Vehicle Dynamics

Dieter Schramm · Manfred Hiller
Roberto Bardini

Vehicle Dynamics

Modeling and Simulation

Second Edition

Springer

Dieter Schramm
University Duisburg-Essen
Duisburg
Germany

Roberto Bardini
BMW Group
München
Germany

Manfred Hiller
University Duisburg-Essen
Duisburg
Germany

ISBN 978-3-662-54482-2 ISBN 978-3-662-54483-9 (eBook)
DOI 10.1007/978-3-662-54483-9

Library of Congress Control Number: 2017939320

Printed on acid-free paper

This Springer imprint is published by Springer Nature
The registered company is Springer-Verlag GmbH Germany
The registered company address is: Heidelberger Platz 3, 14197 Berlin, Germany

Preface 2nd edition

The focus of this book is on the fundamentals of “Vehicle Dynamics” and the mathematical modeling and simulation of motor vehicles. The range of applications encompasses basic single track models as well as complex, spatial multibody systems. The reader will be enabled to develop own simulation models, supported to apply successfully commercial programs, to choose appropriate models and to understand and assess simulation results. The book describes in particular the modeling process from the real vehicle to the mathematical model as well as the validation of simulation results by means of selected applications.

The book aims at students and postgraduates in the field of engineering sciences who attend lectures or work on their thesis. To the same extent, it addresses development engineers and researches working on vehicle dynamics or apply associated simulation programs. The modeling of Vehicle Dynamics is primarily based on mathematical methods used throughout the book. The reader should therefore have a basic understanding of mathematics, e.g., from the first three semesters’ study course in engineering or natural sciences.

This second edition of the book is the English version of the third German edition.

The authors thank all persons who contributed to this edition of the book. Amongst all persons who contributed by giving hints and sometimes, simply asking the right questions we want to highlight in particular the indispensable contributions of Frederic Kracht for diligent proofreading and the solution of unsolvable problems incident to the secrets of contemporary word processor software.

Duisburg, Germany — Dieter Schramm
Duisburg, Germany — Manfred Hiller
München, Germany — Roberto Bardini
March 2017

Preface

The main focus of this book is on the fundamentals of "Vehicle Dynamics" and the mathematical modeling and simulation of motor vehicles. The range of applications encompasses basic single track models as well as complex, spatial multibody systems. The reader will be enabled to develop own simulation models, supported to apply successfully commercial programs, to choose appropriate models and to understand and assess simulation results. The book describes in particular the modeling process from the real vehicle to the mathematical model as well as the validation of simulation results by means of selected applications.

The book is aimed at students and postgraduates in the field of engineering sciences who attend lectures or work on their thesis. To the same extent it addresses development engineers and researches working on vehicle dynamics or apply associated simulation programs.

The modeling of Vehicle Dynamics is primarily based on mathematical methods used throughout the book. The reader should therefore have a basic understanding of mathematics, e.g., from the first three semesters' study course in engineering or natural sciences.

This edition of the book is the English version of the second German edition.

The authors thank all persons who contributed to this edition of the book. Amongst all persons who contributed by giving hints and sometimes simply asking the right questions we want to highlight in particular the indispensable contributions of Stephanie Meyer, Lawrence Louis and Michael Unterreiner who contributed with translation and proof reading of some chapters. We also thank Frederic Kracht for diligent proofreading and the solution of unsolvable problems incident to the secrets of contemporary word processor software.

Duisburg
May 2014

Dieter Schramm
Manfred Hiller
Roberto Bardini

Contents

Nomenclature and Definitions

Variables and Physical Quantities

The name of variables and physical quantities are in general written in italic letters. The notations of locations (points), components and names of coordinate systems, numbers as well as mathematical standard functions, such as e.g. "sin" or "cos" are not written in italic letters.

In addition, the following applies for vectors and tensors as well as matrices:

- Vectors are represented by bold lower case letters, tensors and matrices by bold upper case letters.
- Dots over the respective quantity indicate time derivatives.

Special Notation for Physical Vectors

The subscription of vectors and tensors is made according to the following rules:

- An index on the lower right side represents a denotation and numbering. It denotes, e.g. the body or the coordinate system of the respective quantity.
- For quantities which are described with respect to other quantities a lower left index denotes the reference body or the reference coordinate system. A void index indicates the inertial system as reference system.
- In case that a physical vector is represented by coordinates, the coordinate system is indicated by a left upper index. If no index is present, a physical vector or tensor is given without indicating a specific coordinate system.
- Operators, like inversion, transposing and raising to power as well as differentiation with respect to other variables as time are indicated by a respective right upper index.

- Differentiation with respect to time is indicated by a dot over the respective variable. At this position also other indications like vinculi "–" or tildes "$\sim$" can be present.

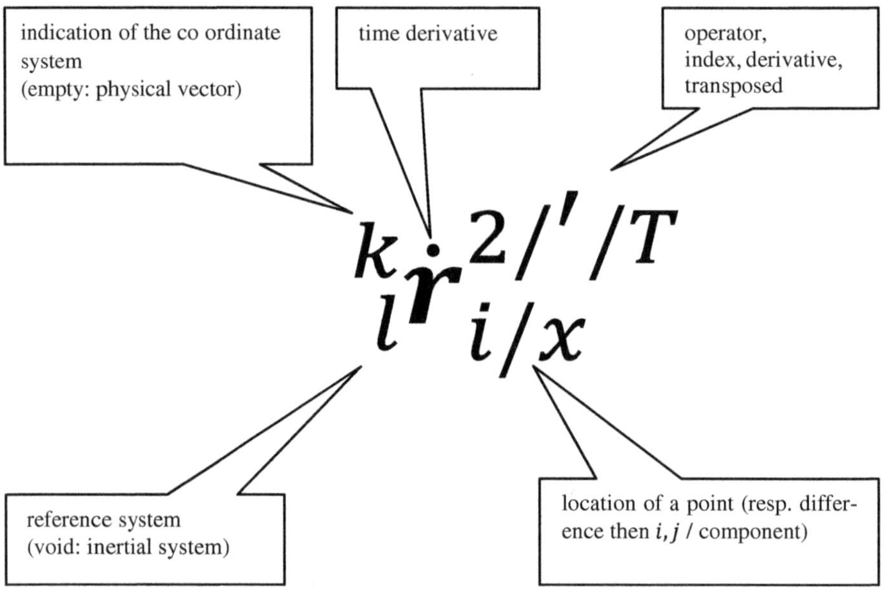

Examples for Subscriptions

$\dot{\boldsymbol{r}}_i$	Absolute velocity of point P_i
$\dot{\boldsymbol{r}}_{i,j}$	Absolute velocity (absolute variation with time) of difference vector $\boldsymbol{r}_j - \boldsymbol{r}_i$
${}_k\dot{\boldsymbol{r}}_i$	Relative velocity of "P_i" with respect to reference system "k"
${}_k\dot{\boldsymbol{r}}_{i,j}$	Relative velocity ${}_k\dot{\boldsymbol{r}}_j - {}_k\dot{\boldsymbol{r}}_i$
${}^i_k\boldsymbol{v}_j$	Coordinate representation of the absolute velocity of point P_j with respect to coordinate system "k", described in coordinates of coordinate system "i"
${}^j\boldsymbol{T}_i$	Rotation tensor, transforming the coordinate representation of vector "$\boldsymbol{a}$" in coordinate system "i" to coordinate system "j": "${}^j\boldsymbol{a} = {}^j\boldsymbol{T}_i {}^i\boldsymbol{a}$"

Partial derivatives of a m-dimensional vectorial function

$$f(x) = \begin{bmatrix} f_1(x_1, \cdots, x_n) \\ \vdots \\ f_m(x_1, \cdots, x_n) \end{bmatrix}$$

with respect to coordinates of a m-dimensional vector x are arranged in a (m, n) - dimensional functional- or JACOBIAN-Matrix:

$$\frac{\partial \boldsymbol{f}(\boldsymbol{x})}{\partial \boldsymbol{x}} = \begin{bmatrix} \frac{\partial f_1(\boldsymbol{x})}{\partial x} \\ \vdots \\ \frac{\partial f_m(\boldsymbol{x})}{\partial x} \end{bmatrix} = \begin{bmatrix} \frac{\partial f_1(\boldsymbol{x})}{\partial x_1} & \cdots & \frac{\partial f_1(\boldsymbol{x})}{\partial x_n} \\ \vdots & \ddots & \vdots \\ \frac{\partial f_m(\boldsymbol{x})}{\partial x_1} & \cdots & \frac{\partial f_m(\boldsymbol{x})}{\partial x_n} \end{bmatrix}.$$

Examples for "Physical" Vectors and Their Representation

$\boldsymbol{e}_{x_i}, \boldsymbol{e}_{y_i}, \boldsymbol{e}_{z_i}$	Unity vectors for coordinate systems
$\boldsymbol{u}_i$	Normalized orientation vector (joint axes)
$\boldsymbol{r}_i$	Position vector to reference point O_i of an "object" (body) "i"
$\boldsymbol{r}_{\bar{i}}$	Position vector to predecessor of reference point O_i
$\boldsymbol{s}_i$	Position Vector to center of gravity S_i
$\boldsymbol{p}_i$	Position vector to "point of interest" P_i (e.g. application point of a force)
$\boldsymbol{r}_{i,j} = \boldsymbol{r}_j - \boldsymbol{r}_i$	Vector difference between two reference points P_i, P_j
$\boldsymbol{v}_i, \dot{\boldsymbol{v}}_i, \boldsymbol{a}_i$	Velocities, accelerations
$\boldsymbol{\omega}_i, \dot{\boldsymbol{\omega}}_i, \boldsymbol{\alpha}_i$	Angular velocity, angular acceleration
$\boldsymbol{F}_i$	Force
$\boldsymbol{L}_i$, $\boldsymbol{T}_i$	Torque
$\boldsymbol{\Theta}_{S_i}, \theta_{S_i}$	Tensor of inertia, moment of inertia
$\boldsymbol{T}_i$	Rotation tensor
$(x, y, z)_i$	Coordinate system (K_i)
$K_i = \{O_i; x_i, y_i, z_i\}$	Coordinate system (K_i), alternative notation
x_i, y_i, z_i	Coordinate axes
ξ_i, η_i, ζ_i	Coordinate axes

Scalars

m_i	Mass
$i, j, k, \ldots$	Indices

N, n	Number of elements, components (e.g. n_β, N_β)
α	Angle
f	Number of degrees of freedom (DoF) (also f_i)

Vectors and Matrices

$\boldsymbol{E}, \boldsymbol{I}$	Unity matrix or unity tensor
$\boldsymbol{g}$	"Vector" of implicit constraint equations
$\boldsymbol{q}$	"Vector" of generalized coordinates
$\boldsymbol{M}$	Mass matrix
$\boldsymbol{b}$	"Vector" of generalized zentripedal- and CORIOLIS forces
$\boldsymbol{Q}$	Generalized forces
$\boldsymbol{\beta}$	Relative or natural joint coordinates
$\boldsymbol{e}^{(i)}$	$\left[0, 0, \ldots \underbrace{, 1,}_{i^{th}\ position} \ldots, 0, 0\right]$
$\boldsymbol{w}$	Position coordinates
$\boldsymbol{Z}$	Reaction forces

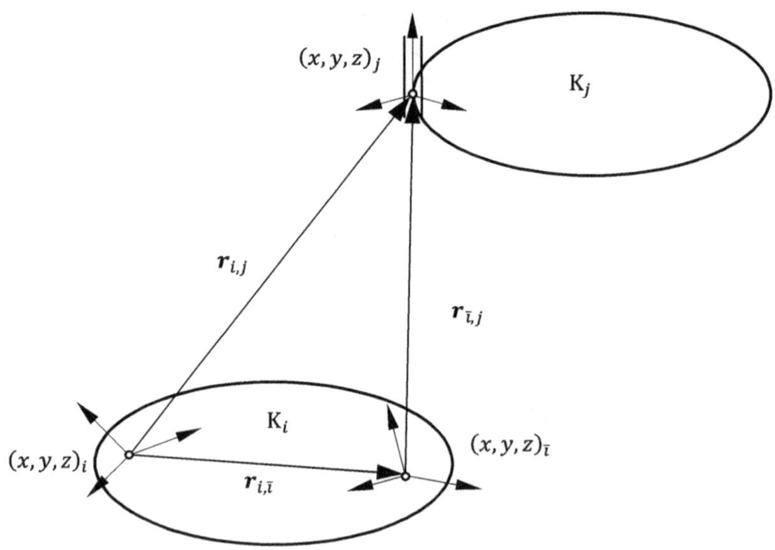

$$\dot{\boldsymbol{r}}_j = \dot{\boldsymbol{r}}_i + \dot{\boldsymbol{r}}_{i,j}$$

$$\dot{\boldsymbol{r}}_{i,j} = \boldsymbol{\omega}_i \times \boldsymbol{r}_{i,j} + \dot{\boldsymbol{r}}_{i,j}$$

$$_i\dot{\boldsymbol{r}}_{i,j} = {}_i\dot{\boldsymbol{r}}_j - {}_i\dot{\boldsymbol{r}}_i$$

(without components!)

Note

$$\boldsymbol{r}_{i,j} = {}_i\,\boldsymbol{r}_j$$

$$\dot{\boldsymbol{r}}_{i,j} \neq {}_i\,\dot{\boldsymbol{r}}_j$$

Trigonometric Functions

Due to space requirements "cos φ" and "sin φ" are, where appropriate, replaced by the short forms "cφ" and "sφ" respectively.

Chapter 1
Introduction

This book addresses the fundamentals, mathematical description and simulation of the dynamics of automobiles. In this context different levels of complexity will be presented, starting with basic single track models up to complex three-dimensional multibody models. A particular focus is the process of establishment of mathematical models from real cars and the validation of the simulation results. The methods presented will be explained in detail based on selected application scenarios.

The intention of this book is to enable the reader to develop his own simulation models and to use them for his daily work, to apply commercial simulation tools in an efficient and dedicated form. In particular the reader will be enabled to choose the appropriate model for a give technical task and to validate the results of simulations.

1.1 Problem Definition

Vehicle dynamics is a branch of vehicle mechanics that deals with the motional actions necessary for moving road vehicles and their resulting forces under consideration of the natural laws. Reference to vehicle dynamics is found in many areas of development of motor vehicles, vehicle systems and their components.

In this chapter an overview of the modeling methods, the fundamental definitions related to vehicle dynamics and the embedding of vehicle dynamics in the development of vehicles will be given.

The use of complex mathematical vehicle models to simulate and develop vehicle systems and their applications, such as in the development of vehicle dynamics control systems or braking systems, has gained significance especially over the last years. The reasons are, on the one hand, economical:

- The effort involved in vehicle testing and measurement has been increasing along with the complexity of the vehicle systems and the prescribed testing conditions. This has a corresponding influence on the development budget available.

D. Schramm et al., Vehicle Dynamics,
DOI 10.1007/978-3-662-54483-9_1

- The increasingly competitive automotive market is forcing manufacturers and suppliers to also contain the costs in the development stage, by replacing prototypes and tests with simulations and virtual prototypes.

On the other hand, many reasons can be attributed to the technology of the new systems. The majority of these are mechatronic systems, ref. e.g. (Isermann 2008), whose typical increase in functionality and optimized product value are based on the function- and hardware oriented combination of mechanical, electrical and electronic components and subsystems, as well as their respective operating systems and functional software.

The interaction of these individual systems, which are derived from different technological domains, on the one hand results in never before seen functional range and product quality and on the other hand in cost efficient solutions, by integrating mechanical, electrical and electronic hardware into modules. The design and testing of such systems with their enormous functional diversity requires high standards in methods of design and testing programs and, as a result, modeling and simulation techniques:

- Vehicle models are the basis for the design and development of vehicle systems and components.
- Vehicle maneuvers can be simulated repeatedly under predefined parameters and conditions.
- Critical maneuvers can be replaced by safe simulations.
- The continuous shortening of product cycles for new models requires shorter developmental phases. This can only be achieved through the implementation of simulations and virtual prototypes.

Based on these requirements, the fields of application for the method of multibody systems in the development of vehicle systems, which is presented in this book, can be deduced:

- Kinematics and dynamics of the chassis and the steering.
- Vehicle dynamics of the entire vehicle.
- Ride comfort of the entire vehicle.
- Analysis of accidents.

The goal in each case is a mathematical description of the relevant areas and functions of the vehicle that can be variably applied for the design, development and evaluation of vehicle dynamics. The numerical simulation of vehicle handling, which is based on these mathematical models, has recently gained enormous significance. It allows the simple, quick and efficient investigation of maneuvers without the need for elaborate testing. The simulation allows for a variation of parameters and conditions in a way that is not possible in actual testing. Since, however, the results generated by numerical simulations are only approximations and their accuracy is dependent on the exactness of the models and the reliability of the system data, great care has to be put into the modeling of these systems.

The driving characteristics of passenger vehicles are influenced by several factors. The wheel locations, which are supposed to conduct predefined motions relative to the chassis, play an important role. By choosing beneficial geometrical parameters in the construction of a wheel suspension system, for example, the stability of the vehicle whilst cornering or changing lanes is guaranteed. Modern wheel suspensions are typically multibody systems with closed kinematic loops. In addition, the handling can be influenced through elastic bearings in the wheel suspensions. For example, the longitudinal flexibility of the wheel location can be achieved through a soft bearing of the transverse link.

The complicated systems and the wish for a reproduction of real events that is as accurate as possible make the development of simulation models a comprehensive and challenging task. Setting up equations efficiently is of vital importance in order to limit the modeling effort and minimize the computation time required for the simulation. Thus the goal of this book is to present an efficient way of creating realistic simulation models of a vehicle. To this end, an overview of the basic mechanical and mathematical processes will be provided, in which the topological structure of the vehicle will be described in detail using fundamentals such as the methodology of kinematic differentials and that of the characteristic joint pairs. Based on this, the modeling of the subsystems and components

- chassis, wheel suspensions,
- wheels and tires,
- force elements,
- drivetrain

and finally the entire system will be dealt with.

1.1.1 Modeling Technical Systems

There are two fundamentally different methods to describe the dynamic behavior of a real process using mathematical models, appropriate to the task at hand:

- In theoretical modeling, the mathematical models are derived from physical laws.
- In experimental modeling a specific model structure, which in most cases is also mathematically formulated, is used as the foundation. On the basis of this model structure, the individual parameters are identified using input and output measurements. A special form of this method is called (model-) identification.

This book almost exclusively deals with theoretical modeling. Physical parameters will be assumed to be known or at least assessable. Typical sources of parameters in vehicle technology are:

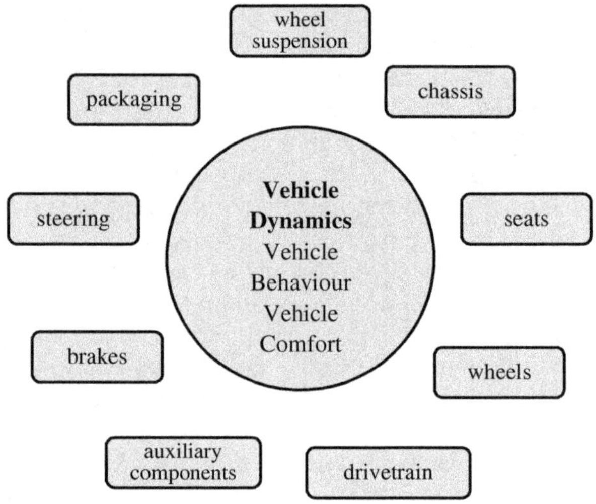

Fig. 1.1 Vehicle Dynamics: Environment and related components

- Computer-aided design models for measurements, masses and moments of inertia.
- Direct measurements of masses, moments of inertia, spring and damper characteristics, and, if possible, friction coefficients.
- Assumptions, estimations and, where applicable, identification methods for other, more difficult or vague characteristics such as friction effects, elasticity in bearings, etc.
- Identification of parameters and characteristic maps through other methods of calculation and simulation such as the finite-element method and of calculating electric and magnetic fields etc.

The identification methods are often used in this context to determine parameters of theoretical models that are either unknown or difficult to measure. Examples are tire models (Chap. 7) or characteristics of force elements (Chap. 9), such as rubber bearings or dampers.

The aim of modeling is to obtain a mathematical-analytical description of the respective system which allows for an investigation of the relevant aspects of the system behavior and the influence of the system components on it, (Fig. 1.1). It is possible to develop models of varying complexity and validity. On the one hand, the more complex a model is, the more accurate the simulation of the system behavior is. On the other hand, however, this will invariably result in complex and mostly nonlinear model equations as well as a need for better computing performance. Additionally, the number of model parameters that have to be determined increases along with the complexity of a model. Most of the time, the effort to procure the parameters required will outweigh the effort in creating the model equations by far. Therefore it is always necessary to critically evaluate whether an increase in model complexity is still adequate to its aims.

1.1.2 Definition of a System

The term system will often be used in this book. Therefore it is important to briefly define what is meant by a system in this book (Hiller 1983):

A system is identified as a set of elements (parts, components), that influence each other through internal functional relationships and physical laws (interaction), on which external influences act (inputs) and the effects of which are commuted to the outside (outputs). Thus, the most important properties of a system are its changeability (motion) and its controllability through a suitable choice of inputs (forces and applied motion). The relationship between external influences on the system and the resulting changes of the system state is defined as system dynamics.

A major part of this book deals with the investigation and analysis as well as the prediction of the dynamic behavior of the vehicle system and its subsystems and components. The following subtasks can be identified:

- Modeling: modeling always involves idealizations and abstractions (Sect. 1.1.4).
- Model investigation: deals with, primarily numeric, solutions to the equations of motion.
- Selection of controlling inputs: Examples in a motor vehicle are steering angle, accelerator and brake pedal position as well as the characteristics of the road surface, but also actuator forces, such as the active anti-roll bars which are examined in Chap. 14.
- Simulation of the system characteristics (Chap. 12).

1.1.3 Simulation and Simulation Environment

Every simulation aims at describing the observed system as accurate as possible in order to be able to deduce the behavior of the real system from the behavior of the model. In this book a vehicle or part of a vehicle as well as, if necessary, a part of its environment will be referred to as a system. Below, solely the simulation of mathematical models on one (or several connected) computers will be examined. The models will be purely mathematical in nature. The simulation is thus equivalent to the running of software, combined, if necessary, with hardware components which are connected via suitable interfaces. The latter are usually referred to as hardware-in-the-loop (HiL) simulations. It is necessary to run the simulation in real time in order to provide the hardware with data.

If one visualizes the vehicle as a mechatronic system, in which, for example, the aforementioned vehicle dynamics control systems and driver assistance systems play an ever increasing role, the simulation of the dynamics of the vehicle components or the entire vehicle as a tool in the process of mechatronic development (VDI-Guideline-2206 2004).

1.1.4 Vehicle Models

The models described in this book are to make it possible to represent the dynamic behavior of real vehicles as realistically as possible. To accomplish this, the models have to meet at least the following criteria:

- Complete spatial kinematics and kinetics of the entire vehicle and, if required, also its subsystems.
- Nonlinear kinematics of the wheel suspension.
- Nonlinear and, where required, also dynamic representations of the force elements.
- Dynamic tire forces.

On the other hand the models have to remain manageable. This is especially important if the simulation models are to be implemented in a hardware-in-the-loop test rig or a drive simulator. In this case the computation time has to remain suitable. This also holds for the use of the models for optimization tasks.

To simulate the handling of a vehicle, different types of vehicle models are possible, depending on the desired level of detail and the task at hand, ref. Table 1.1.

If one assumes the vehicle chassis to be rigid, then the chassis has six degrees of freedom in space, which can, however, be reduced through further assumptions, such as those found in single track and twin track vehicle models. To simulate the vehicle longitudinal motion, it may be sufficient to define just one degree of freedom. Then the other degrees of freedom of the body have to be constrained by using so-called constraint or boundary conditions. Even for a simple model that describes the lateral dynamics, a minimum of two degrees of freedom, for the lateral motion and the yaw motion, is required.

In twin track and especially in complex multibody system models, further degrees of freedom to describe the motion of the components of the wheel suspensions and the drivetrain are required. Furthermore, the subsystems of the vehicle such as the drivetrain, brakes and steering have to be modeled. Systems such as the ABS, ESP, driver assistance systems and other mechatronic systems can also be integrated into the model.

Table 1.1 Modeling of motor vehicles

Model type	Degrees of freedom	Chapter
Single track model, linear	2	10
Single track model, nonlinear	3–7	10
Twin track model	14–30	11
Complex multibody system model	>20	12 and 13, subsystems: 5–9
Finite-Element-model	>500	Not dealt with
Hybrid model	>500	Not dealt with

The use of simulation programs for the development of technical systems in vehicles has gained significance in recent years and is currently state of the art. The prediction of the dynamic behavior of the vehicle allows for conclusions about driving stability, driving safety and comfort of new vehicle systems. Furthermore, such models can show the influence of control systems and actuators on the vehicle's handling and are often prerequisite for an efficient development of such systems. The mechanical components that occur in these systems can be modeled and simulated using the following approaches (Schiehlen and Eberhard 2004).

Multibody Systems (MBS) Multibody systems are suitable for the description of mechanical systems, which consist of bodies that are mostly rigid and are connected via bearings and joints. A multibody system usually consists of rigid bodies with mass, which are subject to concentrated forces and moments at discrete points (Schiehlen and Eberhard 2004). Some of the symbols, commonly used for a typical multibody system, are represented, along with a corresponding example from vehicle technology, in Fig. 1.2.

A rigid body of a multibody system is characterized by its mass and moment of inertia. Characteristic points of a rigid body are the center of gravity S as well as a finite number of node-points P_i, at which concentrated forces and moments act or other bodies are connected via corresponding joints. Elasticity and damping are represented as massless force elements and their typical symbols are shown in Fig. 1.2 along with the corresponding component typically found in a vehicle. Respectively, through depiction of force laws and constraint motions, drivetrains

Fig. 1.2 Typical elements of a multibody system

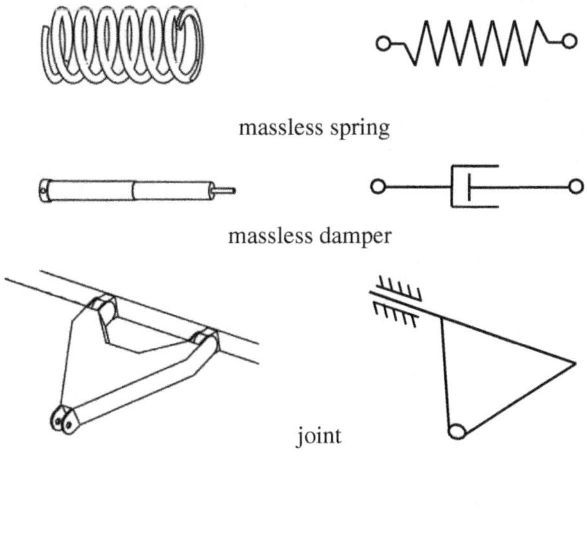

and the actuators can be represented. Of major interest in modeling and simulation are the motion variables of the bodies and, sometimes the forces and moments acting on the bearings and drivetrains. The mathematical description of the kinematics and kinetics of the multibody system results, depending on the modeling and formalization, in ordinary differential or differential algebraic systems of equations with relatively small degrees of freedom and will be dealt with in Chaps. 3 and 4. Here, "kinematics" refers to the description of the possible motions of mechanical systems, while "kinetics" refers to the motion of mechanical systems under the influence of forces.

Finite-Element-Method (FEM) This method is primarily being used to give a mathematical description of the elastic and, where applicable, plastic characteristics of mechanical systems, in which mass and elasticity are distributed continuously throughout the body. The model consists of many finite elements with a simple geometry, whose principle deformation options are constrained by so-called elementary functions. The method is primarily used to examine the effect of external forces on the deformation and stress distribution of a body. The mathematical formulation of the finite-element-method leads to ordinary differential equations with many degrees of freedom.

Continuous Systems (COS) Continuous systems are used for the depiction of elastic characteristics of mechanical systems, in which mass and elasticity, as well as plasticity are distributed continuously throughout the body. The mathematical formulation of continuous systems leads to a description using partial differential equations with infinite number of degrees of freedom. The respective field of application of these structurally different substitute systems is mainly dependent on the geometry and the distribution of stiffness of the initial mechanical system, the goal of the investigation and, thus implicitly, the aspired area of validity of the simulation model. The method of finite elements and the continuous systems are primarily suitable for mechanical systems or bodies with evenly distributed elasticity.

Multibody systems are ideally suited for complex models that help describe vehicle dynamics. However it is also possible to create a vehicle model using the finite-elements-method. This has its advantages especially, when structural deformation and stress distribution have to be determined along with vehicle kinematics and kinetics.

Hybrid mechanical systems which require the modeling of both rigid and elastic bodies can be represented through a combination of multibody systems and the finite-elements-method for example. These are called hybrid systems (Louis and Schramm 2011).

When choosing a suitable method for the simulation of a vehicle, the following aspects have to be considered as well:

- For most of tasks in vehicle dynamics it is sufficient to examine a very limited frequency spectrum, Table 1.2, (Bürger and Dödelbacher 1988; Frik 1994). Hence, it is possible to limit the model to depict a spectrum between 0 and ca. 30 Hz.

Table 1.2 Typical frequencies in vehicle subsystems

Oscillation mode	Frequency (Hz)
Body motion	1–2
Longitudinal vehicle oscillations	4–10
Motor jerk	10–13
Wheel suspensions, deflection	10–15
Steering oscillations	10–16
Body oscillations	30–40

- An exception from the limitation outlined above is for example the simulation of vehicle dynamics control systems. Because of the relatively short time constants of the hydraulics, higher frequency vibrations can occur. These oscillations occur within the region of the natural frequency of the wheel suspensions—due to the bearing elasticity in individual joints—between 15 and 30 Hz as well as the natural frequency of the radials at around 50 Hz. Since these oscillations influence the signals detected by the sensors of the vehicle control systems and lie within the sensor sampling rate, they must be included in the simulation of the regulated vehicle maneuvers.
- In every modeling approach it may be very difficult and in some cases even impossible to obtain the required model data. This is especially the case with friction and damping characteristics, bearing elasticity and tire parameters.

1.2 Complete Vehicle Model

Below, a complete vehicle model is considered to consist of the subsystems chassis, drivetrain, wheel suspensions, wheels, brakes and steering. Inputs to this model are the brake pedal and accelerator position, steering wheel angle, the engaged gear or the position of the automatic lever defined by the driver. The environment acts on the vehicle through the predefined environmental conditions, such as side and head wind, frictional connection coefficient of the road, road inclination and road bumps (Fig. 1.3).

An example of a complex vehicle model that has been modeled as a multibody system is shown in Fig. 1.4. It is not always necessary required to model an entire vehicle. It is possible to divide the vehicle model into its subsystems, which can be examined individually and, if necessary, combined to a complete model afterwards. In Fig. 1.4, this is shown using a complex vehicle model which has been split into its typical subsystems

- vehicle structure (chassis, underbody),
- drivetrain,
- wheel suspensions,
- wheels,
- brakes,
- steering system.

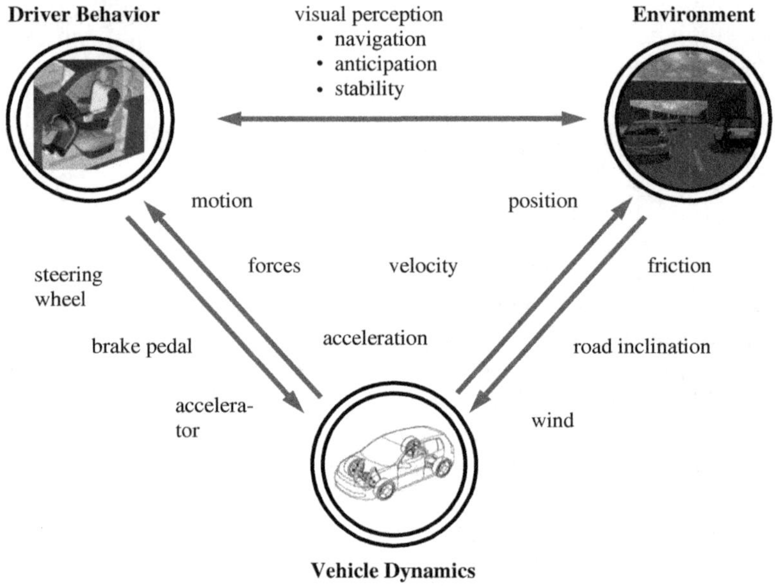

Fig. 1.3 Interaction of a vehicle with the driver and the environment

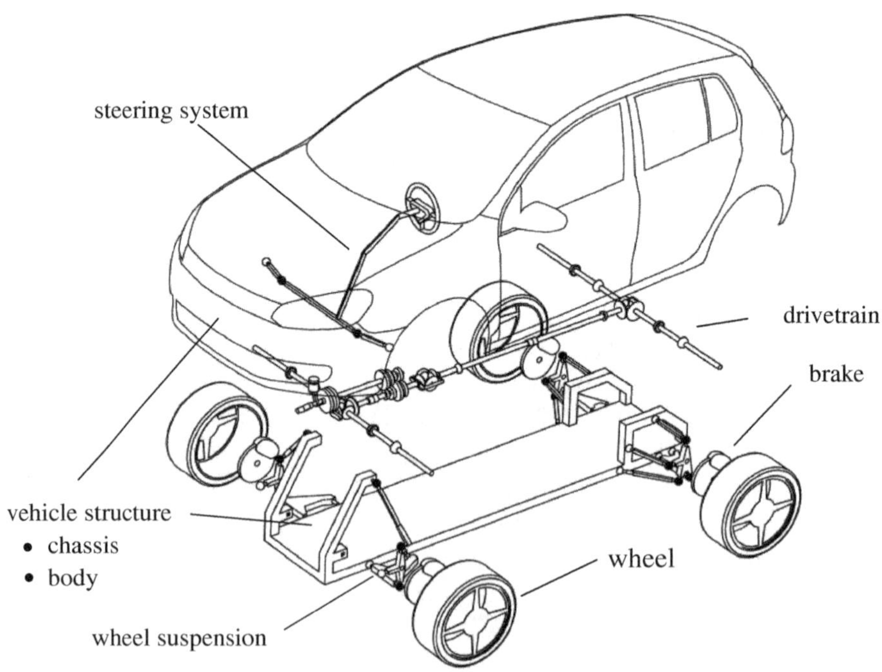

Fig. 1.4 Example of a complex vehicle model

Table 1.3 Overview of vehicle models

	Single track model		Two track model		Multibody systems model		
	linear	nonlinear	without kinematics	with kinematics	reduced	complex	combined with finite-element method
			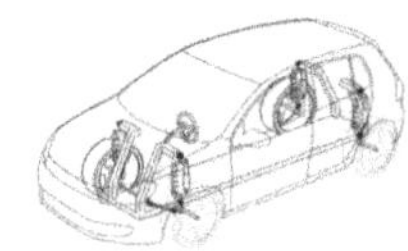		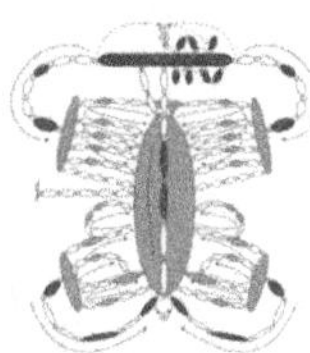		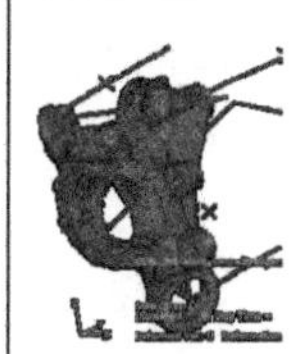
Type of motion	planar translation, yaw						
			roll-, pitch-, vertical dynamics				
					Component motion		
					Special Applications, Rollover, Accident, Crash		
Degrees of freedom	2	3–7	14–25		20– …		20– …
Frequency spectrum	0–2 Hz		0–5 Hz		0–30 Hz		0–200 Hz

1.2.1 Vehicle Models and Application Areas

Depending on the required application, different vehicle models can be used. The fundamental models shown in Tables 1.3 and 1.4 will be discussed in detail in this book. Table 1.4 also indicates whether the respective model is suitable for offline (O) or real-time (E) applications.

1.2.2 Commercial Vehicle Simulation Systems

Nowadays, a multitude of simulation programs and even entire simulation environments with toolsets are available to simulate vehicle dynamics. In this section, a few of these commercial vehicle simulation software systems will be mentioned. The list is however neither complete, nor is the fact that a program has been mentioned or has been omitted an indication of its quality.

MBS Examples of frequently used software solutions for general multibody systems are listed in Table 1.5. They support the development of models through elementary libraries, which contain general as well as application specific elements and usually include graphic user interfaces for model creation (preprocessor) and evaluation (postprocessor). The systems mentioned are either useful to support

Table 1.4 Simulation speed (O: offline, E: real-time)

		Single track		Twin track		MBS-Model	
		Linear	Nonlinear	W/o kinematics	With kinematics	Reduced	Complex
Layout	Components	O/E	O/E	O/E	O/E	O/E	O
	Functions	O/E	O/E	O/E	O/E	O/E	O
Concept design and test	Software	O/E	O/E	O/E	O/E	O/E	O
	Hardware	E	E	E			
	Hardware-in-the loop (HiL)	E	E	E			
	Software-in-the-loop (SiL)	E	E	E			
Vehicle simulator	Human machine interface	E	E	E	E		
	Investigation of algorithm and functions	E	E	E	E		
	Acceptance	E	E	E	E		
Integration in algorithm	Series application	E					

Lation speed (O: offline, E: real-time)

Table 1.5 Programs for the simulation of multibody systems (MBS)

Program	Manufacturer	Type	Reference
Adams/Adams car	MSC. Software Corporation, Santa Ana, USA	MBS	MSC Software (2010)
Simpack	Simpack AG, Gilching, Deutschland	MBS	Simpack AG (2010)
Virtual. Lab vehicle motion	LMS International, Leuven, Belgium	MBS	LMS International (2010a)
Pro/ENGINEER Mechanica	PTC, Needham, USA	MBS	Parametric Technology Corporation (2010)
Simulia Abaqus	Simulia	Hybrid (FEM/MBS)	Simulia (2010)
Dymola	Dassault systemes, Vélizy-Villacoublay, France	MBS	Dassault Systems (2010)
DADS	LMS	MBS	LMS International (2010b)

Table 1.6 Special programs for the simulation of vehicles

Program	Manufacturer	Topic	Literature reference
CarMaker	IPG	MBS, real-time, HiL	IPG (2010)
CarSim	Mechanical solution	MBS, real-time	Carsim (2010)
veDyna	TESIS Dynaware	MBS, real-time	Tesis (2003)
Proracingsim	ProRacing Sim	Motor simulation	Proracing Sim (2010)
AVL Advisor	AVL	MBS, real-time, drivetrain	AVL Advisor (2010)
AVL Cruise	AVL	MBS, real-time, drivetrain	AVL Cruise (2010)
ASM	dSPACE	HiL, embedded	dSpace (2010)
VDMS	Milliken research	MBS	Milliken Research (2010)
SwRi raptor	Southwest research		Southwest Research (2010)
Madymo	TASS	MBS, FEM	Tass (2010)

general mechanical applications or they are derived from the specialization of such systems. There are also simulation environments dedicated to the modelling of vehicle dynamics. These systems, generally, do not only allow the simulation of the vehicle, but also provide the simulation of road profiles and the consideration of the driver (driver model). Examples of such systems can be found in Table 1.6.

1.3 Outline of the Book

In the introduction (Chap. 1), an overview of the different modeling methods and simulation programs has been given. The method of multibody systems will be used in this book to describe the dynamics of vehicles, as it is particularly suited to describe important phenomena occurring in vehicle dynamics.

The required mathematical and kinematical fundamentals will be handled in detail in Chap. 2. Motor vehicles are examples of multibody systems with very complex, spatial kinematics, especially in the area of vehicle suspensions. The characteristic feature is the occurrence of closed kinematic chains, whose mathematical treatment is very difficult. Therefore, the fundamentals of the kinematics of multibody systems will be dealt with in detail in Chap. 3. For years, the authors have successfully implemented concepts of kinematic differentials and transformations as well as methods to incorporate kinematic loops (which frequently occur with wheel suspensions) in development projects. These concepts and methods will be presented in particular below. On the basis of the presented kinematic methods, the equations of motion of multibody systems will be derived in Chap. 4.

Based on the methodology that is described in the first four chapters, the subsystems body, suspensions, tire-road-contact and drivetrain of vehicles will be modeled and simulated in the following Chaps. 5–9. The description of the subsystems will be supplemented with an overview and the mathematical description of the forces acting on the vehicle.

Chapters 10 and 11 deal with basic single and twin track models which are, however, sufficient to describe essential phenomena of vehicle dynamics. The corresponding equations of motion will be given to allow the reader to create his own models and to perform the corresponding simulations. Finally, in Chap. 12 complex, spatial, complete vehicle models will be created and described.

For the different models described in the various chapters, exemplary parameters will be provided which are mainly based on an average middle-class vehicle, in Chap. 13. Furthermore, the kinematics of the chassis of typical vehicles with different drive concepts will be analyzed. Chapter 14 contains application examples of the methods developed in this book.

1.4 Webpage of the Book

A webpage has been created for the book, in which readers have access to additional information and supplements to the contents of the book. The internet address of the page is:

www.imech.de/msdk

References

AVL Advisor (2010) http://www.avl.com. (ed.)
AVL Cruise (2010) http://www.avl.com. (ed.)
Bürger K-H and Dödelbacher G (1988) Verbesserung des Fahrzeugschwingungsverhaltens durch Strukturoptimierung in der Konzeptphase. *VDI-Berichte* 699.
Carsim (2010) http://www.carsim.com. (ed.)
Dassault Systems (2010) http://www.3ds.com/home. (ed.)
dSpace (2010) http://www.dspace.com. (ed.)
Frik S (1994) Untersuchungen zur erforderlichen Modellkomplexität bei der Fahrdynamiksimulation [Dr.-Ing.]. Dissertation, Universität—Gesamthochschule Duisburg
Hiller M (1983) Mechanische Systeme: Eine Einführung in die analytische Mechanik und Systemdynamik. Springer, Berlin u.a. ISBN 978-3-540-12521-1.
IPG (2010) http://www.ipg.de. (ed.)
Isermann R (2008) Mechatronische Systeme Grundlagen. Springer Berlin Heidelberg, Berlin, Heidelberg.—ISBN 978-3-540-32512-3
LMS International (2010a) http://www.lmsintl.com. (ed.)
LMS International (2010b) http://www.lmsintl.com/dads. (ed.)
Louis L and Schramm D (2011) Nonlinear State Estimation of Tire-Road Contact Forces using a 14 DoF Vehicle Model. In *Proceedings Regional Conference on Automotive Research (ReCAR)*, Malaysia. ISBN 978-3-03813-740-5
Milliken Research (2010) http://millikenresearch.com/vdms.html. (ed.)
MSC Software (2010) http://www.mscsoftware.com. (ed.)
Parametric Technology Corporation (2010) http://www.ptc.com. (ed.)
Proracing Sim (2010) http://www.proracingsim.com. (ed.)
Schiehlen W and Eberhard P (2004) Technische Dynamik- Modelle für Regelung und Simulation; 2., neubearb. und erg. Aufl. (ed.), Teubner, Stuttgart u.a. ISBN 978-3-519-12365-1
Simpack AG (2010) http://www.simpack.com. (ed.)
Simulia (2010) http://www.simulia.com/solutions/automotive.html. (ed.)
Southwest Research (2010) http://www.swri.org. (ed.)
Tass (2010) http://www.tass-safe.com. (ed.)
Tesis (2003) ve-DYNA 3.7 Model Overview.
VDI-Guideline-2206 (2004) Design methodology for mechatronic systems. (ed.), Beuth-Verlag GmbH, Berlin.

Chapter 2
Fundamentals of Mathematics and Kinematics

The terms vector and tensor play a central role in mechanics. Below, the difference between a vector as a physical entity and its mathematical representation by means of vector decomposition will be illustrated. The decomposition of vectors into their components requires algorithms and coordinate systems.

2.1 Vectors

2.1.1 Elementary Algorithms for Vectors

The introduction of the term vector is illustrative, as the vector always has a geometrical or physical meaning. The following elementary algorithms are applicable (Fig. 2.1):

Sum and product:

$$\boldsymbol{a} + \boldsymbol{b} = \boldsymbol{b} + \boldsymbol{a} \qquad \text{commutative law,} \tag{2.1}$$

$$\boldsymbol{a} + (\boldsymbol{b} + \boldsymbol{c}) = (\boldsymbol{a} + \boldsymbol{b}) + \boldsymbol{c} \qquad \text{associative law,} \tag{2.2}$$

$$\boldsymbol{a} \cdot (\boldsymbol{b} + \boldsymbol{c}) = \boldsymbol{a} \cdot \boldsymbol{b} + \boldsymbol{a} \cdot \boldsymbol{c} \qquad \text{distributive law.} \tag{2.3}$$

Scalar product:

$$\boldsymbol{a} \cdot \boldsymbol{b} = |\boldsymbol{a}||\boldsymbol{b}| \cos \sphericalangle(\boldsymbol{a}, \boldsymbol{b}); \; ||: \text{EUCLIDEAN Norm} \tag{2.4}$$

$$\boldsymbol{a} \cdot \boldsymbol{b} = \boldsymbol{b} \cdot \boldsymbol{a} \qquad \text{commutative law} \tag{2.5}$$

Vector product or cross product:

$$\boldsymbol{a} \times \boldsymbol{b} \perp (\boldsymbol{a}, \boldsymbol{b}) \tag{2.6}$$

D. Schramm et al., Vehicle Dynamics,
DOI 10.1007/978-3-662-54483-9_2

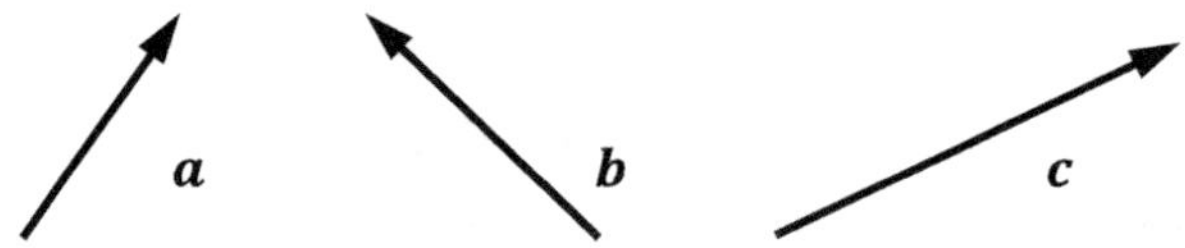

Fig. 2.1 Vectors in the 3-dimensional, Euclidean space $\mathbb{R}^3$

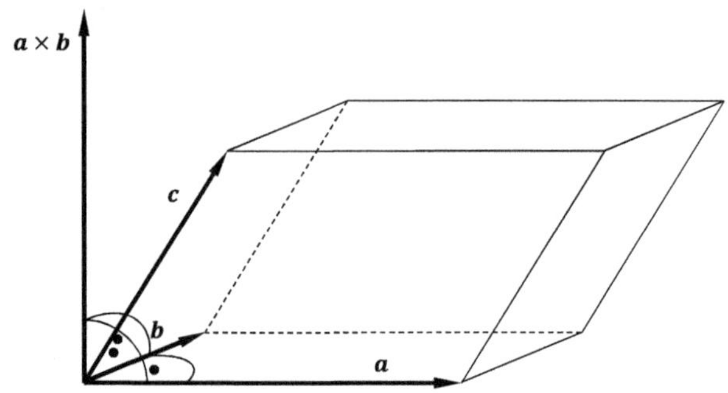

Fig. 2.2 Scalar triple product or parallelepiped of $\boldsymbol{a}, \boldsymbol{b}, \boldsymbol{c}$

$$|\boldsymbol{a} \times \boldsymbol{b}| = |\boldsymbol{a}||\boldsymbol{b}| \sin \varphi \qquad \text{magnitude of cross product.} \tag{2.7}$$

Scalar triple product (Fig. 2.2):

$$\begin{gathered} V = (\boldsymbol{a} \times \boldsymbol{b}) \cdot \boldsymbol{c} \\ (\text{Volume}|V|, \text{ derived from the vectors } \boldsymbol{a}, \boldsymbol{b}, \boldsymbol{c}) \end{gathered} \tag{2.8}$$

Vector triple product:

$$\boldsymbol{a} \times (\boldsymbol{b} \times \boldsymbol{c}) = \boldsymbol{b} \cdot (\boldsymbol{a} \cdot \boldsymbol{c}) - \boldsymbol{c} \cdot (\boldsymbol{a} \cdot \boldsymbol{b}) \tag{2.9}$$

Vector quadruple product (LAGRANGE's identity):

$$(\boldsymbol{a} \times \boldsymbol{b}) \cdot (\boldsymbol{c} \times \boldsymbol{d}) = (\boldsymbol{a} \cdot \boldsymbol{c})(\boldsymbol{b} \cdot \boldsymbol{d}) - (\boldsymbol{b} \cdot \boldsymbol{c})(\boldsymbol{a} \cdot \boldsymbol{d}) \tag{2.10}$$

2.1.2 Physical Vectors

The most important quantities used to describe technical systems, in particular mechanical systems, are physical vectors. The processing and interpretation of these quantities require, as a rule, their representation in a coordinate system. This is represented in the following problem statement:

Given: A vector $\boldsymbol{a}$ is given as a physical quantity, which means that it is independent of its coordinate (component) representation.
Examples:

Displacement	$\boldsymbol{s}$
Velocity	$\boldsymbol{v}$
Angular velocity	$\boldsymbol{\omega}$
Force	$\boldsymbol{F}$
Moment or torque	$\boldsymbol{L}$

To find: Representation of the vector $\boldsymbol{a}$ for analysis for physical-technical applications.
Approach: Introduction of coordinate systems and decomposition of the vectors into components.
Problem: Derivation of the correlation between multiple coordinate systems, i.e. multiple representations for the same vector $\boldsymbol{a}$.

2.2 Coordinate Systems and Components

2.2.1 Coordinate Systems

Coordinate systems $\mathbf{K}_i$ are introduced as orthonormal systems, where the coordinate axes x_i, y_i, z_i are defined as being perpendicular to each other. Unit or base vectors $\boldsymbol{e}_{x_i}, \boldsymbol{e}_{y_i}, \boldsymbol{e}_{z_i}$, having a length of 1, belong to their respective axis (Fig. 2.3).

2.2.2 Component Decomposition

1. Coordinate system $\mathbf{K}_1 = \{\mathrm{O}_1; \boldsymbol{e}_{x_1}, \boldsymbol{e}_{y_1}, \boldsymbol{e}_{z_1}\}$ (Fig. 2.4):

   A decomposition of $\boldsymbol{a}$ in the direction of the unit vectors $\boldsymbol{e}_{x_1}, \boldsymbol{e}_{y_1}, \boldsymbol{e}_{z_1}$ results in:

$$\boldsymbol{a} = {}^1a_x \boldsymbol{e}_{x_1} + {}^1a_y \boldsymbol{e}_{y_1} + {}^1a_z \boldsymbol{e}_{z_1} \tag{2.11}$$

with the projections: ${}^1a_x = \boldsymbol{a} \cdot \boldsymbol{e}_{x_1}$; ${}^1a_y = \boldsymbol{a} \cdot \boldsymbol{e}_{y_1}$; ${}^1a_z = \boldsymbol{a} \cdot \boldsymbol{e}_{z_1}$.
   Practical grouping of the components of $\boldsymbol{a}$ with respect to $\mathbf{K}_1$:

$${}^1\boldsymbol{a} = \left[{}^1a_x, {}^1a_y, {}^1a_z\right]^T \text{ or } {}^1\boldsymbol{a} = {}^1\left[a_x, a_y, a_z\right]^T. \tag{2.12}$$

   Here, $\boldsymbol{a}$ is no longer a physical vector, but rather a column matrix.

2. Coordinate System $\mathbf{K}_2 = \{\mathrm{O}_2; \boldsymbol{e}_{x_2}, \boldsymbol{e}_{y_2}, \boldsymbol{e}_{z_2}\}$:

   A decomposition of $\boldsymbol{a}$ along the unit vectors $\boldsymbol{e}_{x_2}, \boldsymbol{e}_{y_2}, \boldsymbol{e}_{z_2}$ results in

$$\boldsymbol{a} = {}^2a_x \boldsymbol{e}_{x_2} + {}^2a_y \boldsymbol{e}_{y_2} + {}^2a_z \boldsymbol{e}_{z_2} \tag{2.13}$$

with the projections ${}^2a_x = \boldsymbol{a} \cdot \boldsymbol{e}_{x_2}$; ${}^2a_y = \boldsymbol{a} \cdot \boldsymbol{e}_{y_2}$; ${}^2a_z = \boldsymbol{a} \cdot \boldsymbol{e}_{z_2}$.

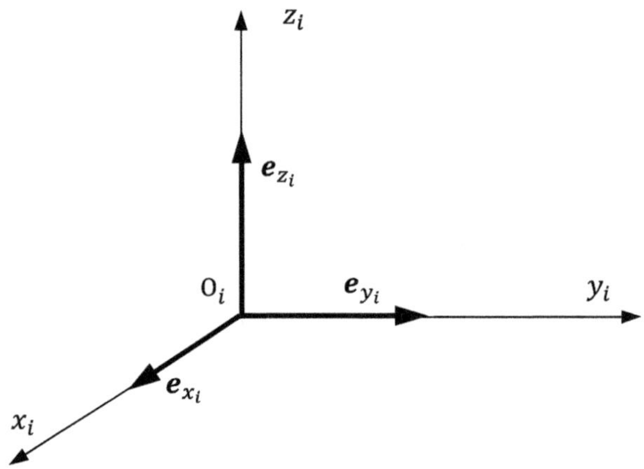

Fig. 2.3 Orthonormal coordinate system $\mathbf{K}_i = \{O_i; e_{x_i}, e_{y_i}, e_{z_i}\}$

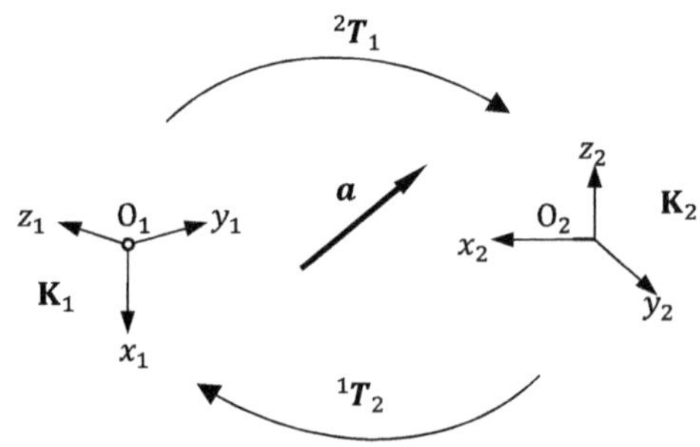

Fig. 2.4 Component decomposition

This results in the components of $\boldsymbol{a}$ with respect to $\mathbf{K}_2$:

$$^2\boldsymbol{a} = \left[^2a_x, {}^2a_y, {}^2a_z\right]^T \text{ or } {}^2\boldsymbol{a} = {}^2\left[a_x, a_y, a_z\right]^T. \tag{2.14}$$

2.2.3 Relationship Between Component Representations

The relationship between the two component representations can be described as the following problem:

Given: Coordinate systems

$$\mathbf{K}_1 = \left\{O_1; \boldsymbol{e}_{x_1}, \boldsymbol{e}_{y_1}, \boldsymbol{e}_{z_1}\right\},$$
$$\mathbf{K}_2 = \left\{O_2; \boldsymbol{e}_{x_2}, \boldsymbol{e}_{y_2}, \boldsymbol{e}_{z_2}\right\}.$$

Component decomposition of vector $\boldsymbol{a}$ in $\mathbf{K}_2 : {}^2\left[a_x, a_y, a_z\right]^T$.

To find: Component decomposition of vector $\boldsymbol{a}$ in $\mathbf{K}_1 : {}^1\left[a_x, a_y, a_z\right]^T$.

The following invariance property can be applied: a physical vector $\boldsymbol{a}$ is independent of its particular coordinate representation; this means that it can be constructed from vector parts from either coordinate systems $\mathbf{K}_1$ or $\mathbf{K}_2$:

$$\boldsymbol{a} \stackrel{!}{=} {}^1a_x\boldsymbol{e}_{x_1} + {}^1a_y\boldsymbol{e}_{y_1} + {}^1a_z\boldsymbol{e}_{z_1} \stackrel{!}{=} {}^2a_x\boldsymbol{e}_{x_2} + {}^2a_y\boldsymbol{e}_{y_2} + {}^2a_z\boldsymbol{e}_{z_2}, \tag{2.15}$$

the symbol "$\stackrel{!}{=}$", in this context, means "corresponds with the mathematical vector."

The consecutive multiplication of Eq. (2.15) with $\boldsymbol{e}_{x_1}, \boldsymbol{e}_{y_1}, \boldsymbol{e}_{z_1}$ yields:

$$\begin{gathered}\cdot\boldsymbol{e}_{x_1} : {}^1a_x \underbrace{\boldsymbol{e}_{x_1} \cdot \boldsymbol{e}_{x_1}}_{1} + {}^1a_y \underbrace{\boldsymbol{e}_{y_1} \cdot \boldsymbol{e}_{x_1}}_{0} + {}^1a_z \underbrace{\boldsymbol{e}_{z_1} \cdot \boldsymbol{e}_{x_1}}_{0} \stackrel{!}{=} \\ {}^2a_x\boldsymbol{e}_{x_2} \cdot \boldsymbol{e}_{x_1} + {}^2a_y\boldsymbol{e}_{y_2} \cdot \boldsymbol{e}_{x_1} + {}^2a_z\boldsymbol{e}_{z_2} \cdot \boldsymbol{e}_{x_1}\end{gathered}$$

etc.

From these equations, one can derive the needed components of the vector $\boldsymbol{a}$ in $\mathbf{K}_1$:

$$\begin{aligned}\cdot\boldsymbol{e}_{x_1} &\to {}^1a_x = {}^2a_x\boldsymbol{e}_{x_2} \cdot \boldsymbol{e}_{x_1} + {}^2a_y\boldsymbol{e}_{y_2} \cdot \boldsymbol{e}_{x_1} + {}^2a_z\boldsymbol{e}_{z_2} \cdot \boldsymbol{e}_{x_1}, \\ \cdot\boldsymbol{e}_{y_1} &\to {}^1a_y = {}^2a_x\boldsymbol{e}_{x_2} \cdot \boldsymbol{e}_{y_1} + {}^2a_y\boldsymbol{e}_{y_2} \cdot \boldsymbol{e}_{y_1} + {}^2a_z\boldsymbol{e}_{z_2} \cdot \boldsymbol{e}_{y_1}, \\ \cdot\boldsymbol{e}_{z_1} &\to {}^1a_z = {}^2a_x\boldsymbol{e}_{x_2} \cdot \boldsymbol{e}_{z_1} + {}^2a_y\boldsymbol{e}_{y_2} \cdot \boldsymbol{e}_{z_1} + {}^2a_z\boldsymbol{e}_{z_2} \cdot \boldsymbol{e}_{z_1}.\end{aligned} \tag{2.16}$$

The matrix notation for this transformation is given in the following:

$$\begin{bmatrix} {}^1a_x \\ {}^1a_y \\ {}^1a_z \end{bmatrix} = \begin{bmatrix} \boldsymbol{e}_{x_2} \cdot \boldsymbol{e}_{x_1} & \boldsymbol{e}_{y_2} \cdot \boldsymbol{e}_{x_1} & \boldsymbol{e}_{z_2} \cdot \boldsymbol{e}_{x_1} \\ \boldsymbol{e}_{x_2} \cdot \boldsymbol{e}_{y_1} & \boldsymbol{e}_{y_2} \cdot \boldsymbol{e}_{y_1} & \boldsymbol{e}_{z_2} \cdot \boldsymbol{e}_{y_1} \\ \boldsymbol{e}_{x_2} \cdot \boldsymbol{e}_{z_1} & \boldsymbol{e}_{y_2} \cdot \boldsymbol{e}_{z_1} & \boldsymbol{e}_{z_2} \cdot \boldsymbol{e}_{z_1} \end{bmatrix} \begin{bmatrix} {}^2a_x \\ {}^2a_y \\ {}^2a_z \end{bmatrix} \tag{2.17}$$

or by

$${}^1\boldsymbol{T}_2 = \begin{bmatrix} \boldsymbol{e}_{x_2} \cdot \boldsymbol{e}_{x_1} & \boldsymbol{e}_{y_2} \cdot \boldsymbol{e}_{x_1} & \boldsymbol{e}_{z_2} \cdot \boldsymbol{e}_{x_1} \\ \boldsymbol{e}_{x_2} \cdot \boldsymbol{e}_{y_1} & \boldsymbol{e}_{y_2} \cdot \boldsymbol{e}_{y_1} & \boldsymbol{e}_{z_2} \cdot \boldsymbol{e}_{y_1} \\ \boldsymbol{e}_{x_2} \cdot \boldsymbol{e}_{z_1} & \boldsymbol{e}_{y_2} \cdot \boldsymbol{e}_{z_1} & \boldsymbol{e}_{z_2} \cdot \boldsymbol{e}_{z_1} \end{bmatrix}, \tag{2.18}$$

$${}^1\boldsymbol{a} = {}^1\boldsymbol{T}_2\, {}^2\boldsymbol{a}. \tag{2.19}$$

The transformation of the components (coordinates) of a vector $\boldsymbol{a}$ with respect to the coordinate system $\mathbf{K}_2$ in components with respect to $\mathbf{K}_1$ is performed with help of the transformation matrix ${}^1\boldsymbol{T}_2$ (note: transformation from $\mathbf{K}_2$ to $\mathbf{K}_1$). This allows the transformation matrix to be clearly interpreted.

$$^1\boldsymbol{T}_2 \begin{cases} \text{Columns: coordinate representation of} \\ \quad \boldsymbol{e}_{x_2}, \boldsymbol{e}_{y_2}, \boldsymbol{e}_{z_2} \text{ in } \mathbf{K}_1 \\ \text{Rows: coordinate representation of} \\ \quad \boldsymbol{e}_{x_1}, \boldsymbol{e}_{y_1}, \boldsymbol{e}_{z_1} \text{ in } \mathbf{K}_2 \end{cases}$$

2.2.4 *Properties of the Transformation Matrix*

The transformations are defined in both directions:

$$^1\boldsymbol{T}_2 : {}^2\boldsymbol{a} \rightarrow {}^1\boldsymbol{a}. \tag{2.20}$$

Correspondingly the inverse transformation is given

$$\begin{aligned} {}^1\boldsymbol{a} \rightarrow {}^2\boldsymbol{a} : {}^2\boldsymbol{T}_1 &= \left({}^1\boldsymbol{T}_2\right)^{-1} (\text{Inverse}), \\ {}^2\boldsymbol{a} = {}^2\boldsymbol{T}_1\, {}^1\boldsymbol{a} &= \left({}^1\boldsymbol{T}_2\right)^{-1}\, {}^1\boldsymbol{a} \end{aligned} \tag{2.21}$$

Analogue to the derivation of the matrix $^1\boldsymbol{T}_2$ (Eqs. 2.17 and 2.18) it follows

$$^2\boldsymbol{T}_1 = \left({}^1\boldsymbol{T}_2\right)^{-1} = \begin{bmatrix} \boldsymbol{e}_{x_1} \cdot \boldsymbol{e}_{x_2} & \boldsymbol{e}_{y_1} \cdot \boldsymbol{e}_{x_2} & \boldsymbol{e}_{z_1} \cdot \boldsymbol{e}_{x_2} \\ \boldsymbol{e}_{x_1} \cdot \boldsymbol{e}_{y_2} & \boldsymbol{e}_{y_1} \cdot \boldsymbol{e}_{y_2} & \boldsymbol{e}_{z_1} \cdot \boldsymbol{e}_{y_2} \\ \boldsymbol{e}_{x_1} \cdot \boldsymbol{e}_{z_2} & \boldsymbol{e}_{y_1} \cdot \boldsymbol{e}_{z_2} & \boldsymbol{e}_{z_1} \cdot \boldsymbol{e}_{z_2} \end{bmatrix}. \tag{2.22}$$

Comparing the equation above to the Eq. (2.17), one obtains Eq. (2.22) by replacing the rows with the columns in Eq. (2.17), i.e. the transpose of matrix $^1\boldsymbol{T}_2$. Hence, the following condition of orthogonality holds:

$$^2\boldsymbol{T}_1 = \left({}^1\boldsymbol{T}_2\right)^{-1} = \left({}^1\boldsymbol{T}_2\right)^T. \tag{2.23}$$

One obtains the inverse transformation $^2\boldsymbol{T}_1$ by transposing the matrix $^1\boldsymbol{T}_2$.

2.3 Linear Vector Functions and Second Order Tensors

Let the vectors $\boldsymbol{x}$ and $\boldsymbol{y}$ be connected through the vector function

$$\boldsymbol{y} = \boldsymbol{T}(\boldsymbol{x}). \tag{2.24}$$

The following mathematical laws are assumed to be valid:

$$T(\mu x) = \mu T(x) \qquad \text{associative law (homogeneity),} \tag{2.25}$$

$$T(x_1 + x_2) = T(x_1) + T(x_2) \quad \text{distributive law (additivity).} \tag{2.26}$$

Consequently the vector function is defined to be homogeneous and linear and can be written as:

$$y = Tx. \tag{2.27}$$

In this linear mapping, where the vector $\boldsymbol{y}$ is assigned to the vector $\boldsymbol{x}$, the quantity $\boldsymbol{T}$ is designated as a second order tensor. The tensor $\boldsymbol{T}$ is just a mapping and nothing has been said about the structure of the above tensor. The properties of $\boldsymbol{T}$ will be determined later.

Examples for Second Order Tensors

1. The Dyadic Product or Tensor Product

The following relation is given:

$$y = (b \cdot x)a = a(b \cdot x). \tag{2.28}$$

From this relation, one can extract the vector $\boldsymbol{x}$ through the following notation:

$$y = (a \circ b)x \quad \text{with} \quad T = a \circ b = ab^T. \tag{2.29}$$

Here, $\boldsymbol{a} \circ \boldsymbol{b}$ is termed as the dyadic or tensor product and it provides a new relationship between the vectors $\boldsymbol{a}$ and $\boldsymbol{b}$.

From the mathematical laws given by Eqs. (2.25) and (2.26), it can easily be shown that

$$T = a \circ b \tag{2.30}$$

is a second order tensor, as defined previously. Consequently the dyadic product in components of a coordinate system (x, y, z) is

$$a = \begin{bmatrix} a_x \\ a_y \\ a_z \end{bmatrix} \quad \text{and} \quad b = \begin{bmatrix} b_x \\ b_y \\ b_z \end{bmatrix},$$
$$T = ab^T = \begin{bmatrix} a_x \\ a_y \\ a_z \end{bmatrix} \begin{bmatrix} b_x, b_y, b_z \end{bmatrix} = \begin{bmatrix} a_xb_x & a_xb_y & a_xb_z \\ a_yb_x & a_yb_y & a_yb_z \\ a_zb_x & a_zb_y & a_zb_z \end{bmatrix}. \tag{2.31}$$

2. The Vector Product or Cross Product

$$\boldsymbol{b} = \boldsymbol{\omega} \times \boldsymbol{a} \tag{2.32}$$

is likewise a relationship between the vectors $\boldsymbol{a}$ and $\boldsymbol{b}$, for which the laws (2.25) and (2.26) are valid as well. Therefore, the vector product can be rewritten as:

$$\boldsymbol{b} = \boldsymbol{\Omega a} = \boldsymbol{\omega} \times \boldsymbol{a} = \tilde{\omega} a \tag{2.33}$$

with $\boldsymbol{\Omega} = \tilde{\omega} = \boldsymbol{\omega} \times \cdots$ corresponding to the operator of the vector product, where $\boldsymbol{\omega}$ represents the rotational velocity.

In component form:

$$\boldsymbol{\Omega} = \begin{bmatrix} 0 & -\omega_z & \omega_y \\ \omega_z & 0 & -\omega_x \\ -\omega_y & \omega_x & 0 \end{bmatrix}, \tag{2.34}$$

$$\boldsymbol{b} = \begin{bmatrix} -\omega_z a_y + \omega_y a_z \\ \omega_z a_x - \omega_x a_z \\ -\omega_y a_x + \omega_x a_y \end{bmatrix}. \tag{2.35}$$

Another important example for a second order tensor is the rotational tensor $\boldsymbol{T}$, which is introduced in the next section.

Note: For the complete definition of a second order tensor, further statements must be made about its transformation properties. The reader is advised to refer to (Klingbeil 1966) for further information.

2.4 Free Motion of Rigid Bodies

2.4.1 General Motion of Rigid Bodies

The general motion of a rigid body in space is characterized by six independent degrees of freedom (DoF). As will be shown later through exemplary physical interpretations, it is best to describe the general motion of a rigid body by superimposing the translational motion of an arbitrarily chosen point (commonly the center of mass) and a spatial rotational motion with respect to this point. This definition of the spatial motion of a rigid body is sufficient for the statements made within the framework of this book, see Fig. 2.5.

Description of the pose In the following, the motion of a rigid body is to be defined as a superposition of a translational and a rotational motion. Thus, one distinguishes between:

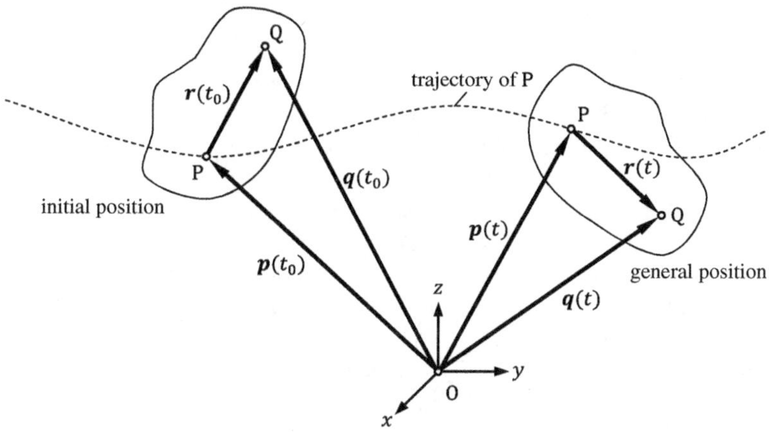

Fig. 2.5 General motion of a rigid body

- Trajectory $\boldsymbol{p}(t)$ of an arbitrary reference point P inside the body: characterized by three translational degrees of freedom.
- Rotational motion of the body about the reference point P: characterized by three rotational degrees of freedom.

The position vector of an arbitrary point Q inside the body (see Fig. 2.5), is given as:

$$\boldsymbol{q}(t) = \boldsymbol{p}(t) + \boldsymbol{r}(t) = \boldsymbol{p}(t) + \boldsymbol{T}(t)\boldsymbol{r}(t_0) \tag{2.36}$$

$\boldsymbol{T}(t)$ is the previously mentioned tensor of rotational motion. The quantities $\boldsymbol{p}(t)$ and $\boldsymbol{T}(t)$ combined contain six independent position coordinates:

$$\boldsymbol{w}(t) = [w_1, \ldots, w_6]^T. \tag{2.37}$$

If three parameters of translation (e.g. the three components of the position vector $\boldsymbol{p}$ to the arbitrary reference point P) and three parameters of rotation (e.g. three independent parameters for the rotational motion, expressed by the rotational tensor $\boldsymbol{T}$) are known, the trajectory of an arbitrary point Q of the rigid body $\boldsymbol{q}(t)$ can be calculated by Eq. (2.36).

Remarks about the rotational tensor $\boldsymbol{T}(t)$

- The rotational tensor is a second-order tensor and is, in general, a function of time.
- If the rotational tensor is written in component form, then it corresponds to a [3 × 3] matrix with nine elements, only three of which are independent. Hence,

there are further dependencies between the elements, which are more specifically discussed in Sect. 2.5.3.

- There exists a close correlation between the rotational tensor $\boldsymbol{T}(t)$ and transformation matrix ${}^{i}\boldsymbol{T}_{j}$, which was introduced in Sect. 2.2 (Eq. 2.18 through 2.23). This correlation will be briefly discussed here.

Relationship between rotational tensor and transformation matrix The relationship between the rotational tensor and the coordinate transformation should first be clarified through a planar example. One considers the transformation between two body-fixed coordinate systems for a planar fixed-point rotation about the axis of rotation $\boldsymbol{e}_z$ with the angle of rotation β (Fig. 2.6). The pair $(\boldsymbol{e}_z, \beta)$, later referred to as the rotation vector, is more closely examined in Sect. 2.5. The body-fixed coordinate systems are on the one hand the system (x_0, y_0, z_0) of the initial position of the rotating body (described by the vector $\boldsymbol{r}_0$) and on the other hand the system of the rotated position (x_1, y_1, z_1) (described by the vector $\boldsymbol{r}_1 = \boldsymbol{r}(t)$).

The representation of the rotation in terms of the rotational tensor $\boldsymbol{T}(\boldsymbol{e}_z, \beta)$ is

$$\boldsymbol{r}_1 = \boldsymbol{T}(\boldsymbol{e}_z, \beta)\boldsymbol{r}_0. \tag{2.38}$$

For the component representation of Eq. (2.38) in the initial coordinate system (x_0, y_0, z_0) the following applies:

$${}^{0}\boldsymbol{r}_1 = {}^{0}\boldsymbol{T}([0,0,1]^T, \beta)\, {}^{0}\boldsymbol{r}_0, \tag{2.39}$$

$$\begin{bmatrix} {}^{0}r_{1x} \\ {}^{0}r_{1y} \\ {}^{0}r_{1z} \end{bmatrix} = \begin{bmatrix} \cos\beta & -\sin\beta & 0 \\ \sin\beta & \cos\beta & 0 \\ 0 & 0 & 1 \end{bmatrix} \begin{bmatrix} {}^{0}r_{0x} \\ {}^{0}r_{0y} \\ {}^{0}r_{0z} \end{bmatrix}. \tag{2.40}$$

If one compares this representation with a corresponding planar transformation between two planar coordinate systems, it is obvious that it corresponds directly to the matrix ${}^{0}\boldsymbol{T}_1$ of the coordinate transformation.

$$(x_1, y_1, z_1) \rightarrow (x_0, y_0, z_0).$$

Or in coordinate representation

$${}^{0}\boldsymbol{r}_1 = {}^{0}\boldsymbol{T}_1\, {}^{1}\boldsymbol{r}_1 \text{ and } {}^{0}\boldsymbol{r}_0 = {}^{1}\boldsymbol{r}_1. \tag{2.41}$$

Thus the following is generally true: the rotational tensor of the rotation $\boldsymbol{r}_0 \rightarrow \boldsymbol{r}_1$, expressed in the system (x_0, y_0, z_0) corresponds to the matrix of the transformation in the opposite direction $(x_1, y_1, z_1) \rightarrow (x_0, y_0, z_0)$.

Description of velocity and motion twist The derivative of Eq. (2.36) with respect to time yields:

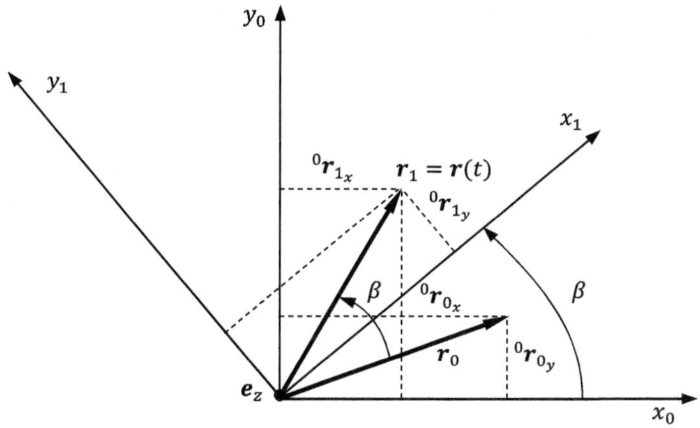

Fig. 2.6 Relationship between rotational tensor and coordinate transformation

Fig. 2.7 Motion twist

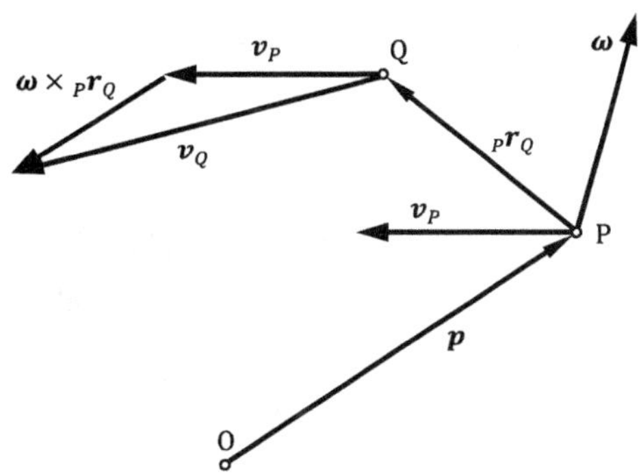

$$\dot{q}(t) = \dot{p}(t) + \dot{r}(t) = \dot{p}(t) + \dot{T}(t)r(t_0). \tag{2.42}$$

With the transformation

$$\dot{T}(t)r(t_0) = \underbrace{\dot{T}(t)T^{-1}(t)}_{\omega\times\cdots}r(t) = \dot{T}(t)T^T(t)r(t) = \omega(t) \times r(t) \tag{2.43}$$

and with the relationship $\boldsymbol{\Omega}(t) = \dot{T}(t)T^T(t)$ (POISON-equation) it follows:

$$\dot{q}(t) = \dot{p}(t) + \omega(t) \times r(t). \tag{2.44}$$

Note: The motion twist describes the state of motion of a rigid body as a whole and consists in accordance to Fig. 2.7 for a specified point P of

Fig. 2.8 Motion screw

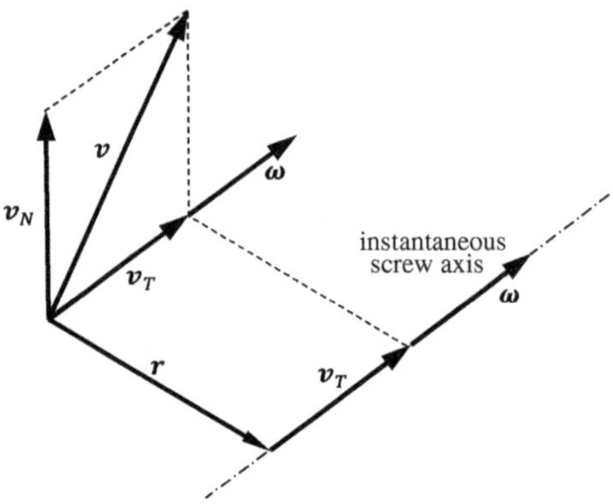

$$\left.\begin{array}{ll}\text{translational velocity} & \boldsymbol{v}_P \\ \text{angular velocity} & \boldsymbol{\omega}\end{array}\right\} \quad (\boldsymbol{v}_P, \boldsymbol{\omega}) - \text{motion twist}$$

If the motion twist $\boldsymbol{v}_P, \boldsymbol{\omega}$ for a point P is given, then for another reference point Q in the body it applies according to Eq. (2.44):

$$\boldsymbol{v}_Q = \boldsymbol{v}_P + \boldsymbol{\omega} \times {}_P\boldsymbol{r}_Q. \tag{2.45}$$

The instantaneous angular velocity $\boldsymbol{\omega}$ of a rigid body is constant everywhere, while the translational velocity depends on the reference point. By systematically translating the reference point, the motion twist can be reduced to a motion screw. A motion screw is characterized by the translational and rotational velocities being parallel to each other along the instantaneous axis of the screw motion see Fig. 2.8.

Calculation of acceleration The second time derivative of Eq. (2.44) yields:

$$\ddot{\boldsymbol{q}}(t) = \ddot{\boldsymbol{p}}(t) + \dot{\boldsymbol{\omega}}(t) \times \boldsymbol{r}(t) + \boldsymbol{\omega}(t) \times \underbrace{\dot{\boldsymbol{r}}(t)}_{\boldsymbol{\omega}(t) \times \boldsymbol{r}(t)}, \tag{2.46}$$

respectively

$$\ddot{\boldsymbol{q}}(t) = \ddot{\boldsymbol{p}}(t) + \dot{\boldsymbol{\omega}}(t) \times \boldsymbol{r}(t) + \boldsymbol{\omega}(t) \times [\boldsymbol{\omega}(t) \times \boldsymbol{r}(t)]. \tag{2.47}$$

2.4.2 Relative Motion

Relative motion plays a crucial role in mechanics, which is particularly true for the mechanical problems related to vehicle dynamics that will be analyzed in this book.

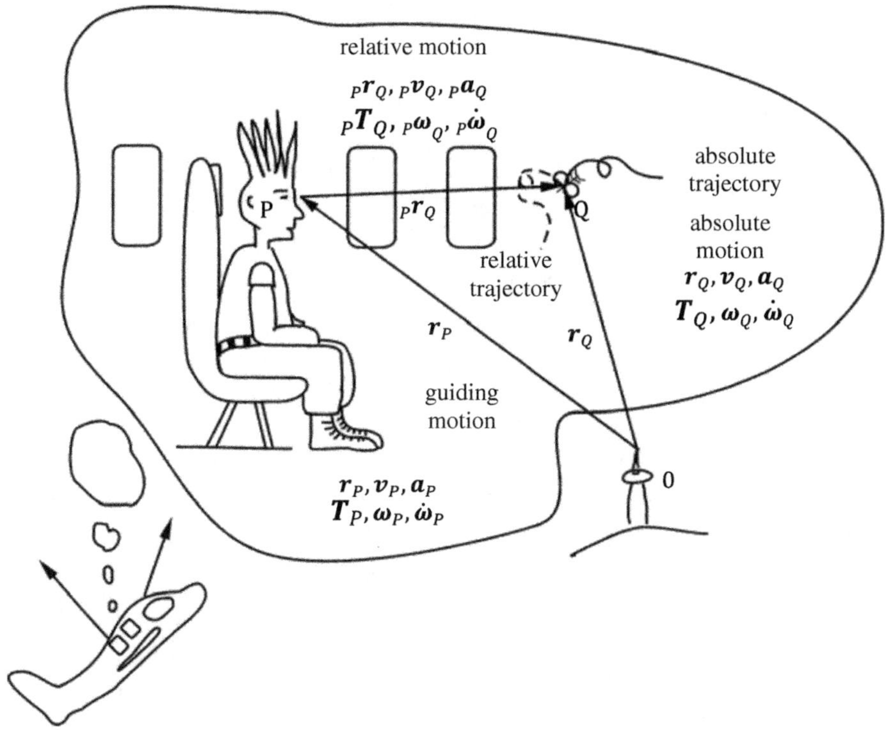

Fig. 2.9 Guiding motion P, relative motion Q with respect to P, and absolute motion Q

Therefore, some fundamental relationships from elementary mechanics will be summarized in this section, and illustrated with the help of an example where a plane, flying relative to the earth (inertial frame 0), is carrying a motionless passenger (system P). Also inside the airplane, a fly Q is moving relative to the passenger and the plane. The airplane together with the passenger acts as the so called “guiding motion” P. Additionally, the fly travels with a relative motion Q with respect to P (Fig. 2.9) (Hiller et al. 1986–1988). Using this, the following problem may be stated:
Given: Guiding motion P and relative motion Q with respect to P.
To find: Absolute motion Q of the fly with respect to inertial system 0.

In particular it applies:
Pose $\boldsymbol{r}_Q(t)$ Translation (expressed by position vectors):

$$\boldsymbol{r}_Q = \boldsymbol{r}_p + {_P\boldsymbol{r}_Q}\,, \tag{2.48}$$

rotation (expressed by rotational tensors):

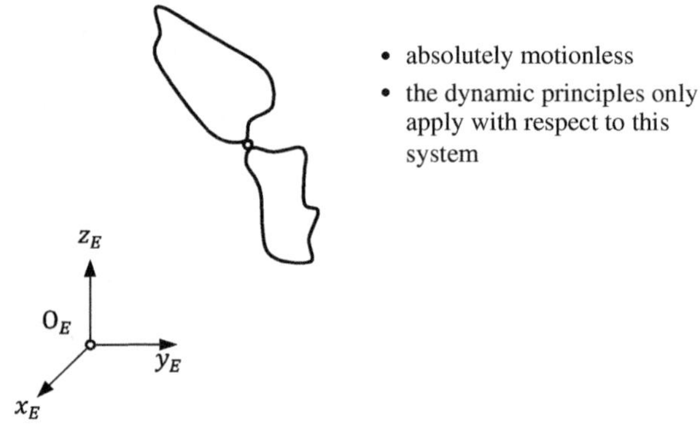

Fig. 2.10 Inertial system

$$\boldsymbol{T}_Q = {}_P\boldsymbol{T}_Q \boldsymbol{T}_P, \tag{2.49}$$

or alternatively by transformation matrices:

$$\boldsymbol{T}_Q = \boldsymbol{T}_P {}^P\boldsymbol{T}_Q. \tag{2.50}$$

Velocity $\boldsymbol{v}_Q(\boldsymbol{t})$:

$$\boldsymbol{v}_Q = \boldsymbol{v}_P + {}_P\boldsymbol{v}_Q + \boldsymbol{\omega}_P \times {}_P\boldsymbol{r}_Q, \tag{2.51}$$

$$\boldsymbol{\omega}_Q = \boldsymbol{\omega}_P + {}_P\boldsymbol{\omega}_Q. \tag{2.52}$$

Acceleration $\boldsymbol{a}_Q(\boldsymbol{t})$: After some calculation we find that:

$$\boldsymbol{a}_Q = \boldsymbol{a}_P + p\boldsymbol{a}_Q + \dot{\boldsymbol{\omega}}_P \times {}_P\boldsymbol{r}_Q + 2\boldsymbol{\omega}_P \times {}_P\boldsymbol{v}_Q + \boldsymbol{\omega}_P \times (\boldsymbol{\omega}_P \times {}_P\boldsymbol{r}_Q) \tag{2.53}$$

$$\dot{\boldsymbol{\omega}}_Q = \dot{\boldsymbol{\omega}}_P + {}_P\dot{\boldsymbol{\omega}}_Q + \boldsymbol{\omega}_P \times {}_P\,\boldsymbol{\omega}_Q. \tag{2.54}$$

Special terminology:

$\dot{\boldsymbol{\omega}}_P \times {}_P\boldsymbol{r}_Q$	EULER-acceleration
$2\boldsymbol{\omega}_P \times {}_P\boldsymbol{v}_Q$	CORIOLIS-acceleration
$\boldsymbol{\omega}_P \times (\boldsymbol{\omega}_P \times {}_P\boldsymbol{r}_Q)$	centripetal acceleration

For supplementary information, refer to (Magnus and Müller-Slany 2005).

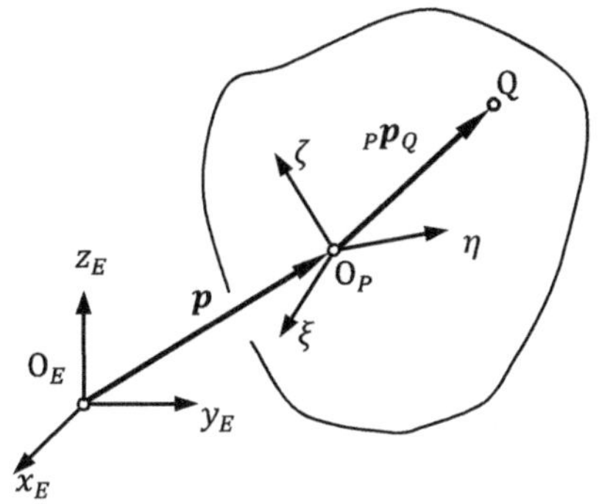

- relative positions of all mass particles of the body are fixed
- the physical characteristics of the body are given with respect to this system (moment of inertia, body - fixed points)

Fig. 2.11 Body-fixed reference frame

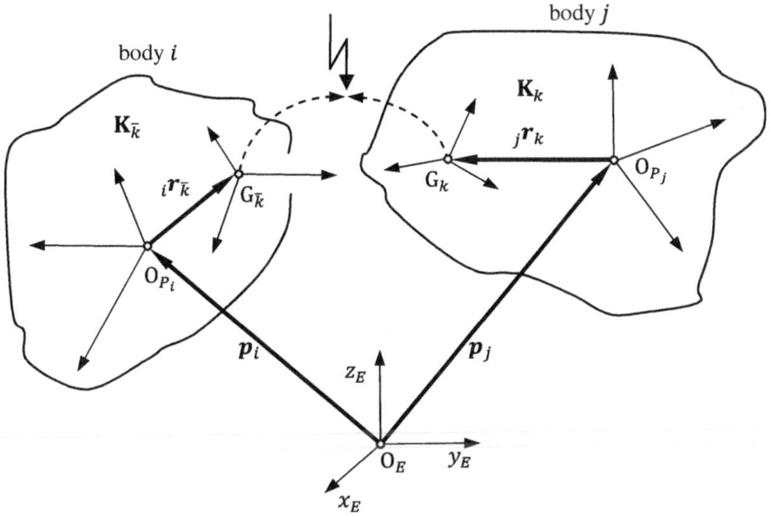

Fig. 2.12 Reference frames for interactions (constraints, forces)

2.4.3 *Important Reference Frames*

- Inertial system $\mathbf{K}_E = \{O_E; x_E, y_E, z_E\}$ (Fig. 2.10)
- Body-fixed reference frame $K_P = \{O_P; \xi_P, \eta_P, \zeta_P\}$ (Fig. 2.11)
- Reference frames for interactions (constraints, forces) (Fig. 2.12)
- Additional coordinate systems $\mathbf{K}_k$, $\mathbf{K}_{\bar{k}}$, which are defined with respect to the body-fixed systems $\mathbf{K}_i$ and $\mathbf{K}_j$.
- With respect to these systems, the relative position and motion of the body "i" and "j" are represented by simple conditions (constraints). This also applies to the interactions (forces).

2.5 Rotational Motion

This section discusses different representations of the rotational motion of rigid bodies. Initially, only position and velocity will be considered. Regarding the position, one must differentiate between infinitesimal and finite rotation (displacement). The fundamental quantity used to describe the finite rotation is the rotational tensor $\boldsymbol{T}$, about which nothing has been said regarding its parameterization until now. In direct correlation with the infinitesimal rotation, the angular velocity $\boldsymbol{\omega}$ serves as an important quantity for its description.

2.5.1 *Spatial Rotation and Angular Velocity in General Form*

General rotational motion can also be treated as the motion of a rigid body around a fixed point (fixed-point rotation). Using this, along with the considerations from Sect. 2.4.1, it follows for the rotational motion (Fig. 2.13):

Pose With the rotational tensor $\boldsymbol{T}(t)$ from Sect. 2.4.1 it follows:

$$\boldsymbol{r}(t) = \boldsymbol{T}(t)\boldsymbol{r}(t_0). \tag{2.55}$$

The mathematical structure of $\boldsymbol{T}(t)$ depends on the chosen rotational parameters (see Sect. 2.5.3). Furthermore, the inverse relationship also exists:

$$\boldsymbol{r}(t_0) = \boldsymbol{T}^{-1}(t)\boldsymbol{r}(t). \tag{2.56}$$

Velocity Here, we get:

$$\dot{\boldsymbol{r}}(t) = \dot{\boldsymbol{T}}(t)\boldsymbol{r}(t_0) = \dot{\boldsymbol{T}}(t)\boldsymbol{T}^{-1}(t)\boldsymbol{r}(t) = \boldsymbol{\omega}(t) \times \boldsymbol{r}(t) \tag{2.57}$$

with

$$\dot{\boldsymbol{T}}(t)\boldsymbol{T}^{-1}(t)\boldsymbol{r}(t) = \boldsymbol{\omega}(t) \times \boldsymbol{r}(t) = \boldsymbol{\Omega}(t)\boldsymbol{r}(t) = \tilde{\boldsymbol{\omega}}\boldsymbol{r}(t). \tag{2.58}$$

Here, $\boldsymbol{\Omega}$ is a skew-symmetric tensor which performs the operation of a vector product (seen also in Sect. 2.3). In a coordinate system (x, y, z), the component representation of $\boldsymbol{\Omega}$ can be read as:

$$\boldsymbol{\Omega} = \tilde{\omega} = \begin{bmatrix} 0 & -\omega_z & \omega_y \\ \omega_z & 0 & -\omega_x \\ -\omega_y & \omega_x & 0 \end{bmatrix}. \tag{2.59}$$

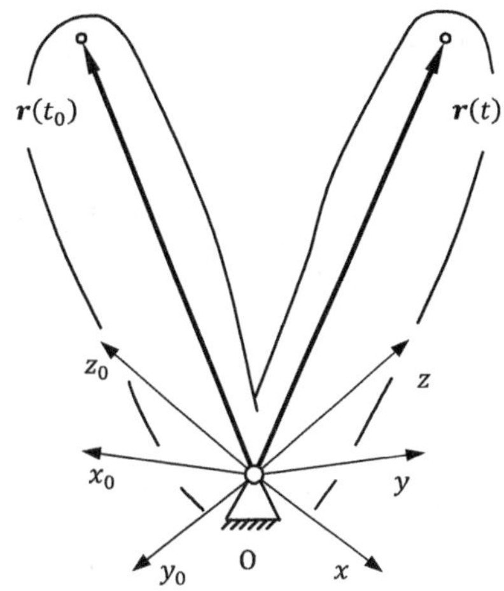

Fig. 2.13 Fixed-point rotation

However, there is still nothing stated about which rotational parameters describe the rotational motion.

2.5.2 *Parameterizing of Rotational Motion*

From a technical perspective, there are two well established physical principles to represent spatial rotational motion. The first is the realization of the spatial rotation as a single rotation about a spatial rotational axis, to be appropriately chosen, along with the respective angle of rotation. This principle was already applied in Sect. 2.4.1 without detailed explanations. The second principle assumes that a spatial rotation can be represented through a combination of three consecutive rotations; so-called "elementary (planar) rotations" about different rotational axes. Therefore, it is obvious that in the case of the spatial fixed-point rotation from the previous section the parameters of the fixed-point rotation may now consist of the three components of the axis vector $\boldsymbol{u}(t)$ and the associated scalar angle of rotation $\beta(t)$, altogether four parameters for the description of the spatial rotation. This pair

$$\vec{\boldsymbol{\beta}} = (\boldsymbol{u}(t), \beta(t)) \tag{2.60}$$

is called the "rotational displacement pair" $\vec{\boldsymbol{\beta}}$, and is comprised of the axis of rotation $\boldsymbol{u}(t)$ and a corresponding angle of rotation $\beta(t)$, (Hiller 1981). As a unit vector with two independent components is sufficient to define an axis vector $\boldsymbol{u}(t)$, a further secondary condition is given in the following equation:

Table 2.1 Parameterization of rotational motion

Parameter types	Number of parameters
Pair of rotation $\vec{\boldsymbol{\beta}} = (\boldsymbol{u}, \beta)$	4
EULER angles θ, ψ, φ (rotation sequence: $z \to x \to z$)	3
CARDAN angles ψ, θ, φ (rotation sequence: $z \to y \to x$)	3
quaternions and EULER parameters $q_0,\ q_1,\ q_2,\ q_3$	4
RODRIGUES parameters $\lambda_1,\ \lambda_2,\ \lambda_3$	3

$$u_x^2 + u_y^2 + u_z^2 = 1. \tag{2.61}$$

When a spatial rotation is described as being about three planar elementary rotations, the corresponding parameters are the angles of elementary rotation, e.g. θ, ψ and φ, just as it is with EULER or CARDAN angles (Hiller 1983).

In the following sections, the method of the rotational displacement pair $\vec{\boldsymbol{\beta}} = (\boldsymbol{u}(t), \beta(t))$ will be discussed as well as the method of the CARDAN angles, an especially popular approach in automotive engineering. Table 2.1 gives a summary of possible parameterizations, each having either three or four parameters.

The geometric interpretation of the EULER parameters and the RODRIGUES parameters as spatial rotations also requires the use of the rotational displacement pair, namely the spatial axis of rotation $\boldsymbol{u}$ and angle of rotation β (Hiller 1981; Hiller and Kecskeméthy 1987; Hiller et al. 1986 – 1988).

2.5.3 The Rotational Displacement Pair and Tensor of Rotation

In this section, the rotation of a rigid body is described with the help of an axis of rotation $\boldsymbol{u}$ and a corresponding angle of rotation β, see Fig. 2.14. One should note that for every finite spatial rotation, a different axis of rotation $\boldsymbol{u}$ and angle of rotation β exist. The following notation is used:

$$\left.\begin{array}{l}\text{axis of rotation } \boldsymbol{u} \\ \text{angle of rotation } \beta\end{array}\right\} = \text{rotational displacement pair } \vec{\boldsymbol{\beta}} = (\boldsymbol{u}, \beta), \tag{2.62}$$
$$\text{initial position } \boldsymbol{r}_0 = \boldsymbol{r}(t_0).$$

The derivation of the tensor of rotation for these rotational parameters can be carried out using various methods, however, they will not be discussed in great detail here (Hiller et al. 1986 – 1988; Klingbeil 1966).

One arrives at the following result using Eq. (2.55):

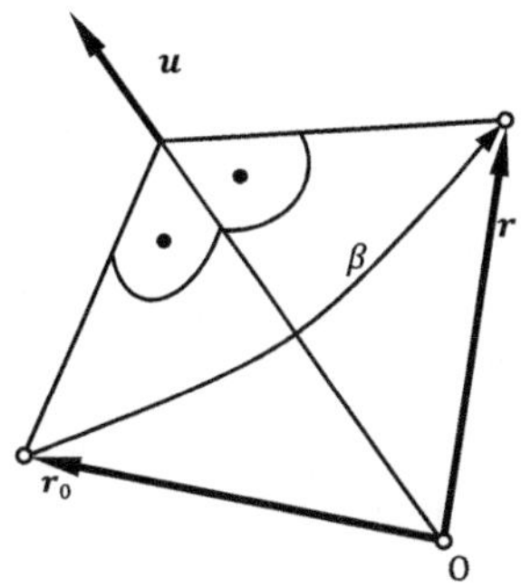

Fig. 2.14 Axis of rotation $\boldsymbol{u}$ and angle of rotation $\boldsymbol{\beta}$ of the rotational displacement pair

$$\boldsymbol{r} = \boldsymbol{T}(\boldsymbol{u}, \beta)\boldsymbol{r}_0 \tag{2.63}$$

where the tensor of rotation is given by

$$\boldsymbol{T}(\boldsymbol{u}, \beta) = \cos\beta\boldsymbol{I} + (1 - \cos\beta)\boldsymbol{u} \circ \boldsymbol{u} + \sin\beta\, \tilde{\boldsymbol{u}}. \tag{2.64}$$

Here $\boldsymbol{u} \circ \boldsymbol{u}$ stands for the dyadic product (see Eqs. 2.28–2.31) and $\tilde{\boldsymbol{u}} = \boldsymbol{U}$ again for the cross product operation with the axis vector $\boldsymbol{u}$ (compare Eq. 2.61). Here $\boldsymbol{I}$ denotes the identity matrix. Another form of this equation is given by

$$\boldsymbol{U}^2 = \boldsymbol{u} \circ \boldsymbol{u} - \boldsymbol{I}, \tag{2.65}$$

$$\boldsymbol{T}(\boldsymbol{u}, \beta) = \boldsymbol{I} + (1 - \cos\beta)\boldsymbol{U}^2 + \sin\beta\boldsymbol{U}. \tag{2.66}$$

The component representation of the tensor of rotation $\boldsymbol{T}(\boldsymbol{u}, \beta)$ in the system (x, y, z) is given below, where "s" = sin and "c" = cos:

$$\boldsymbol{T}(\boldsymbol{u}, \beta) = \begin{bmatrix} (1 - \mathrm{c}\beta)u_x^2 + \mathrm{c}\beta & (1 - \mathrm{c}\beta)u_x u_y - u_z \mathrm{s}\beta & (1 - \mathrm{c}\beta)u_x u_z + u_y \mathrm{s}\beta \\ (1 - \mathrm{c}\beta)u_y u_x + u_z \mathrm{s}\beta & (1 - \mathrm{c}\beta)u_y^2 + \mathrm{c}\beta & (1 - \mathrm{c}\beta)u_y u_z - u_x \mathrm{s}\beta \\ (1 - \mathrm{c}\beta)u_z u_x - u_y \mathrm{s}\beta & (1 - \mathrm{c}\beta)u_z u_y + u_x \mathrm{s}\beta & (1 - \mathrm{c}\beta)u_z^2 + \mathrm{c}\beta \end{bmatrix} \tag{2.67}$$

Some important properties of the tensor of rotation

Decomposing into a symmetric and an anti-symmetric part In general, the symmetric (s) and skew-symmetric (a) parts of a tensor with respect to a matrix are obtained as:

$$\boldsymbol{T}^s = \frac{1}{2}\left(\boldsymbol{T} + \boldsymbol{T}^T\right), \quad \boldsymbol{T}^s = \boldsymbol{T}^{sT}, \tag{2.68}$$

$$\boldsymbol{T}^a = \frac{1}{2}\left(\boldsymbol{T} - \boldsymbol{T}^T\right), \quad \boldsymbol{T}^a = -\boldsymbol{T}^{aT}. \tag{2.69}$$

Hence the following applies for the tensor of rotation $\boldsymbol{T}(\boldsymbol{u},\beta)$:

$$\boldsymbol{T}^s(\boldsymbol{u},\beta) = \boldsymbol{I} + (1-\cos\beta)\boldsymbol{U}^2 \text{ symmetric,} \tag{2.70}$$

$$\boldsymbol{T}^a(\boldsymbol{u},\beta) = \boldsymbol{U}\sin\beta \quad \text{anti-symmetric.} \tag{2.71}$$

Note: The unit-tensor $\boldsymbol{I}$ and the tensor $\boldsymbol{U}^2$ are both symmetric; the tensor $\boldsymbol{U}$ is already a priori anti-symmetric.

Transposed tensor of rotation Using Eq. (2.64) and the relations $\cos\beta = \cos(-\beta)$ and $\sin\beta = -\sin(-\beta)$, the following is valid for the transposed tensor of rotation:

$$\boldsymbol{T}^T(\boldsymbol{u},\beta) = \boldsymbol{T}(\boldsymbol{u},-\beta). \tag{2.72}$$

Inverse tensor of rotation and orthogonality For the inverse of the tensor of rotation, it is valid that

$$\boldsymbol{T}^{-1}(\boldsymbol{u},\beta) = \boldsymbol{T}(\boldsymbol{u},-\beta) = \boldsymbol{T}^T(\boldsymbol{u},\beta). \tag{2.73}$$

It should also be noted that the inverse is equal to the transposed! This follows from the property of orthogonality, i.e. the rows and columns of the tensor of rotation are mutually perpendicular.

2.5.4 Rotational Displacement Pair and Angular Velocity

The relationship between the tensor of rotation and angular velocity is given in Eq. (2.58). Specifically for the rotational displacement pair it follows:

$$\boldsymbol{\Omega} = \dot{\boldsymbol{T}}(\boldsymbol{u},\beta)\boldsymbol{T}^{-1}(\boldsymbol{u},\beta) = \dot{\boldsymbol{T}}(\boldsymbol{u},\beta)\boldsymbol{T}(\boldsymbol{u},-\beta). \tag{2.74}$$

By evaluation of Eq. (2.74) using Eq. (2.64), the following relationship is obtained:

$$\boldsymbol{\omega} = \dot{\beta}\boldsymbol{u} + [\sin\beta\boldsymbol{I} + (1-\cos\beta)\boldsymbol{U}]\dot{\boldsymbol{u}}. \tag{2.75}$$

For small rotations, where $\beta \ll 1$, it is simplified to

$$\boldsymbol{\omega} = \dot{\beta}\boldsymbol{u} + \beta\dot{\boldsymbol{u}} \tag{2.76}$$

and especially for fixed axes of rotation (i.e. $\dot{\boldsymbol{u}} = 0$), one finally obtains

$$\boldsymbol{\omega} = \dot{\beta}\boldsymbol{u}. \tag{2.77}$$

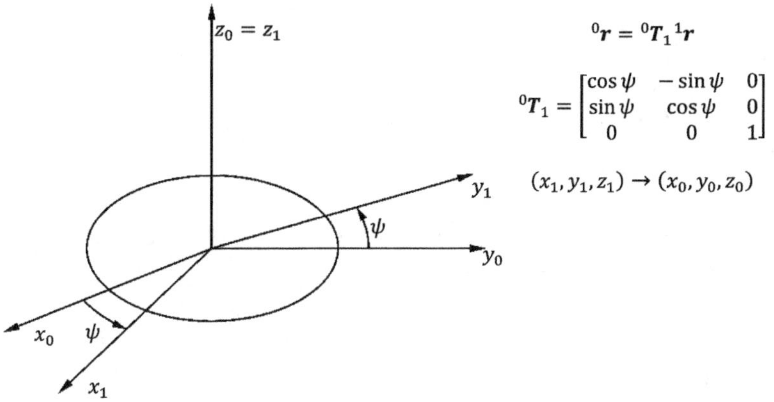

Fig. 2.15 Definition of the first CARDAN rotation

2.5.5 *CARDAN (BRYANT) Angles*

CARDAN angles and the transformation matrix As already mentioned in Sect. 2.5.2, there exists another fundamental parameterization of spatial rotations that is based on three planar, consecutive, elementary rotations about defined coordinate axes. The CARDAN angles (in the English literature often named as BRYANT angles)

$$\psi, \theta, \varphi \tag{2.78}$$

have established themselves in vehicle dynamics, and will be described in the upcoming section. The partial rotations, as well as the resulting complete rotation will be described with the help of transformation matrices. The initial coordinate system (x_0, y_0, z_0) can be regarded as an inertial system and the target system (x_3, y_3, z_3) as the body-fixed system. The transformation of a vector $\boldsymbol{r}$ from the body-fixed coordinate system into the inertial system is described subsequently.

- partial rotation around the z_0-axis: angle of rotation ψ (Fig. 2.15),
- partial rotation around the y_1-axis: angle of rotation θ (Fig. 2.16),
- partial rotation around the x_2-axis: angle of rotation φ (Fig. 2.17).

The resulting transformation from the body-fixed system (x_3, y_3, z_3) into spatially-fixed system (x_0, y_0, z_0) is read as:

$${}^0\boldsymbol{r} = {}^0\boldsymbol{T}_1 \, {}^1\boldsymbol{T}_2 \, {}^2\boldsymbol{T}_3 \, {}^3\boldsymbol{r}. \tag{2.79}$$

Transformation matrix (body-fixed $\rightarrow$ spatially-fixed):

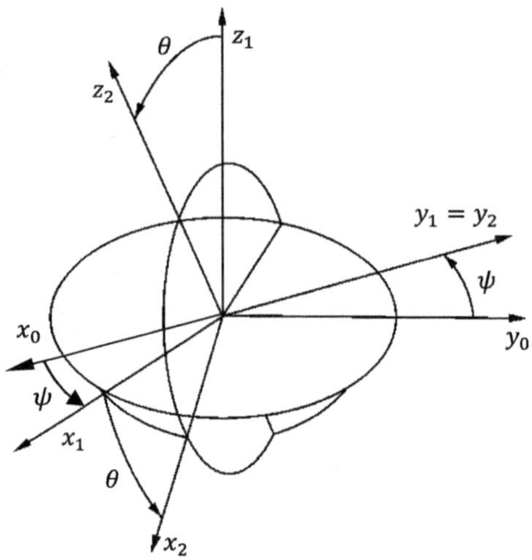

$$^{1}\boldsymbol{r} = {}^{1}\boldsymbol{T}_2\,{}^{2}\boldsymbol{r}$$

$$^{1}\boldsymbol{T}_2 = \begin{bmatrix} \cos\theta & 0 & \sin\theta \\ 0 & 1 & 0 \\ -\sin\theta & 0 & \cos\theta \end{bmatrix}$$

$$(x_2, y_2, z_2) \rightarrow (x_1, y_1, z_1)$$

Fig. 2.16 Definition of the second CARDAN rotation

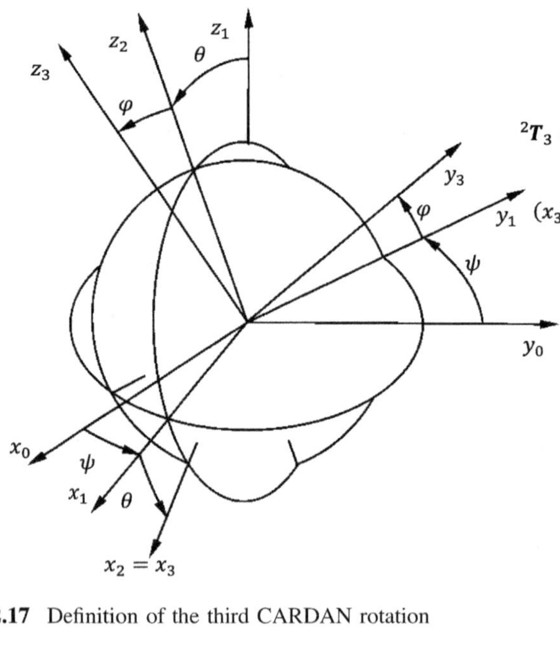

$$^{2}\boldsymbol{r} = {}^{2}\boldsymbol{T}_3\,{}^{3}\boldsymbol{r}$$

$$^{2}\boldsymbol{T}_3 = \begin{bmatrix} 1 & 0 & 0 \\ 0 & \cos\varphi & -\sin\varphi \\ 0 & \sin\varphi & \cos\varphi \end{bmatrix}$$

$$(x_3, y_3, z_3) \rightarrow (x_2, y_2, z_2)$$

Fig. 2.17 Definition of the third CARDAN rotation

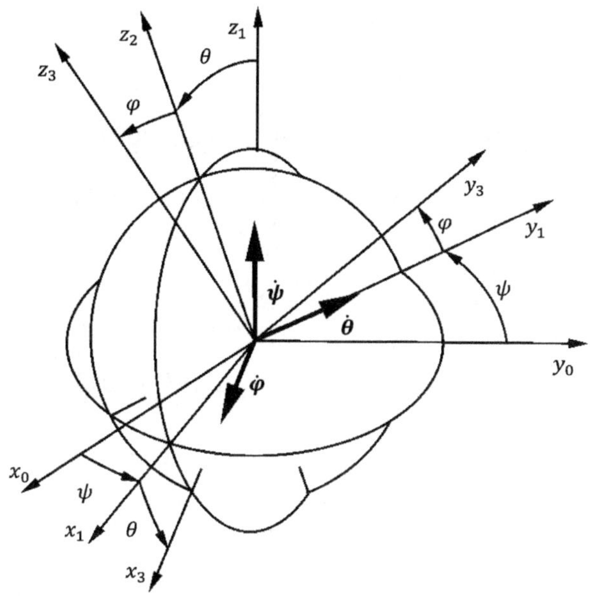

Fig. 2.18 Kinematic CARDAN equations

$$(x_3, y_3, z_3) \rightarrow (x_0, y_0, z_0), \tag{2.80}$$

$$^0\boldsymbol{T}_3 = \begin{bmatrix} c\psi c\theta & -s\psi c\varphi + c\psi s\theta s\varphi & s\psi s\varphi + c\psi s\theta c\varphi \\ s\psi c\theta & c\psi c\varphi + s\psi s\theta s\varphi & -c\psi s\varphi + s\psi s\theta c\varphi \\ -s\theta & c\theta s\varphi & c\theta c\varphi \end{bmatrix}. \tag{2.81}$$

The transformation from the spatially-fixed system into a body-fixed system results from the inversion of Eq. (2.81). It yields (spatially-fixed $\rightarrow$ body-fixed):

$$(x_0, y_0, z_0) \rightarrow (x_3, y_3, z_3), \tag{2.82}$$

$$^3\boldsymbol{T}_0 = \begin{bmatrix} c\psi c\theta & s\psi c\theta & -s\theta \\ -s\psi c\varphi + c\psi s\theta s\varphi & c\psi c\varphi + s\psi s\theta s\varphi & c\theta s\varphi \\ s\psi s\varphi + c\psi s\theta c\varphi & -c\psi s\theta + s\psi s\theta c\varphi & c\theta c\varphi \end{bmatrix}. \tag{2.83}$$

Kinematic CARDAN equations The kinematic CARDAN equations provide the relationship between the angular velocity $\boldsymbol{\omega}$ of the body-fixed system (x_3, y_3, z_3) relative to the initial system (x_0, y_0, z_0) and the time derivatives of the CARDAN angles: $\dot{\psi}, \dot{\theta}, \dot{\varphi}$. From Fig. 2.18, the resulting angular velocity $\boldsymbol{\omega}$ is composed of the partial angular velocities $\dot{\boldsymbol{\psi}}, \dot{\boldsymbol{\theta}}, \dot{\boldsymbol{\varphi}}$, where

$$\boldsymbol{\omega} = \dot{\boldsymbol{\psi}} + \dot{\boldsymbol{\theta}} + \dot{\boldsymbol{\varphi}}. \tag{2.84}$$

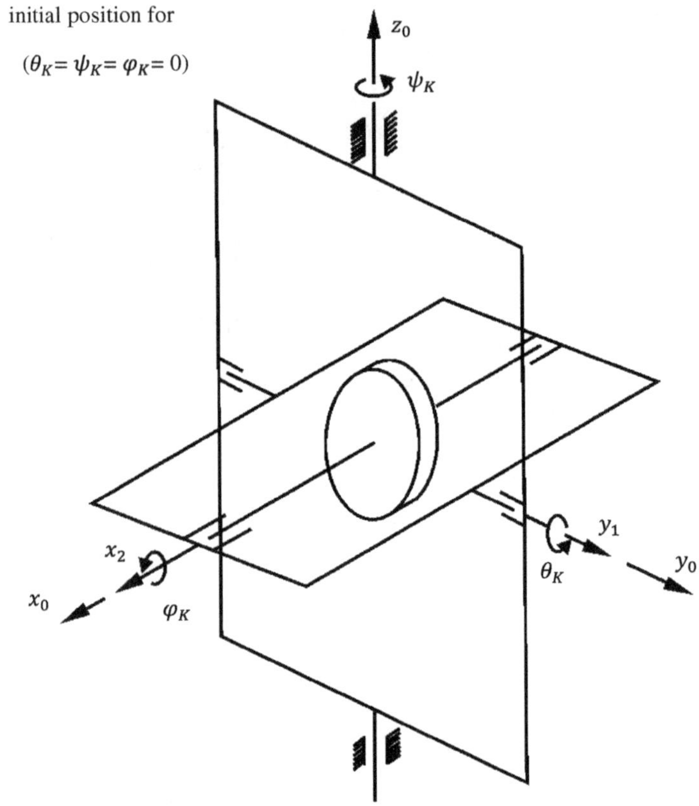

Fig. 2.19 CARDAN angles of a gimbaled gyroscope

This vector equation can be interpreted in the initial system (x_0, y_0, z_0) as

$$^0\boldsymbol{\omega} = \dot{\psi}\,^0\boldsymbol{e}_{z_0} + \dot{\theta}\,^0\boldsymbol{e}_{y_1} + \dot{\varphi}\,^0\boldsymbol{e}_{x_2}. \tag{2.85}$$

The components of $\boldsymbol{\omega}$ in the spatially-fixed system (x_0, y_0, z_0) are extracted from this in the following:

$$\begin{bmatrix} ^0\omega_x \\ ^0\omega_y \\ ^0\omega_z \end{bmatrix} = \begin{bmatrix} 0 & -\sin\psi & \cos\psi\cos\theta \\ 0 & \cos\psi & \sin\psi\cos\theta \\ 1 & 0 & -\sin\theta \end{bmatrix} \begin{bmatrix} \dot{\psi} \\ \dot{\theta} \\ \dot{\varphi} \end{bmatrix}. \tag{2.86}$$

The components of $\boldsymbol{\omega}$ in body-fixed systems (x_3, y_3, z_3), commonly referred to as the kinematic CARDAN equations, read as follows:

$$\begin{bmatrix} {}^3\omega_x \\ {}^3\omega_y \\ {}^3\omega_z \end{bmatrix} = \begin{bmatrix} -\sin\theta & 0 & 1 \\ \cos\theta\sin\varphi & \cos\varphi & 0 \\ \cos\theta\cos\varphi & -\sin\varphi & 0 \end{bmatrix} \begin{bmatrix} \dot{\psi} \\ \dot{\theta} \\ \dot{\varphi} \end{bmatrix}. \tag{2.87}$$

Singular Position: $\theta = \frac{\pi}{2}, \sin\theta = 1, \cos\theta = 0.$

When $\theta = \frac{\pi}{2}$ the angles ψ and φ are no longer distinguishable; and only difference angle $(\psi - \varphi)$ still appears. The reduced transformation matrix from the spatially-fixed system (x_0, y_0, z_0) into the body-fixed system (x_3, y_3, z_3) is

$${}^3\boldsymbol{T}_0 = \begin{bmatrix} 0 & 0 & -1 \\ -\sin(\psi - \varphi) & \cos(\psi - \varphi) & 0 \\ \cos(\psi - \varphi) & \sin(\psi - \varphi) & 0 \end{bmatrix}. \tag{2.88}$$

The initial position of the CARDAN angles can be illustrated by means of a gimbaled gyroscope (Fig. 2.19).

References

Hiller M (1981) Analytisch-numerische Verfahren zur Behandlung räumlicher Übertragungsmechanismen [habil.]. Habilitation, Universität Stuttgart, Düsseldorf: VDI-Verlag

Hiller M (1983) Mechanische Systeme: Eine Einführung in die analytische Mechanik und Systemdynamik. Springer, Berlin u.a. - ISBN 978-3-540-12521-1

Hiller M and Kecskeméthy A (1987) A computer-oriented approach for the automatic generation and solution of the equations of motion for complex mechanisms. In *Proceedings of the 7th. World Congress on Theory of Machines and Mechanisms*, Vol. 1, S 425-30, IFToMM, Pergamon Press., Sevilla, Spain

Hiller M, Kecskeméthy A and Woernle C (1986–1988) Computergestützte Kinematik und Dynamik für Fahrzeuge, Roboter und Mechanismen. In *Carl-Cranz-Kurs*, Vol. 1.16, Carl-Cranz-Gesellschaft, Oberpfaffenhofen

Klingbeil E (1966) Tensorrechnung für Ingenieure. Bibliographisches Institut, Mannheim

Magnus K and Müller-Slany HH (2005) Grundlagen der technischen Mechanik. 7., durchges. und erg. Aufl. (ed.), Teubner, Wiesbaden. - ISBN 978-3-8351-0007-7

Chapter 3
Kinematics of Multibody Systems

In this chapter, the fundamental kinematics for multibody systems are collected and applied to model vehicle components and the entire vehicle. For further details, refer to (Hiller et al. 1986, 1986–1988; Woernle 1988; Hiller and Kecskeméthy 1989; Kecskemethy 1993; Hiller 1995).

3.1 Structure of Kinematic Chains

The combination of rigid bodies interconnected to one another in a multibody system is known as a kinematic chain (this term for linkages is discussed in more detail in Sect. 3.2). A multibody system that is kinematically non-connected can be, if required, converted into a kinematically connected system through implementation of virtual joints that contain six DoF, i.e. $f_{G_i} = 6$ (Fig. 3.1).

3.1.1 Topological Modelling

Two topological principles can be distinguished in kinematic chains: open and closed.

Open kinematic chains–tree structure The path from one body to each other body in an open kinematic chain with tree structure is unique. Within this tree, a preceding body together with its preceding joint, respectively, is assigned to each body. Choosing one body in the tree as a reference body, one gets

$$n_G = n_B \tag{3.1}$$

D. Schramm et al., Vehicle Dynamics,
DOI 10.1007/978-3-662-54483-9_3

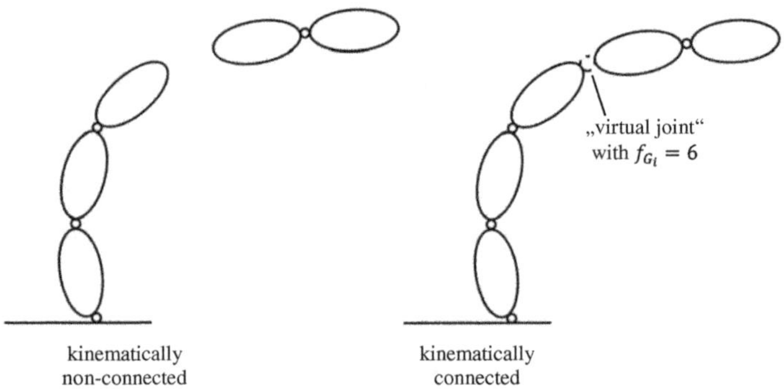

Fig. 3.1 Kinematic chains

with

n_G number of joints
n_B number of bodies (not including the reference body)

Closed kinematic chains—kinematic loops Assuming an open kinematic chain with tree structure, a single independent kinematic loop is obtained by introducing an additional joint in each case. Using Eq. (3.1), the number of independent loops n_L in such a kinematic structure is given by (Fig. 3.2)

$$n_L = n_G - n_B. \tag{3.2}$$

Furthermore, partially and completely closed kinematic chains can be differentiated. A system with kinematic loops forms a partially closed kinematic chain, when

- single partial systems form open chains or
- multiple closed partial systems are connected to each other in an open chain (Fig. 3.3).

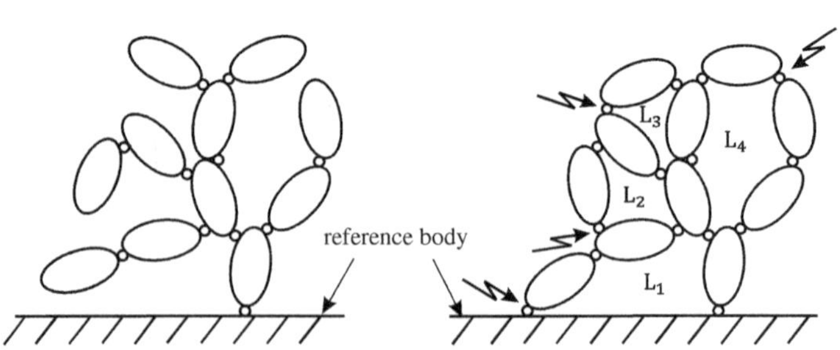

Fig. 3.2 Open kinematic chain and closed kinematic chain

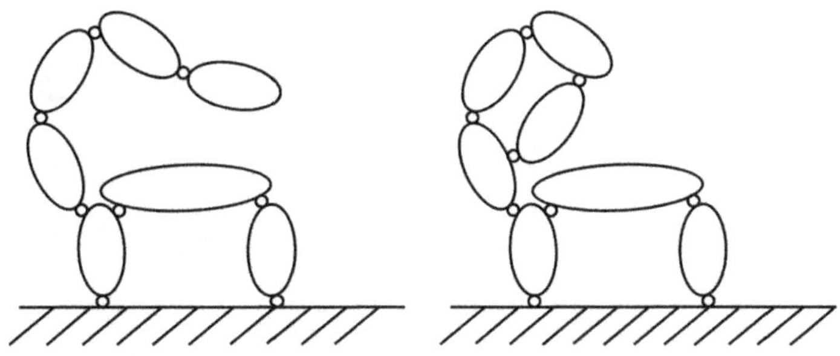

Fig. 3.3 Partially closed kinematic chains

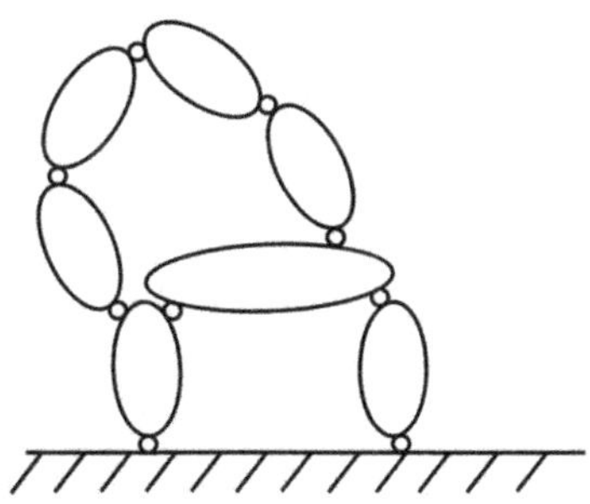

Fig. 3.4 Completely closed kinematic chain (kinematic modeling)

A chain can be considered completely closed when

- each body is a part of a multibody loop and
- each loop has at least one body that is connected to another loop.

A mechanism, by definition, must be a partially or completely closed kinematic chain (for more details see Sect. 3.4.1) (Fig. 3.4).

3.1.2 Kinematic Modelling

Kinematic chains can be grouped into three distinct motion categories:

Planar kinematic chains In a planar kinematic chain all body points move inside or parallel to a reference motion plane (Fig. 3.5). Because of this, the motion of each body in the system has one rotational and two translational motion components. Relative motions between joints must be either translational displacements, that are parallel to the reference motion plane, or rotations, that are normal to the motion plane.

Spherical kinematic chains In a spherical kinematic chain, all body points move on concentric spherical surfaces around a fixed point O in the center (Fig. 3.6). The bodies have three rotational motion components and no translational components.

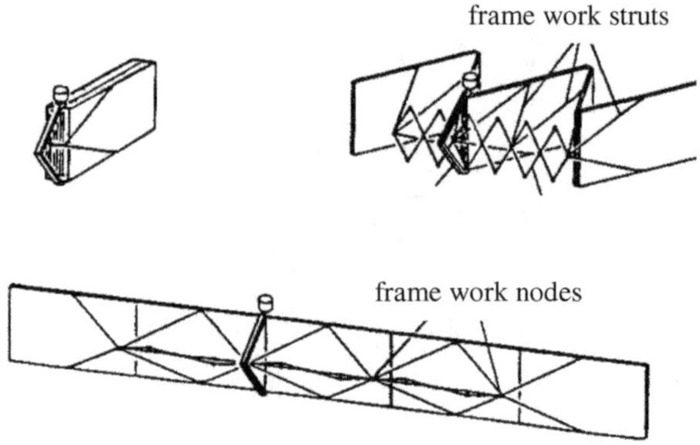

Fig. 3.5 Example for a planar kinematic chain: satellite antenna

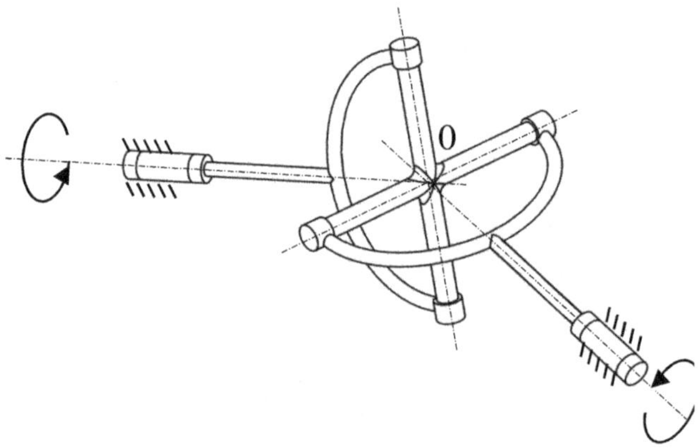

Fig. 3.6 Example for spherical kinematic chains: transmission linkage with CARDAN joint

The relative motions of the bodies can only rotate around the axes going through the fixed point.

Spatial kinematic chains In spatial kinematic chains, the motion of the bodies can be described as general spatial. In general there are three translational and three rotational motion components. The relative motion of the bodies in the joints—depending on the joint—is as well arbitrary. This is illustrated with the complex example of a five-point wheel suspension in Fig. 3.7.

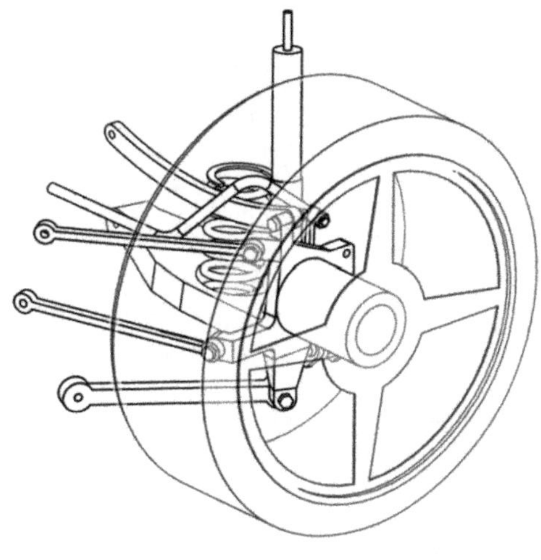

Fig. 3.7 Example for spatial kinematic chains: five-point wheel suspension

3.2 Joints in Kinematic Chains

A joint is defined as a connection between two adjacent bodies within a kinematic chain. Depending on the degrees of freedom f_{G_i} of a joint, it generates $6 - f_{G_i}$ geometric constraints between the two adjacent bodies.

3.2.1 *Joints in Spatial Kinematic Chains*

The relative motion between two bodies in a kinematic chain which are connected by a joint can be described by so-called natural or, respectively, relative coordinates β_i. In Tables 3.1 and 3.2, useful joints for a large class of technically applicable cases are presented. A detailed analysis of the corresponding spatial kinematic chains will be given in subsequent chapters. The natural joint coordinates of these technically relevant joints are angles of rotation $\beta_i = \theta_i$, as well as translational displacements $\beta_i = s_i$. The kinematically most general joint in this overview of Table 3.1 is the screw or helical joint (H) with one degree of freedom. It contains a rotational part of motion R (natural joint coordinate θ_i), simultaneously combined with a translational motion T (natural joint coordinate s_i). The relation between rotation and translation is given by the so-called pitch p, which results in

$$s_i = p \cdot \theta_i. \tag{3.3}$$

Directly related to this are the rotational joint (R) (pitch $p = 0$) and the prismatic joint (P) (pitch $p = \infty$), respectively. The joints with multiple degrees of freedom,

Table 3.1 Joints in spatial kinematic chains

Joint	Symbol	Representation	f	Natural joint coordinates
Screw joint (helical)	H	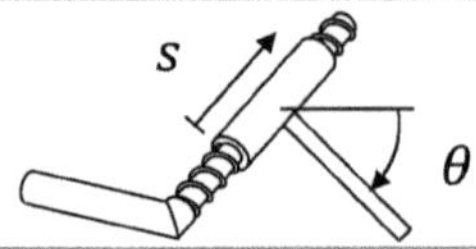	1	$\beta = \theta$ $s = p \cdot \theta$ $p :=$ pitch
Revolute joint	R	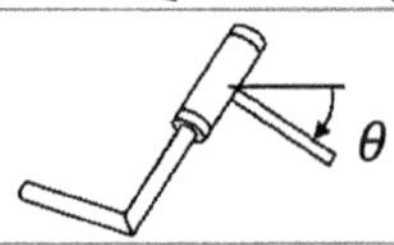	1	$\beta = \theta$
Prismatic joint (slider)	P	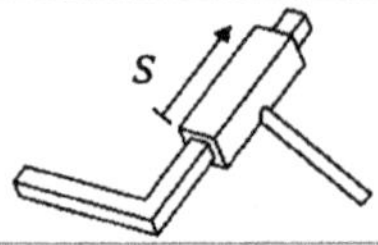	1	$\beta = s$
Cylindrical joint	C	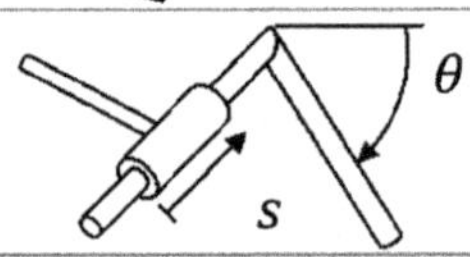	2	$\beta_1 = \theta$ $\beta_2 = s$
CARDAN joint (universal)	T	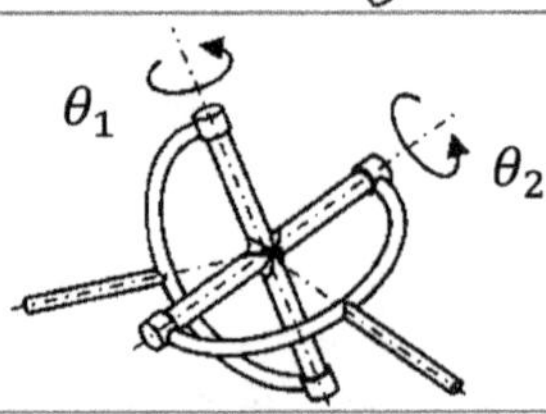	2	$\beta_1 = \theta_1$ $\beta_2 = \theta_2$
Planar joint	E	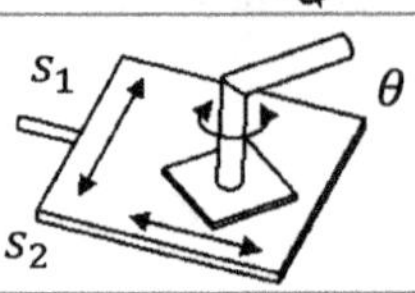	3	$\beta_1 = \theta$ $\beta_2 = s_1$ $\beta_3 = s_2$
Spherical joint	S	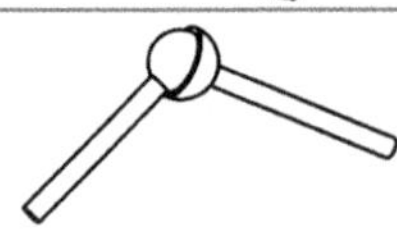	3	e.g. EULER angles $\beta_1 = \psi$ $\beta_2 = \theta$ $\beta_3 = \varphi$

such as the cylindrical joint (C), the CARDAN-joint (T), the spherical joint (S) and the planar joint (E), can on the other hand be constructed by a series connection or a parallel connection of revolute and prismatic joints respectively (see Tables 3.1 and 3.2).

Table 3.2 Construction of multiple d.o.f joints from revolute and prismatic joints

CARDAN joint	Revolute and prismatic joint – cylindrical joint	Spherical joint	Planar joint
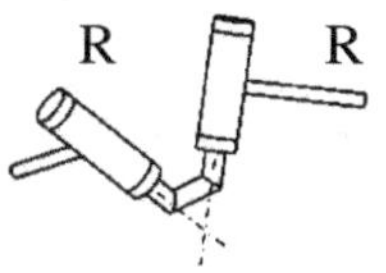	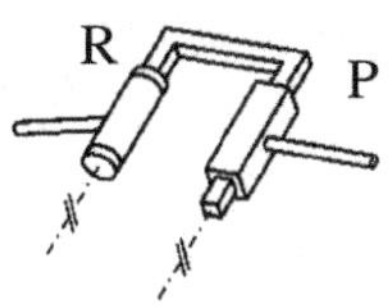	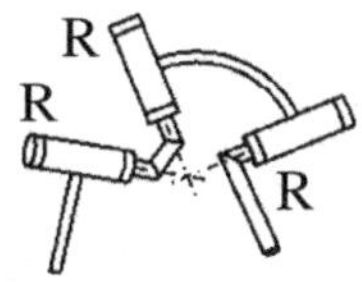	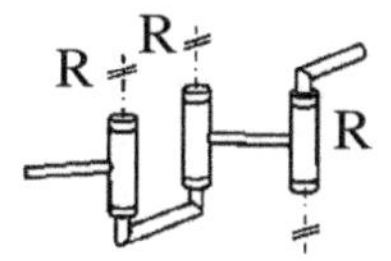 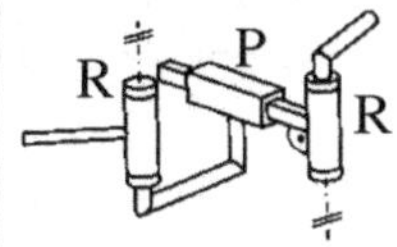 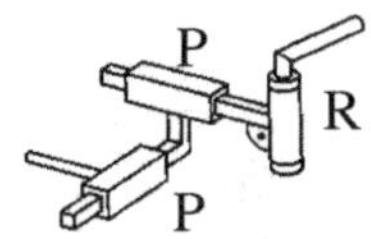
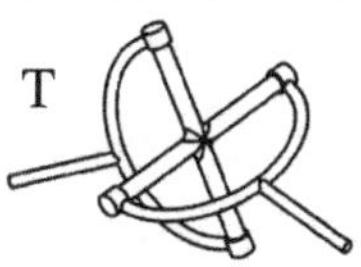	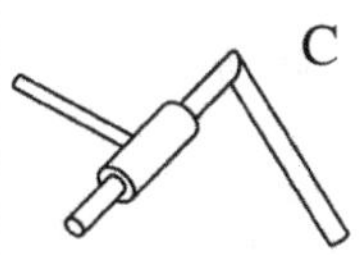	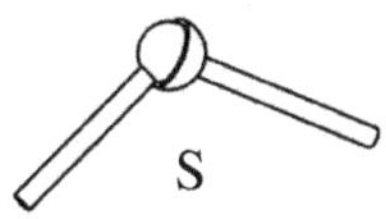	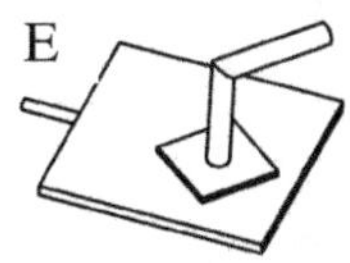

3.2.2 *Joints in Planar Kinematic Chains*

In a planar kinematic chain the individual body in general moves with only one rotational and two translational DoF. As a consequence, only the prismatic joint (displacement axis in the motion plane) and the revolute joint (rotation axis perpendicular to the motion plane) of the joints in Table 3.1 remain (Fig. 3.8).

3.2.3 *Joints in Spherical Kinematic Chains*

In a spherical kinematic chain the individual body in general moves with three rotational degrees of freedom. As a consequence, the spherical motion is comprised of three revolute joints with intersecting axes in the fixed point O of the kinematic chain (Fig. 3.9).

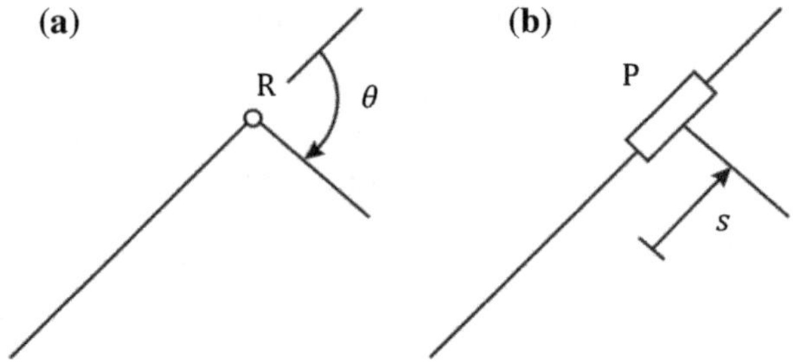

Fig. 3.8 Joints in planar kinematic chains. **a** Revolute joint (planar), **b** Prismatic joint (planar)

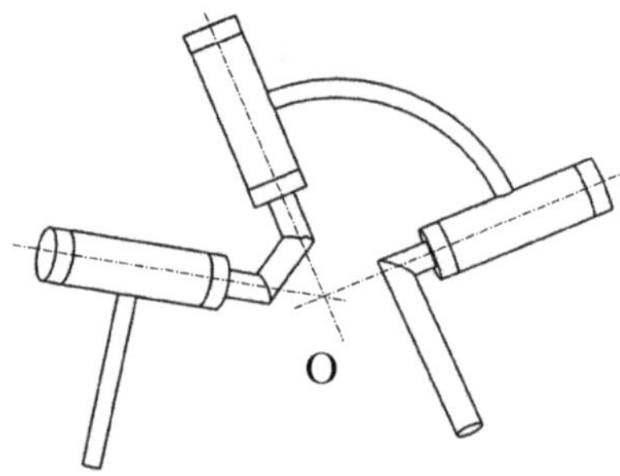

Fig. 3.9 Revolute joints in a spherical kinematic chain

3.2.4 Classification of Joints

In mechanisms and gear trains there is a difference between standard joints (so-called lower kinematic pairs) and complex joints (higher kinematic pairs) (Reuleaux 1875):

- In standard joints, the bodies have surface contact. One distinguishes the following six standard joints (Table 3.3).
- Complex joints have contact along a body line or at a point (Fig. 3.10).

Contact between two bodies can occur on two non-physical, spatially fixed surfaces or rather body fixed surface, spur surface or rather pin surface, along the instantaneous axis of rotation or rather screw axis, which indicate the instantaneous motion state.

In the suspending motion of a five-link wheel suspension, for example, the motion of the wheel carrier relative to the chassis can be represented as a screw motion of the wheel carrier-fixed pin surface with respect to the chassis-fixed spur surface. The contact line is the instantaneous screw axis of the spatial motion of the wheel carrier with respect to the chassis (Fig. 3.11).

Table 3.3 Standard joints

Joint		Contact surface
Screw joint	(H)	Screw surface
Revolute joint	(R)	Rotational surface
Prismatic joint	(P)	Barrel surface of a prism
Cylindric joint	(C)	Cylindric barrel
Dpherical joint	(S)	Spherical surface
Planar joint	(E)	Plane

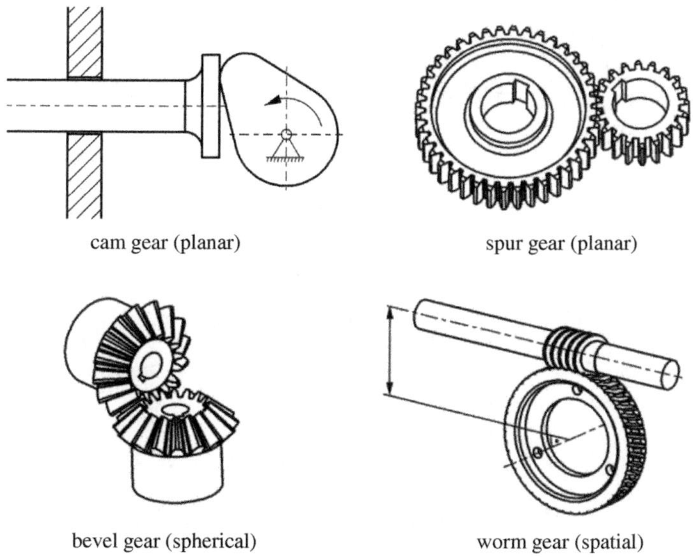

Fig. 3.10 Examples for complex joints with physical contact surfaces

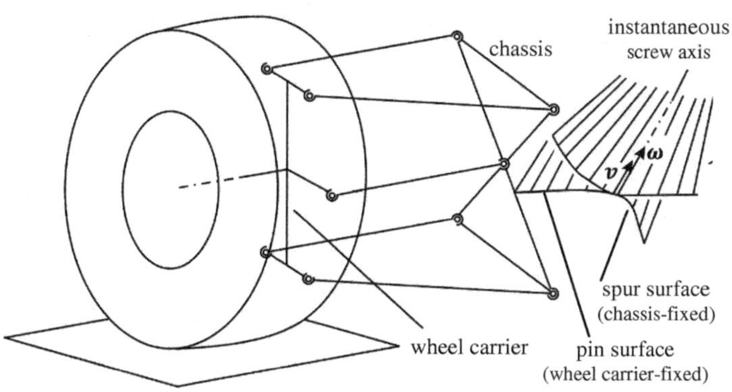

Fig. 3.11 Pin and spur surfaces of a five-point wheel suspension

3.3 Degrees of Freedom and Generalized Coordinates

3.3.1 Degrees of Freedom of Kinematic Chains

An arbitrary, spatial kinematic chain, with or without considerations of kinematics consists of

- n_B bodies (without a reference body),
- n_G joints, each having
- f_{G_i} joint DoF,

where the individual body, in the case of a free motion, can have six independent degrees of freedom in space. The total degree of freedom f (in mechanisms and gear trains in German also referred to as "Laufgrad") relative to the reference body are given by

$$f = 6n_B - \sum_{i=1}^{n_G}(6 - f_{G_i})\text{or} \tag{3.4}$$

$$f = 6(n_B - n_G) + \sum_{i=1}^{n_G} f_{G_i} \tag{3.5}$$

with the number of kinematic loops being

$$n_L = n_G - n_B, \tag{3.6}$$

the following results from Eq. (3.5) for the determination of the degree of freedom:

$$f = \sum_{i=1}^{n_G} f_{G_i} - 6n_L. \tag{3.7}$$

In mechanisms and gear trains, Eq. (3.7) is identified as criterion of KUTZBACH, or with respect to the special case $f = 1$ as criterion of GRÜBLER. Multibody systems in the context of kinematic chains with $f = 1$ are also identified as mechanisms. One distinguishes therefore:

- $f = 1$ mechanism,
- $f = 0$ statically well–defined structure (framework),
- $f < 0$ statically over constrained structure.

It is also shown, that through this classification in mechanics a "natural" transition between kinematics and statics can be produced. Furthermore, Eqs. (3.5) and (3.7) in certain cases (e.g. in so-called over-constrained mechanisms) can produce false results (Möller 1992).

In planar and spherical kinematic chains, the individual bodies in the chain only have three degrees of freedom. The corresponding relations to Eqs. (3.5) and (3.7) then are

$$f = 3(n_B - n_G) + \sum_{i=1}^{n_G} f_{G_i}, \tag{3.8}$$

$$f = \sum_{i=1}^{n_G} f_{G_i} - 3n_L. \tag{3.9}$$

3.3.2 *Examples from Road Vehicle Suspension Kinematics*

Two independent wheel suspensions that are widely used in automotive engineering are presented subsequently as examples for closed kinematic chains with different topological complexities.

The five-link suspension The five-link suspension (in German "Raumlenkerachse") embodies a closed kinematic chain with a very high degree of inner coupling of the inherent kinematic loops (Fig. 3.12).

Because in the assignment of degrees of freedom of the system, the isolated rotation of the suspension arms around their longitudinal axes are neglected, the spherical joints at the chassis end are modeled as CARDAN joints, without loss of generality.

Double wishbone wheel suspension In contrary to the five-link wheel suspension, the double wishbone suspension (Fig. 3.13) possesses only a weak kinematic coupling between the inherent two kinematic loops. Thus the associated kinematic analysis can be stated and solved explicitly, as shown in Sect. 3.5.5 and Chap. 6.

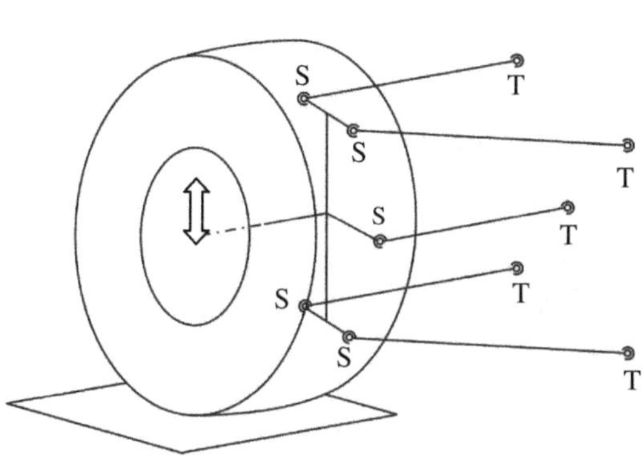

$n_B = 6$ bodies (without wheel)
$n_G = 10$ joints
$f_{G_i} = 3$ spherical joints, S
$f_{G_i} = 2$ CARDAN joints, T
$n_L = 4$ kinematic loops
$f = 1$ DoF (vertical suspension mode)

Fig. 3.12 Five-link wheel suspension (Raumlenkerachse)

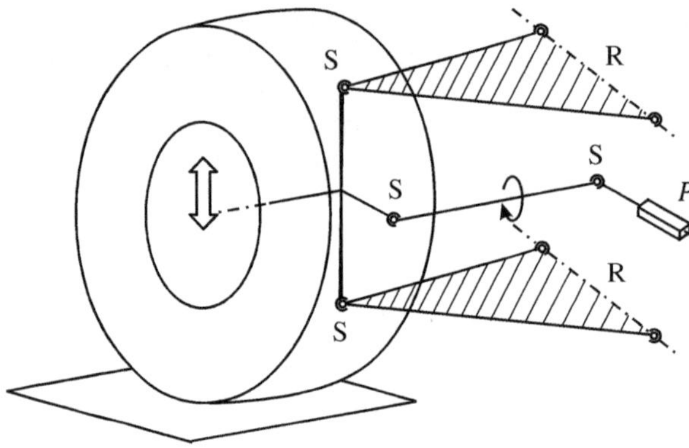

Fig. 3.13 Double wishbone wheel suspension

3.3.3 *Generalized Coordinates*

The position of all the bodies in a kinematic chain with f total degrees of freedom can uniquely be defined through f independent coordinates, the generalized coordinates.

Examples

Kinematic chains with tree structure (planar) In Fig. 3.14 the natural joint coordinates β_i (revolute joints) can directly be selected as the generalized coordinates.

Kinematic chains with loops (planar) For the $f = 4$ degrees of freedom of the kinematic chain shown in Fig. 3.1, $f = 4$ independent joint coordinates have to be chosen as the generalized coordinates. The other joint coordinates in the system have to be calculated as a function of the independent joint coordinates (input coordinates) (Fig. 3.15).

Fig. 3.14 Planar kinematic chain with tree structure

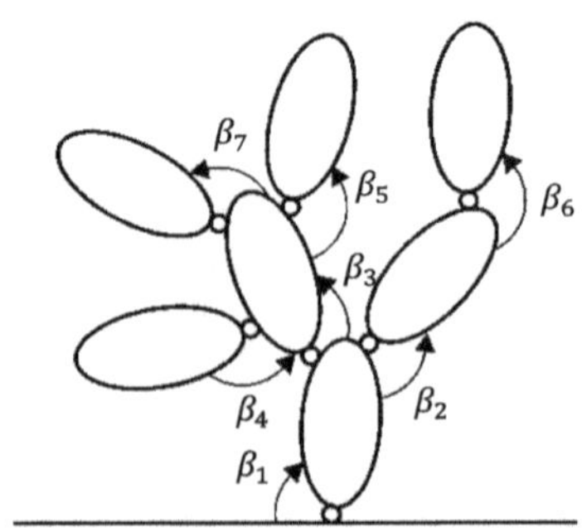

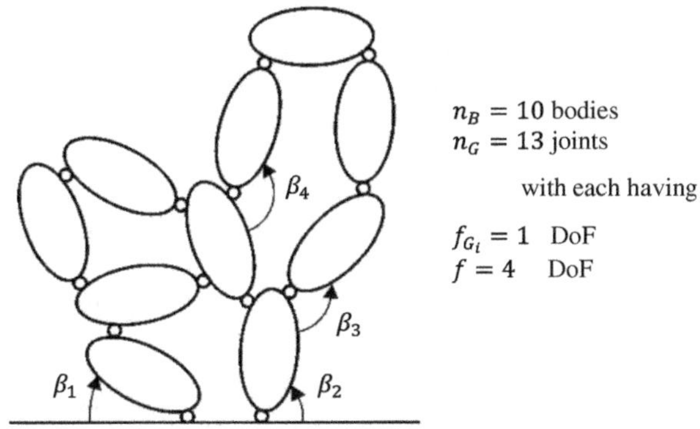

Fig. 3.15 Planar kinematic chain with closed loops

3.4 Basic Principles of the Assembly of Kinematic Chains

For the kinematic analysis of complex multibody systems with kinematic loops, there exist various approaches for stating the equations of motion. Regarding an automatic assembly of multibody systems, there are mainly three basic approaches which will be demonstrated in the following. They are used as an example for the double wishbone steering axle which is based on the double wishbone wheel suspension of the previous section (Fig. 3.16).

It is necessary to have a deep understanding of the topological structure of this suspension (like many other complex multibody systems), as the technical figure alone does not suffice in achieving a well-rounded comprehension of the structure. It can be explained as follows: The system is composed of two coupled kinematic loops and, altogether, has two degrees of freedom – one for the spring deflection in the suspension mode and the other one for the steering motion (compare

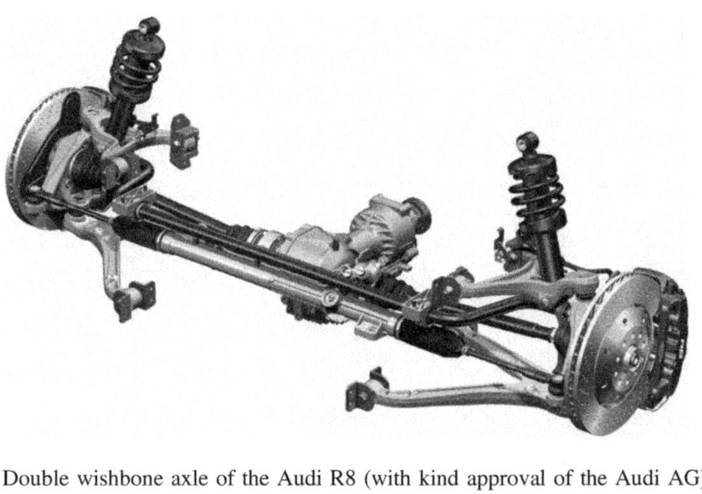

Fig. 3.16 Double wishbone axle of the Audi R8 (with kind approval of the Audi AG)

Sect. 3.3.2). In order to properly survey the structure of a system with kinematic loops, a graphical representation using the symbols shown in Fig. 3.17 is introduced. With help of these symbols, the topological structure of the mechanical system can be represented as follows (Schnelle 1990):

- Decomposition of all (complex) joints with more than one degree of freedom into standard joints (revolute and prismatic) (see Tables 3.1 and 3.2).
- Sketching of all bodies and joints symbolically as described in Fig. 3.17.

As already seen in the previous example (Fig. 3.18), the double wishbone suspension is composed of two revolute joints R_1 and R_2, a prismatic joint P and four spherical joints $S_i, i = 1, \ldots, 4$. The three spherical joints can be decomposed, using two virtual links for each. Using this symbolism leads to the joint-body representation of the double wishbone wheel suspension, as shown in Fig. 3.18. Now the two kinematic loops L_1 and L_2 are discernable as well as the coupling between the revolute joint R_1 and the spherical joint S_1. The DoF of the system—suspension, steering and the isolated DoF from the tie rod, which is not important for further consideration—are also discernable.

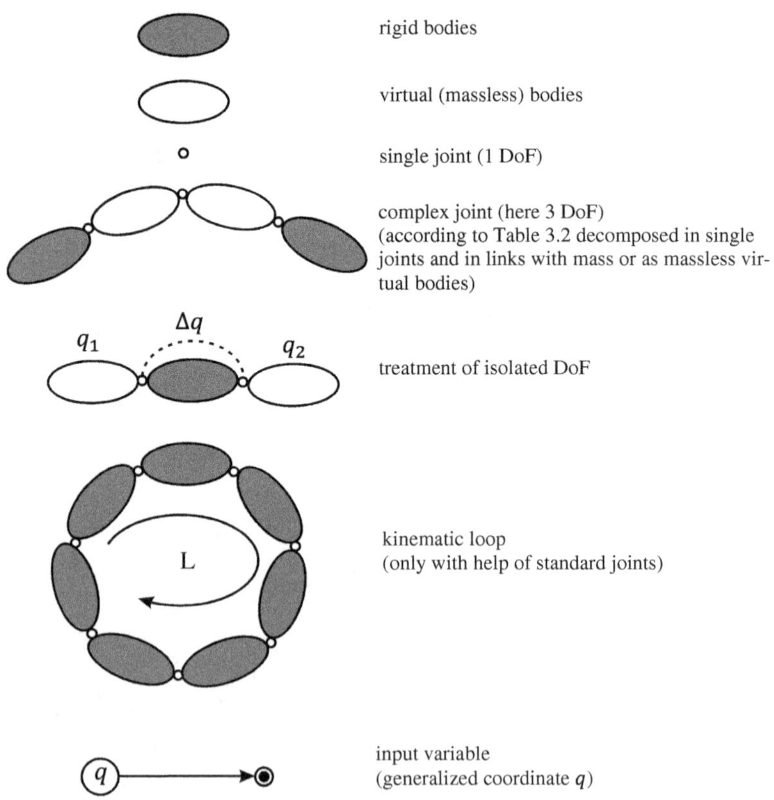

Fig. 3.17 Symbols for the joint-body representation in multibody systems

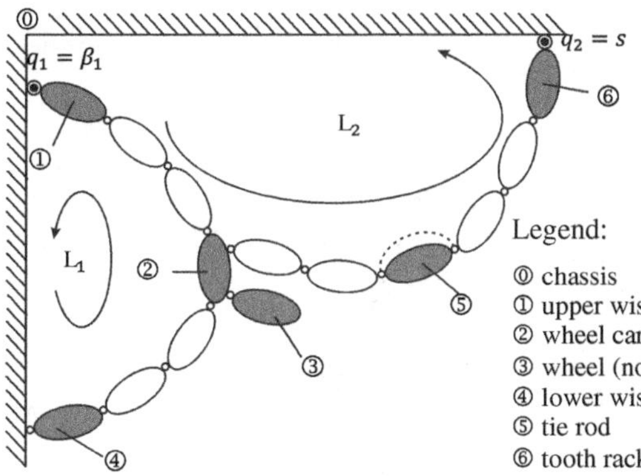

Fig. 3.18 Topological structure of the double wishbone wheel suspension

3.4.1 Sparse-Methods: Absolute Coordinates Formulation

Principle: disconnection of all joints
Constraints: matching of all joint parameters

Here, one first of all assumes that all bodies can move freely. The six position variables of the body-fixed reference systems (three translational and three rotational) are introduced as the description variables of the kinematics for each body "$\mathbf{K}_i$", and they are collected in the vector $\boldsymbol{w}_i$ (compare with Sect. 2.4.1). For a multibody system with n_B bodies the $6n_B$ description variables are introduced and collected in a $[6n_B \times 1]$-vector $\boldsymbol{w} = \left(\boldsymbol{w}_1^T \ldots \boldsymbol{w}_{n_B}^T\right)^T$.

The dependencies between the redundantly introduced position variables are implicitly stated through the formulation of the corresponding closure constraints for all joints. On a joint with f_{G_i} DoF, $6 - f_{G_i}$ invariant joint parameters—independent from their reference system must match. As a suitable common reference system, the inertial system will be chosen. After the choice of (Eq. (3.5))

$$f = 6(n_B - n_G) + \sum_{i=1}^{n_G} f_{G_i}$$

independent, generalized coordinates $\boldsymbol{q} = (q_1 \ldots q_f)$ as input variables, the function $\boldsymbol{w}(\boldsymbol{q})$ is implicitly defined through the system of equations

$$z_i(w, \boldsymbol{q}) = 0, \quad i = 1, \ldots, 6n_B. \tag{3.10}$$

The advantage of this approach lies in the relatively simple formulation of the corresponding constraint equations, as well as in the weak-coupled structure of the whole system of constraint equations (only a few variables per equation, collected into a matrix with many zeroes, the so-called Sparse matrices). A disadvantage is

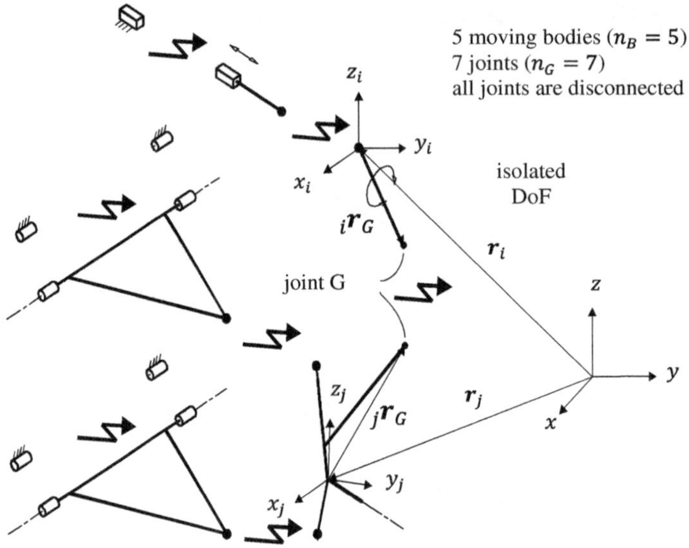

Fig. 3.19 Double wishbone wheel suspension with disconnected joints

the fact that the formulation of the function $\boldsymbol{w}(\boldsymbol{q})$ is profoundly implicit. On the one side, this has a high numerical complexity for the solution of the system of Eq. (3.10); on the other side it can also frequently give numerical singularities during the numerical simulation, which are not necessarily caused by mechanically interpretable singularities (singular positions).

Example (Fig. 3.19).

Double wishbone wheel suspension The assembly yields two vectorial constraint equations for each joint:

$$\boldsymbol{r}_i + {}_i\boldsymbol{r}_G(\beta_i) = \boldsymbol{r}_j + {}_j\boldsymbol{r}_G(\beta_j). \tag{3.11}$$

The total balance for the degrees of freedom is given as follows:

5×6	Equations of motion	30
$-(3\times5+4\times3)$	Constraints	−27
	DoF (one isolated DoF (Fig. 3.19))	3

3.4.2 Vector Loop Methods ("LAGRANGE" Formulation)

Principle: disconnection of only one joint from each kinematic loop; transformation of a system with kinematic loops into a system with tree structure.

Constraints: closure conditions for the kinematic loops.

Here, the multibody system is transformed by the disconnection of suitable joints into a system with a tree structure. The kinematic description variables are introduced as

$$n_\beta = \sum_{i=1}^{n_G} f_{G_i} \tag{3.12}$$

auxiliary variables β_j, which describe the relative position of two adjacent bodies. The dependencies between the auxiliary variables are implicitly stated by formulation of the closure conditions only at the disconnected joints. Here, the absolute values of the fixed joint parameters with respect to the inertial system must be given with the help of the auxiliary variables.

The advantage of this method is that the constraint equations are heavily reduced. A disadvantage is that the computation of the joint parameters using forward kinematics is very complex; one example for this is the non-sparse structure of the complete system of equations (many variables per equation). The numerical complexity of the solution of the constraint equations is comparable with the numerical complexity of the sparse method.

Example: Double wishbone wheel suspension After disconnecting the joints G_1 and G_2, a system with a tree structure and nine relative joint coordinates β_i serves as the auxiliary variables to the motion description.

Assembly: one obtains three constraint equations from each joint, e.g.:

$$\boldsymbol{a}_1 + \boldsymbol{a}_2(\boldsymbol{\beta}) = \boldsymbol{a}_3(\boldsymbol{\beta}) + \boldsymbol{a}_4(\boldsymbol{\beta}). \tag{3.13}$$

The final balance for the degrees of freedom results in

9	Auxiliary variables	9
-2×3	Constraints	−6
	DoF (one isolated DoF (Fig. 3.20))	3

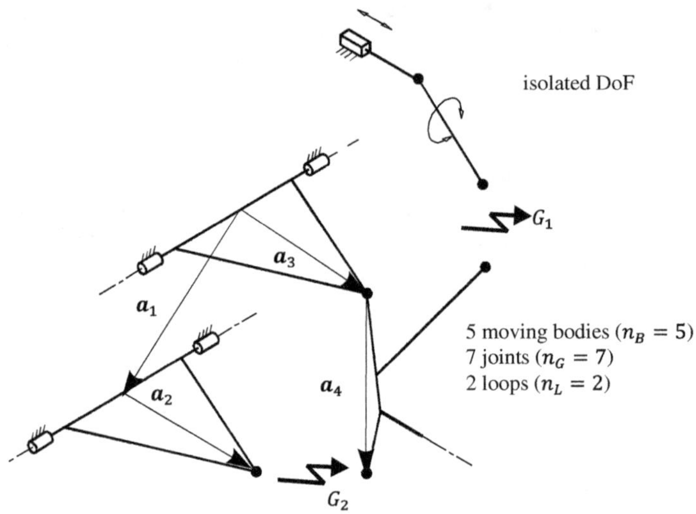

Fig. 3.20 Double wishbone wheel suspension with disconnected kinematic loops

3.4.3 Topological Methods: Formulation of Minimum Coordinates

Principle: Kinematic loops are treated as kinematic transmission elements, which in the following are characterized as the so-called kinematic transformers (Hiller et al. 1986, 1986–1988).

Constraints: The constraint equations, which have to be stated, are composed of two different parts:

- the non-linear, local transmission equations inside of each kinematic loop,
- the linear equations that couple the kinematic loops.

In this method, complete multibody loops are isolated and treated as local transmission elements. As describing kinematic variables the interior, relative coordinates for all loops are introduced. The interdependencies between the describing variables result initially from the local relations inside of each individual loop (transmission functions). Finally, the individually treated transmission elements (loops) are assembled using linear coupling equations and can be illustrated in block diagrams (Sect. 3.5).

The advantage of this method is that the global relations are clearly described (also easily interpretable), and the complete system of equations can often be explicitly solved. A disadvantage is that the initialization in the topological format can sometimes become complex.

In the example of the double wishbone suspension, the system consists of the two coupled kinematic loops L_1 and L_2, where these loops are coupled by the input angle β_1 of the upper trailing arm, as well as by the three angles $\beta_2, \beta_3, \beta_4$ of the spherical joint S_1 from Fig. 3.21. The entire system for the relative kinematics can be combined as follows:

Example

Double wishbone wheel suspension

The system consists of 2 kinematic transmission elements (kinematic transformers) L_1 and L_2

3 linear coupling equations

The DoF are determined by:

2	Kinematic loops, one with two DoF and the other with five DoF	7
3	Coupling equations	−3
1	Branching	−1
	Degrees of freedom (including one isolated DoF (Fig. 3.21))	3

The corresponding block diagram can be found in Fig. 3.21 and will be discussed in further detail in the next section.

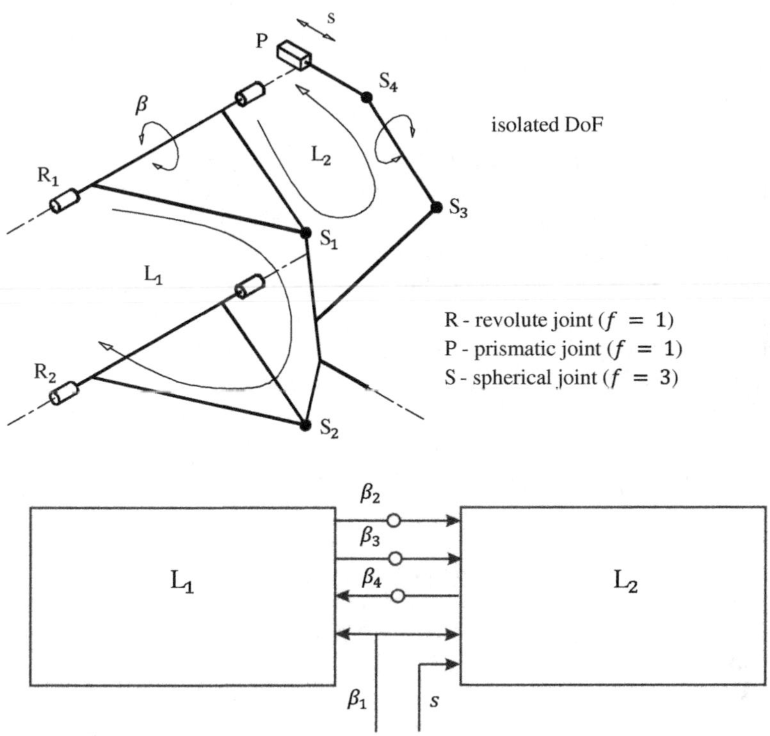

Fig. 3.21 Kinematic transformers

Notes

- The kinematic analysis of a multibody system with kinematic loops, which is based on the topological method, will be analyzed using the example of the double wishbone axle in Sect. 3.5.5.
- Besides the three principles for the formulation of the kinematic equations of complex multibody systems introduced in Sect. 3.4, there are further approaches, which, among other details, use particular properties of the topological structure of the system, leading to an implicit structure of the constraint equation system, which can be numerically solved. This subject is treated in more detail in Chap. 6.

3.5 Kinematics of a Complete Multibody System

3.5.1 Basic Concept

In complex multibody systems which also imply mechanisms, the movements of the individual bodies in the system are strongly connected to each other. The mathematical description of these connections is characterized by a bulky system of highly nonlinear equations that makes the mutual dependencies in the displacements of the bodies not instantly apparent.

The reduction of the system of constraint equations by introducing relative coordinates as description variables was already presented in Sect. 3.4.3. One can further show that when the problem of relative kinematics is solved, all absolute coordinates of the individual bodies are explicitly ascertainable through corresponding forward kinematics in the kinematic network (Hiller et al. 1986–1988). Therefore, the central problem of kinematics of complex multibody systems with closed kinematic loops is the solution of the relative kinematics. In the case under consideration, the closed kinematic loops are considered as non-linear kinematic transmission elements—so called kinematic transformers—which is based upon the idea already introduced in the previous Sect. 3.4.3. The kinematic transformers are then associated with (typically linear) coupling equations to kinematic networks, represented by block diagrams. Finally, it should also be noted that particularly in technical applications (e.g. wheel suspensions), and thereby utilized joint types (compare in Sect. 3.2) through a suitable analysis of the individual kinematic loops, the non-linear constraint equations inside the loop can be recursively (or as far as possible recursively) and explicitly solved. The subsequent absolute kinematics is dealt with in Sect. 3.5.4.

From this approach it follows that through observation of the topological structure of a complex multibody system with kinematic loops, it is also possible to decouple the global interdependencies of local non-linearities. For this, the modeling of the individual multibody loops as kinematic transformers is again regarded.

It can be shown that the kinematic transformers can be coupled using linear equations and that a large class of technical multi-loop multibody systems can be commonly lead back to linearly coupled transformers. The multi-loop multibody system allows itself then to be represented as a block diagram, which again illustrates the global kinematic relations in an understandable form. This concept of block diagrams will be discussed further in the later Chaps. 6 and 12, with practical examples from independent wheel suspensions (Hiller et al. 1986–1988).

3.5.2 Block Wiring Diagram and Kinematic Networks

Individual multibody loops are commonly series-connected to transmission mechanisms (linkages) in order to generate definite transmission functions over definite ways. As a typical spatial transmission element, the spatial four-link mechanism is often used in mechanism design, because it is a simple but non-linear transmission element which already appeared in the double wishbone suspension in the section before (Figs. 3.22 and 3.23).

The spatial four-link mechanism usually functions as a transmission element, so that the rotations of an input crank $\boldsymbol{r}$ are transmitted in the rotations of the output lever $\boldsymbol{s}$. The nonlinear transmission is being generated over the coupler $\boldsymbol{d}$. Whether the spatially-fixed or vehicle-fixed axes of rotation of both levers $\boldsymbol{r}$ and $\boldsymbol{s}$ lie parallel or skew to each other, one distinguishes between planar and spatial four-bar mechanisms. The input angle β (rotation angle of the crank $\boldsymbol{r}$) as well as the output angle γ (rotation angle of the output lever $\boldsymbol{s}$) function as the relevant coordinates for the motion of the mechanism. After the corresponding transmission function $\gamma(\beta)$ is defined, the design parameters of the mechanism can be ignored. The four-link mechanism can from now on be considered as a kinematic transformer with a non-linear input-output behavior. It transforms the input angle β into the output angle γ. This concept can be expanded further, as shown in Sect. 3.5.4.

After the coupling equations between different kinematic transformers are determined, the individual kinematic transformers, which comprise the non-linear–but local–transmission equations of the disconnected multibody loops can be assembled over a block diagram to a kinematic network (compare Fig. 3.21), whose detailed kinematic analysis will be developed in the next section.

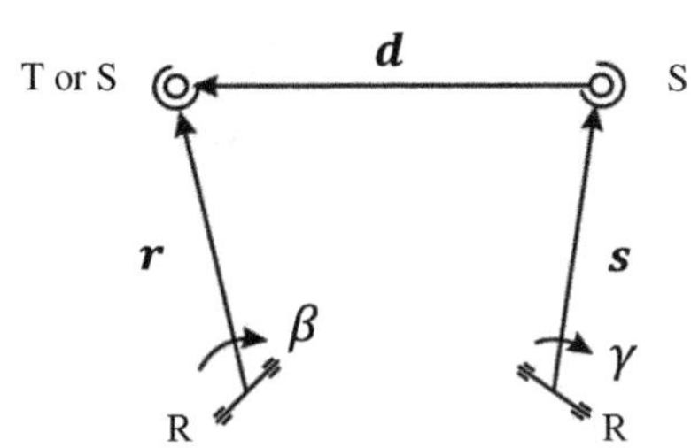

Fig. 3.22 Spatial four-bar mechanism

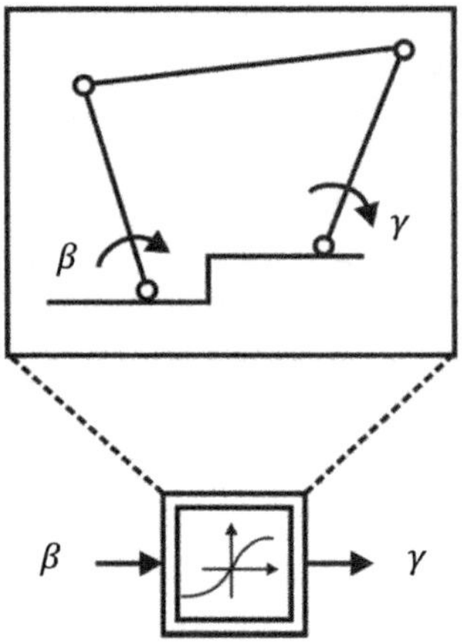

Fig. 3.23 Schematic representation of the spatial four-bar mechanism as transmission element—kinematic transformer

The global constraint equations, as the result of the global assembly of the mechanism in modular form, are now available for further steps of the analysis. The structure based on this approach is extremely advantageous for the handling of complex multibody systems with kinematic loops: Every module, i.e. every kinematic transformer, includes only a limited number of local nonlinear constraint equations, which are manageable and can be solved independently from other transformers in the system: the structure of these constraint equations is only dependent on the geometric parameters inside the local loop. In contrast, the global kinematic interdependencies between the kinematic transformers are linear and, except for the sign and a constant, exclusively given through the topological structure of the multibody system. Therefore, they are (almost) independent from geometric parameters. The complete system can then be combined in one transmission block, whose describing variables are the relative coordinates established above (see Sect. 3.5.3).

The assembly of the complete system can be supported with computer aid through suitable algorithms. The global constraint equations can also be represented with use of nonlinear transmission blocks for the individual loops, as well as linear summing points for the coupling equations, well-arranged with help from block diagrams. This common approach, taken from control theory, is exemplified in Sect. 3.5.5, again using the example of the double-wishbone suspension from automotive engineering introduced earlier in this chapter (Fig. 3.24).

3.5.3 *Relative Kinematics of the Spatial Four-Link Mechanism*

For the position analysis of the nonlinear transmission motion, e.g. the rotation of the input crank (angle β_1) can be chosen as the independent input coordinate

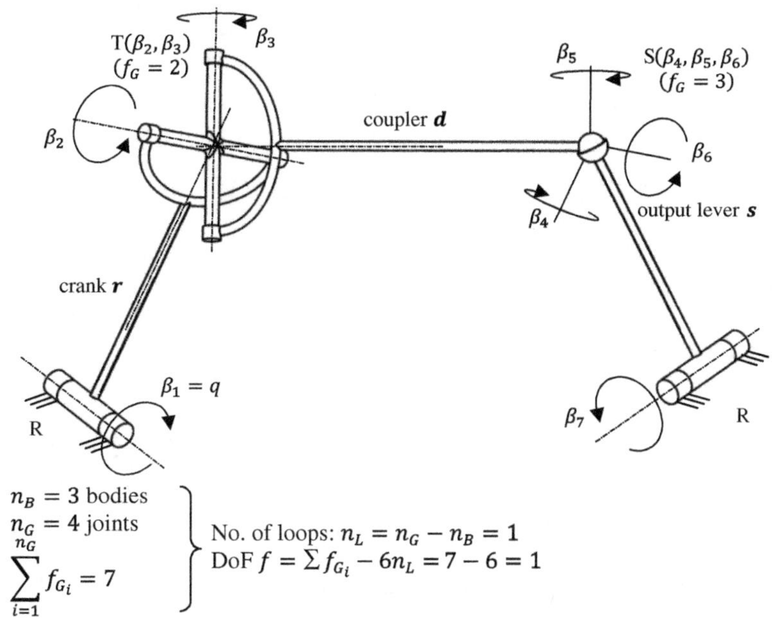

Fig. 3.24 Relative kinematics of the spatial four-link mechanism

$q = \beta_1$. The deflection of the right output lever (angle β_7) such as the motion of the connecting coupling bar ***d*** can be calculated. In particular, one can show that the output variable, which is to be calculated in the first step,

$$\beta_7 = \beta_7(\beta_1, \text{ geometry of the initial position}), \tag{3.14}$$

can be found explicitly through the solution of the equation

$$A\cos\beta_7 + B\sin\beta_7 + C = 0. \tag{3.15}$$

For this equations, there exist, based on three coefficients A, B, and C, which are dependent on β_1 and on the geometry of the initial position of the mechanism, normally two real solutions, only one of which fulfills the constraints of the initial position of the kinematic system. The unknown variables in the four-link mechanism, which only need to be calculated if necessary, are the two angles of the CARDAN-joint β_2, β_3 and the three angles of the spherical joint $\beta_4, \beta_5, \beta_6$. Assuming these two joints, corresponding to Table 3.2, can be constructed from revolute joints, then these angles can be recursively calculated. For further details, see (Woernle 1988). The overall solution structure of the non-linear position analysis of the spatial four-link mechanism can be represented as follows:

$$\left.\begin{array}{l} g_1(\beta_1, \beta_7) = 0, \\ g_2(\beta_1, \beta_7, \beta_2) = 0, \\ g_3(\beta_1, \beta_7, \beta_2, \beta_3) = 0, \\ g_4(\beta_1, \beta_7, \beta_2, \beta_3, \beta_4) = 0, \\ g_5(\beta_1, \beta_7, \beta_2, \beta_3, \beta_4, \beta_5) = 0, \\ g_6(\beta_1, \beta_7, \beta_2, \beta_3, \beta_4, \beta_5, \beta_6) = 0. \end{array}\right\} \tag{3.16}$$

The distinguishable recursive solution structure of the system of constraint equations in Eq. (3.16) somewhat becomes clearer when one observes the JACOBIAN-Matrix $\boldsymbol{J}_\beta$ of the constraint equations, which is also the basis for the velocity-analysis:

$$\boldsymbol{J}_\beta = \frac{\partial \boldsymbol{g}}{\partial \boldsymbol{\beta}} = \begin{array}{c} \left[\begin{array}{cccccc} \times & & & & & \\ \times & \times & & & & \\ \times & \times & \times & & 0 & \\ \times & \times & \times & \times & & \\ \times & \times & \times & \times & \times & \\ \times & \times & \times & \times & \times & \times \end{array}\right] \\ \begin{array}{cccccc} \beta_7 & \beta_2 & \beta_3 & \beta_4 & \beta_5 & \beta_6 \end{array} \end{array} \begin{array}{c} g_1 \\ g_2 \\ g_3 \\ g_4 \\ g_5 \\ g_6 \end{array} . \tag{3.17}$$

This JACOBIAN matrix $\boldsymbol{J}_\beta$ possesses a lower-triangle structure and corresponds to a recursively explicit solution of the position analysis of the four-link mechanism, as shown in Li (1990).

The complete kinematic transmission characteristics of the four-link mechanism on position level, as well as on the velocity and acceleration level can be combined in the non-linear transmission element, which represents now the complete version of the already introduced kinematic transformer (Fig. 3.25).

For more details on the kinematics of four-link mechanisms see the following references (Hiller 1981; Woernle 1988; Kecskemethy 1993).

3.5.4 Relative, Absolute and Global Kinematics

The kinematics of a complex multibody system with kinematic loops has been subdivided in order to give a promising description of the constraint equations in relative kinematics (expressed in relative coordinates) and absolute kinematics

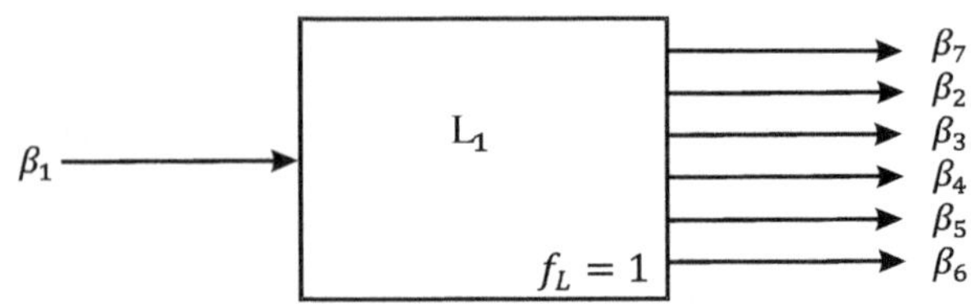

Fig. 3.25 The spatial four-link mechanism as kinematic transformer

(expressed in body coordinates or absolute coordinates). The analysis of the relative kinematics has shown that it contains the problem of the constraint equations. From these considerations it also becomes clear that the major part of the complexity is connected with the derivation and solution of the constraint equations, i.e. the relative kinematics. The remaining part of the analysis, which subsequently still needs to be done, can be carried out with the help of a straight forward approach.

The relative kinematics of the complex multibody system can be combined on the position level, the velocity level, and the acceleration level as follows:

$$\boldsymbol{g}(\boldsymbol{\beta},\boldsymbol{q}) = 0, \tag{3.18}$$

$$\boldsymbol{J}_\beta \dot{\boldsymbol{\beta}} = \boldsymbol{V}\ddot{\boldsymbol{q}},\ \boldsymbol{J}_\beta = \frac{\partial \boldsymbol{g}}{\partial \boldsymbol{\beta}}, \tag{3.19}$$

$$\boldsymbol{J}_\beta \ddot{\boldsymbol{\beta}} + \dot{\boldsymbol{J}}_\beta \dot{\boldsymbol{\beta}} = \boldsymbol{V}\ddot{\boldsymbol{q}}. \tag{3.20}$$

Therein, $\boldsymbol{\beta}$ represents the vector $[n_B \times 1]$ for the auxiliary variables (relative or natural (joint) coordinates), and $\boldsymbol{q}$ the vector $[f \times 1]$ for the (independent) generalized coordinates (f = number of DoF). The Eq. (3.18) includes the implicit system of the constraint equations, which is—after a topological analysis—available in a modular structure. It can be solved explicitly in case of kinematically and loop-wise decomposable multibody systems (Hiller et al. 1986–1988, 1986). The system JACOBIAN matrix $\boldsymbol{J}_\beta$ of the time derivatives of Eqs. (3.19) and (3.20) is produced through the assembly of the local JACOBIAN matrices of the individual loops along the diagonals. The system distribution matrix $\boldsymbol{V}$ assigns the generalized coordinates of the selected input variables of the system. The $[n_B \times 1]$-Vektor $\dot{\boldsymbol{J}}_\beta \dot{\boldsymbol{\beta}}$ appearing in Eq. (3.20) is produced block wise from the corresponding expressions of the kinematic transformers, as it can be shown.

The complete relative kinematics can now be assembled in a black box, which possesses the generalized coordinates $\boldsymbol{q}$ along with their time derivatives $\dot{\boldsymbol{q}}, \ddot{\boldsymbol{q}}$ as inputs and the auxiliary (intermediate) variables $\boldsymbol{\beta}$ along with their time derivatives $\dot{\boldsymbol{\beta}}, \ddot{\boldsymbol{\beta}}$ as outputs (Fig. 3.26).

As soon as the relative kinematics is modularly available, the absolute motion of every arbitrary body in the system can be recursively calculated with help from elementary kinematic expression from the forward kinematics (compare Sect. 3.4.1).

Putting together the six absolute position coordinates of the body K_i in vector $\boldsymbol{w}_i$, as well as all the $6n_B$ absolute position coordinates of the bodies in vector $\boldsymbol{w}$, one can calculate in recursive form, starting from the absolute reference system, the

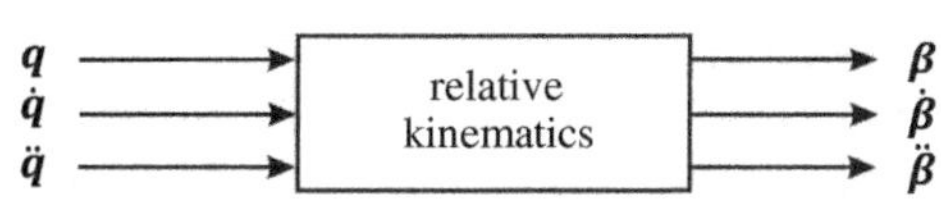

Fig. 3.26 Module for relative kinematics of a complex multibody system

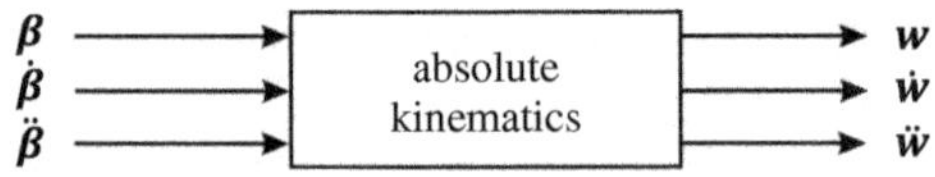

Fig. 3.27 Module for absolute kinematics of a complex multibody system

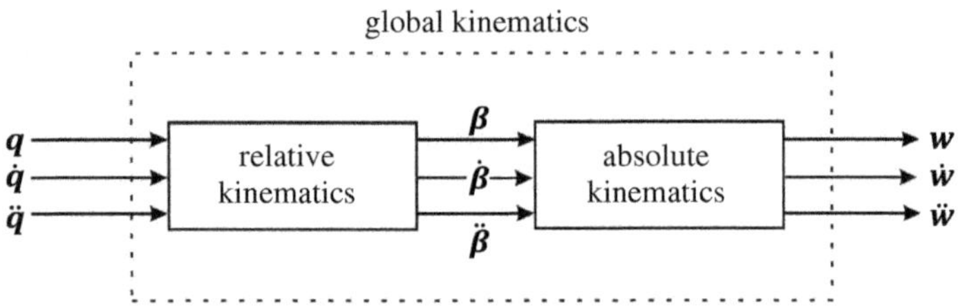

Fig. 3.28 The global kinematics of complex multibody systems

position kinematics, as well as their time derivatives as indirect functions of the relative kinematics:

$$\boldsymbol{w}_i = \boldsymbol{w}_i(\boldsymbol{w}_1, \ldots, \boldsymbol{w}_{i-1}, \boldsymbol{\beta}, t), \tag{3.21}$$

$$\dot{\boldsymbol{w}}_i = \dot{\boldsymbol{w}}_i\left(\boldsymbol{w}, \dot{\boldsymbol{w}}_1, \ldots, \dot{\boldsymbol{w}}_{i-1}, \boldsymbol{\beta}, \dot{\boldsymbol{\beta}}, t\right), \tag{3.22}$$

$$\ddot{\boldsymbol{w}}_i = \ddot{\boldsymbol{w}}_i\left(\boldsymbol{w}, \dot{\boldsymbol{w}}, \ddot{\boldsymbol{w}}_1, \ldots, \ddot{\boldsymbol{w}}_{i-1}, \boldsymbol{\beta}, \dot{\boldsymbol{\beta}}, \ddot{\boldsymbol{\beta}}, t\right). \tag{3.23}$$

The Eqs. (3.21)–(3.23) represent in recursive form the most general type of dependency of the absolute motion, where an appropriate numbering of the bodies (starting from the reference system) is adopted for reasons of clarity. Actually, it is usually possible in many cases to represent the absolute motion of a body in dependency of the absolute motion of the predecessor as well as the relative motion between both bodies.

For the derivation of the equations of motion, the necessary partial derivatives are calculated using the so-called kinematic differentials, which do not require knowledge of the exact structure of the interdependencies in Eqs. (3.21)–(3.23) (Sect. 4.5). In the following, it is sufficient to represent the results of absolute kinematics again as a black box, whose inputs have the meaning of relative coordinates as well as their time derivatives and their outputs having the meaning of absolute coordinates together with their time derivatives (Fig. 3.27).

The modules relative kinematics and absolute kinematics can now be combined into one block, called global kinematics, which can be regarded again as a black box and which represents the complete kinematic analysis of a complex multibody system with kinematic loops (Fig. 3.28).

With help from the global kinematics, one can determine, based on given generalized coordinates as well as their time derivatives as inputs, the motion of each arbitrary body in the multibody system. This enables the automatic (computerized) and very efficient derivation of the motion equations, in a numerical or symbolic manner, as shown in Sect. 4.6.

3.5.5 Example: Double Wishbone Suspension

On the example of the double wishbone suspension, which plays an important role in automotive engineering, as already discussed in Sect. 3.4.3 under the title "topological methods", the kinematic analysis of multibody systems with kinematic loops will be carried out further (Figs. 3.29 and 3.30).

The topological structure of the system consists of two independent kinematic loops L_1 and L_2, which are coupled over the revolute joint R_1 (angle of rotation β_1) and the spherical joint S_1 (spherical angles β_2, β_3, β_4) of the upper wishbone, respectively. One recognizes that the loop L_1, which also embodies the wheel carrier, corresponds to the spatial four-link mechanism, introduced in Sects. 3.5.2 and 3.5.3, where the wheel carrier is connected with, both upper and lower, wishbones over the corresponding spherical joints. This is required, because later, the steering motion s which is independent from the spring deflection β_1, can be introduced into the system. Having a separate glance on the spatial four-link mechanism, one can recognize that this loop possesses two degrees of freedom, one for the spring deflection β_1 of the wheel mount and a second independent DoF, which corresponds to an isolated rotation of the wheel carrier about the connecting straight line between upper and lower spherical links of the wishbone, which is required for the guidance of the steering motion over the steering mechanism referred to in Fig. 3.29.

In a similar manner, the second kinematic loop L_2 of the double wishbone axle can be prepared and successively interrelated with the loop L_1 over the linear equations in the variables $\beta_2, \beta_3, \beta_4$ of the spherical joint S_1 (Figs. 3.29 and 3.30),

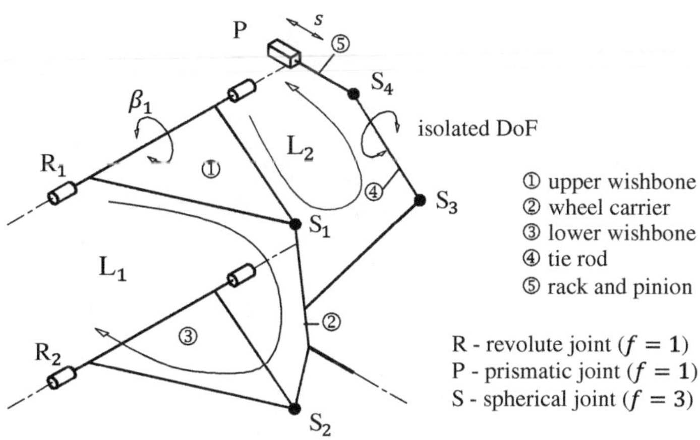

Fig. 3.29 Double wishbone suspension—kinematic structure

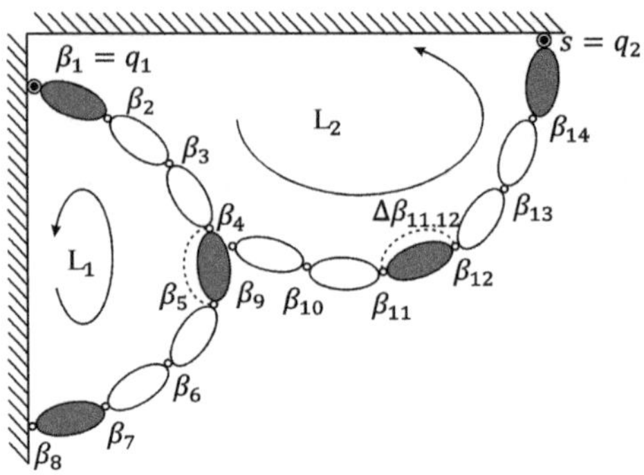

Fig. 3.30 Relative kinematics of the double wishbone suspension—topology

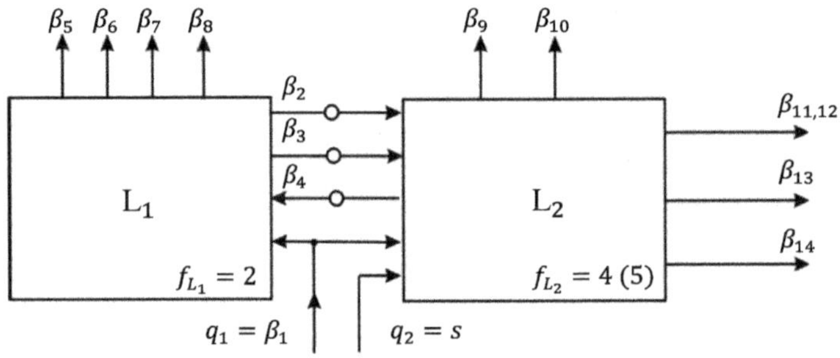

Fig. 3.31 Block diagram of the relative kinematics of the double wishbone suspension

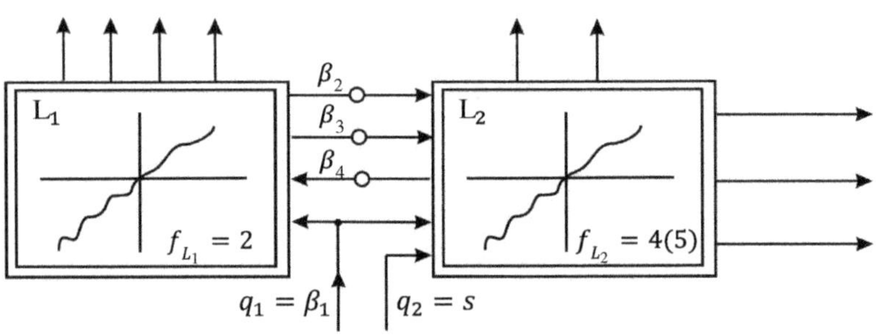

Fig. 3.32 Relative kinematics of the double wishbone suspension

corresponding to the coupling of the loops via these joints, to the already shown block diagram for the calculation of the relative kinematics (Fig. 3.31). An important advantage of this approach, which can be applied in its basic ideas to a large number of further technically interesting kinematic structures, is the minimized kinematic calculation amount: Only those kinematic variables must be calculated, which are necessary for the further continuation of the calculation. For example, in most practical cases the angles of the spherical joints are not required. It should still be emphasized once more that in the preceding example the calculation for the complete system was carried out in explicit-recursive form (Hiller et al. 1986–1988), (Woernle 1988) and (Schnelle 1990).

Note: In the later applications (especially in Chaps. 6 and 12), the simplified symbolism already demonstrated in Fig. 3.23 is used for the block diagrams, in order to emphasize the nonlinear transmission behavior inside each kinematic loop. In the case of the double wishbone suspension (Fig. 3.31) will be replaced by (Fig. 3.32).

References

Hiller M (1981) Analytisch-numerische Verfahren zur Behandlung räumlicher Übertragungsmecha nismen [habil.]. Habilitation, Universität Stuttgart, Düsseldorf: VDI-Verlag

Hiller M (1995) Multiloop Kinematic Chains, and Dynamics of Multiloop Systems in Kinematics and Dynamics of Multi-body Systems

Hiller M and Kecskeméthy A (1989) Equations of motion of complex multibody systems using kinematical differentials. In *Transactions of the Canadian Society of Mechanical Engineers*, Vol. 13, S 113–121

Hiller M, Kecskeméthy A and Woernle C (1986–1988) Computergestützte Kinematik und Dynamik für Fahrzeuge, Roboter und Mechanismen. In *Carl-Cranz-Kurs*, Vol. 1.16, Carl-Cranz-Gesellschaft, Oberpfaffenhofen

Hiller M, Kecskeméthy A and Woernle C (1986) A loop-based kinematical analysis of complex mechanisms. In *Presented at the Design Engineering Technical Conference*, Columbus, Ohio

Kecskemethy A (1993) Objektorientierte Modellierung der Dynamik von Mehrkörpersystemen mit Hilfe von Übertragungselementen [Dr.-Ing.]. Dissertation, Universität Duisburg, Düsseldorf: VDI-Verlag

Li H (1990) Ein Verfahren zur vollständigen Lösung der Rückwärtstransformation für Industrieroboter mit allgemeiner Geometrie Dissertation, Universität – Gesamthochschule – Duisburg

Möller M (1992) Verfahren zur automatischen Analyse der Kinematik mehrschleifiger räumlicher Mechanismen [Dr.-Ing.]. Dissertation, Universität Stuttgart

Reuleaux F (1875) Theoretische Kinematik: Grundzüge einer Theorie des Maschinenwesens. Vol. 1, F. Vieweg und Sohn - ISBN 978-3-8364-0352-8

Schnelle K-P (1990) Simulationsmodelle für die Fahrdynamik von Personenwagen unter Berücksichtigung der nichtlinearen Fahrwerkskinematik Dissertation, Universität Stuttgart, Stuttgart

Woernle C (1988) Ein systematisches Verfahren zur Aufstellung der geometrischen Schließbedingungen in kinematischen Schleifen mit Anwendung bei der Rückwärtstransformation für Industrieroboter [Dr.-Ing.]. Dissertation, Universität Stuttgart, Düsseldorf: VDI-Verlag

Chapter 4
Equations of Motion of Complex Multibody Systems

In this chapter several methods from classical mechanics to state equations of motion of mechanical systems are briefly explored (Sect. 4.1–4.4). Principally, all these methods are also suited for the modeling of the vehicles regarded in this book, which can be considered as complex multibody systems. Concerning the practical application of a method, this decision is mainly dependent on the associated modeling and calculation effort, according to the complexity of the system. Consequently, the equations of motion for this kind of large scale systems with many bodies and many kinematic loops (and therefore many constraints) will be developed numerically or symbolically with the help of the computer. The computational cost is dependent on the number of describing coordinates, including its geometrical significance; further being defined by the formulation of the constraint equations, i.e. through the modeling of the kinematics of the system, which is reflected directly by the number of required mathematical operations.

A particular approach is further detailed in Sect. 4.5. Based on d'ALEMBERT's principle for rigid bodies, a method which is suitable for the development of the equations of motion for mechanisms and complex multibody systems is derived. This method is based on the kinematic structure of the system. With an adequate formulation of the constraint equations (see Chap. 3), compared to other methods, the number of the necessary mathematical operations can be significantly reduced. The implementation of this method, as well as its practical application by means of a computer is further illustrated in Sect. 4.6 of this chapter, see also (Hiller 1983, 1995; Hiller et al. 1986–1988, 1986; Kecskemethy 1993).

4.1 Fundamental Equation of Dynamics for Point Mass Systems

Given is a system composed of

$$N \text{ mass points } m_i, \ i = 1, \ \ldots, N.$$

D. Schramm et al., Vehicle Dynamics,
DOI 10.1007/978-3-662-54483-9_4

which have fundamentally been introduced for holonomic systems. It is necessary to articulate that JOURDAIN's principle in Sect. 4.2 applies to holonomic and non-holonomic systems.

4.2 JOURDAIN'S Principle

JOURDAIN's Principle Similar to the fundamental equation from Sect. 4.1, the following equation applies to a mechanical system composed of point masses, (Schiehlen and Eberhard 2004):

$$\sum_{i=1}^{N}[\boldsymbol{F}_i - m_i\boldsymbol{a}_i] \cdot \delta\boldsymbol{v}_i = 0. \tag{4.6}$$

This relation can be extended to systems consisting of rigid bodies. It articulates that the virtual power of the reaction forces of this system disappears. Consequently a strong similarity with the fundamental equation of dynamics (Eq. 4.5), as well as to d'ALEMBERT's principle can be recognized. Further, the virtual velocities of the point masses replace the corresponding displacements. In order to calculate the reaction forces, the implicitly performing holonomic constraints have to be supplemented with the associated expressions for the non-holonomic constraints.

4.3 LAGRANGE Equations of the First Kind for Point Mass Systems

Starting with the fundamental equation of dynamics (Eq. 4.5), the LAGRANGE equations of the first kind are successively specified for point mass systems. A system is described as having

- N point masses m_i, $\boldsymbol{r_i}$,
- g geometric constraints

$$f_\alpha(t;\boldsymbol{r}_1,\ldots,\boldsymbol{r}_N) = 0, \quad \alpha = 1,\ldots,g, \tag{4.7}$$

- k kinematic constraints

$$\phi_\beta = \sum_{i=1}^{N}\boldsymbol{l}_{i\beta}(t;\boldsymbol{r}_1,\ldots,\boldsymbol{r}_N)\cdot\boldsymbol{v}_i + d_\beta(t;\boldsymbol{r}_1,\ldots,\boldsymbol{r}_N) = 0, \qquad \beta = 1,\ldots,k. \tag{4.8}$$

The system possesses $f = 3N - g - k$ degrees of freedom (DoF). By introducing the LAGRANGE multipliers λ_α, μ_β, the LAGRANGE equations of the first kind can now be read as:

$$\begin{aligned} m_i \boldsymbol{a}_i &= \boldsymbol{F}_i + \sum_{\alpha=1}^{g} \lambda_\alpha \frac{\partial f_\alpha}{\partial \boldsymbol{r}_i} + \sum_{\beta=1}^{k} \mu_\beta \boldsymbol{l}_{i\beta}, \quad i = 1, \ldots N, \\ & f_\alpha(t; \boldsymbol{r}_1, \ldots, \boldsymbol{r}_N) = 0, \quad \alpha = 1, \ldots, g, \\ & \sum_{i=1}^{N} \boldsymbol{l}_{i\beta} \cdot \boldsymbol{v}_i + d_\beta = 0, \quad \beta = 1, \ldots, k. \end{aligned} \tag{4.9}$$

This is a system composed of $(3N + g + k)$ equations for the $(3N + g + k)$ unknowns:

$$x_i, y_i, z_i, \lambda_\alpha, \mu_\beta. \tag{4.10}$$

The reaction forces $\boldsymbol{R}_i$ of the system are obtained from Eq. (4.9) as

$$\boldsymbol{R}_i = \sum_{\alpha=1}^{g} \lambda_\alpha \frac{\partial f_\alpha}{\partial \boldsymbol{r}_i} + \sum_{\beta=1}^{k} \mu_\beta \boldsymbol{l}_{i\beta}, \quad i = 1, \ldots, N. \tag{4.11}$$

Note: Introducing for each body the pair of instantaneous velocities $(\boldsymbol{v}_i, \boldsymbol{\omega}_i)$ which together represent the instantaneous kinematic state of the system, the dynamic equations for the conservation of linear and angular momentum for rigid bodies, also denominated as NEWTON-EULER equations, can be formulated. Subsequently, the LAGRANGE equations of the first kind for systems of rigid bodies can also be composed.

Advantages of the LAGRANGE equations of the first kind:

- They are valid for holonomic and non-holonomic systems.
- The reaction forces can be calculated directly and easily.

Disadvantages of the LAGRANGE equations of the first kind:

- Although the system possesses only $f = 3N - g - k$ degrees of freedom, $3N + g + k$ equations need to be stated and solved. This also holds true when only the motion of the system is of interest, and the reaction forces do not have to be calculated.

4.4 LAGRANGE Equations of the Second Kind for Rigid Bodies

By introducing the independent generalized coordinates (corresponding to the number of degrees of freedom)

$$q_1, q_2, \ldots, q_f \tag{4.12}$$

the LAGRANGE equations of the second kind for holonomic systems can be stated as follows:

$$\frac{d}{dt}\left(\frac{\partial T}{\partial \dot{q}_j}\right) - \frac{\partial T}{\partial q_j} = Q_j, \quad j = 1,\ldots,f, \tag{4.13}$$

where

T is the kinetic energy of the system,
Q_j are the generalized (applied) forces

The kinetic energy of the rigid body is then

$$T = T_{trans} + T_{rot} = \frac{1}{2} m \boldsymbol{v}_s^2 + \frac{1}{2} \boldsymbol{\omega}^T \boldsymbol{\Theta}_S \boldsymbol{\omega}, \tag{4.14}$$

where

$\boldsymbol{v}_s$ velocity of the center of mass S,
$\boldsymbol{\omega}$ angular velocity of the rigid body,
$\boldsymbol{\Theta}_s$ tensor of inertia with respect to S.

If the chosen body-fixed coordinate system is a principle-axis system, then the rotational energy can be written as:

$$T_{rot} = \frac{1}{2}\left(A\omega_x^2 + B\omega_y^2 + C\omega_z^2\right), \tag{4.15}$$

with

$$\boldsymbol{\Theta}_S = \begin{pmatrix} A & 0 & 0 \\ 0 & B & 0 \\ 0 & 0 & C \end{pmatrix}$$

and the principle moments of inertia A, B and C.

Advantages of the LAGRANGE equations of the second kind:

- Smallest possible number f of equations.
- For the formation, only the kinetic energy T and the generalized forces Q_j are required.

Disadvantage of the LAGRANGE equations of the second kind:

- High complexity of the expression for the kinetic energy in mechanisms and other complex multibody systems.

Note: The equations of motion for constraint systems can also be stated, where the LAGRANGE equations of the second kind in a first step are written for a free system that is then expanded by the introduction of LAGRANGE multipliers, using the constraint equations of the system. The result is an expanded system of $(f + g)$ equations, comparable to the LANGRANGE equations of the first kind. This approach builds the theoretical basis for several computer programs that analyze multibody systems.

4.5 D'ALEMBERT's Principle

Expanding upon the fundamental equation of dynamics for point mass systems, one considers the term $m_i \boldsymbol{a}_i$ in Eq. (4.5) to stand for the inertial forces in the system, which already represents d'ALEMBERT's principle for point mass systems. The expansion into systems of rigid bodies follows easily, as again the pair of instantaneous velocities $(\boldsymbol{v}_i, \boldsymbol{\omega}_i)$ for each rigid body in the system to formulate the conservation of linear and angular momentum as before is utilized. For a system composed of n_B rigid bodies, one can then derive the following equation, which represents d'ALEMBERT's principle for systems of rigid bodies:

$$\sum_{i=1}^{n_B} [(m_i \ddot{\boldsymbol{r}}_{S_i} - \boldsymbol{F}_i) \cdot \delta \boldsymbol{r}_{S_i} + (\boldsymbol{\Theta}_{S_i} \dot{\boldsymbol{\omega}}_i + \boldsymbol{\omega}_i \times \boldsymbol{\Theta}_{S_i} \boldsymbol{\omega}_i - \boldsymbol{T}_i) \cdot \delta \boldsymbol{\varphi}_i] = 0, \tag{4.16}$$

with

$m_i, \boldsymbol{\Theta}_{S_i}$ mass and tensor of inertia for body i,
$\ddot{\boldsymbol{r}}_{S_i}$ acceleration of the center of mass for body i,
$\boldsymbol{\omega}_i, \dot{\boldsymbol{\omega}}_i$ angular velocity and acceleration for body i,
$\boldsymbol{F}_i, \boldsymbol{T}_i$ applied forces and torques on the body i,
$\delta \boldsymbol{r}_{S_i}, \delta \boldsymbol{\varphi}_i$ virtual displacements of translation and rotation of body i.

In Eq. (4.16) appear the absolute virtual displacements of the bodies i, which are not independent due to the constraints of the system. Using Eq. (4.16), it is now the goal to derive the subsequent relation through introduction of f independent generalized coordinates $q_1, \ldots, q_f$ (corresponding to the total number of degrees of freedom of the system) with respect to their independent virtual displacements $\delta q_1, \ldots, \delta q_f$

$$\boldsymbol{M}\ddot{\boldsymbol{q}} + \boldsymbol{b} = \boldsymbol{Q}. \tag{4.17}$$

In this case there are:

$\boldsymbol{q}$ $[f \times 1]$-vector of the generalized coordinates,
$\boldsymbol{M}$ $[f \times f]$-generalized mass matrix,
$\boldsymbol{b}$ $[f \times 1]$-vector: of generalized gyroscopic forces,
$\boldsymbol{Q}$ $[f \times 1]$-vector of generalized applied forces

In a succeeding step it is necessary to state the kinematic relationship between the absolute and dependent coordinates and the independent generalized coordinates. Further, the terms belonging to the translational motion and the rotational motion of the body i are separately analyzed. In order to facilitate the relationship between the vectorial terms of translation and rotation with the independent coordinates, a coordinate representation has to be introduced. The inertial system $\mathbf{K}_E = \{O_E; x_E, y_E, z_E\}$ will serve as the reference system for the upcoming considerations. Using this, the following general relationships are derived:

$$\begin{aligned} \delta_E \boldsymbol{r}_{S_i} &= {}_E\boldsymbol{J}_{S_i} \delta \boldsymbol{q}, \quad {}_E\ddot{\boldsymbol{r}}_{S_i} = {}_E\boldsymbol{J}_{S_i} \ddot{\boldsymbol{q}} + {}_E\boldsymbol{a}_{S_i}, \\ \delta_E \boldsymbol{\varphi}_i &= {}_E\boldsymbol{J}_{\varphi_i} \delta \boldsymbol{q}, \quad {}_E\dot{\boldsymbol{\omega}}_i = {}_E\boldsymbol{J}_{\varphi_i} \ddot{\boldsymbol{q}} + {}_E\boldsymbol{\alpha}_{\varphi_i} \end{aligned} \tag{4.18}$$

with the JACOBIAN matrices (expressed in system $\mathbf{K}_E$)

$${}_E\boldsymbol{J}_{S_i} = \frac{\partial_E \boldsymbol{r}_{S_i}}{\partial \boldsymbol{q}}, {}_E\boldsymbol{J}_{\varphi_i} = \frac{\partial_E \boldsymbol{\varphi}_i}{\partial \boldsymbol{q}}. \tag{4.19}$$

Thus, the individual elements of the equations of motion can be read as follows:

- mass matrix

$$\boldsymbol{M} = \sum_{i=1}^{n_B} \left[{}_E\boldsymbol{J}_{S_i}^T {}_E\boldsymbol{J}_{S_i} m_i + {}_E\boldsymbol{J}_{\varphi_i}^T {}_E\boldsymbol{\Theta}_{S_i} {}_E\boldsymbol{J}_{\varphi_i} \right]. \tag{4.20}$$

- generalized gyroscopic forces

$$\boldsymbol{b} = \sum_{i=1}^{n_B} \left\{ {}_E\boldsymbol{J}_{S_i}^T {}_E\boldsymbol{a}_{S_i} m_i + {}_E\boldsymbol{J}_{\varphi_i}^T \left[{}_E\boldsymbol{\Theta}_{S_i} {}_E\boldsymbol{\alpha}_{\varphi_i} + {}_E\boldsymbol{\omega}_i \times \left({}_E\boldsymbol{\Theta}_{S_i} {}_E\boldsymbol{\omega}_i \right) \right] \right\}. \tag{4.21}$$

- generalized applied forces and torques

$$\boldsymbol{Q} = \sum_{i=1}^{n_B} \left[{}_E\boldsymbol{J}_{S_i}^T {}_E\boldsymbol{F}_i + {}_E\boldsymbol{J}_{\varphi_i}^T {}_E\boldsymbol{T}_i \right]. \tag{4.22}$$

The unknowns in these equations are still the JACOBIAN matrices

$${}_E\boldsymbol{J}_{S_i} = \frac{\partial_E \boldsymbol{r}_{S_i}}{\partial \boldsymbol{q}}, \; {}_E\boldsymbol{J}_{\varphi_i} = \frac{\partial_E \boldsymbol{\varphi}_i}{\partial \boldsymbol{q}} \tag{4.23}$$

and the generalized acceleration terms

$$ {}_{E}\boldsymbol{a}_{S_i} = \sum_{j=1}^{f}\sum_{k=1}^{f} \frac{\partial_E^2 \boldsymbol{r}_{S_i}}{\partial q_j \partial q_k} \dot{q}_j \dot{q}_k, \tag{4.24} $$

$$ {}_{E}\boldsymbol{\alpha}_{\varphi_i} = \sum_{j=1}^{f}\sum_{k=1}^{f} \frac{\partial_E^2 \boldsymbol{\varphi}_i}{\partial q_j \partial q_k} \dot{q}_j \dot{q}_k. \tag{4.25} $$

It is therefore necessary to discuss in the succeeding sections the evaluation of the mass matrix, the generalized gyroscopic forces, and the generalized applied forces. However, the analysis will not be carried out through the expressions given in Eqs. (4.19), (4.24) and (4.25) as their analytical calculation is very complex. In particular, the computing time of the second derivatives (Eqs. 4.24 and 4.25) grows rapidly with the increasing complexity of the multibody system. This is mainly due to the cumulative complexity of the constraint equations due to the inherent kinematic loops. Using instead the concept of global kinematics of complex multibody systems with kinematic loops (consisting of relative kinematics and absolute kinematics, see Sect. 3.5), the number of required calculation operations can be drastically reduced when compared to traditional methods. Further, it has to be emphasized that the evaluation of all the complex terms mentioned in the equations above is not fixed to a reference frame that has to be selected a priori. Here, a choice of favorable reference frames which enable a calculation with a minimum number of operations is possible. Conversely, for the derivation of Eqs. (4.19), (4.24) and (4.25), a reference frame has to be introduced, which may include the disadvantages mentioned above.

4.6 Computer-Based Derivation of the Equations of Motion

The difficulty in deriving the dynamic equations of complex multibody systems lies mainly in the incorporation of the constraint equations in a general form. Again, kinematics plays the central role. Thus, based on the complex topology of the system, the kinematic analysis has to be appropriately prepared. Further issues like the calculation of the applied forces as well as the reaction forces can also be simplified using an appropriate approach for the kinematics. This will, however, not be investigated in this context.

4.6.1 Kinematic Differentials of Absolute Kinematics

Before stating the dynamic equations, it is suitable to complete the kinematic analysis with relations for the determination of partial derivatives. In particular, the relationships between the first and second differentials of the generalized coordinates $\boldsymbol{q}$ and the corresponding differentials of the absolute coordinates of the bodies i, collected in the vector

$$\boldsymbol{w}_i = \begin{bmatrix} \boldsymbol{r}_{S_i} \\ \boldsymbol{\varphi}_i \end{bmatrix}, \tag{4.26}$$

are required. Considering the function

$$\boldsymbol{w}_i = \boldsymbol{w}_i(\boldsymbol{q}), \tag{4.27}$$

the first and second time derivatives follow as

$$\dot{\boldsymbol{w}}_i = \boldsymbol{J}_{w_i}\dot{\boldsymbol{q}}, \tag{4.28}$$

$$\ddot{\boldsymbol{w}}_i = \boldsymbol{J}_{w_i}\ddot{\boldsymbol{q}} + \boldsymbol{a}_{w_i}. \tag{4.29}$$

The [6×f]-JACOBIAN matrices $\boldsymbol{J}_{w_i}$ as well as the $[6 \times 1]$–vectors $\boldsymbol{a}_{w_i}$ are still unknown. They can be determined from analytical functions $\boldsymbol{w}_i(\boldsymbol{q})$, each with the corresponding partial derivatives:

$$\boldsymbol{J}_{w_i} = \frac{\partial \boldsymbol{w}_i}{\partial \boldsymbol{q}}, \tag{4.30}$$

$$\boldsymbol{a}_{w_i} = \sum_{j=1}^{f}\sum_{k=1}^{f} \frac{\partial^2 \boldsymbol{w}_i}{\partial q_j \partial q_k}\dot{q}_j\dot{q}_k. \tag{4.31}$$

The preparation of analytical derivatives is extremely complex due to the highly implicit nature of the constraint equations of complex multibody systems. Even applying mathematical tools for a symbolic formula manipulation, like MATHEMATICA or MAPLE, can be problematic because the intermediate data are so extensive that these terms subsequently can no longer be effectively handled. Therefore, an alternative solution is presented, where analytical derivatives are replaced by kinematic expressions, needing much less operations.

First derivatives of the absolute coordinates of a rigid body. In order to state the global kinematics for a general complex multibody system where the position $\boldsymbol{w}_i$ of all bodies i is given, the time derivatives $\dot{\boldsymbol{w}}_i$ of the absolute coordinates of all bodies for arbitrary values of the generalized velocities $\dot{\boldsymbol{q}}$ can be specified using elementary-kinematic expressions (see Sect. 3.5). Pseudo input velocities $\tilde{\dot{\boldsymbol{w}}}_i^{(j)}$ can be defined for unique, dimensionless pseudo velocities of the generalized coordinates

$$\tilde{\dot{q}}^{(j)} = e^{(j)}, e^{(j)} = \left[0, \ldots, 0, \underbrace{1}_{j^{th}\ column}, 0, \ldots, 0\right], \tag{4.32}$$

where the $[f \times 1]$-unit vectors possess a "1" as j^{th} element, otherwise only zeroes. As the actual time derivatives $\dot{\boldsymbol{w}}_i$ are linear combinations of the generalized velocities $\dot{q}_j$, which again are independent from each other, the following relation holds:

$$\dot{\boldsymbol{w}}_i = \sum_j \tilde{\dot{\boldsymbol{w}}}_i^{(j)} \dot{q}_j. \tag{4.33}$$

The comparison of Eq. (4.33) with Eq. (4.28) yields the simple rule for the determination of the corresponding JACOBIAN matrix:

$$j\text{th column } |\boldsymbol{J}_{w_i}| = \tilde{\dot{\boldsymbol{w}}}_{i(j)}. \tag{4.34}$$

This means that the individual columns of the required JACOBIAN matrices can be effectively and systematically extracted from the expressions of the previously performed kinematic analysis (see in particular Sect. 3.5). This method enables a particularly efficient calculation of the needed partial differentiations for the JACOBIAN matrices. One designates it as the kinematic differentials of the first kind, (Kecskemethy 1993).

Second derivatives. Assuming similarly position and velocity of the system to be given, the acceleration $\ddot{\boldsymbol{w}}_i$ of all the bodies can also be calculated for arbitrary values of the generalized accelerations $\ddot{\boldsymbol{q}}$ by simple kinematic means (see Sect. 3.5). In particular, the desired pseudo-input accelerations $\tilde{\ddot{\boldsymbol{w}}}_i$ can be directly determined with the generalized input acceleration $\tilde{\ddot{\boldsymbol{q}}} = 0$, and one obtains the following from Eq. (4.29):

$$\boldsymbol{a}_{w_i} = \tilde{\ddot{\boldsymbol{w}}}_i. \tag{4.35}$$

The Eqs. (4.34) and (4.35) contain all required relations between the differentials of the generalized coordinates and the absolute coordinates of the bodies. They are definable alone by the elementary kinematic expressions (mainly with the use of the laws of relative kinematics). Therefore, they are denominated as the kinematic differentials of the second kind, (Kecskemethy 1993).

Kinematic differentials. The time derivatives of the collected absolute coordinates can now be split up into their translational and rotational components $\dot{\boldsymbol{r}}_{S_i}$, $\ddot{\boldsymbol{r}}_{S_i}$ and $\boldsymbol{\omega}_i$, $\dot{\boldsymbol{\omega}}_i$ respectively. The corresponding relations are

$$\dot{\boldsymbol{r}}_{S_i} = \sum_j \tilde{\dot{\boldsymbol{r}}}_{S_i}^{(j)} \dot{q}_j, \quad \ddot{\boldsymbol{r}}_{S_i} = \sum_j \tilde{\dot{\boldsymbol{r}}}_{S_i}^{(j)} \ddot{q}_j + \tilde{\ddot{\boldsymbol{r}}}_{S_i}, \tag{4.36}$$

$$\boldsymbol{\omega}_i = \sum_j \tilde{\boldsymbol{\omega}}_i^{(j)} \dot{q}_j, \quad \dot{\boldsymbol{\omega}}_i = \sum_j \tilde{\boldsymbol{\omega}}_i^{(j)} \ddot{q}_j + \tilde{\dot{\boldsymbol{\omega}}}_i. \tag{4.37}$$

From the description in the Eqs. (4.36) and (4.37), another advantage becomes evident: While the relations between the differentials using the analytical path are only ascertainable through the derivatives of a defined component representation of all vectors, here it is possible to represent the relations by using only physical vectors (meaning kinematic quantities), which are independent from the selection of the coordinate system. The transition into component representation can be arbitrarily delayed, i.e. the selection of the coordinate system can be adjusted to the reduction of the computational complexity of every term that needs to be evaluated. This allows the formulation of the equations of motion of general, complex multibody systems in an extremely compact and efficient manner.

4.6.2 Equations of Motion

For the dynamics of a system composed of n_B rigid bodies, d'ALEMBERT's principle was given in Sect. 4.5 as:

$$\sum_{i=1}^{n_B} [(m_i \ddot{\boldsymbol{r}}_{S_i} - \boldsymbol{F}_i) \cdot \delta \boldsymbol{r}_{S_i} + (\boldsymbol{\Theta}_{S_i} \dot{\boldsymbol{\omega}}_i + \boldsymbol{\omega}_i \times \boldsymbol{\Theta}_{S_i} \boldsymbol{\omega}_i - \boldsymbol{T}_i) \cdot \delta \boldsymbol{\varphi}_i] = 0, \tag{4.38}$$

with the variables

$m_i, \boldsymbol{\Theta}_{S_i}$	mass and tensor of inertia for body i,
$\ddot{\boldsymbol{r}}_{S_i}$	acceleration of the center of mass for body i,
$\boldsymbol{\omega}_i, \dot{\boldsymbol{\omega}}_i$	angular velocity and acceleration for body i,
$\boldsymbol{F}_i, \boldsymbol{T}_i$	applied forces and torques on the body i,
$\delta \boldsymbol{r}_{S_i}, \delta \boldsymbol{\varphi}_i$	virtual displacements of translation and rotation of body i

In order to set up the equations of motion from Eq. (4.38), the dependent virtual displacements $\delta \boldsymbol{r}_{S_i}, \delta \boldsymbol{\varphi}_i$ as well as the accelerations $\ddot{\boldsymbol{r}}_{S_i}, \dot{\boldsymbol{\omega}}_i$ must be expressed in dependency of the virtual displacements and of the accelerations of the independent generalized coordinates. Keeping in mind that the transformation rules apply equally for the virtual displacements as they do for the velocities, one can now gather from Eqs. (4.36) and (4.37):

$$\delta \boldsymbol{r}_{S_i} = \sum_{j=1}^{f} \tilde{\dot{\boldsymbol{r}}}_{S_i}^{(j)} \delta q_j, \quad \ddot{\boldsymbol{r}}_{S_i} = \sum_{j=1}^{f} \tilde{\dot{\boldsymbol{r}}}_{S_i}^{(j)} \ddot{q}_j + \tilde{\ddot{\boldsymbol{r}}}_{S_i}, \tag{4.39}$$

$$\delta\boldsymbol{\varphi}_i = \sum_{j=1}^{f} \tilde{\boldsymbol{\omega}}_i^{(j)} \delta q_j, \quad \dot{\boldsymbol{\omega}}_i = \sum_{j=1}^{f} \tilde{\boldsymbol{\omega}}_i^{(j)} \ddot{q}_j + \tilde{\dot{\boldsymbol{\omega}}}_i. \tag{4.40}$$

Substituting Eqs. (4.39) and (4.40) into Eq. (4.38) provides, with consideration of the independency of the virtual displacements δq_j, the equations of motion in the minimal form:

$$\boldsymbol{M}\ddot{\boldsymbol{q}} + \boldsymbol{b} = \boldsymbol{Q}. \tag{4.41}$$

The coefficients of the generalized $[f \times f]$-mass matrix $\boldsymbol{M}$, the $[f \times 1]$-vector of the generalized gyroscopic and centrifugal forces $\boldsymbol{b}$ and of the $[f \times 1]$-vector of the generalized applied forces $\boldsymbol{Q}$ can be calculated as follows:

$$\left.\begin{aligned} M_{j,k} &= \sum_{i=1}^{n_B} \left[m_i \tilde{\boldsymbol{r}}_{S_i}^{(j)} \cdot \tilde{\boldsymbol{r}}_{S_i}^{(k)} + \tilde{\boldsymbol{\omega}}_i^{(j)} \cdot \left(\boldsymbol{\Theta}_{S_i} \tilde{\boldsymbol{\omega}}_i^{(k)} \right) \right], \\ b_j &= \sum_{i=1}^{n_B} \left[m_i \tilde{\boldsymbol{r}}_{S_i}^{(j)} \cdot \tilde{\ddot{\boldsymbol{r}}}_{S_i} + \tilde{\boldsymbol{\omega}}_i^{(j)} \cdot \left(\boldsymbol{\Theta}_{S_i} \tilde{\dot{\boldsymbol{\omega}}}_i + \boldsymbol{\omega}_i \times \boldsymbol{\Theta}_{S_i} \boldsymbol{\omega}_i \right) \right], \\ Q_j &= \sum_{i=1}^{n_B} \left[\tilde{\boldsymbol{r}}_{S_i}^{(j)} \cdot \boldsymbol{F}_i + \tilde{\boldsymbol{\omega}}_i^{(j)} \cdot \boldsymbol{T}_i \right]. \end{aligned}\right\} \tag{4.42}$$

The Eq. (4.42) can be directly evaluated by known global kinematics referred to in Sect. 3.5. The evaluation can on the one hand be carried out on a purely numerical basis, which means that the kinematic differentials are defined directly through corresponding repetitions of the global kinematics with adequate inputs, without consideration of redundant operations. On the other hand, the redundant operations can be eliminated by the construction of the kinematic differentials through analytical formation. Then the equations of motion are achieved in compact, symbolic form. The numeric relationship has the advantage of a very fast implementation (for individual simulations), while the symbolic approach provides an optimized version of the motion equations (e.g. for the design of variant constructions). Both representations have in common that the structure of Eq. (4.42) is independent from coordinate representations. Depending on the application, the individual terms can be evaluated using appropriately chosen component representations (Fig. 4.1).

4.6.3 Dynamics of a Spatial Multibody Loop

The subsequently exemplarily analyzed spatial multibody loop (Fig. 4.2) possesses a spherical joint and four revolute joints with the unit vectors $\boldsymbol{u}_1, \boldsymbol{u}_2, \boldsymbol{u}_3$ and $\boldsymbol{u}_4$ for each rotational axis. All bodies, with exception of the coupler body (center of mass S_m) and the resting frame, are modeled as massless bodies, in order to simplify the dynamics. The spatial kinematic loop contains seven natural coordinates: the four

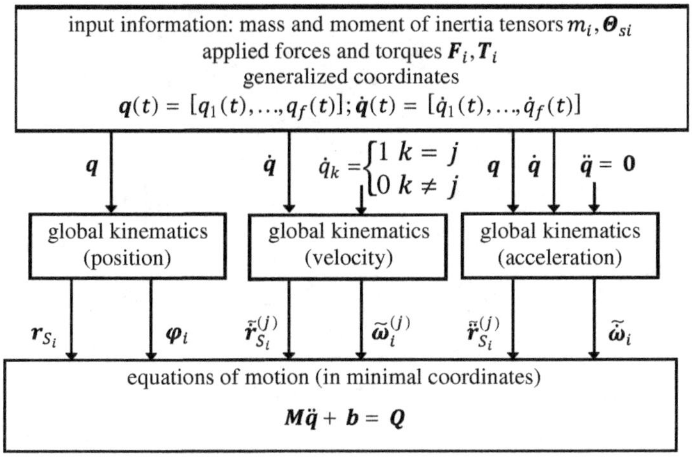

Fig. 4.1 Equations of motion of complex multibody systems applying kinematic differentials

relative angles $\beta_1, \beta_2, \beta_3, \beta_4$ of the four revolute joints and three more angles in the spherical joint S_1, which are, however, not required in this context. The system possesses one degree of freedom, for which one scalar equation of motion can be formulated. Basically, every coordinate can be selected as a generalized coordinate. Because of the proximity of this example to a technical application, as well as, in order to circumvent singularities, a suitable choice for the input motion is the coordinate $\beta_1 = q$. The kinematics of the system, which means the nonlinear equations for position, velocity, and acceleration are functions of this input variable β_1 (which also holds for its derivatives), are formulated in the following, (Hiller 1995).

Core system of implicit equations. In order to get an implicit core system of minimal order out of the constraint equations, a characteristic joint pair is chosen having a maximum number of degrees of freedom, (Woernle 1988). In this example, the two revolute joints R_2, R_3 with the two perpendicular axes $\boldsymbol{u}_2$ and $\boldsymbol{u}_3$ can be interpreted as a CARDAN joint with two degrees of freedom. The second joint of the characteristic joint pair is the spherical joint S_1 with three joint coordinates. The CARDAN joint and the spherical joint compose the characteristic joint pairing of the loop, eliminating in this first step five unknowns. The corresponding (scalar) characteristic constraint parameter is the distance d (Fig. 4.2)

$$\boldsymbol{d} \cdot \boldsymbol{d} = d^2. \tag{4.43}$$

The vector $\boldsymbol{d}$ can be represented on one side with respect to the so-called lower segment as $\boldsymbol{d} = \boldsymbol{r}_0 + \boldsymbol{r}_1$, and on the other side with respect to the upper segment as $\boldsymbol{d} = \boldsymbol{r}_3 + \boldsymbol{r}_4$. From this, two expressions for the distance d^2 are gained, which must be identical. The square of the corresponding terms with the projections

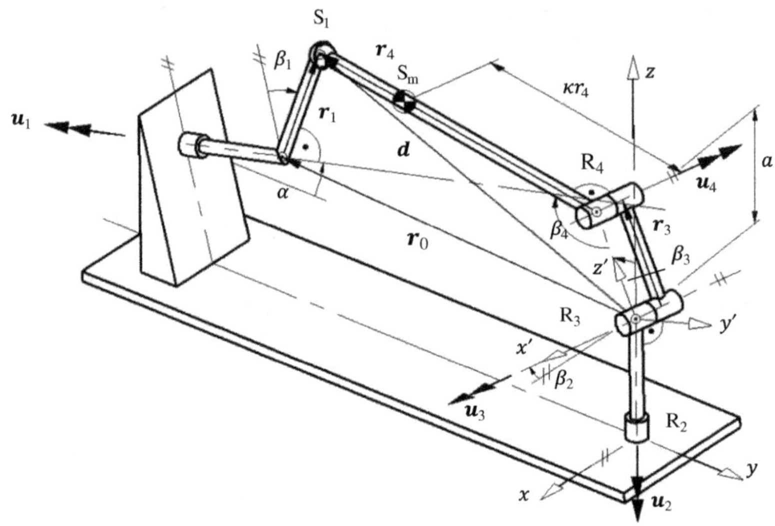

Fig. 4.2 Spatial multibody loop

$$\boldsymbol{r}_0 \cdot \boldsymbol{r}_1 = (r_1 a \cos\alpha)\cos\beta_1, \quad \boldsymbol{r}_3 \cdot \boldsymbol{r}_4 = -(r_3 r_4)\cos\beta_4, \tag{4.44}$$

provide an explicit relation for the angle β_4

$$\cos\beta_4 = -C_1\cos\beta_1 + C_2, \tag{4.45}$$

$$C_1 = \frac{r_1 a \cos\alpha}{r_3 r_4}, \quad C_2 = \frac{r_3^2 + r_4^2 - r_0^2 - r_1^2}{2 r_3 r_4}. \tag{4.46}$$

The Eq. (4.45) yields two symmetrical solutions for the function $\beta_4(\beta_1)$, only one of which is compatible with the initial position of the kinematic loop.

Complementary angles. The complementary variables are, in this case, the angles β_2 and β_3, which can be calculated in simple manner using the projections

$$\cos\beta_2 = \boldsymbol{e}_x \cdot \boldsymbol{u}_3, \quad \sin\beta_2 = \boldsymbol{e}_z \cdot \left(\boldsymbol{u}_3' \times \boldsymbol{e}_x\right), \tag{4.47}$$

$$\cos\beta_3 = -\boldsymbol{u}_2 \cdot \boldsymbol{e}_z', \quad \sin\beta_3 = \boldsymbol{u}_3 \cdot \left(\boldsymbol{e}_z' \times \boldsymbol{u}_2\right) \tag{4.48}$$

Here, the unit vectors $\boldsymbol{u}_3$ and $\boldsymbol{e}_z'$ can be expressed by using the already known vectors: For $\boldsymbol{u}_3$ there are two possible solutions:

$$\boldsymbol{u}_3 = \pm\frac{\boldsymbol{d} \times \boldsymbol{u}_2}{|\boldsymbol{d} \times \boldsymbol{u}_2|}. \tag{4.49}$$

For $\boldsymbol{e}_z'$ one gets

$$\boldsymbol{e}_z' = \frac{(\boldsymbol{d}\cdot\boldsymbol{e}_z')\boldsymbol{d} - \left[\boldsymbol{u}_3\cdot(\boldsymbol{d}\times\boldsymbol{e}_z')\right]\cdot(\boldsymbol{d}\times\boldsymbol{u}_3)}{d^2}, \tag{4.50}$$

where the projections $\boldsymbol{d}\cdot\boldsymbol{e}_z'$ and $\boldsymbol{u}_3\cdot(\boldsymbol{d}\times\boldsymbol{e}_z')$ can be defined in the system $(x,y,z)'$. The explicit evaluation of Eqs. (4.47)–(4.50) provides

$$\cos\beta_2 = \pm\frac{b\cos\alpha + r_1\cos\beta_1\sin\alpha}{K_1(\beta_1)}, \tag{4.51}$$

$$\sin\beta_2 = \pm\frac{r_1\sin\beta_1}{K_1(\beta_1)}, \tag{4.52}$$

$$\cos\beta_3 = K_5(\beta_1,\beta_2,\beta_4), \tag{4.53}$$

$$\sin\beta_3 = -K_3(\beta_1,\beta_2,\beta_4)\sin\beta_2 - K_4(\beta_1,\beta_2,\beta_4)\cos\beta_2, \tag{4.54}$$

with the abbreviations

$$K_1(\beta_1) = \sqrt{\left[(b\cos\alpha + r_1\cos\beta_1\sin\alpha)^2 + r_1^2\sin\beta_1^2\right]}, \tag{4.55}$$

$$K_2(\beta_4) = r_4^2 + r_3^2 - 2r_3r_4\cos\beta_4, \tag{4.56}$$

$$\begin{aligned} K_3(\beta_1,\beta_2,\beta_4) &= \frac{r_1\sin\beta_1(r_4\cos\beta_4 - r_3)}{K_2(\beta_4)} \\ &+ \frac{r_4\sin\beta_4}{K_2(\beta_4)}(a\sin\beta_2 - b\sin\alpha\sin\beta_2 + r_1\sin\beta_2\cos\beta_1\cos\alpha), \end{aligned} \tag{4.57}$$

$$\begin{aligned} K_4(\beta_1,\beta_2,\beta_4) &= \frac{(r_4\cos\beta_4 - r_3)(b\cos\alpha + r_1\cos\beta_1\sin\alpha)}{K_2(\beta_4)} \\ &+ \frac{r_4\sin\beta_4}{K_2(\beta_4)}(a\cos\beta_2 - b\sin\alpha\cos\beta_2 + r_1\cos\beta_2\cos\alpha), \end{aligned} \tag{4.58}$$

$$\begin{aligned} K_5(\beta_1,\beta_2,\beta_4) &= \frac{(r_3 - r_4\cos\beta_4)(a - b\sin\alpha + r_1\cos\beta_1\cos\alpha)}{K_2(\beta_4)} \\ &+ \frac{r_4\sin\beta_4}{K_2(\beta_4)}(b\cos\alpha\cos\beta_2 + r_1\sin\beta_1\sin\beta_2 + r_1\cos\beta_1\cos\beta_2\sin\alpha). \end{aligned} \tag{4.59}$$

All vectors can now be represented with respect to each arbitrarily chosen coordinate system. This completes the position kinematics of this spatial multibody loop.

Time derivatives of the relative angles. For the angular velocity $\dot{\beta}_4$, it follows from the derivation of Eqs. (4.45) and (4.46):

$$\dot{\beta}_4 = D_{41}\dot{\beta}_1, \quad D_{41} = -C_1 \frac{\sin\beta_1}{\sin\beta_4}. \tag{4.60}$$

Here, the coefficient D_{41} becomes singular for $\sin\beta_4 = 0$, which means that the vectors $\boldsymbol{r}_3$ and $\boldsymbol{r}_4$ are parallel. This can, however, be excluded in the given case. The angular velocities $\dot{\beta}_2$ and $\dot{\beta}_3$ can be defined through the time derivation of Eqs. (4.47)–(4.48). It is easier, however, to define these values with help of the velocity of the vector $\boldsymbol{d}$. Two defining equations for this vector were already formulated in the previous step. From the equality of the corresponding derivatives it follows

$$(\boldsymbol{u}_1 \times \boldsymbol{r}_1)\dot{\beta}_1 = (\boldsymbol{u}_2 \times \boldsymbol{d})\dot{\beta}_2 + (\boldsymbol{u}_3 \times \boldsymbol{d})\dot{\beta}_3 + (\boldsymbol{u}_4 \times \boldsymbol{r}_4)\dot{\beta}_4. \tag{4.61}$$

In Eq. (4.61), the two unknown values $\dot{\beta}_2$ and $\dot{\beta}_3$ are contained. Note that the unity vectors $\boldsymbol{u}_3$ and $\boldsymbol{u}_4$ are normal to the unity vector $\boldsymbol{u}_2$, and one gets two independent scalar equations for $\dot{\beta}_2$ and $\dot{\beta}_3$ from Eq. (4.61) through the projection on $\boldsymbol{u}_3$ and on $\boldsymbol{u}_2$ respectively. The corresponding scalar products lead to

$$\dot{\beta}_2 = D_{21}\dot{\beta}_1, \quad \dot{\beta}_3 = D_{31}\dot{\beta}_1 + D_{34}\dot{\beta}_4 \tag{4.62}$$

with the coefficients

$$\begin{aligned} D_{21} &= \frac{\boldsymbol{u}_3 \cdot (\boldsymbol{u}_1 \times \boldsymbol{r}_1)}{\boldsymbol{u}_3 \cdot (\boldsymbol{u}_2 \times \boldsymbol{d})}, \\ D_{31} &= \frac{\boldsymbol{u}_2 \cdot (\boldsymbol{u}_1 \times \boldsymbol{r}_1)}{\boldsymbol{u}_2 \cdot (\boldsymbol{u}_3 \times \boldsymbol{d})}, \\ D_{34} &= -\frac{\boldsymbol{u}_2 \cdot (\boldsymbol{u}_4 \times \boldsymbol{r}_4)}{\boldsymbol{u}_2 \cdot (\boldsymbol{u}_3 \times \boldsymbol{d})} \end{aligned} \tag{4.63}$$

or expressed in scalar quantities

$$\begin{aligned} D_{21} &= \frac{r_1(\cos\beta_1\cos\beta_2 + \sin\beta_1\sin\beta_2\sin\alpha)}{r_1(\cos\beta_1\cos\beta_2\sin\alpha + \sin\beta_1\sin\beta_2) + b\cos\beta_2\sin\alpha}, \\ D_{31} &= \frac{r_1\sin\beta_1\cos\alpha}{r_3\sin\beta_3 + r_4\sin(\beta_4 - \beta_3)}, \\ D_{34} &= \frac{r_4\sin(\beta_3 - \beta_4)}{r_3\sin\beta_3 + r_4\sin(\beta_4 - \beta_3)}. \end{aligned} \tag{4.64}$$

These coefficients become singular for $\boldsymbol{u}_2 \cdot (\boldsymbol{u}_3 \times \boldsymbol{d}) = 0$, which means when the distance vector $\boldsymbol{d}$ is parallel to $\boldsymbol{u}_2$. Again, this special configuration can also be excluded, however, in the given case. If the input variable chosen to be β_1, the simulation of the kinematics and dynamics of the multibody loop is then free of singularities. Because of Eqs. (4.62), no further independent conditions for the angular velocities $\dot{\beta}_2$ and $\dot{\beta}_3$ can be formulated. With the projection of Eq. (4.61) in the direction of vector $\boldsymbol{d}$, one gets the condition

$$\boldsymbol{d} \cdot (\boldsymbol{u}_1 \times \boldsymbol{r}_1)\dot{\beta}_1 = \boldsymbol{d} \cdot (\boldsymbol{u}_4 \times \boldsymbol{r}_4)\dot{\beta}_4, \tag{4.65}$$

which is equivalent to Eq. (4.60). This can be shown, after substituting of $\boldsymbol{d} = \boldsymbol{r}_0 + \boldsymbol{r}_1$ into the left side of Eq. (4.61), as well as the substitution of $\boldsymbol{d} = \boldsymbol{r}_3 + \boldsymbol{r}_4$ into the right side of Eq. (4.61). For the angular acceleration, one obtains, using the time derivative of Eq. (4.60):

$$\ddot{\beta}_4 = D_{41}\,\ddot{\beta}_1 - E_4, \quad E_4 = \frac{C_1 \cos\beta_1 \left(\dot{\beta}_2\right)^2 + \cos\beta_4 \left(\dot{\beta}_4\right)^2}{\sin\beta_4}. \tag{4.66}$$

The angular accelerations $\ddot{\beta}_2$ and $\ddot{\beta}_3$ could be defined in correspondence with the derivation of Eq. (4.62). However, the derivation of Eq. (4.61) leads to a much simpler solution, by repeating the projections which were already used to define the angular velocities $\dot{\beta}_2$ and $\dot{\beta}_3$. The differentiation of Eq. (4.61) leads to:

$$\begin{aligned}
&(\boldsymbol{u}_1 \times \boldsymbol{r}_1)\ddot{\beta}_1 + (\boldsymbol{u}_1 \times \dot{\boldsymbol{r}}_1)\dot{\beta}_1 \\
&\quad = (\boldsymbol{u}_2 \times \boldsymbol{d})\,\ddot{\beta}_2 + (\boldsymbol{u}_2 \times \dot{\boldsymbol{d}})\dot{\beta}_2 + (\boldsymbol{u}_3 \times \boldsymbol{d})\,\ddot{\beta}_3 \\
&\qquad + (\dot{\boldsymbol{u}}_3 \times \boldsymbol{d})\dot{\beta}_3 + (\boldsymbol{u}_3 \times \dot{\boldsymbol{d}})\dot{\beta}_3 + (\boldsymbol{u}_4 \times \boldsymbol{r}_4)\,\ddot{\beta}_4 \\
&\qquad + (\dot{\boldsymbol{u}}_4 \times \boldsymbol{r}_4)\dot{\beta}_4 + (\boldsymbol{u}_4 \times \dot{\boldsymbol{r}}_4)\dot{\beta}_4.
\end{aligned} \tag{4.67}$$

The scalar product of Eq. (4.67) with $\boldsymbol{u}_3$, as well as $\boldsymbol{u}_2$, provides:

$$\begin{aligned}
\ddot{\beta}_2 &= D_{21}\,\ddot{\beta}_1 + E_2, \\
\ddot{\beta}_3 &= D_{31}\,\ddot{\beta}_1 + D_{34}\,\ddot{\beta}_4 + E_3.
\end{aligned} \tag{4.68}$$

The coefficients E_2 and E_3 depend on triple products executed with position vectors and velocities. After the elimination of the vanishing terms $\boldsymbol{u}_2 \cdot (\dot{\boldsymbol{u}}_3 \times \boldsymbol{d})$ and $\boldsymbol{u}_2 \cdot (\dot{\boldsymbol{u}}_4 \times \boldsymbol{r}_4)$ yields:

$$\begin{aligned}
E_2 &= \frac{\boldsymbol{u}_3 \cdot (\boldsymbol{u}_1 \times \dot{\boldsymbol{r}}_1)\dot{\beta}_1 - \boldsymbol{u}_3 \cdot (\dot{\boldsymbol{u}}_4 \times \boldsymbol{r}_4)\dot{\beta}_4 - \boldsymbol{u}_3 \cdot (\dot{\boldsymbol{u}}_3 \times \boldsymbol{d})\dot{\beta}_3 - \boldsymbol{u}_3 \cdot (\boldsymbol{u}_2 \times \dot{\boldsymbol{d}})\dot{\beta}_2}{\boldsymbol{u}_3 \cdot (\boldsymbol{u}_2 \times \boldsymbol{d})}, \\
E_3 &= \frac{\boldsymbol{u}_2 \cdot (\boldsymbol{u}_1 \times \dot{\boldsymbol{r}}_1)\dot{\beta}_1 - \boldsymbol{u}_2 \cdot (\boldsymbol{u}_4 \times \dot{\boldsymbol{r}}_4)\dot{\beta}_4 - \boldsymbol{u}_2 \cdot (\boldsymbol{u}_3 \times \dot{\boldsymbol{d}})\dot{\beta}_3}{\boldsymbol{u}_2 \cdot (\boldsymbol{u}_3 \times \boldsymbol{d})}.
\end{aligned} \tag{4.69}$$

The arising velocities can be expressed in dependency of the relative velocities $\dot{\beta}_1, \dot{\beta}_2, \dot{\beta}_3$ and $\dot{\beta}_4$:

$$\dot{\boldsymbol{r}}_1 = (\boldsymbol{u}_1 \times \boldsymbol{r}_1)\dot{\beta}_1, \quad \dot{\boldsymbol{u}}_3 = -\dot{\boldsymbol{u}}_4 = (\boldsymbol{u}_2 \times \boldsymbol{u}_3)\dot{\beta}_2, \tag{4.70}$$

$$\dot{\boldsymbol{r}}_4 = (\boldsymbol{u}_2 \times \boldsymbol{r}_4)\dot{\beta}_2 + (\boldsymbol{u}_3 \times \boldsymbol{r}_4)\left[\dot{\beta}_3 - \dot{\beta}_4\right]. \tag{4.71}$$

The explicit evaluation of the appearing triple products in the corresponding suitable coordinate system provides:

$$\begin{aligned}
\boldsymbol{u}_3 \cdot (\boldsymbol{u}_1 \times \dot{\boldsymbol{r}}_1)\dot{\beta}_1 &= r_1\dot{\beta}_1^2(\sin\beta_1 \cos\beta_2 - \cos\beta_1 \sin\beta_2 \sin\alpha),\\
\boldsymbol{u}_3 \cdot (\dot{\boldsymbol{u}}_4 \times \boldsymbol{r}_4)\dot{\beta}_4 &= -r_4 \cos(\beta_4 - \beta_3)\dot{\beta}_2\dot{\beta}_4,\\
\boldsymbol{u}_3 \cdot (\dot{\boldsymbol{u}}_3 \times \boldsymbol{d})\dot{\beta}_3 &= \dot{\beta}_2\dot{\beta}_3(r_4\cos(\beta_4 - \beta_3)) - r_3\cos\beta_3,\\
\boldsymbol{u}_3 \cdot \left(\boldsymbol{u}_2 \times \dot{\boldsymbol{d}}\right)\dot{\beta}_2 &= r_1\dot{\beta}_1\dot{\beta}_2(\sin\beta_1\cos\beta_2\sin\alpha - \sin\beta_2\cos\beta_1),\\
\boldsymbol{u}_3 \cdot (\boldsymbol{u}_2 \times \boldsymbol{d}) &= -r_1\cos\beta_1\cos\beta_2\sin\alpha - b\cos\beta_2\cos\alpha - r_1\sin\beta_1\sin\beta_2,\\
\boldsymbol{u}_2 \cdot (\boldsymbol{u}_1 \times \dot{\boldsymbol{r}}_1)\dot{\beta}_1 &= \dot{\beta}_1^2 r_1\cos\beta_1\cos\alpha,\\
\boldsymbol{u}_2 \cdot (\boldsymbol{u}_4 \times \dot{\boldsymbol{r}}_4)\dot{\beta}_4 &= r_4\dot{\beta}_4\left(\dot{\beta}_3 + \dot{\beta}_4\right)\cos(\beta_4 - \beta_3),\\
\boldsymbol{u}_2 \cdot \left(\boldsymbol{u}_3 \times \dot{\boldsymbol{d}}\right)\dot{\beta}_3 &= r_1\dot{\beta}_1\dot{\beta}_2(\cos\beta_1\sin\beta_2 - \cos\beta_2\sin\beta_1\sin\alpha),\\
\boldsymbol{u}_2 \cdot (\boldsymbol{u}_3 \times \boldsymbol{d}) &= r_3\sin\beta_3 + r_4\sin(\beta_4 - \beta_3).
\end{aligned}$$

Absolute kinematics. For the derivation of the equations of motion, only the absolute kinematics of the coupler body, the only body with mass in the system, as defined at the beginning of this example, is required. So, only the absolute velocity and acceleration of the center of mass S_m as well as the absolute angular velocity and acceleration of the same body will be considered. The corresponding derivations yield:

$$\boldsymbol{\omega} = \boldsymbol{u}_2\dot{\beta}_2 + \boldsymbol{u}_3\left[\dot{\beta}_3 - \dot{\beta}_4\right], \tag{4.72}$$

$$\dot{\boldsymbol{s}} = \dot{\boldsymbol{d}} - [1-\kappa]\dot{\boldsymbol{r}}_4 = (\boldsymbol{u}_1 \times \boldsymbol{r}_1)\dot{\beta}_1 - [1-\kappa](\boldsymbol{\omega} \times \boldsymbol{r}_4), \tag{4.73}$$

$$\dot{\boldsymbol{\omega}} = \boldsymbol{u}_2\ddot{\beta}_2 + \boldsymbol{u}_3\left[\ddot{\beta}_3 - \ddot{\beta}_4\right] + \boldsymbol{a}_\varphi, \tag{4.74}$$

$$\ddot{\boldsymbol{s}} = (\boldsymbol{u}_1 \times \boldsymbol{r}_1)\ddot{\beta}_1 - [1-\kappa](\dot{\boldsymbol{\omega}} \times \boldsymbol{r}_4) + \boldsymbol{a}_s. \tag{4.75}$$

The acceleration terms $\boldsymbol{a}_\varphi$ and $\boldsymbol{a}_s$ can be described with the previously defined velocities as

$$\boldsymbol{a}_\varphi = (\boldsymbol{u}_2 \times \boldsymbol{u}_3)\left[\dot{\beta}_3 - \dot{\beta}_4\right]\dot{\beta}_2, \tag{4.76}$$

$$\boldsymbol{a}_\mathrm{s} = (\boldsymbol{u}_1 \times \dot{\boldsymbol{r}}_1)\dot{\beta}_1 + [1-\kappa](\boldsymbol{\omega} \times \dot{\boldsymbol{r}}_4). \tag{4.77}$$

For the evaluation of the corresponding vector expressions, it is convenient to do this in components of the inherent vectors. One gets for example for the angular velocity $\boldsymbol{\omega}$ in components:

$$\boldsymbol{\omega} = \begin{bmatrix} \cos\beta_2\left(\dot{\beta}_3 - \dot{\beta}_4\right) \\ -\sin\beta_2\left(\dot{\beta}_3 - \dot{\beta}_4\right) \\ -\dot{\beta}_2 \end{bmatrix}. \tag{4.78}$$

Kinematic differentials. Because the system possesses only one degree of freedom, only a one-column JACOBIAN matrix must be stated. Here, the velocity terms of the former section must again be formulated for the special input $\dot{\tilde{\beta}}_1 = 1$. From the corresponding equations one gets:

$$\left.\begin{aligned} \dot{\tilde{\beta}}_4 &= D_{41}, \\ \dot{\tilde{\beta}}_2 &= D_{21}, \\ \dot{\tilde{\beta}}_3 &= D_{31} + D_{34}\,\dot{\tilde{\beta}}_4, \end{aligned}\right\}, \tag{4.79}$$

$$\left.\begin{aligned} \tilde{\boldsymbol{\omega}} &= \boldsymbol{u}_2\,\dot{\tilde{\beta}}_2 + \boldsymbol{u}_3\left[\dot{\tilde{\beta}}_3 - \dot{\tilde{\beta}}_4\right], \\ \dot{\tilde{\boldsymbol{s}}} &= (\boldsymbol{u}_1 \times \boldsymbol{r}_1) - [1-\kappa]\left(\tilde{\boldsymbol{\omega}} \times \boldsymbol{r}_4\right) \end{aligned}\right\} \tag{4.80}$$

The pseudo accelerations (which means the accelerations for the specially given input acceleration $\ddot{\tilde{q}} = 0$) analogously provide:

$$\left.\begin{aligned} \ddot{\beta}_4 &= -E_4, \\ \ddot{\tilde{\beta}}_2 &= E_2, \\ \ddot{\tilde{\beta}}_3 &= D_{34}\,\ddot{\beta}_4 + E_3, \end{aligned}\right\} \tag{4.81}$$

$$\left.\begin{aligned} \dot{\tilde{\boldsymbol{\omega}}} &= \boldsymbol{u}_2\ddot{\tilde{\beta}}_2 + \boldsymbol{u}_3\left[\ddot{\tilde{\beta}}_3 - \ddot{\tilde{\beta}}_4\right] + \boldsymbol{a}_\varphi, \\ \ddot{\tilde{\boldsymbol{s}}} &= [1-\kappa]\left(\dot{\tilde{\boldsymbol{\omega}}} \times \boldsymbol{r}_4\right) + \boldsymbol{a}_\mathrm{s}. \end{aligned}\right\} \tag{4.82}$$

With these expressions, the equations of motion can now be formulated in closed form.

Equations of motion. From d'ALEMBERT's Principle, the scalar equation of motion follows for this example as:

$$(m\ddot{s} + mg\boldsymbol{e}_z) \cdot \delta s + (\omega \times \boldsymbol{\Theta}_S \omega + \boldsymbol{\Theta}_S \dot{\omega}) \cdot \delta \varphi = 0, \tag{4.83}$$

where $\boldsymbol{\Theta}_S$ indicates the moment of inertia tensor with respect to the center of mass S_m. The virtual displacements in Eq. (4.83) can be represented as linear combinations of the virtual displacement, as well as, of the accelerations of the generalized coordinate $q = \beta_1$ using the kinematic differentials that have been derived in the previous step:

$$\left.\begin{aligned} \delta s &= \tilde{\dot{s}}\, \delta q, \\ \ddot{s} &= \tilde{\dot{s}}\, \ddot{q} + \tilde{\ddot{s}}, \\ \delta \varphi &= \tilde{\omega} \delta q, \\ \dot{\omega} &= \tilde{\omega}\, \ddot{q} + \tilde{\dot{\omega}}. \end{aligned}\right\} \tag{4.84}$$

Substitution of Eq. (4.84) into Eq. (4.83) finally leads to the explicit and analytical form of the scalar equation of motion in this example:

$$M\ddot{\beta}_1 + b = Q. \tag{4.85}$$

The generalized mass M, the generalized centripetal and CORIOLIS forces b, and the generalized forces Q follow as:

$$\left.\begin{aligned} M &= m\tilde{\dot{s}} \cdot \tilde{\dot{s}} + (\boldsymbol{\Theta}_S\, \tilde{\omega}) \cdot \tilde{\omega}, \\ b &= m\tilde{\dot{s}} \cdot \tilde{\ddot{s}} + \left(\boldsymbol{\Theta}_S \tilde{\dot{\omega}} + \omega \times \boldsymbol{\Theta}_S \omega\right) \cdot \tilde{\omega}, \\ Q &= -mg\boldsymbol{e}_z \cdot \tilde{\dot{s}}. \end{aligned}\right\} \tag{4.86}$$

The (numerical) evaluation of Eq. (4.86) can be carried out in an arbitrary coordinate system, because the corresponding expressions only include physical representations of position, velocity, and acceleration vectors. The representation in Eq. (4.86) is also very suitable for physical interpretations.

References

Hiller M (1983) Mechanische Systeme: Eine Einführung in die analytische Mechanik und Systemdynamik. Springer, Berlin u.a. - ISBN 978-3-540-12521-1

Hiller M (1995) Multiloop Kinematic Chains, and Dynamics of Multiloop Systems in Kinematics and Dynamics of Multi-body Systems

Hiller M, Kecskeméthy A and Woernle C (1986–1988) Computergestützte Kinematik und Dynamik für Fahrzeuge, Roboter und Mechanismen. In *Carl-Cranz-Kurs*, Vol. 1.16, Carl-Cranz-Gesellschaft, Oberpfaffenhofen

Hiller M, Kecskeméthy A and Woernle C (1986) A loop-based kinematical analysis of complex mechanisms. In *Presented at the Design Engineering Technical Conference*, Columbus, Ohio

Kecskemethy A (1993) Objektorientierte Modellierung der Dynamik von Mehrkörpersystemen mit Hilfe von Übertragungselementen [Dr.-Ing.]. Dissertation, Universität Duisburg, Düsseldorf: VDI-Verlag

Schiehlen W and Eberhard P (2004) Technische Dynamik- Modelle für Regelung und Simulation; mit 4 Tabellen und 44 Beispielen. 2., neubearb. und erg. Aufl. (ed.), Teubner, Stuttgart u.a. - ISBN 978-3-519-12365-1

Woernle C (1988) Ein systematisches Verfahren zur Aufstellung der geometrischen Schließbedingungen in kinematischen Schleifen mit Anwendung bei der Rückwärtstransformation für Industrieroboter [Dr.-Ing.]. Dissertation, Universität Stuttgart, Düsseldorf: VDI-Verlag

Chapter 5
Kinematics and Dynamics of the Vehicle Body

Based on Chaps. 3 and 4, where methods for the derivation of the equations of motion (kinematics and dynamics) of general complex multibody systems have been presented, in this chapter the particular analytical formulation of the kinematics and dynamics of the vehicle body will be developed. From an analytical point of view the vehicle body, consisting of the chassis and the car body, plays the role of a reference body for the subsequent vehicle components, like front wheel suspensions, rear-wheel suspensions or drivetrain.

5.1 Vehicle-Fixed Reference Frame

The chassis and the car body subsequently will be treated as one single rigid body. Effects like torsion or other deformations of the chassis will not be taken into account. The chassis can freely move in space. To describe its position and orientation in space, a vehicle-fixed reference frame $\mathbf{K}_V = \{O_V; x_V, y_V, z_V\}$ will be introduced. The chassis-fixed reference point O_V is located in the vehicle center plane, between the front wheels, on the level of the wheel center (DIN 1994). The x-axis points into the direction of the longitudinal axis of the vehicle, the y-axis into the lateral direction of the vehicle to the left, and the z-axis upwards in the vertical direction.

For the complete spatial description of the chassis, it is suitable to use for the translational part the three components ${}^E x_V, {}^E y_V, {}^E z_V$ of the position vector $\boldsymbol{r}_V$ expressed in coordinates of the inertial system, as well as the three CARDAN angles ψ_V (yaw angle), θ_V (pitch angle), and φ_V (roll angle). The introduced reference frames are shown in Fig. 5.1. The orientation of the vehicle-fixed system with respect to the inertial system is uniquely determined by the CARDAN angles. The spatial motion can be illustrated, by starting with an initial position, where the orientation of the inertial system and the vehicle system coincide. Then the vehicle

D. Schramm et al., Vehicle Dynamics,
DOI 10.1007/978-3-662-54483-9_5

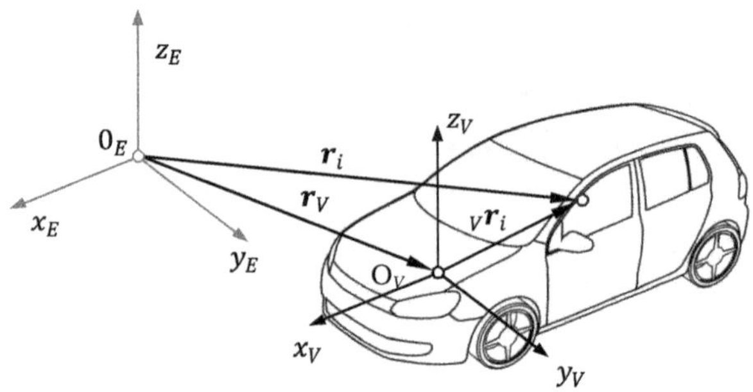

Fig. 5.1 Reference frames for the description of the vehicle body

successively executes the three rotations of the CARDAN angles around the well-defined axes, as follows (Fig. 5.2):

- The system x_E, y_E, z_E urns into system x_1, y_1, z_1 with $z_1 = z_E$, by rotating around the z_E-axis with the yaw angle ψ_V.
- The system x_1, y_1, z_1 turns into system x_2, y_2, z_2 with $y_2 = y_1$, by rotating around the y_1-axis with the pitch angle θ_V.
- The system x_2, y_2, z_2 turns into system x_V, y_V, z_V with $x_V = x_2$, by rotating around the x_2-axis with the roll angle φ_V.

For the transformation of the coordinates of an arbitrary vector ${}_V\boldsymbol{r}_i$, given in components of the vehicle system (Fig. 5.1), into the inertial system, the following relation holds:

$$ {}^E_V\boldsymbol{r}_i = {}^E\boldsymbol{T}_V\,{}^V_V\boldsymbol{r}_i, \tag{5.1}$$

with

${}^E_V\boldsymbol{r}_i$ vector in coordinates of the inertial system,
${}^V_V\boldsymbol{r}_i$ vector in coordinates of the vehicle system,
${}^E\boldsymbol{T}_V$ transformation matrix vehicle $\rightarrow$ inertial system

The calculation of the transformation matrix in Eq. (5.1), based on CARDAN angles as rotational parameters, can be found in (Hiller 1983). Omitting the index "V" at the CARDAN angles one gets:

$$ {}^E\boldsymbol{T}_V = \begin{bmatrix} \mathrm{c}\theta\mathrm{c}\psi & \mathrm{s}\varphi\mathrm{s}\theta\mathrm{c}\psi - \mathrm{c}\varphi\mathrm{s}\psi & \mathrm{c}\varphi\mathrm{s}\theta\mathrm{c}\psi + \mathrm{s}\varphi\mathrm{s}\psi \\ \mathrm{c}\theta\mathrm{s}\psi & \mathrm{s}\varphi\mathrm{s}\theta\mathrm{s}\psi + \mathrm{c}\varphi\mathrm{c}\psi & \mathrm{c}\varphi\mathrm{s}\theta\mathrm{s}\psi - \mathrm{s}\varphi\mathrm{c}\psi \\ -\mathrm{s}\theta & \mathrm{s}\varphi\mathrm{c}\theta & \mathrm{c}\varphi\mathrm{c}\theta \end{bmatrix}. \tag{5.2}$$

For the inverse transformation from the inertial to the vehicle system this follows, using the transposed transformation matrix ${}^V\boldsymbol{T}_E$ (see also Chap. 2). The inverse (transposed) transformation is of great importance here, since all vectors

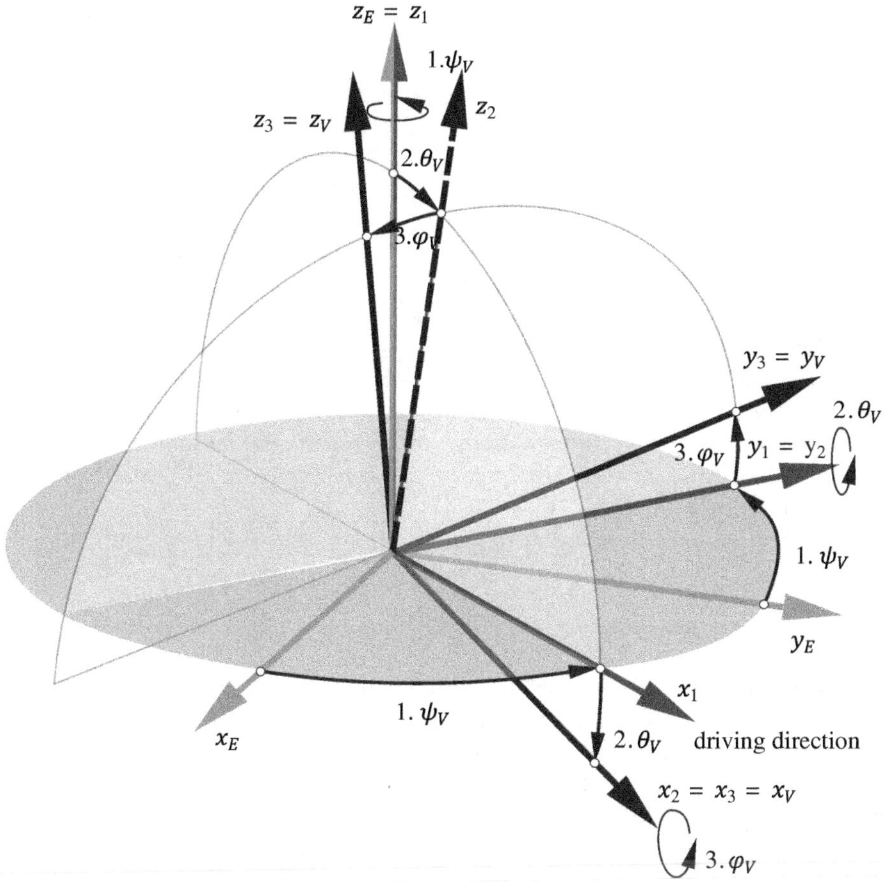

Fig. 5.2 Definition of the CARDAN angles

and tensors, used for the calculation of the equations of motion are given in components of the vehicle system.

Based on CARDAN angles as kinematic parameters for the description of spatial rotations, also the corresponding angular velocities and angular accelerations, respectively, can be expressed with the help of the time derivatives of the CARDAN angles, and one gets the well-known kinematic CARDAN equations (Hiller 1983). In the context of this book, the required kinematic equations for the angular velocity and angular acceleration, respectively, of the vehicle, will be expressed in components of the vehicle system (compare Sect. 2.55):

$$ {}^{V}\boldsymbol{\omega}_V = \begin{bmatrix} -\sin\theta & 0 & 1 \\ \cos\theta\sin\varphi & \cos\varphi & 0 \\ \cos\theta\cos\varphi & -\sin\varphi & 0 \end{bmatrix} \begin{bmatrix} \dot{\psi} \\ \dot{\theta} \\ \dot{\varphi} \end{bmatrix}, \tag{5.3} $$

$$
\begin{aligned}
{}^{V}\dot{\omega}_V = &\begin{bmatrix} -\sin\theta & 0 & 1 \\ \cos\theta\sin\varphi & \cos\varphi & 0 \\ \cos\theta\cos\varphi & -\sin\varphi & 0 \end{bmatrix} \begin{bmatrix} \ddot{\psi} \\ \ddot{\theta} \\ \ddot{\varphi} \end{bmatrix} \\
&+ \begin{bmatrix} -c\theta\dot{\theta} & 0 & 0 \\ -s\theta s\varphi\dot{\theta} + c\theta c\varphi\,\dot{\varphi} & -s\varphi\,\dot{\varphi} & 0 \\ -s\theta c\varphi\dot{\theta} - c\theta s\varphi\,\dot{\varphi} & -c\varphi\,\dot{\varphi} & 0 \end{bmatrix} \begin{bmatrix} \dot{\psi} \\ \dot{\theta} \\ \dot{\varphi} \end{bmatrix}.
\end{aligned}
\tag{5.4}
$$

The upper left index in Eqs. (5.3) and (5.4) indicates the coordinate system in which the components are given, the lower right index "*V*" states that one is actually regarding the motion of the vehicle system. An index on the lower left side indicates in case of a relative motion the reference system; in the case that the reference system is the inertial system, the index is omitted. Using the angular velocity and the angular acceleration of the vehicle, one can now calculate the absolute velocity and the absolute acceleration, respectively, of arbitrary vehicle fixed points:

$$\dot{\boldsymbol{r}}_i = \dot{\boldsymbol{r}}_V + \boldsymbol{\omega}_V \times {}_V\boldsymbol{r}_i, \tag{5.5}$$

$$\ddot{\boldsymbol{r}}_i = \ddot{\boldsymbol{r}}_V + \dot{\boldsymbol{\omega}}_V \times {}_V\boldsymbol{r}_i + \boldsymbol{\omega}_V \times (\boldsymbol{\omega}_V \times {}_V\boldsymbol{r}_i). \tag{5.6}$$

For the statement of the equations of motion of the vehicle model, one has to calculate mainly kinematic expressions, like velocities or accelerations, which are "physical" vectors, to be represented in components in arbitrarily chosen coordinate systems. In case of the vehicle model under consideration, it is suitable to represent all vectors in the vehicle-fixed system. In this system, a large number of position vectors are constant, while others (e.g. in the area of wheel suspensions) can be transformed by simple transformations of the vehicle system.

5.2 Kinematical Analysis of the Chassis

In this section, in a first step it will be discussed how the interrelation of the kinematic topologies of a tree-type structure and a closed-loop structure can be used for the analysis and the statement of the modeling equations of wheel suspensions (see Chap. 3). This will be illustrated with the example of a multi-link rear wheel suspension.

5.2.1 Incorporation of the Wheel Suspension Kinematics

Regarding an individual analysis of wheel suspensions, position, velocity, and acceleration of arbitrarily chosen wheel suspension points in a vehicle at rest have to be determined. This means that velocities and accelerations with respect to the vehicle are determined to be designated with the left lower index "V". Figure 5.3 illustrates the corresponding scheme of interdependencies of the vector ${}_V\boldsymbol{r}_T$ from the vehicle reference system to a selected point T of the wheel carrier, using the body-joint-representation, as introduced in Sect. 3.4.

The objective of the individual analysis of the car body kinematics–discussed in the previous section–has been the calculation of position, velocity, and acceleration of arbitrary chassis points with respect to the inertial system. The corresponding general scheme, which represents a tree-type structure, is shown in Fig. 5.4. The open or tree-type kinematic chain on the left side of this figure represents the free motion (with 6 DoF) of the chassis with respect to inertia. For the analysis of the complete system, the subsystems containing the kinematic loops (i.e. wheel

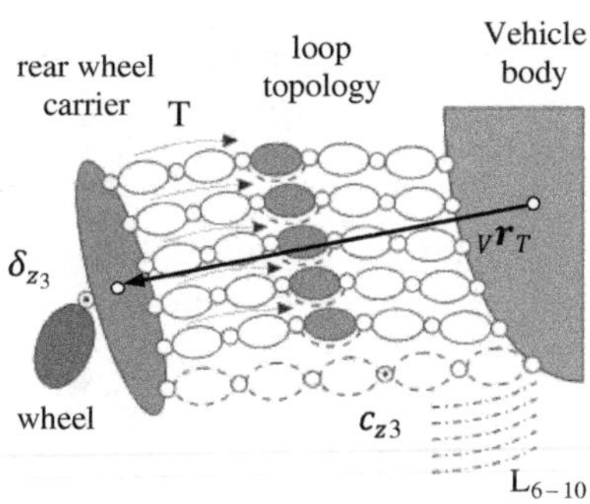

loop analysis

$$
\begin{aligned}
{}_V\boldsymbol{r}_T &= {}_V\boldsymbol{r}_T\,(c_{z3})\\
{}_V\dot{\boldsymbol{r}}_T &= {}_V\dot{\boldsymbol{r}}_T(c_{z3},\dot{c}_{z3})\\
{}_V\ddot{\boldsymbol{r}}_T &= {}_V\ddot{\boldsymbol{r}}_T(c_{z3},\dot{c}_{z3},\ddot{c}_{z3})\\
{}_V\boldsymbol{\omega}_T &= {}_V\boldsymbol{\omega}_T\,(c_{z3},\dot{c}_{z3})\\
{}_V\dot{\boldsymbol{\omega}}_T &= {}_V\dot{\boldsymbol{\omega}}_T\,(c_{z3},\dot{c}_{z3},\ddot{c}_{z3})
\end{aligned}
$$

Fig. 5.3 Individual analysis of a multilink rear wheel suspension

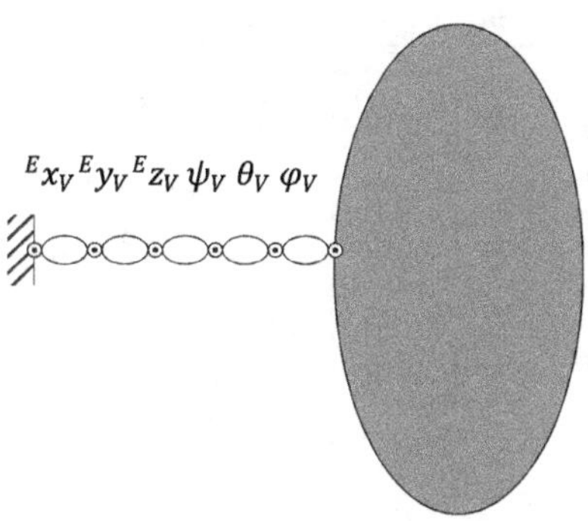

$$
\begin{aligned}
\boldsymbol{r}_V &= \boldsymbol{r}_V({}^E x_V \ldots \varphi_V)\\
\dot{\boldsymbol{r}}_V &= \dot{\boldsymbol{r}}_V({}^E x_V \ldots \varphi_V, {}^E\dot{x}_V \ldots \dot{\varphi}_V)\\
\ddot{\boldsymbol{r}}_V &= \ddot{\boldsymbol{r}}_V({}^E x_V \ldots \varphi_V, {}^E\dot{x}_V .. \dot{\varphi}_V, {}^E\ddot{x}_V \ldots \ddot{\varphi}_V)
\end{aligned}
$$

$$
\begin{aligned}
{}^E\boldsymbol{T}_V &= {}^E\boldsymbol{T}_V(\psi_V,\theta_V,\varphi_V)\\
\boldsymbol{\omega}_V &= \boldsymbol{\omega}_V(\psi_V,\theta_V,\varphi_V,\dot{\psi}_V,\dot{\theta}_V,\dot{\varphi}_V)\\
\dot{\boldsymbol{\omega}}_V &= \dot{\boldsymbol{\omega}}_V(\psi_V,\theta_V,\varphi_V,\dot{\psi}_V,\dot{\theta}_V,\dot{\varphi}_V,\ddot{\psi}_V,\ddot{\theta}_V,\ddot{\varphi}_V)
\end{aligned}
$$

Fig. 5.4 Scheme of the chassis kinematics, inertial system E, chassis system V, calculation of the tree-structure suspensions, etc.) will be attached to the tree-type structure, as shown in Fig. 5.4. One gets position, velocity, and acceleration of the wheel suspension points with respect to the inertial system by applying the well-known equations for the relative motion of two bodies.

Guiding motion, Relative motion,

$$\boldsymbol{r} = \boldsymbol{r}_V \qquad + {}_V\boldsymbol{r}, \tag{5.7}$$

$$\dot{\boldsymbol{r}} = \dot{\boldsymbol{r}}_V + \boldsymbol{\omega}_V \times {}_V\boldsymbol{r} \qquad + {}_V\dot{\boldsymbol{r}}, \tag{5.8}$$

$$\ddot{\boldsymbol{r}} = \ddot{\boldsymbol{r}}_V + \dot{\boldsymbol{\omega}}_V \times {}_V\boldsymbol{r} + \boldsymbol{\omega}_V \times (\boldsymbol{\omega}_V \times {}_V\boldsymbol{r}) \qquad + 2\boldsymbol{\omega}_V \times {}_V\dot{\boldsymbol{r}} + {}_V\ddot{\boldsymbol{r}}, \tag{5.9}$$

$$\boldsymbol{\omega} = \boldsymbol{\omega}_V \qquad + {}_V\boldsymbol{\omega}, \tag{5.10}$$

$$\dot{\boldsymbol{\omega}} = \dot{\boldsymbol{\omega}}_V + \boldsymbol{\omega}_V \times {}_V\boldsymbol{\omega} \qquad + {}_V\dot{\boldsymbol{\omega}}. \tag{5.11}$$

Equations (5.7)–(5.11) show that the kinematics expressed in the inertial system–which is needed for the equations of motion of the complete vehicle–can be obtained in a very simple way from the already known loop-kinematics, together with the very simple kinematics of the tree-type kinematic chain "chassis". In general, one can say that in systems which contain a combination of tree-type and loop topologies, it is recommended to analyze first trees and loops separately, and to fit them together later-on (Fig. 5.5).

For the equations of motion, in addition to the "real" velocities also the so-called pseudo velocities, introduced in Chap. 4, are needed. The idea here is to calculate unknown partial derivatives, which are needed later-on, in an intelligent kinematic

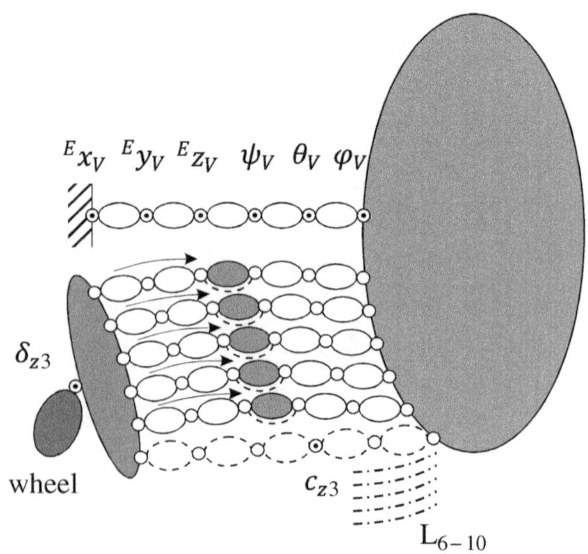

Fig. 5.5 Combination of chassis and wheel suspension kinematics

way (Sect. 4.6.1). In Eqs. (5.8) and (5.9), one can see that the calculation of the pseudo velocities can be simplified, depending if the actually regarded degree of freedom (DoF) in this procedure belongs to the tree-type or the loop-type structure. In case of the tree-type motion, the last term on the right hand side of Eqs. (5.8) and (5.9) becomes "0"; in case of the loop-type motion, left of the last one vanishes, because the corresponding DoF have the velocity "0".

5.2.2 Equations of Motion

For the dynamics of a multibody system of n_B rigid bodies, in Sect. 4.5 d'ALEMBERT's principle has been applied:

$$\sum_{i=1}^{n_B}[(m_i\ddot{\boldsymbol{r}}_{Si} - \boldsymbol{F}_i)\cdot\delta\boldsymbol{r}_{Si} + (\boldsymbol{\Theta}_{Si}\dot{\boldsymbol{\omega}}_i + \boldsymbol{\omega}_i \times \boldsymbol{\Theta}_{Si}\boldsymbol{\omega}_i - \boldsymbol{T}_i)\cdot\delta\boldsymbol{\varphi}_i] = 0, \tag{5.12}$$

with

$m_i, \boldsymbol{\Theta}_{Si}$ mass and tensor of inertia,
$\ddot{\boldsymbol{r}}_{Si}$ acceleration of the center of mass,
$\boldsymbol{\omega}_i, \dot{\boldsymbol{\omega}}_i$ angular velocity and angular acceleration,
$\boldsymbol{F}_i, \boldsymbol{T}_i$ external forces and torques,
$\delta\boldsymbol{r}_{Si}, \delta\boldsymbol{\varphi}_i$ virtual displacements

for each body "*i*".

The tensors of inertia and the external torques have to be expressed with respect to the body centers of mass. For the equations of motion after Eq. (5.12), the dependent virtual displacements $\delta\boldsymbol{r}_{Si}, \delta\boldsymbol{\varphi}_i$, as well as the accelerations $\ddot{\boldsymbol{r}}_{Si}, \dot{\boldsymbol{\omega}}_i$, have to be calculated as a function of the virtual displacements and the accelerations of the independent generalized coordinates. Regarding that for the virtual displacements the same transformation rules are valid as for the velocities, one gets:

$$\delta\boldsymbol{r}_i = \sum_{j=1}^{f}\tilde{\dot{\boldsymbol{r}}}_{S_i}^{(j)}\delta q_j,\ \ddot{\boldsymbol{r}}_i = \sum_{j=1}^{f}\tilde{\dot{\boldsymbol{r}}}_{S_i}^{(j)}\ddot{q}_j + \tilde{\ddot{\boldsymbol{r}}}_{S_i}, \tag{5.13}$$

$$\delta\boldsymbol{\varphi}_i = \sum_{j=1}^{f}\tilde{\boldsymbol{\omega}}_i^{(j)}\delta q_j,\ \dot{\boldsymbol{\omega}}_i = \sum_{j=1}^{f}\tilde{\boldsymbol{\omega}}_i^{(j)}\ddot{q}_j + \tilde{\dot{\boldsymbol{\omega}}}_i. \tag{5.14}$$

After inserting Eqs. (5.13) and (5.14) into Eq. (5.12), and considering the independent virtual displacements δq_j, the equations of motion in minimal form are:

$$\boldsymbol{M}\ddot{\boldsymbol{q}} + \boldsymbol{b} = \boldsymbol{Q}. \tag{5.15}$$

The coefficients of the generalized $[f \times f]$-mass matrix $\boldsymbol{M}$, as well as of the $[f \times 1]$-vector of the generalized gyroscopic and centrifugal forces $\boldsymbol{b}$, and the $[f \times 1]$-vector of the generalized external forces $\boldsymbol{Q}$ are the following:

$$\begin{aligned} \boldsymbol{M}_{j,k} &= \sum_{i=1}^{n_B} [m_i \tilde{\dot{\boldsymbol{r}}}_{S_i}^{(j)} \tilde{\dot{\boldsymbol{r}}}_{S_i}^{(k)} + \tilde{\boldsymbol{\omega}}_i^{(j)} (\boldsymbol{\Theta}_{S_i} \tilde{\boldsymbol{\omega}}_i^{(k)})] \\ \boldsymbol{b}_j &= \sum_{i=1}^{n_B} \left[m_i \tilde{\dot{\boldsymbol{r}}}_{S_i}^{(j)} \tilde{\ddot{\boldsymbol{r}}}_{S_i} + \tilde{\boldsymbol{\omega}}_i^{(j)} \left(\boldsymbol{\Theta}_{S_i} \tilde{\dot{\boldsymbol{\omega}}}_i + \boldsymbol{\omega}_i \times \boldsymbol{\Theta}_{S_i} \boldsymbol{\omega}_i \right) \right], \\ \boldsymbol{Q}_j &= \sum_{i=1}^{n_B} [\tilde{\dot{\boldsymbol{r}}}_{S_i}^{(j)} \boldsymbol{F}_i + \tilde{\boldsymbol{\omega}}_i^{(j)} \boldsymbol{T}_i]. \end{aligned} \tag{5.16}$$

In case of the already solved global kinematics of the system, the Eq. (5.16) can be evaluated, following the procedure of Sect. 4.6.

References

DIN (1994) Fahrzeugdynamik und Fahrverhalten—Begriffe. in *FAKRA* (ed.), Vol. Deutsche Norm DIN 70000 ISO 8855, Deutsches Institut für Normung e.V.: Normenausschuss Kraftfahrzeuge im DIN

Hiller M (1983) Mechanische Systeme: Eine Einführung in die analytische Mechanik und Systemdynamik. Springer, Berlin u.a.—ISBN 978-3-540-12521-1

Chapter 6
Modeling and Analysis of Wheel Suspensions

The wheel guidance or wheel suspension has made a remarkable development during its centennial history. By the implementation of more sophisticated kinematic structures, the generation of accurate and reproducible wheel movements became possible. As the knowledge about the vibration behavior of vehicles and their driving dynamics increased, the subtleties of the spatial kinematics came to the force. Accordingly, today a variety of wheel guidance designs and suspension geometries with different objectives and characteristics is available. Due to the use of computer-aided processes, today the desired design of the chassis is possible. As a result, safe driving behavior and best driving comfort can be achieved at the same time.

6.1 Function of Wheel Suspension Systems

By definition in the original sense of automotive engineering the wheel axle of a vehicle includes the rigid connection of two oppositely and independently rotatable mounted wheels. The wheels are interconnected to the vehicle body through the axle. This definition applies to all rigid and semi-rigid axles (Sect. 6.2) because in these cases the wheels are mounted directly on the axle and move together with it. In the case of independent wheel suspensions this connection no longer exists. The wheel is rotatable mounted on a wheel carrier. The wheel carrier itself functions as a coupling element of a kinematic chain and is connected to the axle with the help of a corresponding number of connecting rods. It is not clearly defined whether the axle includes the connecting rods and the wheel carrier or whether they form a new assembly. In this book, the combination of connecting rods and wheel carrier refers to the wheel suspension.

The essential function of the wheel suspension is to connect the vehicle body with the wheels. Thereby it is possible to carry the body along the drive way and to transmit forces in the horizontal plane. The wheel suspension gives the wheel a primary vertically aligned movement possibility. As a result, the wheel follows a

D. Schramm et al., Vehicle Dynamics,
DOI 10.1007/978-3-662-54483-9_6

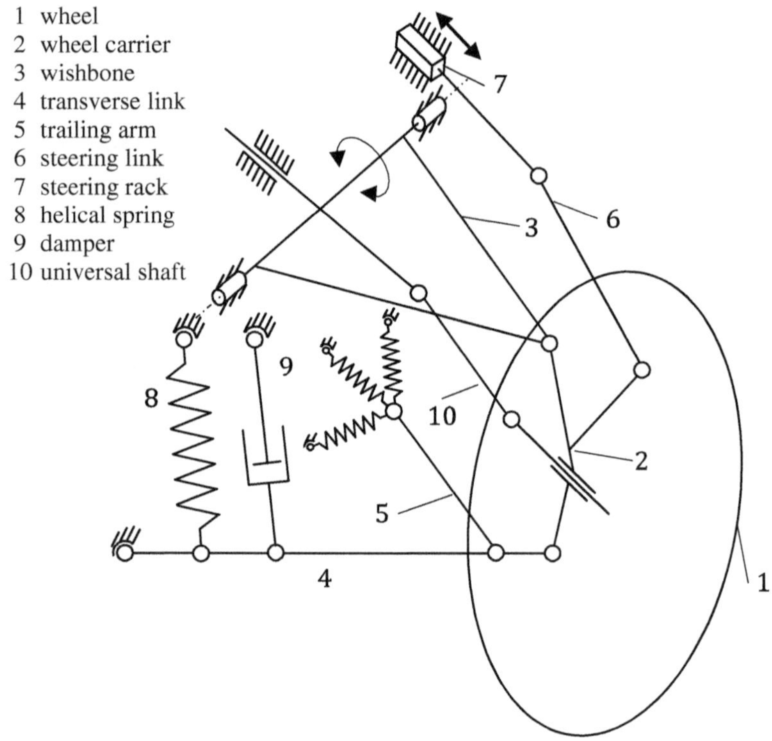

Fig. 6.1 Basic elements of an independent suspension (scheme)

route with uneven road surfaces to a certain extent. By using spring and damping elements, the resulting body movements are reduced and driving safety and comfort are ensured. When it comes to a front suspension the steering movements of the wheel are usually also transmitted over a special connecting rod of the wheel suspension, the so-called tie rod.

Furthermore, the wheel suspension influences the position of the wheel relative to the road by its geometry and the spring rate and damping. This allows a systematic influence on the dynamic driving characteristics of the vehicle. The adjustment of these characteristics takes up a compromise, because the requirements of a good driving behavior and a high comfort are the most time inconsistent with one another. Here, an important role is often played by the rubbery-elastic manufactured bearing of the vehicle-mounted rod joints. In addition, the mechanism of the wheel suspension has a certain elastic behavior. The correct laying of the transverse and longitudinal stiffness (elasto-kinematics) is of particular importance for the influence on the toe angle and camber angle course under load and thus for the adjustment in particular of fast moving cars. Not every wheel suspension system that meets the kinematic requirements of the kinematic behavior of a wheel, offers the same potential for elasto-kinematic adjustment (Sect. 6.2).

In Fig. 6.1 the multilink version of a wheel suspension is shown schematically. Item (1) is the wheel with the tire, which is pivoted at the wheel carrier (2) by the wheel bearing. In general, the wheel carrier also holds the brake mechanism. The wheel carrier also functions as a link of a spatial kinematic chain of wishbone (3) and transverse link (4). The trailing-arm (5) backs the transverse link against the vehicle on an elastic designed bearing that serves as a noise and vibration isolator. By using the tie rod (6) which is connected with the vehicle-mounted steering (7), the orientation of the wheel carrier can be adjusted. The coil spring (8) and the damper (9) back the transverse link against the vehicle body. The universal shaft (10) finally transmits the driving torque to the wheel (Chap. 8).

6.2 Different Types of Wheel Suspension

In automotive engineering a multitude of different wheel suspensions is used. Independent from the type of the respective wheel suspension it is intended that the wheels are lead at the vehicle body in a way in which the potential of the wheels is used optimally in all driving situations. Next to functionality, other aspects such as weight, space requirements as well as cost play an important role in the design of the chassis. Therefore, depending on the car classification and consequentially the requirements regarding comfort, driver safety and cost as well as constructive aspects, propelled, or not propelled, guided or unguided axes are used.

Over the course of decades, the composition of wheel suspensions has developed into three fundamentally different concepts:

- beam axle, the oldest known form of wheel suspension,
- twist-beam rear suspension, that are especially used as rear axles without drive shaft,
- independent suspension in which the following two forms are distinguished:
  - independent suspension with one pivot axis and
  - independent suspension with more than one pivot axis.

Typical independent suspensions with one rotation axis are trailing-arm, wishbone and semi trailing-arm axles. Typical independent suspensions with more than one rotation axis are double wishbone axles as well as suspension struts under the MacPherson principle. Multilink axles for which single rotation axes are not clearly distinguishable are counted towards the group of the independent suspensions with more than one rotation axis.

The following section presents a few practical implementations of the previously named concepts. Attached one can find a short kinematic description with the main characteristics such as the number of kinematic loops, the number of bodies and joints as well as the consequential degree of freedom. Furthermore the advantages and disadvantages of the practical features of the respective concepts are illustrated.

Fig. 6.2 Power-driven rigid front axle with Panhard bar of the Mercedes G-Class (courtesy of Daimler AG 2009)

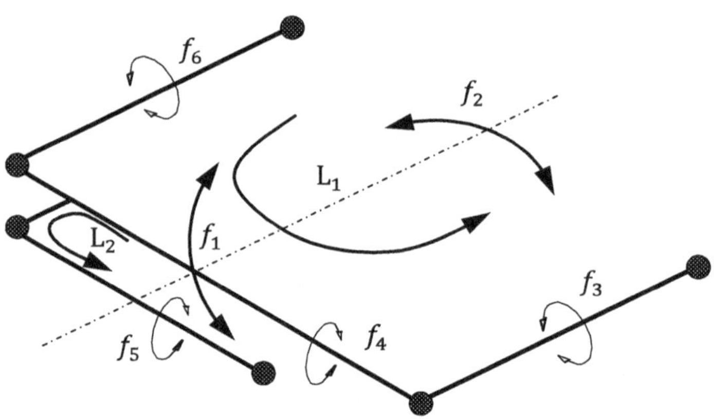

Fig. 6.3 Rigid body model of the front axle of the Mercedes G-Class (without steering mechanism)

6.2.1 *Beam Axles*

The solid connection of two wheels via a lateral axle is called a beam axle (Figs. 6.2 and 6.3). Here, the two opposing wheels can influence each other. This concept originates from horse-drawn carriages. Today, beam axles are only very rarely used as the front axles of automobiles. It's only in all-terrain vehicles that the beam axle has proven itself thanks to its good off-road characteristics. This kind of axle is used more frequently as power-driven rear axle in utility cars (Table 6.1).

Table 6.1 Model characteristics and features of the beam axle

Model	Advantages	Disadvantages
4 bodies	Simple structure	High space requirements
6 spherical joints	Inexpensive	Heavy
	No change in track width high off-road articulation possible	Kinematic coupling of right and left side and therefore large camber variation for unilateral deflection high unsprung mass
$4 \times 6 = 24$	Equations of motion	
$6 \times 3 = 18$	Constraints	
6	DoF of the axles (vertical deflection f_1, roll f_2, 4 isolated rotational DoF f_3, f_4, f_5, f_6)	
2	Kinematic loops L_1 and L_2	

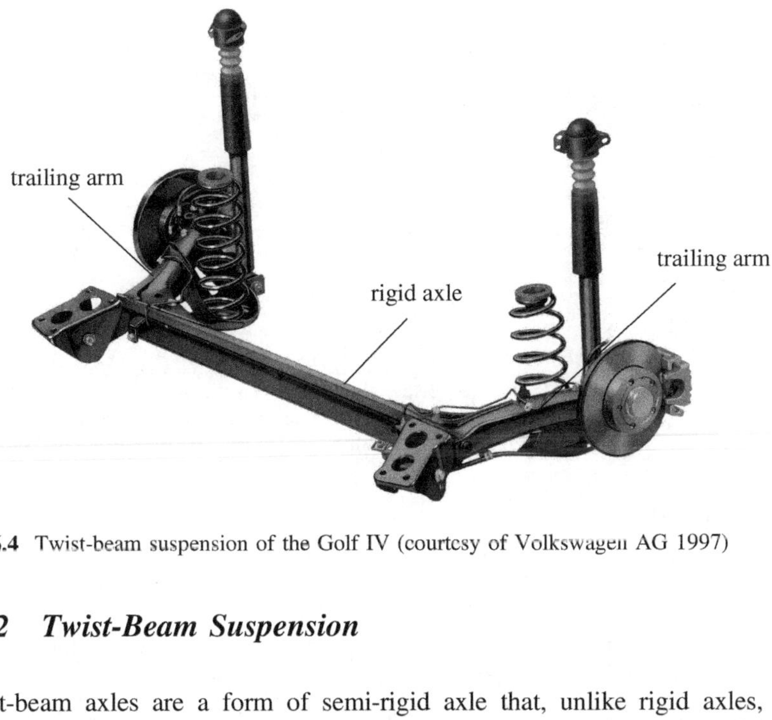

Fig. 6.4 Twist-beam suspension of the Golf IV (courtesy of Volkswagen AG 1997)

6.2.2 *Twist-Beam Suspension*

Twist-beam axles are a form of semi-rigid axle that, unlike rigid axles, allow relative movement between the wheel carriers as a result of the torsion of the axle beam. The axle beam is intentionally given elastic properties to combine the characteristics of beam axles and independent rear suspension. Both wheel carriers are attached to the torsional stiff and deflection-resistant trailing arms, which in turn are transversally connected via a deflection-resistant profile (cross member) with low torsional stiffness (Figs. 6.4 and 6.5), (Table 6.2).

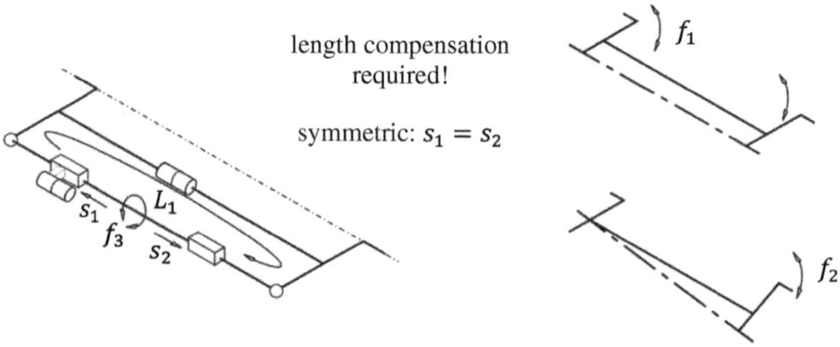

Fig. 6.5 Rigid body model of a twist-beam suspension

Table 6.2 Model characteristics and features of the twist-beam suspension

Model	Advantages	Disadvantages
4 bodies	Low space requirements	Drive difficult to implement
1 revolute joint with torsion spring	Stabilizing effect through cross member	Steering difficult to implement
2 spherical joints	Low weight	Limited potential for optimizing driving dynamics
1 prismatic joint	Good roll compensation	
1 cylindrical joint		
$4 \times 6 = 24$	Equations of motion	
$1 \times 5 + 2 \times 3 + 1 \times 5 + 1 \times 4 = 20$	Constraints	
4	DoF: 2 DoF of the axle (bilateral vertical deflection f_1, unilateral vertical deflection f_2), isolated rotation f_3 and compensation for longitudinal movement s_1, s_2	
1	Kinematic loop L_1	

6.2.3 *Trailing-Arm Axle*

Trailing-arm axles are a form of independent suspension with a single link (or arm) between the wheel carrier and body. The link is rigidly connected to the wheel carrier and via a revolute joint to the chassis. Depending on the alignment of the link's axis of rotation, they are classified as trailing-arm or semi-trailing-arm (Sect. 6.2.4) and trailing-arm suspension (Figs. 6.6 and 6.7).

Trailing-arm axles have only one trailing link per wheel, which absorbs the longitudinal and lateral forces allowing the wheel to swing around a rotational axis perpendicular to the longitudinal axis of the vehicle. The links are usually elastically mounted to the body in a way that allows them to move longitudinally for improved comfort (Table 6.3).

Fig. 6.6 Trailing-arm rear axle of the Mercedes A-Class (Mödinger et al. 1997)

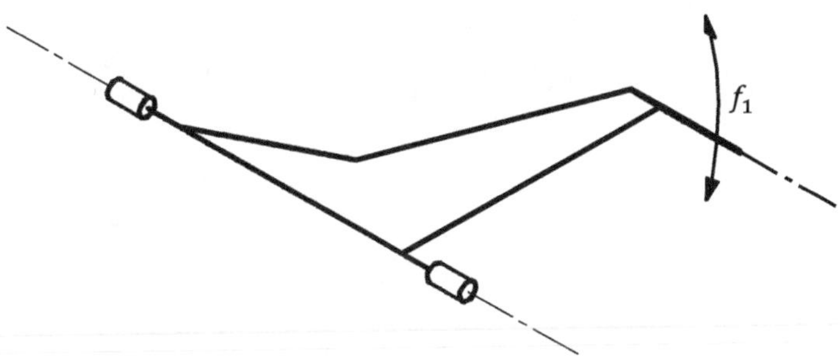

Fig. 6.7 Rigid body model of a trailing-arm wheel suspension

Table 6.3 Model characteristics and features of the trailing-arm wheel suspension

Model	Advantages	Disadvantages
1 body	Little space requirements	Restricted design for axle kinematics
1 revolute joint	Simple construction	No roll compensation
	Small unsprung masses	Low instant center of rotation, therefore strong rolling tendency
	Track width, toe in and camber constant with unilateral suspension (not relative to road surface)	Poor longitudinal elasticity
$1 \times 6 = 6$	Equations of motion	
$1 \times 5 = 5$	Constraints	
1	DoF per wheel suspension (f_1)	
0	Kinematic loops	

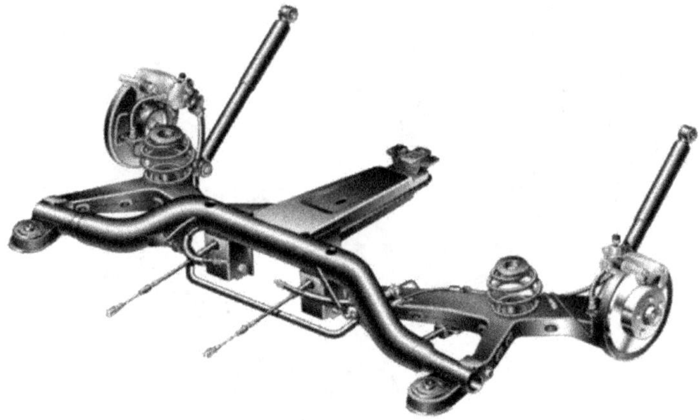

Fig. 6.8 Trailer arm axle of the Volkswagen Sharan (Schuster et al. 1995)

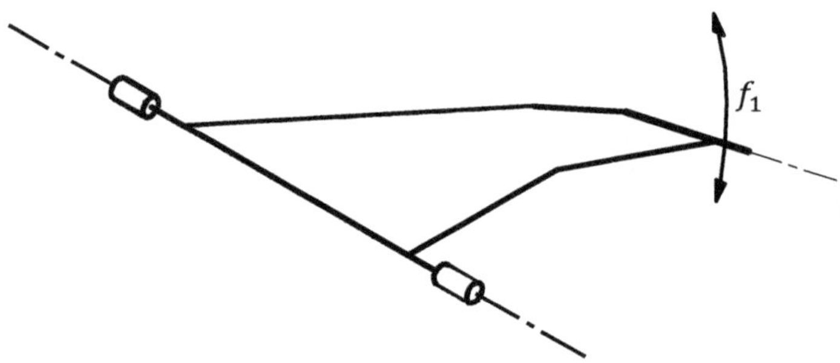

Fig. 6.9 Rigid body model of a trailer arm wheel suspension

6.2.4 Trailer Arm Axle

The skewed assembly of a single rod enables a good absorption of longitudinal and transverse forces if it is connected to the chassis by means of two axes of rotation set out in the longitudinal direction of the car (Figs. 6.8 and 6.9). Through this the advantages of trailing-arm wheel and wishbone wheel suspension can be combined without encompassing large disadvantages (Table 6.4).

6.2.5 Double Wishbone Axles

If a wheel suspension is only guided with laterally placed rods, there needs to be one rod above and one below the center of the wheel. Additionally there needs to be a steering rod in order to prohibit steering or to allow for defined steering

Table 6.4 Model characteristics and features of the trailer arm wheel suspension

Model	Advantages	Disadvantages
1 body	Relatively simple structure	Restricted possibilities of elasto-kinematic set-up
1 revolute joint	Good applicability for powering	
	Roll compensation possible	
$1 \times 6 = 6$	Equations of motion	
$1 \times 5 = 5$	Constraints	
1	DoF per wheel suspension (f_1)	
0	Kinematic loops	

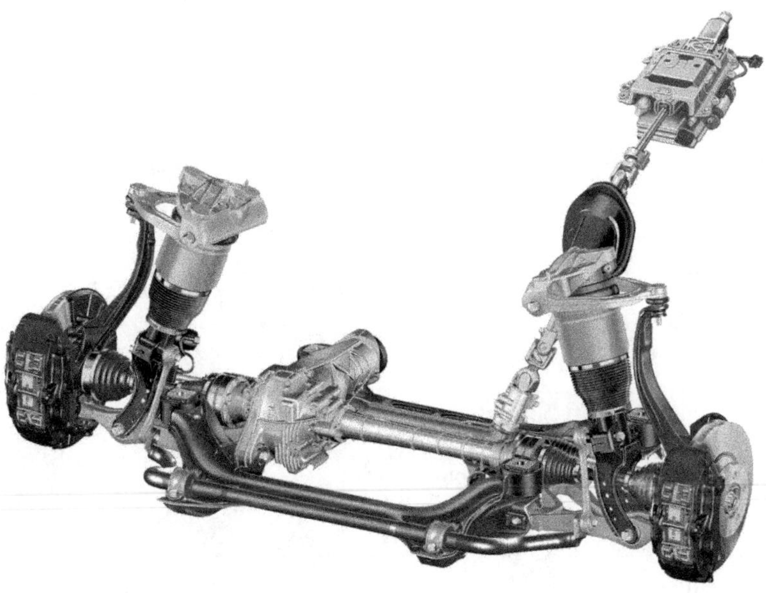

Fig. 6.10 Steered double wishbone axle front wheel suspension of the VW Tuareg (courtesy of Volkswagen AG 2002)

(Fig. 6.10). Such an assembly is called a double wishbone axle wheel suspension. In Fig. 6.11 the tie rod, which is mounted via spherical joints, undertakes the transmission of the steering forces towards the wheel carrier. Hence the rigid body model's isolated DoF f_3 (Table 6.5).

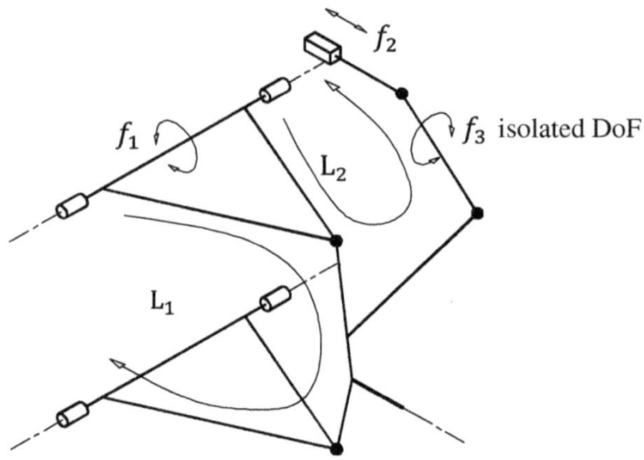

Fig. 6.11 Rigid body model of a double wishbone axle wheel suspension

Table 6.5 Model characteristics and features of the double wishbone axle wheel suspension

Model	Advantages	Disadvantages
5 bodies	Spatial wheel kinematics (track width, toe in, camber) largely freely designable	Large space requirements
2 revolute joints	Very well suitable for powering	Complex structure
1 prismatic joint		Complex longitudinal suspension and elasto-kinematic set-up
4 spherical joints		
$2 \times 5 + 1 \times 5 + 4 \times 3 = 27$	Constraints	
$5 \times 6 = 30$	Equations of motion	
3	DoF per wheel suspension (spring f_1, steer f_2, isolated DoF f_3)	
2	Kinematic loops L_1 and L_2	

6.2.6 *Wheel Suspension Derived from the MacPherson Principle*

In wheel suspensions derived from the MacPherson Principle one differentiates fundamentally between the two designs of spring struts and damper struts. If the damper uses the clearance in the helical spring and at the same time supports both ends of the spring, it is called a spring strut (see Fig. 6.12). If the spring on the other hand is not mounted on the damper but separately, it is called a damper strut.

Essential features of the damper strut and spring strut axles are that a standing damper is connected with the wheel carrier tightly and the piston rod of the damper is mounted at the chassis hinged. Thus a cylindrical joint is created which enables the steering and spring displacement of the wheel carrier. For completion of the wheel suspension two further rods are needed. One serves as a steering rod and is connected to the steering. The other works in the lower tier, usually in the form of a triangular wishbone, for the guidance of the wheel (Fig. 6.13), (Table 6.6).

Fig. 6.12 MacPherson spring strut axle of the BMW Mini with steering system (courtesy of BMW Group 2001)

Fig. 6.13 Rigid body model of a MacPherson spring strut wheel suspension

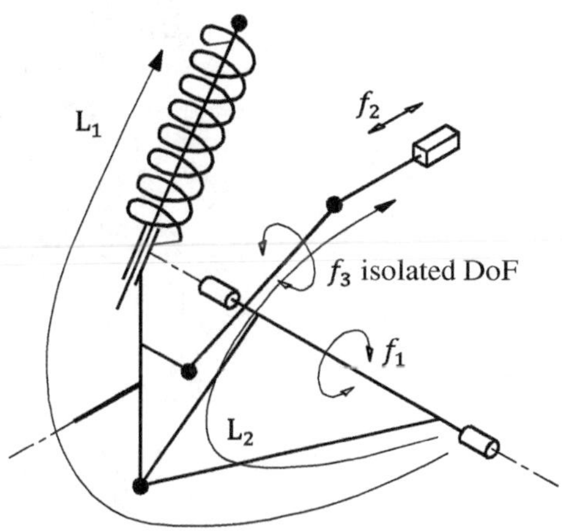

6.2.7 *Multi-Link Axles*

The use of four or more links for the suspension of a wheel is only possible if none of them is connected rigidly to the wheel carrier. Therefore every rod needs to be fixed to the wheel carrier by means of a movable joint. In order to exploit all of the kinematic potential of the layout, five independent links have to be used.

Table 6.6 Model characteristics and features of the MacPherson wheel suspension

Model	Advantages	Disadvantages
5 bodies	Low building effort	Bending torque in damper impairs responding qualities.
1 revolute joint	Low space requirements	
2 prismatic joints	Few unsprung masses	Large design height
4 spherical joints	Well suitable for drive	
$5 \times 6 = 30$	Equations of motion	
$1 \times 5 + 2 \times 5 + 4 \times 3 = 27$	Constraints	
3	DoF per wheel suspension (spring f_1, steer f_2, isolated DoF f_3)	
2	Kinematic loops L_1 and L_2	

Fig. 6.14 Integral-IV wheel suspension of the BMW 5 Series (courtesy of BMW Group 2003)

The installation of a trapezoidal link instead of two lower two-pivot-links leads to a so-called integral-link-suspension (Fig. 6.14). Here, the absorption of the torque created by longitudinal forces is realized through a perpendicular additional link (integral-link) which supports the wheel carrier on the trapezoidal link (Fig. 6.15), (Table 6.7).

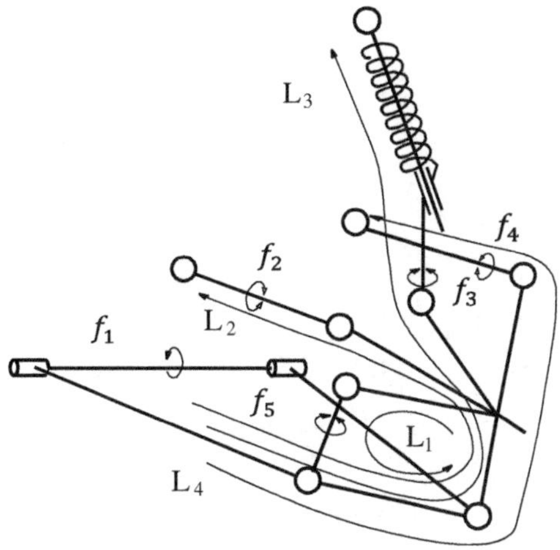

Fig. 6.15 Rigid body model of an integral-IV rear suspension

Table 6.7 Model characteristics and features of the Integral IV-steering axle

Model	Advantages	Disadvantages
7 bodies	Good kinematic construction possibilities	High building expense
1 revolute joint	Precise wheel suspension	High cost
1 prismatic joint	Elasto-kinematics well adjustable	
9 spherical joints	Good lateral and longitudinal suspension possible	
$1 \times 5 + 1 \times 5 + 9 \times 3 = 37$	Constraints	
$7 \times 6 = 42$	Equations of motion	
5	DoF per wheel suspension (spring f_1, 4 isolated DoF f_2,f_3,f_4,f_5)	
4	Kinematic loops L_1, L_2, L_3 and L_4	

6.3 Characteristic Variables of Wheel Suspensions

In order to evaluate the features of wheel suspensions it makes sense to use some universally valid terms and characteristic variables. Concerning this, (DIN 1993) determines quantities that have proven themselves in the description of the position of the wheel in space.

The angle γ between the wheel plane and the plane that is parallel to the middle axis of the vehicle and perpendicular to the ground (Fig. 6.16) is called camber angle. The camber angle is considered positive if the upper part of the wheel is tilted away from the vehicle at the top.

The track width s is the distance between the wheel-road contact points of an axle. A further important variable, especially in braking applications, is the toe difference of the vehicle which describes the difference between the distances $(B - C)$ of the rim flanges (Fig. 6.17).

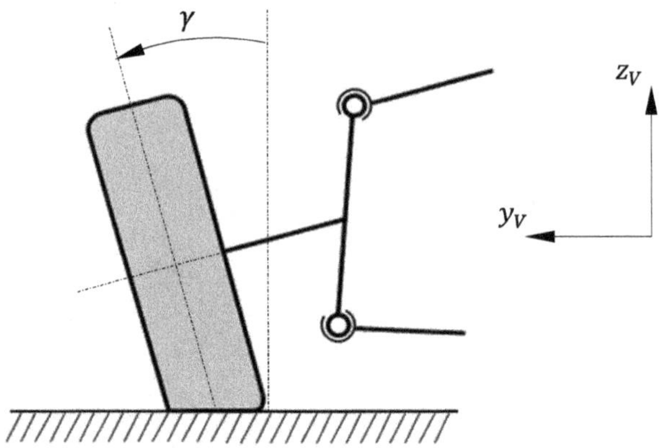

Fig. 6.16 Definition of the camber angle γ

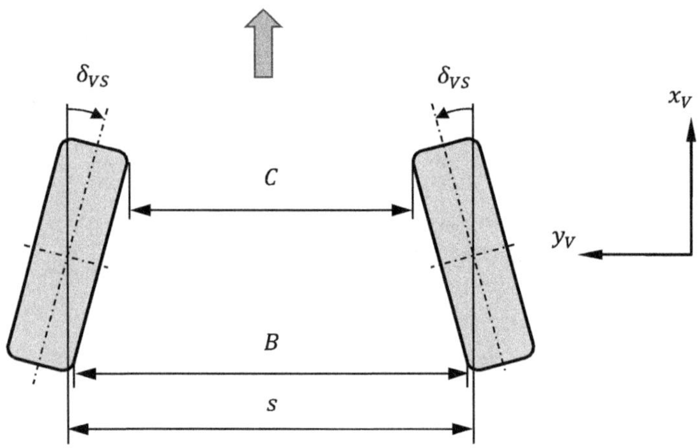

Fig. 6.17 Definition of the track width s

If the wheel is tilted towards the middle of the vehicle in front, a toe-in exists ($B > C$). If ($B < C$), a toe-out exists. The toe-in angle δ_{VS} is the angle between the longitudinal axis of the vehicle and the tire-midplane.

The distance between the cutting line of the tire-midplane and the local road surface from the cutting line of the steering axle with the road surface is called roll radius r_r (Fig. 6.18). If the point of intersection A is out of the tire-midplane, the roll radius is negative.

The angle between one plane perpendicular to the road surface of the vehicle and parallel to the longitudinal axis and the steering axis (instantaneous screw axis of the wheel carrier) is called kingpin angle σ (Fig. 6.19).

The definition of self-steering offset can be extracted from Fig. 6.20. If the point D lays in the direction of movement in front of E, there is no self-steering offset

Fig. 6.18 Definition of roll radius r_r and steering axis

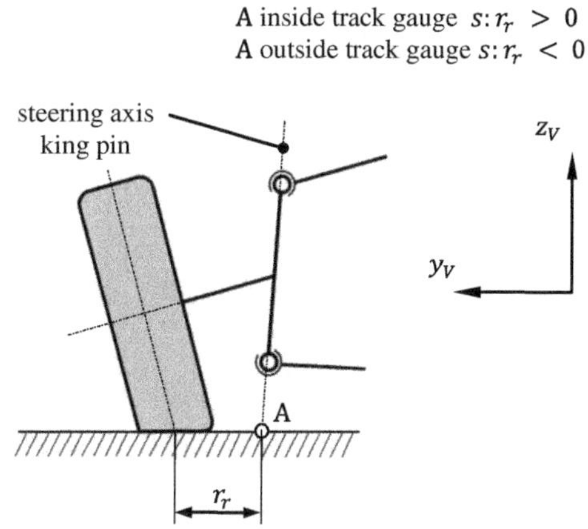

Fig. 6.19 Definition of the kingpin angle σ

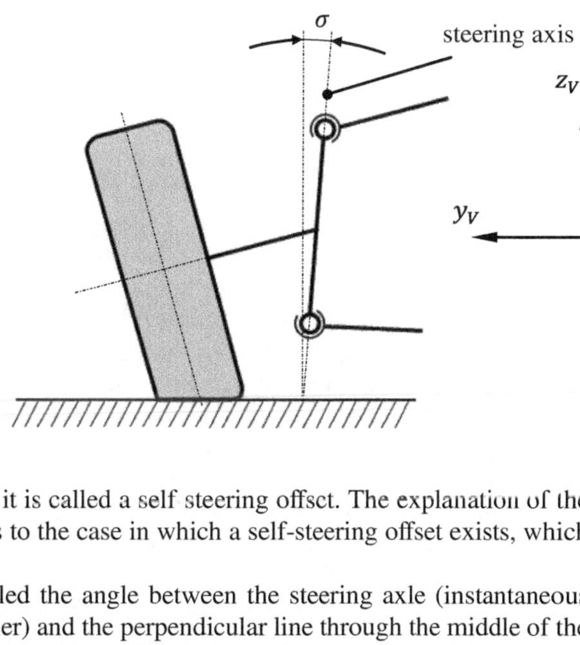

present, if E is in front of D it is called a self steering offset. The explanation of the following definitions applies to the case in which a self-steering offset exists, which is almost always the case.

The caster angle τ is called the angle between the steering axle (instantaneous screw axis of the wheel carrier) and the perpendicular line through the middle of the wheel (Fig. 6.20). The caster trail or self-steering offset length n_l is the distance between the point of intersection of the steering axle and the road surface (point E) and the point of intersection between the local road surface plane and the perpendicular line through the middle of the wheel (point D). Finally, the distance between the line of interception of the steering axle with the horizontal tire-midplane and the perpendicular line through the middle of the wheel is called self-steering offset l_n.

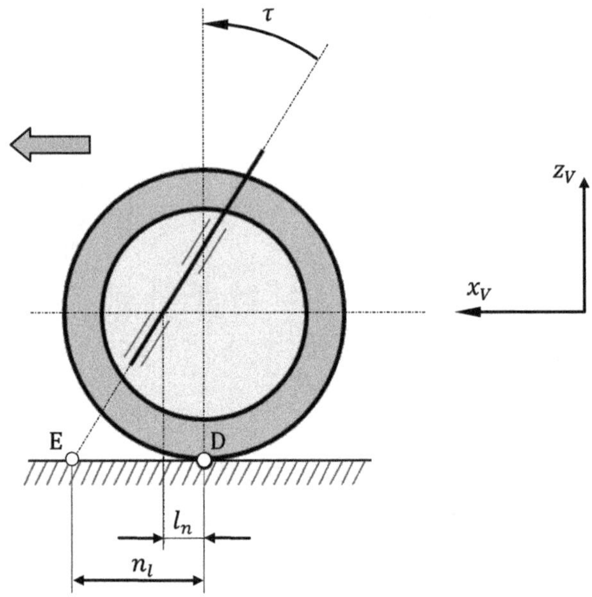

Fig. 6.20 Definition of positive and negative caster

6.4 One Dimensional Quarter Vehicle Models

In order to be able to assess the possibilities and limits of mechanical systems, the use of theoretical mathematical techniques has proven to be of great value. By means of a suitable reduction of the model, the computing time and interpretability of the results can be optimized. Consequently, it can be sufficient to use a simple equivalent mass-spring system in order to theoretically analyze the vehicle vibrations in vertical direction. The influence of vehicle suspension and damping, wheel mass and tire elasticity on comfort, driving safety and range of spring deflection can be determined simply by using quarter vehicle models. The vehicle model hereby consists of a two-body-system, which is linked to a wheel model (Fig. 6.21). This model reduction is acceptable for a passenger car because the coupling masses are usually very small compared to the car body mass. Even though they influence each other's vibration properties only very slightly, the following restrictions need to be considered:

- The effects of pitch- and roll oscillations are disregarded in the evaluation of comfort.
- In the evaluation of driving safety the effects of the wheel load fluctuations to the lateral dynamics of the vehicle cannot be observed. For this, the use of a more complex wheel and tire model is necessary such as the one described in Sect. 7.4.

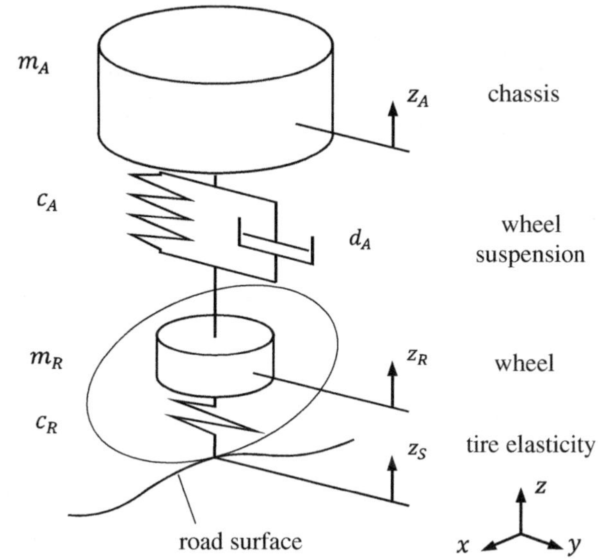

Fig. 6.21 Planar quarter-vehicle model (heave ride model)

Table 6.8 Data of the passive reference model

Parameter		Numerical value	Unit
Wheel mass	m_R	48	kg
Body mass (ratio)	m_A	401	kg
Wheel spring stiffness	c_R	235,000	N/m
Vehicle body spring stiffness	c_A	28,426	N/m
Vehicle body suspension damper	d_A	2,836	Ns/m

The planar quarter vehicle model (heave ride model) is depicted in Fig. 6.21. It consists of the wheel mass m_R and the proportionate body mass m_A. The masses are linked via a spring (spring stiffness c_A) and a damper (damping coefficient d_A), which is placed parallel to the spring.

The contact to the road is established via the wheel, which is modeled through the spring stiffness c_R. The damping of the wheel can usually be disregarded.

The principles of linear momentum for the system in z-direction are:

$$\text{chassis:} \quad m_A\ddot{z}_A + d_A(\dot{z}_A - \dot{z}_R) + c_A(z_A - z_R) = 0, \tag{6.1}$$

$$\text{wheel:} \quad m_R\ddot{z}_R - d_A(\dot{z}_A - \dot{z}_R) - c_A(z_A - z_R) + c_R z_R = c_R z_S. \tag{6.2}$$

For the evaluation of the results, a system of comparison is needed. For example the vehicle data of Table 6.8 can be used, which are aligned with a upper class car.

In order to achieve better results from the quarter-vehicle model, the pro rata chassis mass can be determined by breaking down the half chassis mass onto three mass points $m_{A,v}$ (front axle), $m_{A,h}$ (rear axle) and m_K (coupling mass) (Fig. 6.22).

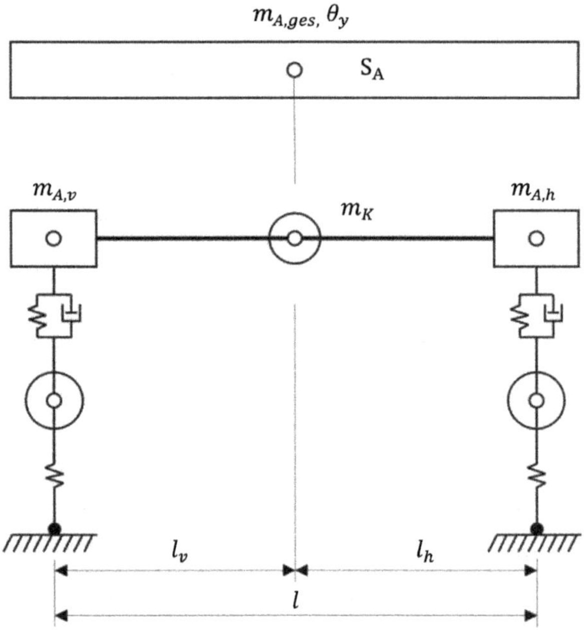

Fig. 6.22 For the derivation of the proportionate vehicle body mass

From the conditions for the preservation of the chassis mass $m_{A,ges}$, of the center of mass S_A and the mass moment of inertia $\theta_y = m_A i_y^2$ (radius of gyration i_y)

$$m_{A,v} + m_{A,h} + m_K = m_{A,ges}, \tag{6.3}$$

$$m_{A,v} l_v - m_{A,h} l_h = 0, \tag{6.4}$$

$$m_{A,v} l_v^2 + m_{A,h} l_h^2 = m_A i_y^2, \tag{6.5}$$

one obtains three equivalent masses

$$m_{A,v} = m_{A,ges} \frac{i_y^2}{l_v l}, \quad m_{A,h} = m_{A,ges} \frac{i_y^2}{l_h l}, \tag{6.6}$$

$$m_k = m_{A,ges} \left(1 - \frac{i_y^2}{l_h l_v} \right). \tag{6.7}$$

For the wheel based, proportional body mass m_A of the quarter-vehicle model, $m_A = \frac{1}{2} m_{A,v}$ (front wheel) respectively $m_A = \frac{1}{2} m_{A,h}$ (rear wheel) is used.

Furthermore it is necessary for an appropriate quarter-vehicle model to consider the kinematic transmission of the wheel suspension forces as shown in Fig. 6.23. The kinematic transmission λ results from the vertical deflection z_r of the wheel suspension as

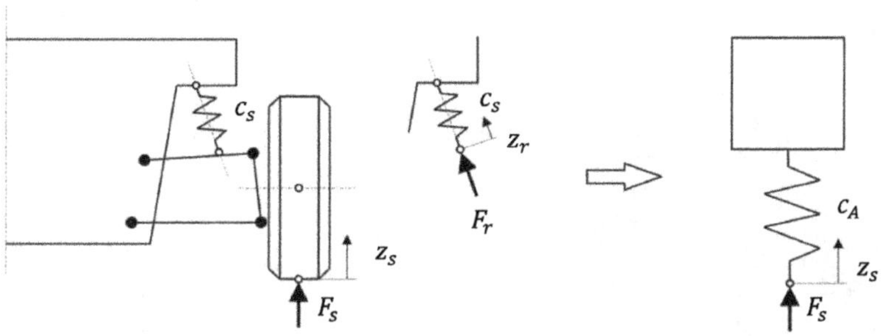

Fig. 6.23 Kinematic transmission of the wheel suspension forces

$$\lambda(z_s) = \frac{dz_r}{dz_s} = \frac{\dot{z}_r}{\dot{z}_s}. \tag{6.8}$$

From the principle of virtual work, the relation between the force F_s at the tire-road contact point and the elastic force F_r can be derived as follows:

$$F_s \delta z_s = F_r \delta z_r = F_r \frac{dz_r}{dz_s} \delta z_s \rightarrow F_s = \frac{dz_r}{dz_s} F_r = \lambda F_r. \tag{6.9}$$

From this, the spring stiffness related to the vertical deflection z_s of the tire contact point is calculated as follows:

$$c_s = \frac{dF_s}{dz_s} = \frac{d(\lambda F_r)}{dz_s} = \frac{d\lambda}{dz_s} F_r + \lambda \frac{dF_r}{dz_s} = \frac{d\lambda}{dz_s} F_r + \lambda \underbrace{\frac{dF_r}{dz_r}}_{c_s} \underbrace{\frac{dz_r}{dz_s}}_{\lambda}, \tag{6.10}$$

$$c_A = \frac{d\lambda}{dz_s} F_r + \lambda^2 c_s. \tag{6.11}$$

6.5 Three-Dimensional Model of a MacPherson Wheel Suspension

The use of wheel guiding damper struts derived from the MacPherson principle is very common in automotive engineering. Damper struts differ from spring struts in having the spring not supported by the damper housing, but by the transverse link. MacPherson wheel suspensions can be found in passenger cars from compact cars up to the upper middle-sized class as well as in vans and pickup trucks (examples: Fiat Punto, Ford Fiesta, VW Golf, Opel Vectra, Porsche Boxter, VW Sharan, Mercedes-Benz Sprinter, etc.). The design distinguishes itself by having small

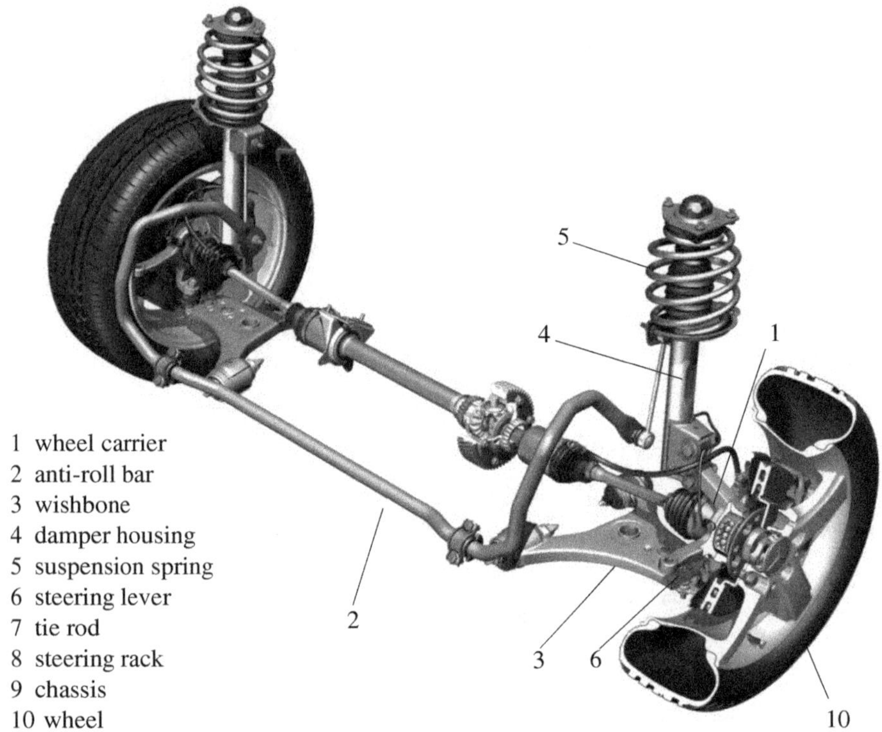

Fig. 6.24 MacPherson-axle of the Mercedes B-Class (courtesy of the Daimler AG 2009). The components tie rod (*7*), steering rack (*8*) and chassis (*9*) are depicted in Figs. 6.25 and 6.26

space requirements and good characteristics in wheel guidance. The kinematic of the MacPherson wheel suspension is the topic of many studies, i.e. (Cronin 1981), or (Schmidt and Wolz 1987). Using the McPherson principle as an example of this wheel suspension, the advantages compared to other approaches become evident. In particular it is possible to present explicitly solvable kinematic equations in a compact form.

6.5.1 Kinematic Analysis

In Fig. 6.24, the MacPherson spring strut axle of the Mercedes B-Class is presented. The wheel carrier (1) on which the wheel (10) is pivot-mounted, is connected tightly with the lower half of the damper housing (4). The piston rod of the damper slides in the wheel carrier fixed tube of the damper and is mounted via a spherical joint to the chassis. The spring (5) supports itself against the vehicle body on the outer shell of the damper guidance. At the lower spring plate, the anti-roll bar

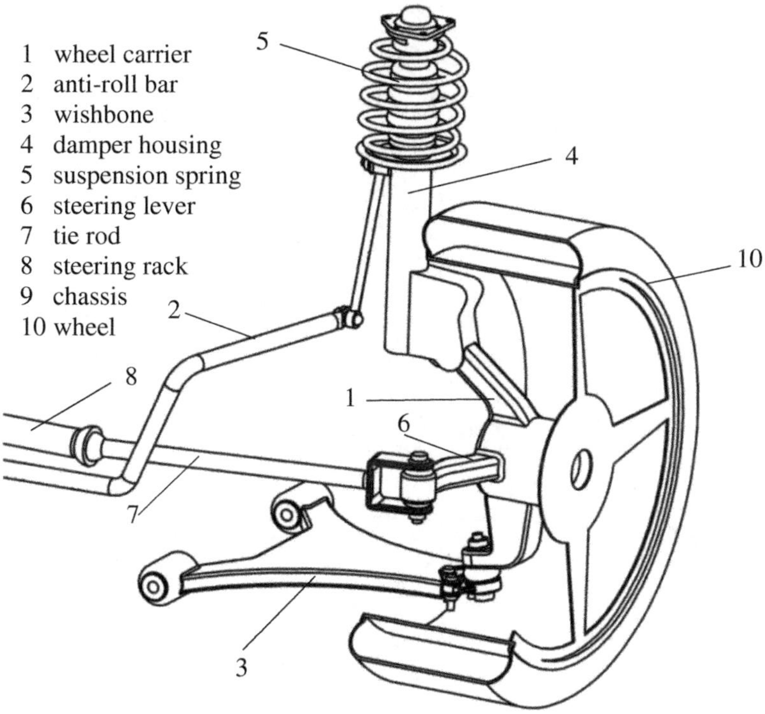

Fig. 6.25 Multibody model of a MacPherson-wheel suspension. The component chassis (*9*) is depicted in Fig. 6.26

(2) comes into place as an additional force element. The guidance of the wheel carrier is additionally supported by the wishbone (3), which is pivot-mounted to the chassis (9). Between the wishbone and the wheel carrier, there is another spherical joint. The wheel carrier can use it in its steering mode to rotate around the connecting line of the spherical joints. The steering motion is transmitted through a tie rod (7) with spherical joints at its ends into the steering lever (6), which is a part of the wheel carrier (1). By the spherical joint of the tie rod on the side of the vehicle, the steering movement can be introduced through a turning motion (worm and roller steering and ball and nut power steering) via a four-bar linkage (or a longitudinal displacement (rack and pinion steering (8))).

Based on the structure of the real system described above, the multibody model shown in Fig. 6.25 is designed. It contains the same elements as the front axle depicted in Fig. 6.24. Additionally, the tie rod (7) and the steering rack (8) were added for the modeling of the steering mechanism. The damper forms the prismatic joint while the spring as a force element is not a part of the kinematic model and is therefore left out.

From the model of the MacPherson wheel suspension depicted in Fig. 6.25 we now derive a substitute system as a multibody model, which only consists of rigid

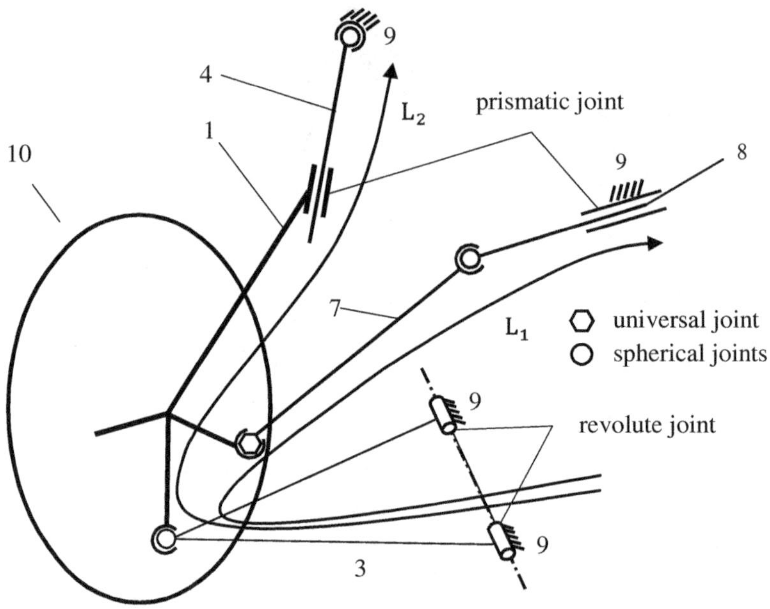

Fig. 6.26 Rigid body model of the MacPherson wheel suspension—topology

bodies and joints and which forms the basis for further kinematic analyses. In the chosen modeling the wishbone bearing on the side of the vehicle is realized as a revolute joint with a gyratory DoF. The result is the rigid body model of Fig. 6.26.

If one replaces the spherical joint on the wheel carrier by a CARDAN joint (two revolute joints with one virtual intermediate body) and the other spherical joint of the model by three revolute joints with two virtual intermediate bodies each in order to eliminate the isolated degrees of freedom of the tie rod, the system topology depicted in Fig. 6.27 is created. The number of independent kinematic loops n_L can be measured by enumerating the available joints and bodies (without base plate), while all joints are expressed through basic joints like revolute joints R ($f = 1$), prismatic joints P ($f = 1$), respectively (see Sect. 3.2).

Without considering the wheel, the following applies:

$$\begin{array}{ll} \text{number of joints } n_G & n_G = 14, \\ \text{number of bodies } n_B & n_B = 12, \\ \text{number of loops } n_L = n_G - n_B & n_L = 2. \end{array} \qquad (6.12)$$

The loop L_1 is directed via the transverse link or wishbone (3), the wheel carrier (1) and the damper housing (4) back to the chassis (9). After combining the two joint coordinates of the isolated degree of freedom (two spherical joints in the loop), 7 coordinates remain, thus one generalized coordinate α as input and six,

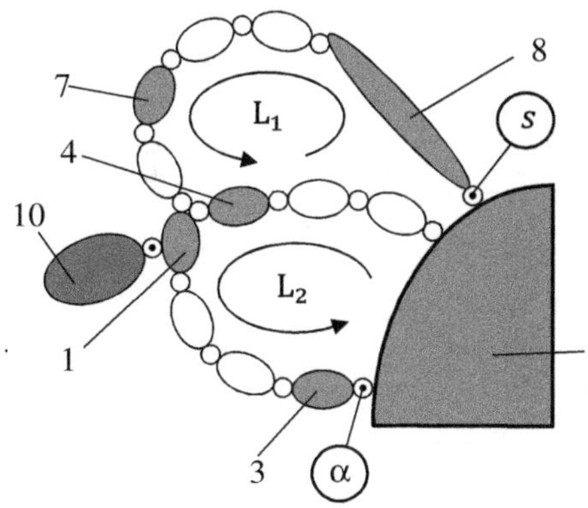

Fig. 6.27 Joint-body-display of the MacPherson wheel suspension

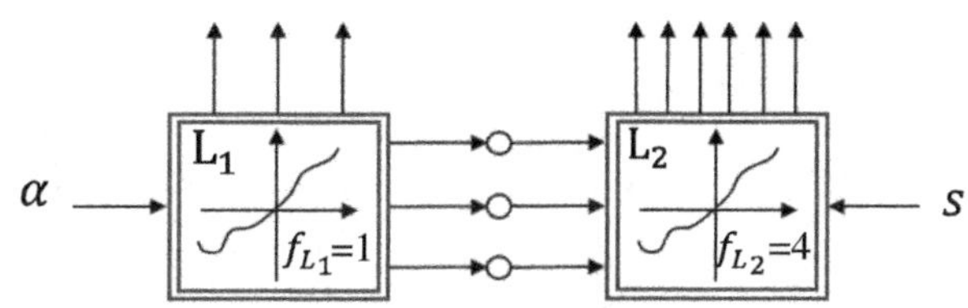

Fig. 6.28 Block diagram of the MacPherson suspension with explicit solution flow

initially unknown coordinates, as output. With this, the kinematic loop L_1 can be solved.

The loop L_2 includes the steering mechanism (7 and 8). It contains (without isolated DoF) 10 joint coordinates, thus four inputs. Three of the joint coordinates (four minus one isolated DoF) are also part of loop L_1, thus already known. The fourth input must therefore be the generalized input variable of the complete system, which in this case is the displacement s of the steering rack. With this, the second loop is solvable as well and the solution flow is defined: One recognizes that this kinematic system is explicitly solvable.

In order to present the previously described explicit solution process clearly, one can use the idea of kinematic transformers that have been introduced in Chap. 3. In Fig. 6.28, the connection between the two kinematic loops for the convenient solution flow is presented in a block diagram. In the following section, the procedure for an explicit solution for the position, velocity, and acceleration of the wheel suspension as a function on the two independent generalized joint coordinates α and s is described. First, the non-linear transmission behavior of the loops L_1 and L_2 using the method of the characteristic pair of joints (Woernle 1988) is analyzed (Chap. 3) for that purpose. The time derivatives—velocity and acceleration—are calculated next.

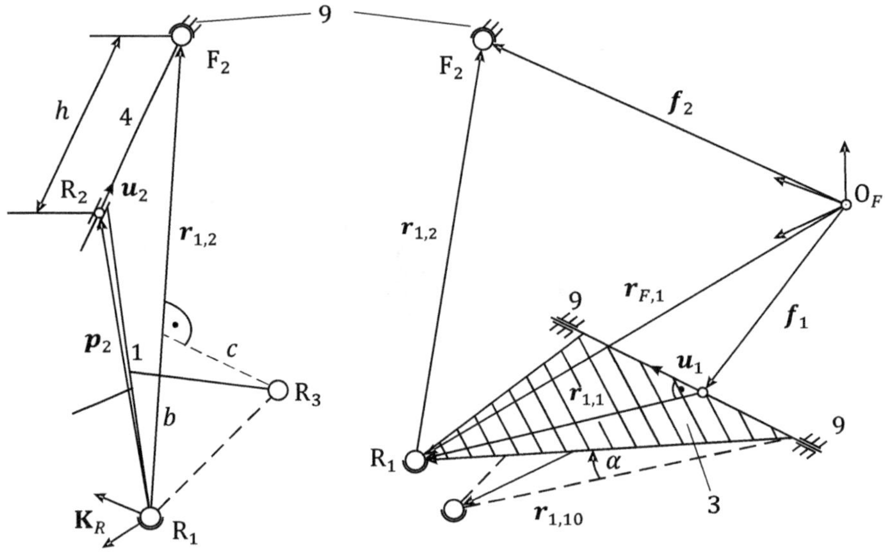

Fig. 6.29 Analysis of the loop L_1 in two segments

6.5.2 Explicit Solution

Position The kinematic loop L_1 is subdivided by the two spherical joints into an upper and a lower segment (Fig. 6.29). The upper segment contains the wheel carrier and the damper strut, the lower segment contains the chassis and the wishbone. The task is now to determine the distance d of the points R_1 and F_2 in both segments. In the lower segment, the chassis-fixed points F_1 and F_2 as well as the transverse link axle $\boldsymbol{u}_1$ are provided. Thereby, the calculation of the connection vector $\boldsymbol{r}_{1,2}$ of the points R_1 and F_2 becomes:

$$\boldsymbol{r}_{1,2} = \boldsymbol{f}_2 - \boldsymbol{r}_{F,1} = \boldsymbol{f}_2 - \boldsymbol{f}_1 - \boldsymbol{r}_{1,1}. \tag{6.13}$$

The vector $\boldsymbol{r}_{1,1}$ is generated from the turning of the vector $\boldsymbol{r}_{1,10}$ with the angle α around the center $\boldsymbol{u}_1$. The vector $\boldsymbol{r}_{1,10}$ describes the wishbone orientation for $\alpha = 0$. This reference orientation can be chosen arbitrarily, the vector only has to include a constructively given angle (here 90°) with the center $\boldsymbol{u}_1$ and has an equally given distance. One obtains

$$\boldsymbol{r}_{1,1} = \boldsymbol{T}(\boldsymbol{u}_1, \alpha)\boldsymbol{r}_{1,10}. \tag{6.14}$$

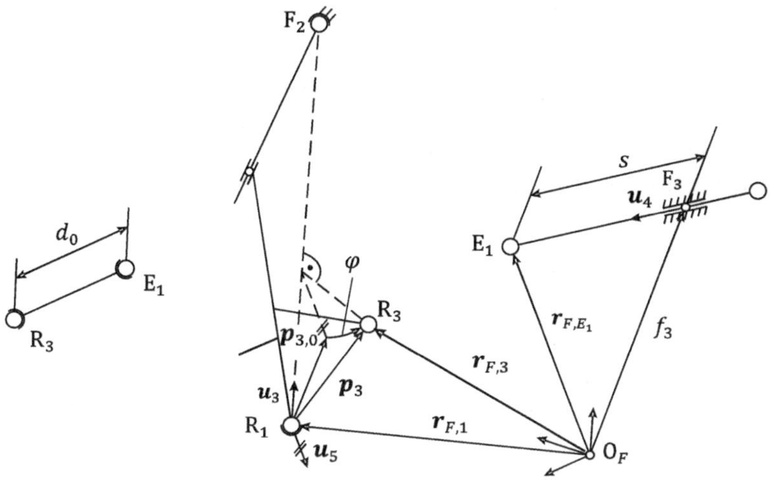

Fig. 6.30 Analysis of the loop L_2 in two segments

All vectors from Eqs. (6.13) and (6.14) can be written in components of the vehicle system. Therefore, the constraint parameter results as the square d_u^2 of the wanted distance in the lower segment:

$$d_u^2 = {}_F\boldsymbol{r}_{1,2\,F}^T\boldsymbol{r}_{1,2}. \tag{6.15}$$

In the upper segment the distance d_0 is depending on the auxiliary variable h. One gets the following vector relation which can be expressed in components of the wheel carrier system K_R:

$$\boldsymbol{r}_{1,2} = \boldsymbol{p}_2 + h\boldsymbol{u}_2, \tag{6.16}$$

$$d_0^2 = {}_R\boldsymbol{r}_{1,2\,R}^T\boldsymbol{r}_{1,2}. \tag{6.17}$$

The equation of the Eqs. (6.15) and (6.17) supplies a relation which can be solved for the unknown joint coordinate h. With this, the displacement of the damper strut in its guidance is known. Furthermore, with the position of the points R_1 and F_2, the current steering axle is to which the unit vector $\boldsymbol{u}_R$ (Fig. 6.30) is assigned. The position of the loop L_1 is therefore known.

The loop L_2 is now also separated into two segments in which the upper segment only consists of the tie rod of which the length is known. With this, the constraint parameter d_0^2 is immediately given. In the lower segment, at first the position of the steering rack end point E_1 as a function of the generalized coordinate s (translation of the steering rack) can be indicated as

$$\boldsymbol{r}_{F,E_1} = \boldsymbol{f}_3 + s\boldsymbol{u}_4. \tag{6.18}$$

The position of the point R_3 results from the rotation of the wheel carrier with the unknown angle φ around the current steering axle $\boldsymbol{u}_3$. Here, the construction of the starting position—given by the vector $\boldsymbol{p}_{3,0}$—is not entirely simple. Inside the wheel carrier fixed system $\boldsymbol{K}_R$, the distances b and c can be calculated after solving the first loop by projection of the wheel carrier point R_3 onto the current steering axle (given in $\boldsymbol{r}_{1,2}$). With this, a vector $\boldsymbol{p}_{3,0}$ can be constructed inside the vehicle system which is consistent with the geometry. At first, one needs to obtain a unit vector $\boldsymbol{u}_5$, which is perpendicular to the steering axle, i.e.

$$\boldsymbol{u}_5 = \frac{\boldsymbol{u}_3 \times \boldsymbol{r}_{F,1}}{\left\|\boldsymbol{u}_3 \times \boldsymbol{r}_{F,1}\right\|} \tag{6.19}$$

this yields the following:

$$\boldsymbol{p}_{3,0} = b\boldsymbol{u}_3 + c\boldsymbol{u}_5, \tag{6.20}$$

$$\boldsymbol{r}_{F,3} = \boldsymbol{r}_{F,1} + \boldsymbol{T}(\boldsymbol{u}_3, \varphi)\boldsymbol{p}_{3,0}, \tag{6.21}$$

$$d_u^2 = \left({}_F\boldsymbol{r}_{F,3} - {}_F\boldsymbol{r}_{F,E_1}\right)^T \left({}_F\boldsymbol{r}_{F,3} - {}_F\boldsymbol{r}_{F,E_1}\right). \tag{6.22}$$

Equalizing d_u^2 with the given quantity d_0^2 results in an explicitly solvable equation for the unknown angle φ. This by itself is not valuable, but becomes important in the identification of the position vector $\boldsymbol{r}_{F,3}$ from Eq. (6.21). With this, the calculation of the position is complete. The position of the points R_1, F_2 and R_3 is now provided in components of the vehicle system as well as the wheel carrier system. The determination of the transformation matrix ${}^F\boldsymbol{T}_R$ of the two systems can be carried out according to Sect. 2.2.

Velocities With known position, every loop is cut open at a spherical joint. Then, the velocity of the hinge point is determined from both sides and equated. The resulting vector equation is solved using convenient projections. In the loop L_1, the velocity ${}_F\boldsymbol{v}_{F_2}$ of the hinge point F_2 (Fig. 6.29) relative to the vehicle shall be established. From the side of the vehicle, the following equation results immediately:

$${}_F\boldsymbol{v}_{F_2} = 0. \tag{6.23}$$

The second approach results in the velocity ${}_F\boldsymbol{v}_{F_2}$ for the relative motion of bodies

$$\begin{aligned} {}_F\boldsymbol{v}_{F_2} &= {}_F\boldsymbol{v}_{R_1} + {}_F\boldsymbol{\omega}_R \times \boldsymbol{r}_{1,2} + {}_R\boldsymbol{v}_{F_2} \\ &= \dot{\alpha}\boldsymbol{u}_1 \times \boldsymbol{r}_{1,1} + {}_F\boldsymbol{\omega}_R \times \boldsymbol{r}_{1,2} + \dot{h}\boldsymbol{u}_2. \end{aligned} \tag{6.24}$$

Combining Eqs. (6.23) and (6.24) and building the scalar product with the vector $\boldsymbol{r}_{1,2}$, the unknown velocity $\dot{h}$ can be determined.

$$\dot{h} = -\frac{\left(\boldsymbol{u}_1 \times \boldsymbol{r}_{1,1}\right)^T \boldsymbol{r}_{1,2}}{\boldsymbol{u}_2 \boldsymbol{r}_{1,2}} \dot{\alpha}. \tag{6.25}$$

The vector product of $\boldsymbol{r}_{1,2}$ with both sides of the Eq. (6.24) results in the part $\left({}_F\boldsymbol{\omega}_R\right)_s$ of the velocity of the wheel carrier which is perpendicular to the vector $\boldsymbol{r}_{1,2}$ (the remaining part cannot be determined from the first loop):

$$\left({}_F\boldsymbol{\omega}_R\right)_s = -\frac{\boldsymbol{r}_{1,2} \times \left(\boldsymbol{u}_1 \times \boldsymbol{r}_{1,1}\right)}{\boldsymbol{r}_{1,2}\boldsymbol{r}_{1,2}} \dot{\alpha} - \frac{\left(\boldsymbol{r}_{1,2} \times \boldsymbol{u}_2\right)}{\boldsymbol{r}_{1,2}\boldsymbol{r}_{1,2}} \dot{h}. \tag{6.26}$$

The calculation of the velocities in the second loop is carried out analogously (see Fig. 6.30). Here, the velocity ${}_F\boldsymbol{v}_{R_3}$ of the hinge point R_3 relative to the chassis is determined and equalized in two approaches. The observation of the wheel carrier motion results in:

$${}_F\boldsymbol{v}_{R_3} = {}_F\boldsymbol{v}_{R_1} + \left(\left({}_F\boldsymbol{\omega}_R\right)_s + \dot{\varphi}\boldsymbol{u}_3\right) \times \boldsymbol{p}_3. \tag{6.27}$$

The velocity ${}_F\boldsymbol{v}_{R_1}$ of the point R_1 and the component $\left({}_F\boldsymbol{\omega}_R\right)_s$ of the angular velocity of the wheel carrier, which is perpendicular to the steering axle, are already known. The angular velocity $\dot{\varphi}$ with which the wheel carrier rotates around the current steering axle is wanted. For the velocity ${}_F\boldsymbol{v}_{R_3}$ the following is also valid if it is determined via the motion of the rack-and-pinion strut and the motion of the tie rod:

$${}_F\boldsymbol{v}_{R_3} = \dot{s}\boldsymbol{u}_4 + {}_F\boldsymbol{\omega}_S \times \left(\boldsymbol{r}_{F,3} - \boldsymbol{r}_{F,E_1}\right). \tag{6.28}$$

Here, the linear velocity $\dot{s}$ of the steering rack is given as an input variable, the angular velocity ${}_F\omega_S$ of the tie rod is unknown, one can again only determine its component $\left({}_F\omega_S\right)_S$ which is oriented perpendicular to the tie rod. After combining Eqs. (6.27) and (6.28), one obtains the following equation, which is analogous to the above:

$$\dot{\varphi} = \frac{\dot{s}\boldsymbol{u}_4 - {}_F\boldsymbol{v}_{R_1} - \left(\left({}_F\boldsymbol{\omega}_R\right)_S \times \boldsymbol{p}_3\right)^T \left(\boldsymbol{r}_{F,3} - \boldsymbol{r}_{F,E_1}\right)}{\left(\boldsymbol{u}_3 \times \boldsymbol{p}_3\right)^T \left(\boldsymbol{r}_{F,3} - \boldsymbol{r}_{F,E_1}\right)}. \tag{6.29}$$

With this, the angular velocity ${}_F\boldsymbol{\omega}_R$ of the wheel carrier relative to the chassis is completely known as

$$ {}_F\boldsymbol{\omega}_R = \left({}_F\boldsymbol{\omega}_R\right)_s + \dot{\varphi}\boldsymbol{u}_3. \tag{6.30}$$

For the angular velocity of the steering tie rod, the following holds:

$$ \left({}_F\boldsymbol{\omega}_S\right)_s = \frac{\left(\boldsymbol{r}_{F,3} - \boldsymbol{r}_{F,E_1}\right) \times \left({}_F\boldsymbol{v}_{R_1} + {}_F\boldsymbol{\omega}_R \times \boldsymbol{p}_3 - \dot{s}\boldsymbol{u}_4\right)}{\left(\boldsymbol{r}_{F,3} - \boldsymbol{r}_{F,E_1}\right)^T \left(\boldsymbol{r}_{F,3} - \boldsymbol{r}_{F,E_1}\right)}. \tag{6.31}$$

With this, the velocities of all bodies relative to the chassis are known. With the relationships of the relative kinematics it is possible—if needed—to determine the velocity with respect to the inertial system. The kinematic differentials (Sect. 4.6.1) which further are needed for the set-up of the equations of motion result immediately from the Eqs. (6.23–6.31) if they are evaluated while using the pseudo input velocities:

$$ a)\ \dot{\alpha} = 1, \quad \dot{s} = 0, \qquad b)\ \dot{\alpha} = 0, \quad \dot{s} = 1. \tag{6.32}$$

Acceleration For the calculation of accelerations, the same procedure is used as the one we have used for the velocities. The cut-open loops are depicted in Figs. 6.29 and 6.30. In loop L_1, the acceleration ${}_F\boldsymbol{a}_{F_2}$ of the hinge point F_2 is determined in two ways:

$$ {}_F\boldsymbol{a}_{F_2} = 0, \tag{6.33}$$

$$ \begin{aligned} {}_F\boldsymbol{a}_{F_2} = {} & \ddot{\alpha}\boldsymbol{u}_1 \times \boldsymbol{r}_{1,1} + \dot{\alpha}^2\boldsymbol{u}_1 \times \left(\boldsymbol{u}_1 \times \boldsymbol{r}_{1,1}\right) + {}_F\boldsymbol{\alpha}_R \times \boldsymbol{r}_{1,2} \\ & + 2\,{}_F\boldsymbol{\omega}_R \times \dot{h}\boldsymbol{u}_2 + {}_F\boldsymbol{\omega}_R \times \left({}_F\boldsymbol{\omega}_R \times \boldsymbol{r}_{1,2}\right) + \ddot{h}\boldsymbol{u}_2. \end{aligned} \tag{6.34}$$

From the Eqs. (6.33) and (6.34), the acceleration $\ddot{h}$ at the damper tube as well as the component $\left({}_F\boldsymbol{\alpha}_R\right)_s$ of the angular acceleration of the wheel carrier which is perpendicular to the steering axle is determined. The acceleration ${}_F\boldsymbol{a}_{R_3}$ of the point R_3 in the second loop can again be obtained through two approaches:

$$ \begin{aligned} {}_F\boldsymbol{a}_{R_3} = {} & \ddot{\alpha}\boldsymbol{u}_1 \times \boldsymbol{r}_{1,1} + \dot{\alpha}^2\boldsymbol{u}_1 \times \left(\boldsymbol{u}_1 \times \boldsymbol{r}_{1,1}\right) + \left(\left({}_F\boldsymbol{\alpha}_R\right)_s + \ddot{\varphi}\boldsymbol{u}_3\right) \\ & \times \boldsymbol{p}_3 + {}_F\boldsymbol{\omega}_R \times \left({}_F\boldsymbol{\omega}_R \times \boldsymbol{p}_3\right), \end{aligned} \tag{6.35}$$

$$ {}_F\boldsymbol{a}_{R_3} = \ddot{s}\boldsymbol{u}_4 + {}_F\boldsymbol{\alpha}_R \times \left(\boldsymbol{r}_{F_3} - \boldsymbol{r}_{F,E_1}\right) + {}_F\boldsymbol{\omega}_S \times \left({}_F\boldsymbol{\omega}_S \times \left(\boldsymbol{r}_{F_3} - \boldsymbol{r}_{F,E_1}\right)\right). \tag{6.36}$$

From the Eqs. (6.35) and (6.36), the angular acceleration of the tie rod can be obtained via the variable $\ddot{\varphi}$, the angular acceleration ${}_F\boldsymbol{\alpha}_R$ of the wheel carrier and the perpendicular component $\left({}_F\boldsymbol{\alpha}_F\right)_s$. With this, the determination of the accelerations is complete. For the set-up of the equations of motion, the relationships (6.33)–(6.36) are evaluated with the special generalized accelerations

$$\ddot{\alpha} = 0, \ddot{s} = 0 \tag{6.37}$$

The kinematic analysis of the MacPherson wheel suspension is thereby completed. It is especially noticeable that the constraint equations of the overall system are explicitly solvable, and an iteration of unknown variables was not necessary. Because of this, the process is fundamentally different from the approach chosen by Schmidt and Wolz (1987). They state the constraint equations by cutting through a body (duplication of the body) and not by observing a characteristic pair of joints (Woernle 1988) and therefore state consistent and well formable, but only numerically solvable constraint equations. Therefore, a much higher computational effort is needed which is undesirable in simulations.

It is furthermore remarkable that all of the equations for the determination of velocities and accelerations are based on vectors. Next to the very clear and transparent presentation this procedure has the advantage, that the transition to reference frames and coordinates, which are needed for the numerical calculation, can be freely decided by the user. Therefore one can choose the frames of reference freely for every term.

6.6 Three-Dimensional Model of a Five-Link Rear Wheel Suspension

The five-link wheel suspension (also called five-point suspension), used for example at the Daimler AG as a rear wheel suspension in the mid-range and top-of-the-range cars, is depicted in Fig. 6.31. This wheel suspension represents a modular unit that is precisely adjustable in its kinematic values. The advantages of such a suspension are especially in the good elasto-kinematic scopes for design, the precise wheel guidance, the elasto-kinematic tune ability as well as the good longitudinal and lateral suspension. The very high expenditure in building and the cost resulting from it are the main disadvantages of this design.

At the example of this wheel suspension we will illustrate the advantages of the implicit approach compared to the one discussed in Sect. 6.5.2.

6.6.1 Kinematic Analysis

As basic module in Fig. 6.31 one can determine the sub-frame (1), which is flexural rigid, yet torsional flexible. This sub-frame is connected to the floor assembly via four rubber bearings. The bearings also partially assume the task of noise reduction. The wheel carrier (2) is now connected to the sub-frame via five spatially arranged transverse links (3–7). Every transverse link is fixed to the chassis as well as the wheel carrier via each one partially elastically mounted revolute-, or spherical joint.

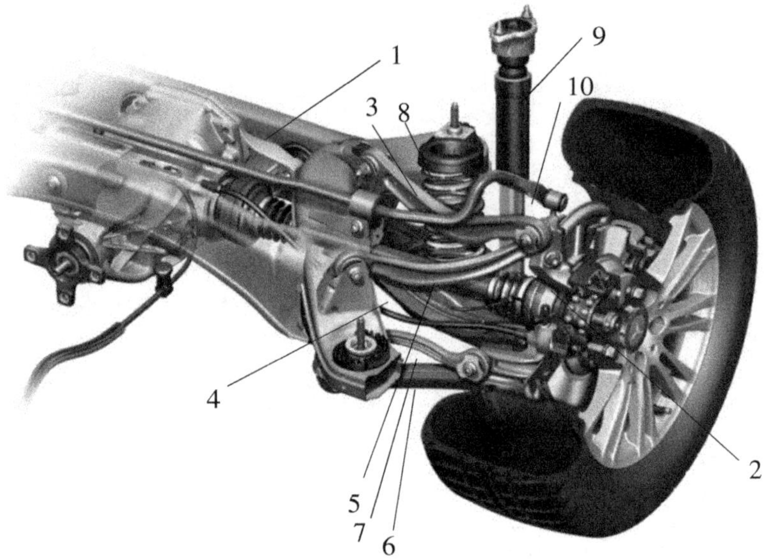

Fig. 6.31 Rear axle of the Mercedes E-Class (courtesy of Daimler AG 2009)

The joints are buffered by the use of rubber sockets against undesirable vibrations. One of those handles, the spring link (4), supports the shock absorber (8) as well as the coil spring (9) while at the same time the anti-roll bar (10) is fixed to it. The particular kinematic characteristics of such a suspension are now achieved by use of the different lengths for the five transverse links. The wheel carrier (2) is guided in vertical direction by the upper link (3) and the spring link (4), while the trailing-arm (5) and the push rod (6) primarily define the longitudinal alignment of the wheel carrier. The kinematic toe-in change can be turned off or determined with the position of the tie rod (7). Concerning the use of the five-link wheel suspension for a steered axle, exactly this link could be connected to the track rod.

Starting from the earlier mentioned structure of the real system, the multibody model depicted in Fig. 6.32 is designed. It contains the same essential elements as the wheel suspension system shown in Fig. 6.31. However, the spring and the absorber are not part of the kinematic model. These can be left out in the step of abstraction from the real model towards the multibody model. For illustration purposes, they were however added to Fig. 6.32.

From the model of the five-link wheel suspension depicted in Fig. 6.32 we now derive a substitute model. This multibody model will only consist of rigid bodies and joints and serves as a basis for all following considerations of the wheel suspension. By the use of appropriate simplification one obtains the rigid body model depicted in Fig. 6.33.

In the rigid body model, the rubber sockets that represent the bearings of the five transverse links are replaced by ideal spherical joints. The transverse links are

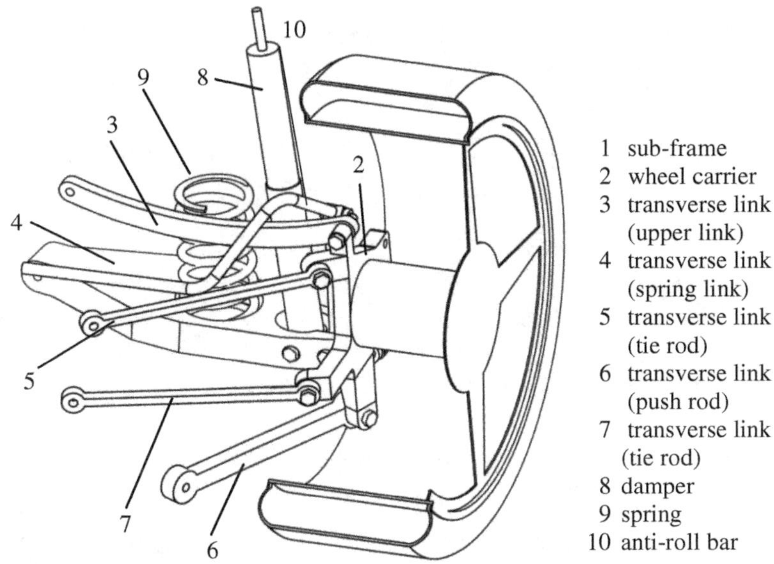

Fig. 6.32 Multibody model of the five-link suspension

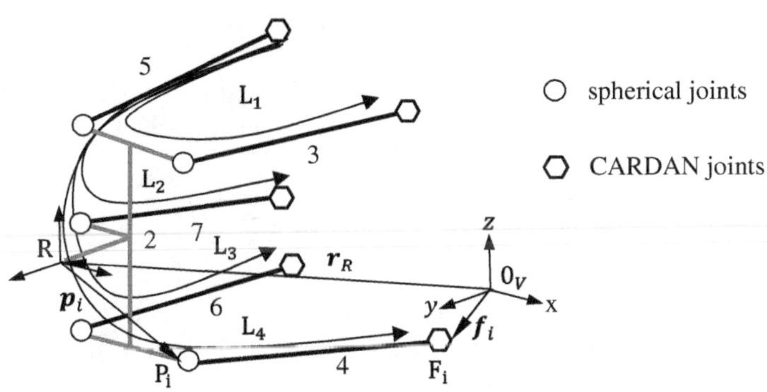

Fig. 6.33 Multibody model of the five-link wheel suspension

modeled as simple rods. The anti-roll bar is completely left out as it does not have an influence on the kinematic behavior of the suspension. The masses of the single components are not considered, respectively added to the wheel carrier.

From the derived rigid body model, we now develop a joint-body-representation. Since the five links are connected to the structure via a spherical joint (three DoF each), they each have a so-called isolated degree of freedom, namely the rotation around its own longitudinal axis. Because this movement has no impact on the behavior of the whole axle, all spherical joints on the side of the vehicle are replaced by CARDAN joints (two degrees of freedom each). As a consequence the global kinematic of the axle does not change since the isolated DoF are eliminated. This results in a joint-body-representation which is apparent from Fig. 6.34.

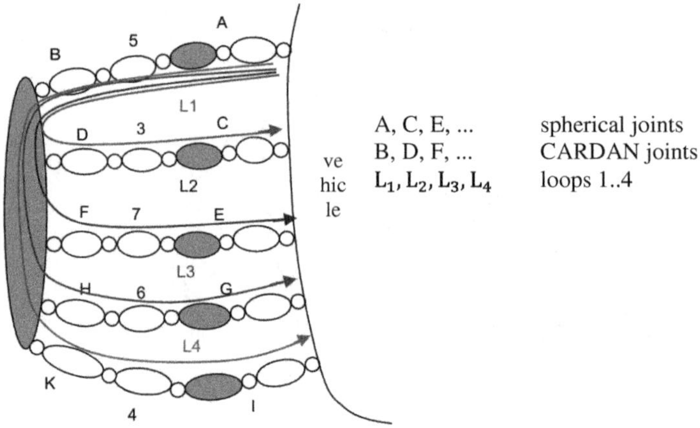

Fig. 6.34 Joint-body-presentation (topology) of the five-link wheel suspension

The number of independent kinematic loops $n_L$can be obtained by counting the available joints and bodies (without base body = vehicle body). One has

$$\begin{array}{ll} \text{number joints} & n_G = 10, \\ \text{number bodies} & n_B = 6, \\ \text{number loops} & n_L = n_G - n_B = 4. \end{array} \tag{6.38}$$

The number of DoF f of the wheel suspension is now calculated using the Grübler-Kutzbach Criterion (Sect. 3.3) (without isolated DoF) as

$$f = \sum_{i}^{n_G} f_{G_i} - 6n_L = 5 \cdot 3 + 5 \cdot 2 - 6 \cdot 4 = 1, \tag{6.39}$$

where f_{G_i} is the number of degrees of freedom of the ith joint. The here observed five-link wheel suspension therefore has exactly one DoF which corresponds with the vertical deflection. To use a generalized coordinate for this degree of freedom, in this example the vertical deflection of the wheel carrier immediately suggests itself. With regard to a convenient process of solution for the set-up of the equations of motion it is however useful to assign the generalized coordinate to a joint coordinate. Therefore, a joint angle is chosen as the generalized coordinate q_{gen} for the derivation of the solution described in the following section.

6.6.2 *Implicit Solution*

In order to obtain a clear description of the process in solving the kinematic loops, we now develop the relative kinematics of the system by stating and linking

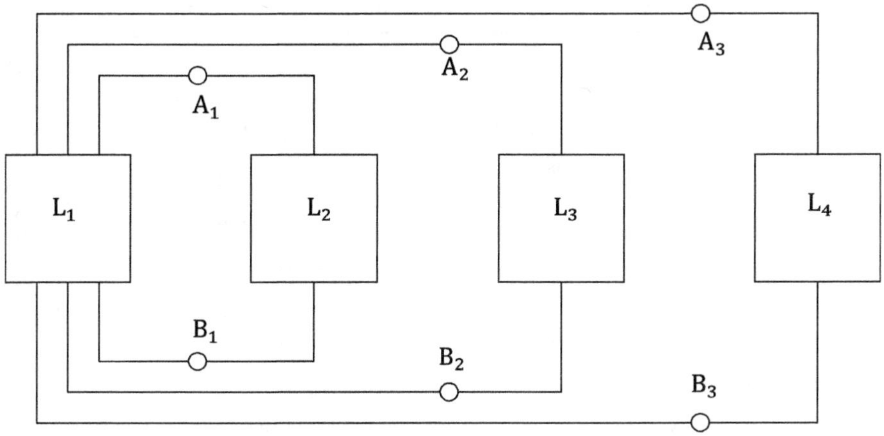

Fig. 6.35 First, simple block diagram

kinematic transformers (see Sect. 3.5). Every loop is represented by a black box, which are then linked with each other correspondingly. For details of this approach (Kecskemethy 1993). The needed local degrees of freedom of the loops f_L calculate as

$$\begin{array}{lll} \text{General} & f_{L_i} & = n_G(\mathrm{L}_i) - 6, \\ \text{Loop 1} & f_{L_1} & = 10 - 6 = 4, \\ \text{Loop 2} & f_{L_2} & = 10 - 6 = 4, \\ \text{Loop 3} & f_{L_3} & = 10 - 6 = 4, \\ \text{Loop 4} & f_{L_4} & = 10 - 6 = 4. \end{array} \tag{6.40}$$

Every kinematic transformer therefore needs four input variables. The difficulty now is to connect these transformers as well as possible so that there will be as few implicit constraint equations as possible. The number of coupling constraints between the four loops at the single joints (Fig. 6.34) is calculated as follows:

Number of constraints at joint G_i:

$$\begin{array}{ll} & p_i = n_L(\mathrm{G}_i) - n_B(\mathrm{G}_i) + 1 \\ p_A = 4 - 2 + 1 = 3, & p_F = 1 - 2 + 1 = 0, \\ p_B = 4 - 2 + 1 = 3, & p_G = 1 - 2 + 1 = 0, \\ p_C = 1 - 2 + 1 = 0, & p_H = 1 - 2 + 1 = 0, \\ p_D = 1 - 2 + 1 = 0, & p_I = 1 - 2 + 1 = 0, \\ p_E = 1 - 2 + 1 = 0, & p_K = 1 - 2 + 1 = 0. \end{array} \tag{6.41}$$

In joint A and B all loops are therefore coupled with each other. Hence we achieve the for the moment simplified block diagram depicted in Fig. 6.35.

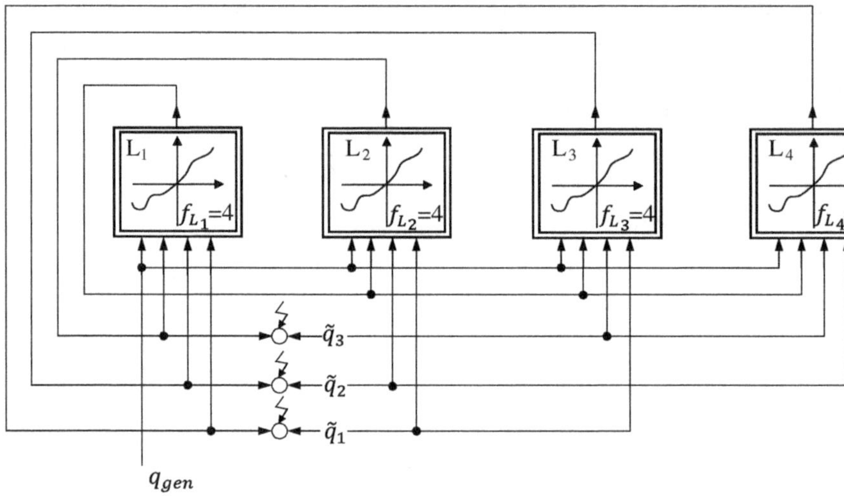

Fig. 6.36 Block diagram of the five-link wheel suspension with 3 implicit equations

By determination of the solution flow and the definition of corresponding pseudo-inputs $\tilde{q}_1$, $\tilde{q}_2$, and $\tilde{q}_3$ including the generalized coordinate q_{gen}, the block diagram depicted in Fig. 6.36 can be derived. Subsequently, the solving of three implicit core equations remains for the solving of this system of equations—next to the solving of the recursive-explicit part.

Position The five-link wheel suspension consists of six rigid bodies (Fig. 6.34). The wheel carrier in a non-guided condition has six degrees of freedom in space. The five links each only need five coordinates since the rotation around their longitudinal axes is not considered. Of the therefore remaining 31 body coordinates however only one is independent (the above mentioned generalized coordinate q_{gen} which is needed for the independent deflection of the wheel suspension). The remaining 30 are calculated by solving the constraint equations which result from the geometrical constraints of the system. When aggregating all 31 body coordinates to a $[31 \times 1]$ vector $\boldsymbol{w}$, the structure of the corresponding implicit system of equations looks as

$$g_i(\boldsymbol{w}, q_{gen}) = 0, \quad i = 1, \ldots, 31. \tag{6.42}$$

One of the 31 equations describes the assignment of the generalized coordinate q_{allg} to the corresponding body coordinate of the wheel carrier, meaning the z-coordinate of the point R in Fig. 6.33. The remaining 30 coordinates are geometrical constraints, which have to be fulfilled. These constraints can in this example be provided from the 25 joint degrees of freedom (compare Fig. 6.34) together with the five predefined lengths of the transverse links.

The aim is now to reduce the order of the implicit system of Eq. (6.42) and to use explicit kinematic relations as far as possible. This is possible through the introduction of n_s auxiliary variables $\beta_i, i = 1,\ldots,n_s$, which is summarized as the vector $\boldsymbol{\beta}$. The auxiliary variables are relative coordinates. They have to be chosen in such a way that all body coordinates w_k, $k = 1,\ldots,31$ can be calculated explicitly from $\boldsymbol{\beta}$. Thus the order of the implicit system of equations that is to be solved reduces to the number n_s of the auxiliary variables. Eq. (6.42) then becomes

$$g_i(\boldsymbol{\beta}, q_{allg}) = 0, \qquad i = 1,\ldots,n_s, \tag{6.43}$$

$$\boldsymbol{w} = \boldsymbol{w}(\boldsymbol{\beta}, t). \tag{6.44}$$

By the use of relative intermediate variables, the global kinematic block is partitioned into two parts. The assignments between the generalized coordinates and auxiliary variables are summarized in the relative kinematic, the explicit dependence $\boldsymbol{w} = \boldsymbol{w}(\boldsymbol{\beta}, t)$ matches the absolute kinematic. Finally, just as it is already described in the block diagram of Fig. 6.36, three core equations remain to be solved iteratively.

Now, we will show a different approach, which is based on Schnelle (1990). Here, the constraint equations account for the special structure of the wheel suspension with five homogenous links $\boldsymbol{l}_i$. One chooses for example the six position coordinates of the wheel carrier—the three components of the position vector $\boldsymbol{r}_R$ (Fig. 6.33) and the three CARDAN-angles ψ, θ and φ of its finite orientation relative to the chassis—as auxiliary variables. Then one arrives at $n_s = 6$.

From this choice of the generalized coordinates the first constraint equation immediately follows

$$g_1(\boldsymbol{\beta}, q_{allg}) = r_{R,z} - q_{allg} = 0. \tag{6.45}$$

The five missing constraint equations are received from the requirement that the distance between the corresponding rod positions are equal to the respective constant length of the link:

$$g_{i+1}(\boldsymbol{\beta}) = \boldsymbol{l}_i(\boldsymbol{\beta})^T \boldsymbol{l}_i(\boldsymbol{\beta}) - l_i^2 = 0 \quad i = 1,\ldots,5. \tag{6.46}$$

The vectors $\boldsymbol{l}_i$ then results explicitly in dependency of the auxiliary variables. From Fig. 6.33, the following applies:

$$\boldsymbol{l}_i(\boldsymbol{\beta}) = \boldsymbol{r}_R(\boldsymbol{\beta}) + \boldsymbol{p}_i(\boldsymbol{\beta}) - \boldsymbol{f}_i. \tag{6.47}$$

The numerical solution of the Eqs. (6.45) and (6.46) finally corresponds to the wanted relation $\boldsymbol{\beta} = \boldsymbol{\beta}(q_{allg})$.

Velocities and accelerations Respectively, the velocity $\dot{\boldsymbol{l}}_i(\boldsymbol{\beta}, \dot{\boldsymbol{\beta}})$ can be calculated explicitly. Here one has

$$\dot{\boldsymbol{l}}_i(\boldsymbol{\beta},\dot{\boldsymbol{\beta}}) = \dot{\boldsymbol{r}}_R(\boldsymbol{\beta},\dot{\boldsymbol{\beta}}) + \boldsymbol{\omega}_{R,F}(\boldsymbol{\beta},\dot{\boldsymbol{\beta}}) \times \boldsymbol{p}_i(\boldsymbol{\beta}). \tag{6.48}$$

The time derivative of the constraint Eq. (6.46) is

$$\dot{g}_{i+1}(\boldsymbol{\beta},\ \dot{\boldsymbol{\beta}}) = 2\dot{\boldsymbol{l}}_i(\boldsymbol{\beta},\dot{\boldsymbol{\beta}})^T \boldsymbol{l}_i(\boldsymbol{\beta}) = 0, \quad i = 1,\ldots,5. \tag{6.49}$$

So, the velocity of the constraint $\dot{g}_{i+1}$ can be determined explicitly from the auxiliary variables and their velocities. This is used in the partial derivative of $\dot{g}_{i+1}$ Eq. (6.49) is for this purpose—according to the principle of the kinematic differentials—from Chap. 4 evaluated with a special $\dot{\tilde{\boldsymbol{\beta}}}^{(j)}$:

$$\tilde{\dot{\boldsymbol{\beta}}}^{(j)} = (0\ 0\ \ldots\ \underset{\substack{\uparrow \\ j^{th}\ column}}{1}\ \ldots\ 0). \tag{6.50}$$

One can then obtain the partial derivative in explicit form

$$\frac{\partial g_{i+1}}{\partial \beta_j} = \tilde{\dot{g}}_{i+1}^{(j)} = \dot{g}_{i+1}^{(j)}\left(\boldsymbol{\beta}, \tilde{\dot{\boldsymbol{\beta}}}^{(j)}\right). \tag{6.51}$$

The constraints of the velocities arising from this can be combined to a linear system

$$\boldsymbol{J}_\beta \dot{\boldsymbol{\beta}} - \boldsymbol{V}\dot{q}_{allg} = 0. \tag{6.52}$$

Here, the JACOBIAN-Matrix $\boldsymbol{J}_\beta$ with $\boldsymbol{J}_{\beta_{i,j}} = \dfrac{\partial g_i}{\partial \beta_j}$ as well as the (constant) distribution matrix $\boldsymbol{V}$ occur. One can proceed with the acceleration accordingly and obtain

$$\boldsymbol{J}_\beta \ddot{\boldsymbol{\beta}} + \dot{\boldsymbol{J}}_\beta \dot{\boldsymbol{\beta}} - \boldsymbol{V}\ddot{q}_{allg} = 0. \tag{6.53}$$

The solving of (6.52) and (6.53) results in

$$\dot{\boldsymbol{\beta}} = \boldsymbol{J}_\beta^{-1} \boldsymbol{V}\dot{q}_{allg}, \tag{6.54}$$

$$\ddot{\boldsymbol{\beta}} = \boldsymbol{J}_\beta^{-1}\left(\boldsymbol{V}\ddot{q}_{allg} - \dot{\boldsymbol{J}}_\beta \dot{\boldsymbol{\beta}}\right). \tag{6.55}$$

With this, the relative kinematic is solved.

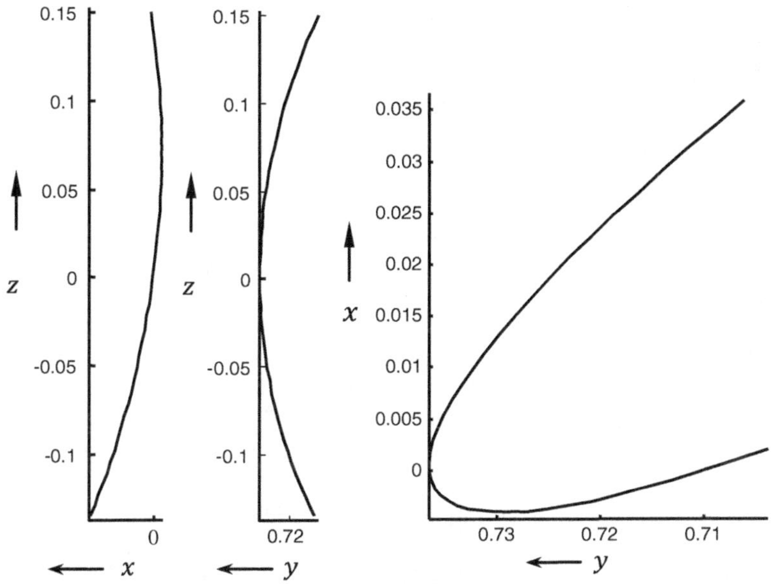

Fig. 6.37 Trajectory of the tire center of the five-link wheel suspension W124 (Mercedes)

6.6.3 Simulation Results of the Three Dimensional Quarter Vehicle Model

Every type of wheel suspension has its, dependent on the respective design, characteristic behavior in the deflecting and hopping of the wheel. The most important kinematic parameters of the wheel motion and hence the position of the wheel towards the road are the camber angle and toe angle changes of the resilient axle. Since the rear axle determines the stability of the vehicle, too extreme alterations of the toe angle result in an uncomfortable tracking effect of the axle. Therefore, the toe angle should stay as neutral as possible. The camber angle should adjust itself in a way in which there is always a possibly beneficial contact between the wheel and the road. With the aid of the earlier mentioned kinematic equations the characteristic quantities of the wheel suspension can now be calculated exemplary for a spatial quarter vehicle with a five link wheel suspension.

Trajectory of the wheel center For the five-link wheel suspension analyzed in the previous sections, the recording of the trajectory of the tire center is helpful. In Fig. 6.37, the wheel trajectory in the three levels $z - x, z - y.$ and $x - y$ is depicted in the vehicle-fixed coordinate system.

Camber- and toe angle graphs The camber and toe angles (see Sect. 6.3) are calculated from the alignment of the wheel normal with respect to the vehicle-fixed coordinate system as depicted in Fig. 6.38.

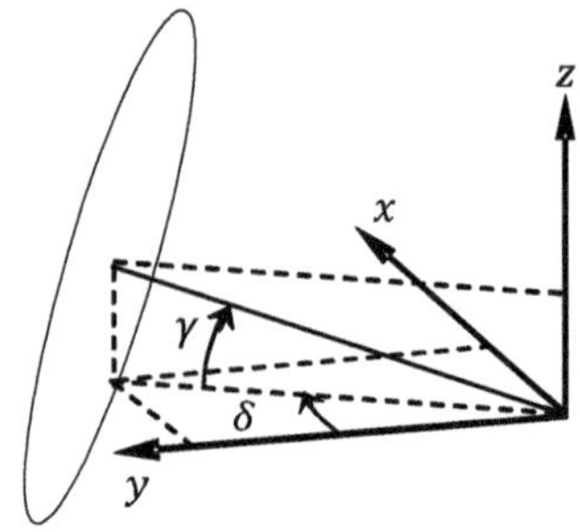

Fig. 6.38 Camber and toe angle definition and calculation

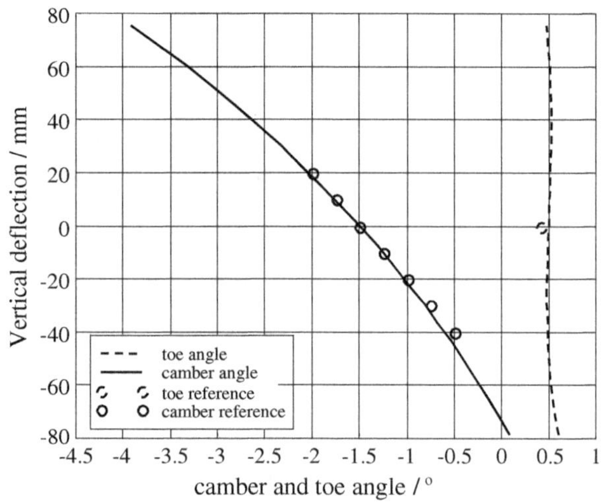

Fig. 6.39 Camber and toe angle graph of a five-link wheel suspension

The camber and toe angles are therefore calculated as such:

$$\text{camber angle:} \qquad \gamma = -\arctan\left(\frac{z}{\sqrt{x^2+y^2}}\right), \tag{6.56}$$

$$\text{toe angle:} \qquad \delta = -\arctan\left(\frac{x}{y}\right). \tag{6.57}$$

In Fig. 6.39, the calculated camber and toe angle graphs of a five-link wheel suspension in correlation with the compression travel are plotted. Additionally, there are also reference points from a corresponding, actually measured camber curve depicted. The deviations between calculation and measurement are caused by the elasto-kinematic, which was neglected in the modeling presented here. The camber- and toe variation at the wheel suspension can be visualized with suitable animation software, as shown in Fig. 6.40.

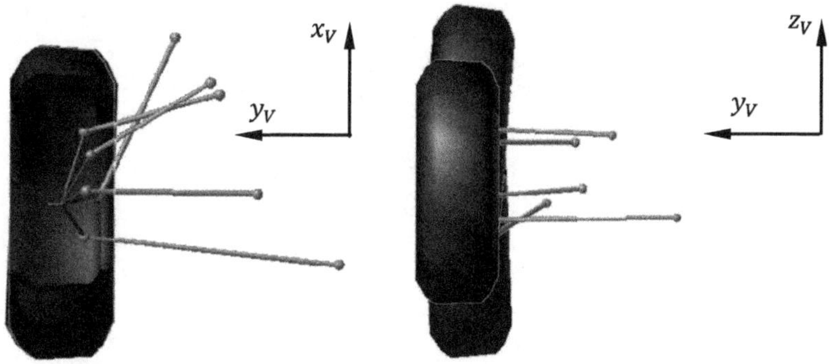

Fig. 6.40 Wheel position at ±80 mm deflection/hopping: (*left* the almost not visible change in toe angle, *right* the change in camber angle)

Fig. 6.41 The quarter vehicle model on the axle test bench

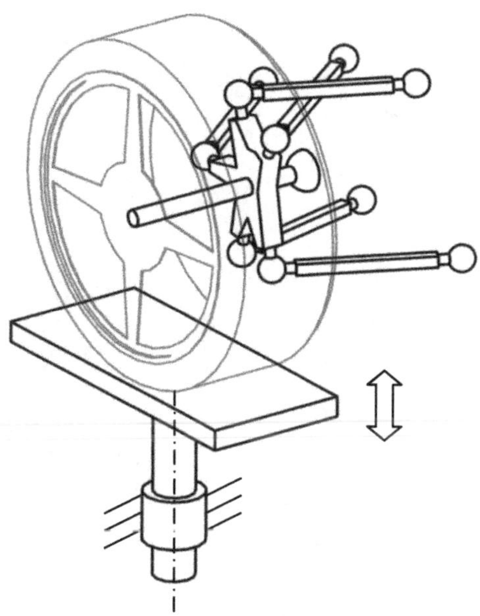

Simulation of the dynamics of an axle test bench Finally, the dynamic simulation of an axle test bench as presented in Fig. 6.41 is shown for exemplary reasons. The wheel is subject to an imposed vibration of the wheel suspension generated by a plate. The amplitude of the wheel movement is larger than the stimulus amplitude, a phase shift between the movement of the ground and the wheel is virtually not identifiable (see Fig. 6.42). The occurring areas during the transient period in which the ground force is not observable, the wheel has lifted off and there is no contact to the ground (compare Figs. 6.43 and 6.44).

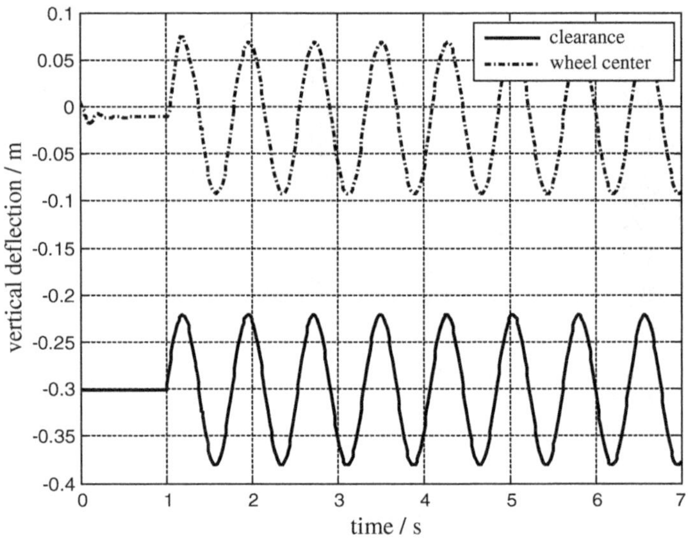

Fig. 6.42 Vertical deflection of the base plate and the wheel carrier

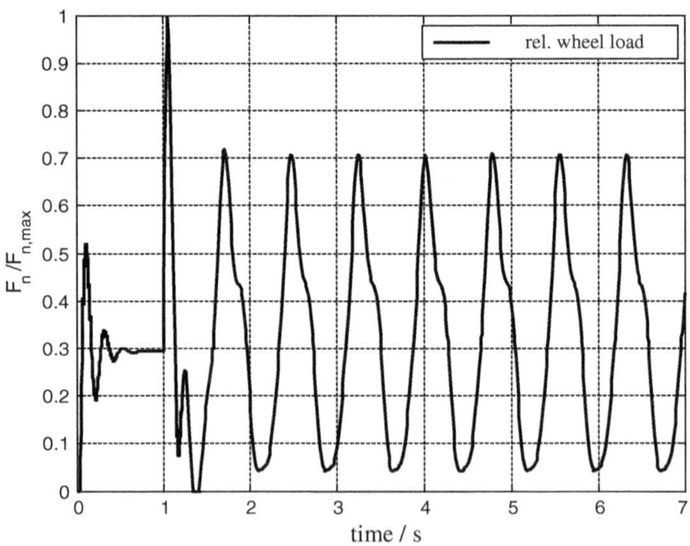

Fig. 6.43 Standardized vertical wheel load (normal force)

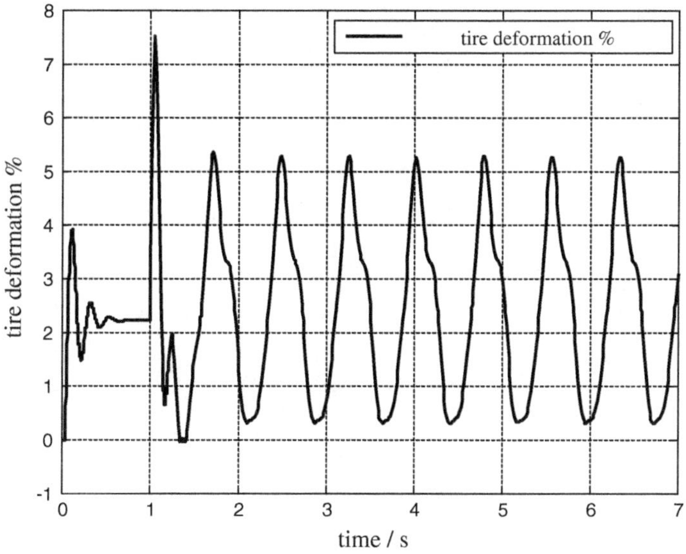

Fig. 6.44 Elastic vertical deformation of the wheel

References

Cronin DL (1981) McPherson Strut Kinematics. *Mechanism and Machine Theory* 16. S 631–44

DIN (1993) DIN_70020-1: Strassenfahrzeuge—Kraftfahrzeugbau—Begriffe von Abmessungen. (ed.), Deutsches Institut für Normung e.V., Berlin

Kecskemethy A (1993) Objektorientierte Modellierung der Dynamik von Mehrkörpersystemen mit Hilfe von Übertragungselementen [Dr.-Ing.]. Dissertation, Universität Duisburg, Düsseldorf: VDI-Verlag

Mödinger W, Bublitz HG, K.-J., Schulz W, Minning M and Braun R (1997) Das Fahrwerk des neuen A-Klasse von Daimler-Benz. *Sonderausgabe ATZ und MTZ* Die neue A-Klasse von Daimler-Benz. S 102–9

Schmidt A and Wolz U (1987) Nichtlineare räumliche Kinematik von Radaufhängungen—kinematische und dynamische Untersuchungen mit dem Programmsystem. *MESA VERDE - Automobilindustrie 1987* 6. S 639–44

Schnelle K-P (1990) Simulationsmodelle für die Fahrdynamik von Personenwagen unter Berücksichtigung der nichtlinearen Fahrwerkskinematik Dissertation, Universität Stuttgart, Stuttgart

Schuster H, Balk A and Oehlschlaeger H (1995) Der Sharan—Die Großraumlimousine von Volkswagen. *ATZ Automobiltechnische Zeitschrift* 97. S 400–15

Woernle C (1988) Ein systematisches Verfahren zur Aufstellung der geometrischen Schließbedingungen in kinematischen Schleifen mit Anwendung bei der Rückwärtstransformation für Industrieroboter [Dr.-Ing.]. Dissertation, Universität Stuttgart, Düsseldorf: VDI-Verlag

Chapter 7
Modeling of the Road-Tire-Contact

A sufficiently accurate knowledge of the interaction between the tires and the road surface is of vital importance to describe and evaluate the dynamics of vehicles. The contact road-tire is, apart from the influence of aerodynamic factors, the only possibility to actively influence the motion of the vehicle. Here all the forces and the torques will be transmitted over a post card sized patch called the tire contact patch.

The wheel is defined as the rotating components, spinning about the wheel spin axis, of the vehicle undercarriage. These include the wheel carrier, which is assumed to be rigid, the rotating parts of the brake, parts of the drive shaft and the drive train as well as the tires. The wheel is fitted to the suspension using a wheel bearing. The wheels have three fundamental properties that, depending on the application, have to be represented in the modeling of the wheel forces:

- the absorption of the wheel loads and the protection of the other vehicle components and the passengers from impact loads,
- the transmission of acceleration and braking forces and
- the lateral forces during cornering.

Physically this means the transfer of forces and torques in all three spatial directions. In the process the wheel is assigned to two functionally separate subsystems of the vehicle. On the one hand, it is part of the suspension and on the other, a part of the drive train. In both these areas, it is the last element of the causal chain and as such is the immediate interface to the road surface.

Modern tires are, from the production point of view, elaborately designed viscoelastic forms and represent, from the modeling point of view in vehicles, force components with complex nonlinear and dynamic characteristics. The degree of complexity of the tire models, for applications in vehicle models, ranges thus from the simple linear to the very complex nonlinear models, depending on the situation and the application. This chapter deals with the suitable mathematical models required to describe the transfer characteristics of the tires, without delving too

D. Schramm et al., *Vehicle Dynamics*,
DOI 10.1007/978-3-662-54483-9_7

much into the physical characteristics of the tire. For this purpose it shall be referred to the comprehensive literature available in this topic. In Ammon (2013) the fundamental mathematical characteristics of the tires are dealt with. In Pacejka (2006) one can find a comprehensive introduction to the mathematical tire models as well as an overview over the methods to measure the required data. The construction, history and the fundamental characteristics of tires and wheels is described in Reimpell and Sponagel (1988) or in Leister (2009). Comprehensive information along with the description of known tire models can be found in Gipser (2010).

While in the case of vehicle dynamics models, the characteristic curves of the tires are often measured first on the test rig and then replicated, as exact as possible, through a model (the empirical method), the physical modeling is based on the knowledge of the exact mechanism, by which the forces are generated. The physical approach requires a lot more computational effort when compared to the models used in vehicle dynamics. Another difference lies in the fact that the vehicle dynamics models are suitable to recreate the stationary and non-stationary tire characteristics in the frequency range up to 20 Hz of vehicle dynamics (modeling of low frequency forces and deformations). Compared to this the comfort models can represent high dynamic situations up to 80 Hz and more (for example: vibrations on uneven surfaces). Through this it is possible to predict the outcome of working points that cannot be measured realistically. Depending on the job description one is to choose the tire model, which offers the best compromise between computational time and performance.

7.1 Tire Construction

The basic construction of a radial tire is shown in Fig. 7.1. From the mathematical description of the vehicle dynamics point of view the following components of the tire are important:

- The tread (1) is made of rubber and contains the tire tread profile consisting of the tread knobs and tread grooves.
- The carcass (2) consists of tensile surfaces covered in rubber, made up of e.g. artificial silk, nylon, and rayon. The carcass along with the tire pressure gives the tire its strength. It runs transversally to the rolling direction, radially from bead ring (4) to bead ring.
- The belt (3) these days is usually a composite layer of steel that rests on the tread surface of the carcass. It encloses the tire from the outside and gives the tread its strength.
- The two bead rings (4) ensure a tight fit of the tire on the wheel and guarantee, along with the enclosed rubber, a seal between the tire and the rim.

While the scope and depth of the tire models available today are very different, all models rely on kinematic variables as input parameters to calculate the tire forces. For this reason, the preparation of these input variables is described in detail in Sect. 7.4.

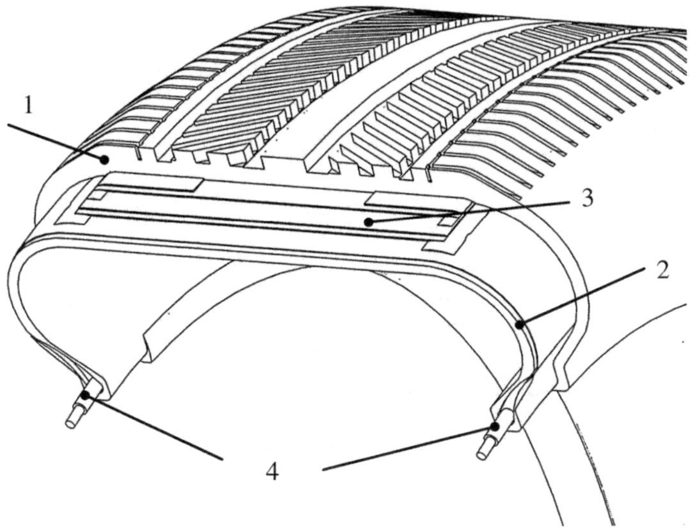

Fig. 7.1 Basic construction of a radial tire

7.2 Forces Between Wheel and Road

The procedure for the integration of the tire forces in a simulation model based on the multi-body system method consists of several steps:

- Position and velocity of the contact point: these quantities are determined from the known kinematic position and velocity of the wheel carrier and the wheel according to Chaps. 5 and 6. To this end the methods discussed in Chaps. 2 and 3 will be applied.
- Kinematic tire model: the slip quantities in both the longitudinal and the transversal directions as well as the boring slip from the kinematic variables using the known road surface data.
- Physical tire model: in the final step, starting from the slip variables and using a physical or mathematical tire model, the forces and, if appropriate, the torques acting on the contact surface of the tire will be calculated.

7.3 Stationary Tire Contact Forces

The force transmission between tire and road surface takes place through the tire contact patch (tire contact area). This is formed due to the tire load in the contact area between the tire and the road surface. It is to be noted here, that depending on the pattern of the tread, only the tread lugs are in direct contact with the road

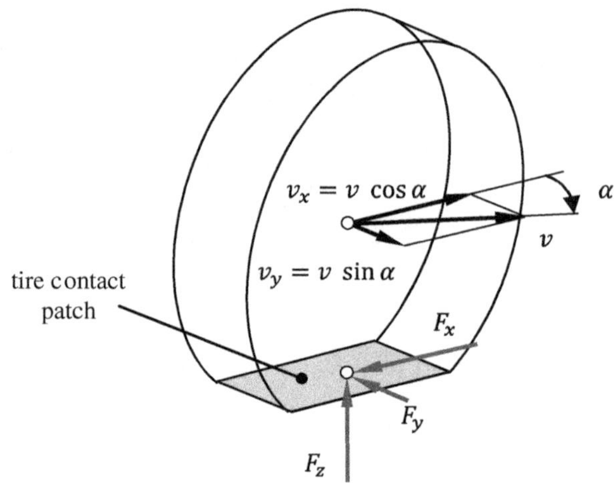

Fig. 7.2 Contact forces between road surface and tire

Table 7.1 Components of the contact forces at the tires

Direction	Force description	Symbol
Longitudinal, circumferential	Longitudinal force, circumferential force	F_x
Transversal, lateral	Transversal force, side force, lateral force	F_y
Vertical	Vertical force, upward force, tire contact force	F_z

surface. As part of the modeling of the vehicle as a multibody system, the forces between the tire and the road surface will be described through resulting (single) contact forces and torques. In this case, it generally deals with applied forces and torques, i.e. they can be determined as a function of position and velocity variables. According to Fig. 7.2 one can decompose the contact force into three components (Table 7.1).

The decomposition of the forces into the different spatial directions is thereby relative to a wheel fixed coordinate system, i.e. the underlying coordinate system does not rotate about the axis of the wheel rotation. Accordingly, the current velocity of the wheel center v is split into the longitudinal component $v_x = \cos\alpha$ and the transversal component $v_y = \sin\alpha$ (Fig. 7.2).

In the z-direction there is a unidirectional force transmission, since only compressive forces are transmitted. The force components in the longitudinal and lateral directions which are parallel to the road surface are generally transmitted through friction. They are therefore, dependent on the material pairing tire/road surface, limited in value.

7.3.1 Tires Under Vertical Loads

The tire acts, physically, predominantly as an air spring in the vertical direction. Through the vertical tire load F_z a contact area is formed at the contact region with the road surface (contact patch). In the case of a standing wheel a static deflection of

$$\Delta r = r_0 - r_{stat} \tag{7.1}$$

results, with the construction radius of the tire r_0 and the so-called static tire radius r_{stat}, Ref. Fig. 7.3. Thereby the vertical surface pressure p_z is distributed symmetrically in the contact patch. From the radii r_{stat} and r_0 a rough approximation of the contact length L can be calculated. According to the Fig. 7.3 with the tire deflection given by $|\Delta r| = |r_0 - r_{stat}| \ll r_0$ one arrives at the equation:

$$\left(\frac{L}{2}\right)^2 = r_0^2 - (r_0 - \Delta r)^2 \approx 2r_0\Delta r \Rightarrow L \approx 2\sqrt{2r_0\Delta r}. \tag{7.2}$$

The wheel load F_z depends primarily on the tire deflection $r_0 - r_{stat}$. The deformation of the tire leads to a weakly progressive characteristic curve of the spring, Ref. (Ammon 2013):

$$F_z = \begin{cases} c_z\Delta r + c_{z,3}\Delta r^3 \text{ für } \Delta r \geq 0 \\ 0 \text{ für } \Delta r < 0 \end{cases}. \tag{7.3}$$

Here the progressive wheel load component is usually small when compared with the linear component. For normal applications, it is thus sufficient to use a linearized form of the Eq. (7.3), linearized about a working point. Care should be

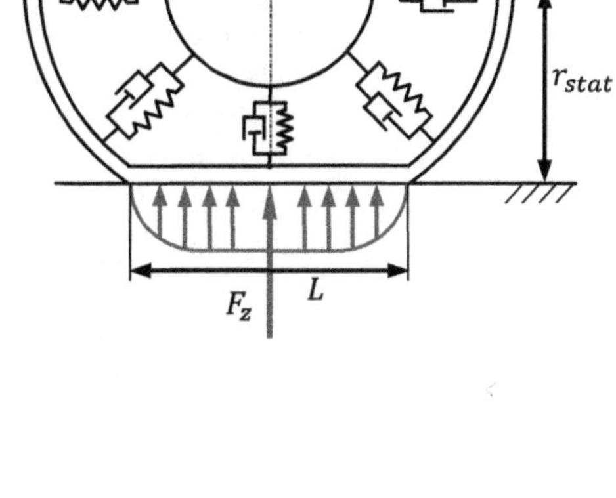

Fig. 7.3 Stationary tire under normal load

taken however to include the case $F_z = 0$ (lifting of the wheel) which can occur and which should therefore be considered.

7.3.2 Rolling Resistance

In the following a simple model will be presented to explain principally the fundamentally relevant effects of a rolling wheel required in vehicle dynamics, Fig. 7.4. To this end the belt of the tire is connected to the wheel carrier through spring and damper elements. The dampers account for the damping characteristics of the wheel rubber. This leads to a velocity dependent increase in normal pressure at the front part of the contact area along the direction of rolling and to a decrease in pressure at the rear part of the contact area. The value of the resulting normal force F_z does not change in any significant effect due to the rolling of the wheel. Hence, even in the case of the rolling wheel, the static tire radius r_{stat} is defined as the distance between the assumed planar tire contact area and the wheel axis of rotation. Due to the forward displacement of the surface pressure, the fictional point at which the normal force acts is also displaced in the direction of rolling by a value e from the center of the contact area. Hence there results a torque balance of the acceleration-less rolling wheel given by

$$F_x r_{stat} - F_z e = 0 \tag{7.4}$$

and the rolling resistance force is given by

$$F_R = F_x = \frac{e}{r_{stat}} F_z. \tag{7.5}$$

This rolling resistance force must be compensated using the vehicle drive torque.

7.3.3 Tires Under Longitudinal (Circumferential) Forces

Adhesion coefficient Fundamentally two physical effects are responsible for the transmission of longitudinal forces in the contact area (Gillespie 1992) (Fig. 7.5):

- Traction (adhesion) through adhesion friction in the contact patch: surface adhesion occurs due to the intermolecular bonding forces between tire rubber and the material of the road surface. While the effect dominates the force generation between the tire and the road surface on dry roads, it significantly reduces on wet roads.
- Hysteresis friction: this leads to (positive) locking through the meshing effects between the tire contact patch (tread) and the road surface. This effect is due to the viscoelastic material characteristic of the tire rubber. The large damping of

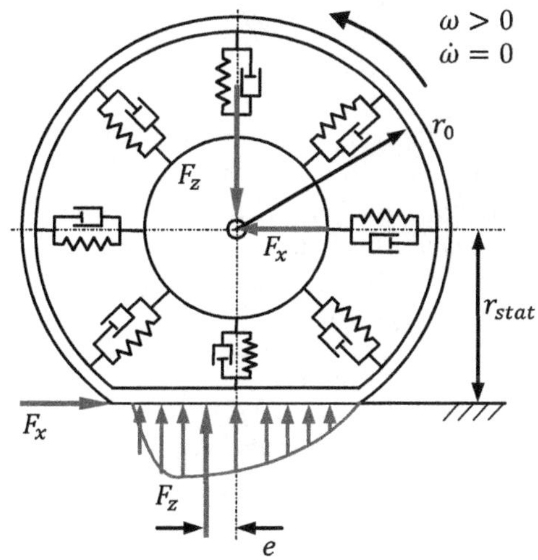

Fig. 7.4 Description of the rolling resistance

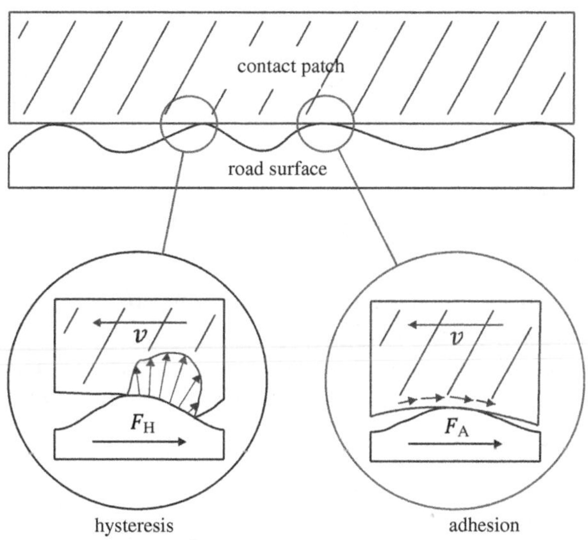

Fig. 7.5 Description of the forces between road and tires

the rubber in the contact area results in a high hysteresis friction coefficient. This effect is much less affected by water on the road surface.

Both effects depend on small relative motion between the contact partners tire tread and the road surface of the contact area. The ratio between the longitudinal force F_x and the normal force F_z is defined as the adhesion coefficient in the longitudinal direction

$$\mu = \frac{F_x}{F_z}. \tag{7.6}$$

The formation of the longitudinal forces can be described through the shear deformation of the tread in conjunction with the friction behavior between the tread and the road. This requires, however, a macroscopic description of this shear deformation, which can be incorporated into, for example, the multibody system formalism. The kinematics of the shear deformation will be described through the term longitudinal slip.

Circumferential slip (longitudinal slip) The circumferential slip is a kinematic quantity, which describes the state of motion of a driven, braked or non-driven rolling wheel. The wheel here is assumed to be a rigid body. This is why it is called also rigid body slip.

In a planar motion of a rigid wheel one distinguishes fundamentally between two states of motion

- pure kinematic rolling without sliding and
- combined rolling and sliding,

Table 7.2. In vehicles the second case is relevant. First the motion of a planar wheel with the variables

- wheel radius r,
- velocity v of the wheel center point,
- velocity v_P of the (fictional) wheel contact point,
- angular velocity ω of the wheel

will be investigated. In an ideal rolling wheel the velocity v_P of the contact patch would be vanishing. In a real wheel (tire) this is not the case. The deviation, relative to the wheel speed is referred to as the slip. The acceleration slips_A and the brake slips_B are differentiated below. To this end the value of the relative velocity of the

Table 7.2 Rigid body slip

$v = \omega r$	$v < \omega r$		$v > \omega r$	
Rolling wheel	Driven wheel	Spinning wheel	Braked wheel	Blocked wheel
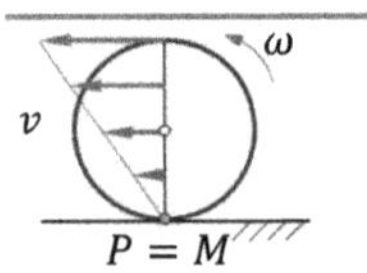	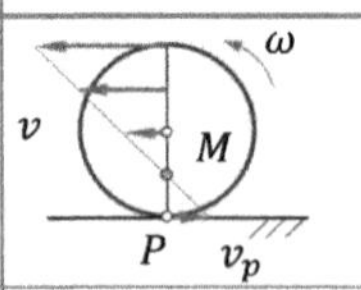	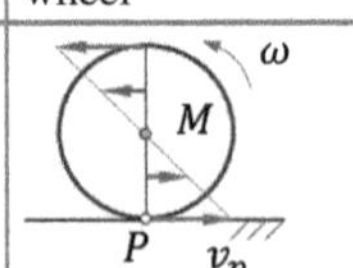	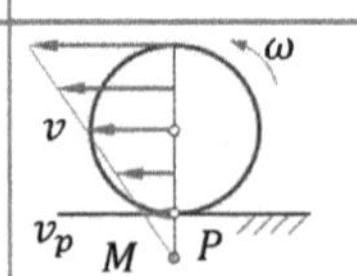	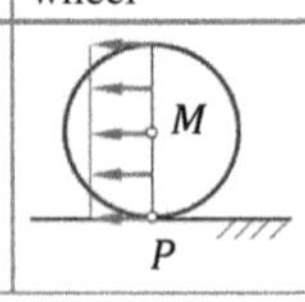
No slip	Drive slip		Brake slip	
$s_A = 0$ $s_B = 0$	$s_A = \frac{v_P}{\omega r} = \frac{\omega r - v}{\omega r}$	$s_A = 1$	$s_B = \frac{v_P}{v} = \frac{v - \omega r}{v}$	$s_B = 1$

fictional tire contact point P is given relative to the larger of the two values v and ωr. For a driven wheel the acceleration slip $(v < \omega r)$ is given by

$$s_A = \frac{v_P}{\omega r} = \frac{\omega r - v}{\omega r} \tag{7.7}$$

and for a decelerating wheel the brake slip $(v > \omega r)$ is given by:

$$s_B = \frac{v_P}{v} = \frac{v - \omega r}{v}. \tag{7.8}$$

Summarizing one can write:

$$s_{A,B} = \frac{|v_P|}{\max(v, \omega r)} = \frac{|v - \omega r|}{\max(v, \omega r)}. \tag{7.9}$$

The slip is mostly given in percentage, i.e. for example $s_A = 90\,\%$ instead of $s_A = 0.9$. The slip definition used here guarantees that the value of the slip in the extreme conditions for blocked wheel $(v \neq 0, \omega = 0)$ and for a spinning wheel $(v = 0, \omega \neq 0)$ takes the value of one. Thus the slip is normalized within the interval [0, 1]. Special considerations require the cases of startup and motion at very low velocities. This is dealt with in detail in Schuster (1999). There can be found several other definitions of slip in the literature, Ref. for example (Rill 1994) and again (Schuster 1999). This is to be considered when using simulation software. This is especially true for the limiting cases of blocked and spinning wheels. All definitions are similar however in that the slip of the rolling wheel is zero.

In the following these definitions of slip will be applied to radial tires. To this end the dynamic wheel radius determinable from the measurements of the rolling circumference will be first defined.

Rolling circumference and dynamic tire radius The rolling motion of a radial-ply tire is significantly different from that of a cross-ply tire, which is no longer relevant in modern passenger vehicles. The belt of a radial tire behaves as a non-elastic band along the outer tire structures due to the non-elastic nature of the steel wires, when compared with the surrounding rubber. Even under the most extreme loading conditions that can be encountered, the longitudinal elongation of the belt is under 1 % (Reimpell and Sponagel 1988). It is therefore justified in assuming the rolling circumference U, in normal operating range, is a constant. The rolling circumference U is unwound on the road during one revolution. This quantity will be customarily measured on a tire towed with a velocity of 60 km/h, without drive or brake forces, and will be given in tire tables (DIN 1986). The rolling circumference U defines the dynamic tire radius

$$r_{dyn} = \frac{U}{2\pi}. \tag{7.10}$$

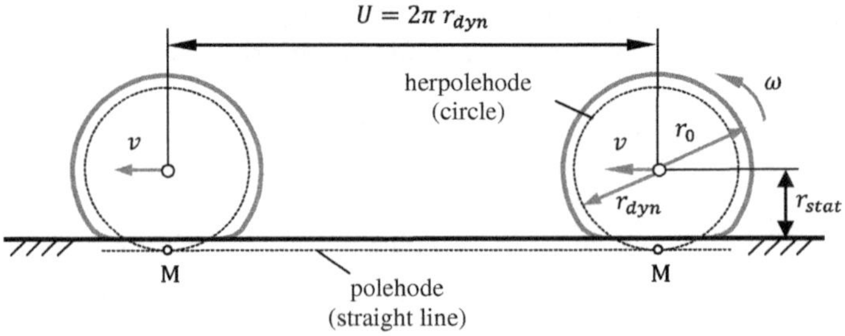

Fig. 7.6 For the definition of the dynamic tire radius

Hence r_{dyn} is the radius of an imaginary wheel-fixed circle, which on rolling over one revolution delivers the measured rolling circumference U. As the longitudinally stiff belt lies, not on the surface, but below the carcass of the tire, the dynamic tire radius r_{dyn} is smaller than the manufacturing radius r_0. In belted tires the dynamic tire radius r_{dyn} is closer to the construction radius r_0 than to the statistical tire radius r_{stat}. The relationship between the radii r_0, r_{stat} and r_{dyn} can be estimated using a simple geometric consideration. To this end, one compares the unwinding length of an imaginary rigid disc, with a radius r_{dyn}, with the half contact patch length $L/2$ (Fig. 7.6):

$$U = r_{dyn}\alpha = r_0 \sin\alpha \rightarrow r_{dyn} = r_0 \frac{\sin\alpha}{\alpha} \approx r_0(1 - \frac{\alpha^2}{6}). \tag{7.11}$$

If one now also considers the geometric relationship (see Fig. 7.3)

$$r_{stat} = r_0 \cos\alpha \approx r_0\left(1 - \frac{\alpha^2}{2}\right) \rightarrow \frac{\alpha^2}{2} \approx (1 - \frac{r_{stat}}{r_0}) \tag{7.12}$$

and substitutes this relationship in Eq. (7.11), one obtains

$$r_{dyn} \approx r_0\left(1 - \frac{\alpha^2}{6}\right) \approx r_0\left(1 - \frac{1}{3} + \frac{r_{stat}}{3r_0}\right) = \frac{2}{3}r_0 + \frac{1}{3}r_{stat}. \tag{7.13}$$

The wheel fixed circle with radius r_{dyn} is hence the herpolehode of the motion. The imaginary road-fixed straight line on which the tire unwinds is called the polehode. The instantaneous contact point M of the herpolehode and the polehode is the instantaneous center of rotation. One should note that the polehode lies under the road surface. In Table 7.3 are a few examples for statistic and dynamic tire radii are given (Reimpell and Sponagel 1988).

The difference between the statistic and dynamical radii is not, or rather is not primarily, due to the expansion of the tire belt due to the centrifugal forces, which are relatively small.

Table 7.3 Examples of statistical and dynamical tire radii

Tires	r_0/mm	r_{stat}/mm	U/mm	r_{dyn}/mm
185/65R15	311	284	1,895	302
195/65R15	318	290	1,935	308
205/65R15	324	294	1,975	315

Circumferential slip of a wheel with belted tires From now on, the rolling of the tire with a radius r_{dyn} will be assumed to be slip-less, i.e. it holds $s_{A,B} = 0$. As such, a compromise will be accepted, that this normalization of the slip is not absolutely correct, as the dynamic tire radius is valid only for the specific velocity, at which the experiment was performed. The advantage is however, that the state slip-free is simple and comprehensible using a single standard experiment, without introducing or considering contact mechanical processes in the contact patch. Furthermore it allows the definition of the slip of a rigid tire:

$$\text{drive Slip } \left(v < \omega r_{dyn}\right): \quad s_A = \frac{\omega r_{dyn} - v}{\omega r_{dyn}} \tag{7.14}$$

$$\text{brake Slip } \left(v > \omega r_{dyn}\right): \quad s_B = \frac{v - \omega r_{dyn}}{v} \tag{7.15}$$

Summarizing one can write:

$$s_{A,B} = \frac{\left|v - \omega r_{dyn}\right|}{\max\left(v, \omega r_{dyn}\right)}. \tag{7.16}$$

This kinematic analogous model of the slip-less rolling wheel is extended such that the rolling of the treads on the road is visible and describable. This helps to visualize and subsequently formalize the emergence of the circumferential forces on the braked, and the driven wheel respectively. The belt, with the tread, is considered here to be a circumferentially stiff closed band with a circumference $U = 2\pi r_{dyn}$, see (Ammon 2013). Comparable to the tread of a tracked vehicle, this band is considered in the upper areas of the rigid wheel disc as a materialized herpolehode with a radius r_{dyn} and in the patch area through an imaginary guidance plate parallel to the road surface (patch length L, entering point E, and exiting point A). Since the wheel disc, in the lower region of the guidance plate, would penetrate the circumferential belt and the road surface, it has to be dematerialized again in this region.

The treads carried by the belt are represented through elastic beam elements and profile stud elements respectively (Brush Model, Fig. 7.7). The wheel center has a velocity v, and the wheel disc has an angular velocity ω. Hence the belt runs with a

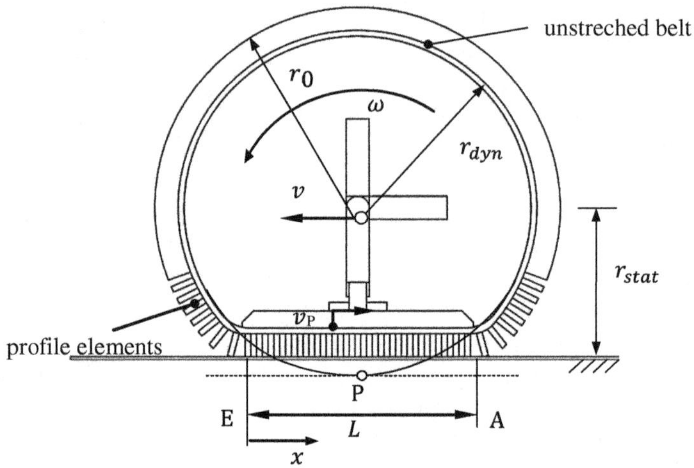

Fig. 7.7 Mechanical analogous model for a belted tire per (Ammon 2013)

longitudinal velocity of ωr_{dyn}. The belt section $\overline{\text{EA}}$ has, due to the assumed stiffness of the belt, in the patch the absolute velocity (sliding velocity):

$$v_P = v - \omega r_{dyn}. \tag{7.17}$$

This slip-less motion of the wheel, with a belted tire, is thus plausibly represented using this model:

- Slip-less motion: $v = \omega r_{dyn} \rightarrow v_P = 0$: The belt section $\overline{\text{EA}}$ is at rest. There occurs no shear or sliding motion of the profile elements.
- Driven wheel $v < \omega r_{dyn} \rightarrow v_P < 0$: There is an acceleration slip as per (7.14). The profile elements in the contact patch are sheared to the front.
- Braked wheel: $v > \omega r_{dyn} \rightarrow v_P > 0$: There is a brake slip as per (7.15). The profile elements will be sheared to the back.

Force transmission in the circumferential direction To explain the principle of the development of the circumferential forces, the motion of a single profile element through the patch area $\overline{\text{EA}}$ will first be investigated. The case of a driven wheel will be assumed exemplary: $v < \omega r_{dyn}$. This view is also valid for a braked wheel (Fig. 7.7).

Motion and shear deformation of a profile element One observes for the time being a single profile element on its path through the contact patch Fig. 7.8. This profile element reaches at time $t_0 = 0$ the entry point E $(x = 0)$ of the contact area $\overline{\text{EA}}$. Accordingly, it will be transported through the patch with the constant

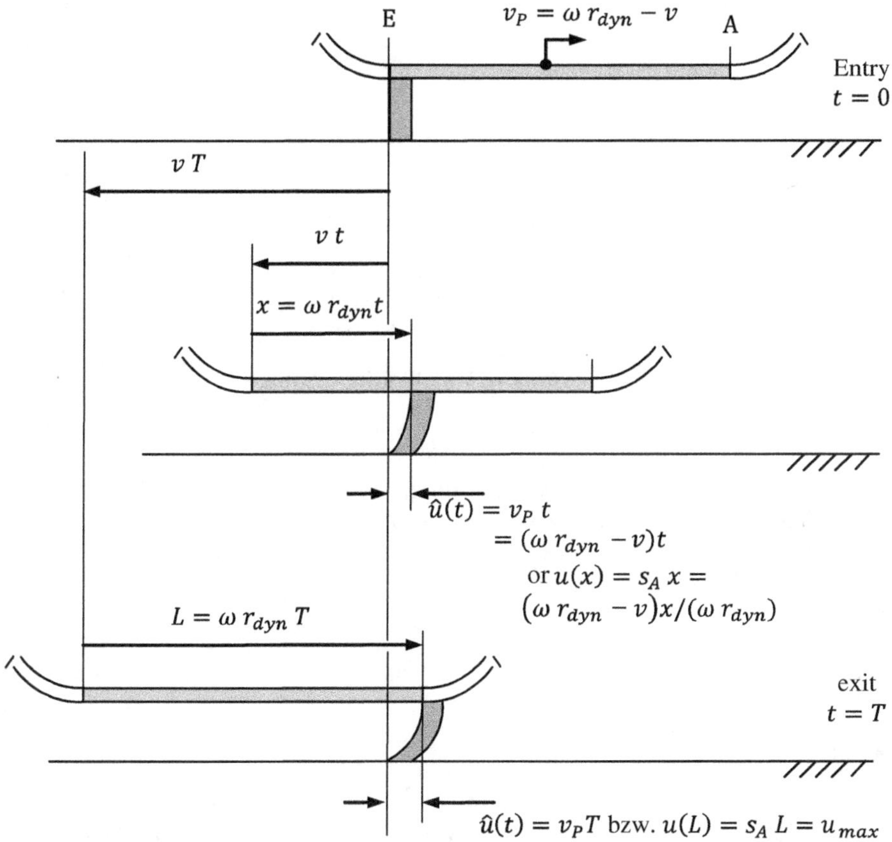

Fig. 7.8 Path of a profile element in a tread

(acceleration-free motion of the wheel) circumferential velocity $\omega r_{dyn} > v$ of the belt. At time t the profile will then be found at the position in the contact patch at:

$$x = \omega r_{dyn} t. \tag{7.18}$$

A probable assumption is now, that the regarded profile element at the entry into the contact surface at time $t = 0$, and also for a certain time after that, is adhered to the road surface. As the belt section $\overline{\text{EA}}$ has the absolute velocity $v_P = \omega r_{dyn} - v$ opposite to the driving direction v, this element will be deformed in the horizontal direction through a stretch $\hat{u}(t)$:

$$\hat{u}(t) = v_P t = \left(\omega r_{dyn} - v\right)t. \tag{7.19}$$

If one now represents the time t through x as in (7.18) and substitutes the driving slip s_A from (7.14), one arrives at a linear relationship with x, with gradient s_A, resulting in an increase of the deformation in the direction of the entry point E:

$$u(x) = \left(\omega r_{dyn} - v\right)\frac{x}{\omega r_{dyn}} = s_A x \tag{7.20}$$

the element reaches the exit point A $(x = L)$ after a time

$$T = \frac{L}{\omega r_{dyn}} \tag{7.21}$$

under the patch and experiences the maximum deformation there

$$u_{max} = u(L) = s_A L. \tag{7.22}$$

Tangential stress on the profile elements If one assumes a linear-elastic behavior of the deformation of the profile element, then the profile element transfers a shear stress in the length dx and breath dy, that is proportional to deformation $u(x)$:

$$\tau_x(x) = ku(x) \tag{7.23}$$

with a corresponding constant k. The constant depends on

- the shear modulus of the elastic material,
- the height of the tread and
- the profile.

The assumption of a continuous stress distribution over the entire path surface does not consider any tire profile and is thus only an approximation.

The resulting circumferential force F_x transferred by all the profile elements in the contact surface A is, under the assumption of a constant patch breadth b,

$$F_x = \int\limits_A \tau_x(x)\,dA = \int\limits_0^L \tau_x(x)b\,dx = \frac{1}{2}kbL^2 s_A. \tag{7.24}$$

The force F_x is proportional to the triangular surface under the tangential stress distribution. The relationships (7.23) and (7.24) are only valid when the profile element is in adhesion over the entire length of the patch L. This is not generally applicable as through the friction behavior certain deviations have to be considered.

Adhesion through COULOMB's friction For the adhesion between the profile elements and the road surface a coulomb friction will be assumed. Let the normal pressure distribution $p_z(x)$ be constant over the contact patch breath b and decreasing to the points E and A. The resulting vertical force (tire load) is then

$$F_Z = \int_A p_Z(x)\, dA = \int_0^L p_Z(x) b\, dx. \tag{7.25}$$

For the tangential stress $\tau_x(x)$ the adhesion condition with the adhesion coefficient μ_H:

$$\tau_x(x) \le \tau_{xH}(x) \quad \text{with} \quad \tau_{xH}(x) = \mu_H\, p_Z(x). \tag{7.26}$$

If the tangential stress surpasses the adhesion boundary at a point x, then the profile element begins to slide at this point. Therefore, instead of an adhesion stress, a sliding stress will occur

$$\tau_{xG}(x) = \mu_G\, p_Z(x) \tag{7.27}$$

with the sliding coefficient μ_G. For a friction pair rubber-asphalt the following rule applies

$$\mu_G < \mu_H \rightarrow \tau_{xG}(x) < \tau_{xH}(x) = \mu_H\, p_Z(x). \tag{7.28}$$

The contact area $\overline{\text{EA}}$ is thus divided into an adhesion zone $\overline{\text{EG}}$ and a sliding zone $\overline{\text{GA}}$. With increasing slip s_A the adhesion boundary G moves from the exit point A to the left. The resulting transferred circumferential force relates to the surface under the resulting tangential stress distribution, see Fig. 7.9.

Slip-force curve The transferred circumferential force F_x can only be described in relation with the circumferential slip $s_{A,B}$. As a general rule, this results in a similar

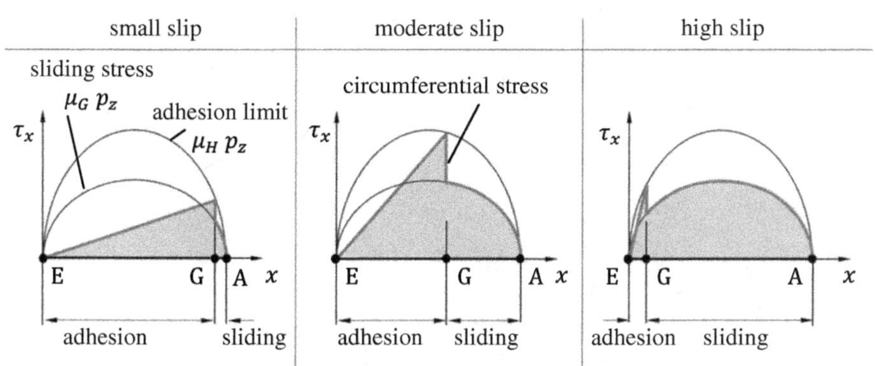

Fig. 7.9 Circumferential stress in the slip area according to Ammon (2013)

curve, that is however different, dependent on the tires, with respect to the maximum and the slope.

The typical curve force of this slip-force curve can be described, based on the preliminary work done in this chapter, as follows:

For small slip values the adhesion region extends almost over the entire patch length. The circumferential force F_x increases linearly thereafter, due to the assumptions made and according to the Eq. (7.24). It is therefore given as

$$F_x = c_s s \tag{7.29}$$

with the circumferential stiffness of the tire

$$c_s = \frac{1}{2} k b L^2. \tag{7.30}$$

If the longitudinal slip increases further, the sliding region increases disproportionately and the resulting force $F_x(s)$ between the road and the tire exceeds the maximum possible force $F_{x,max}$. This maximum value defines the maximum adhesion coefficient (Fig. 7.10).

$$\mu_{max} = \frac{F_{x,max}}{F_Z}. \tag{7.31}$$

This maximum force adhesion coefficient μ_{max} is significantly smaller than the adhesion coefficient μ_H of the material pair rubber and asphalt, that can be as high as 2 under the right conditions. It therefore always holds that:

$$\mu_{max} < \mu_H. \tag{7.32}$$

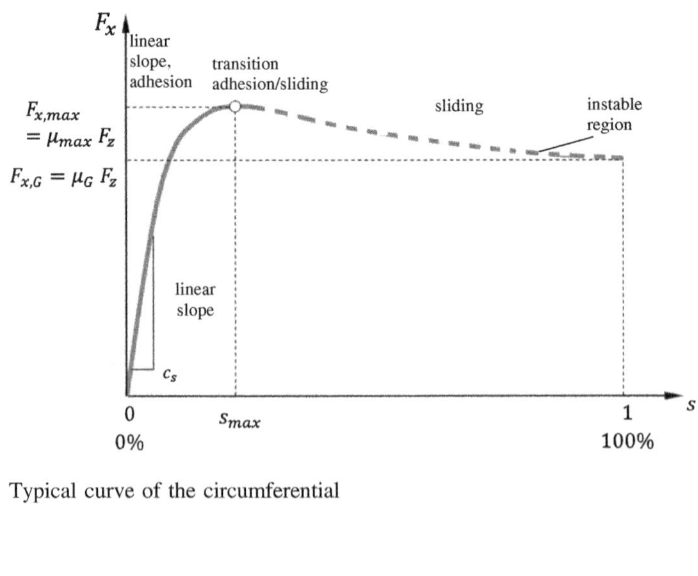

Fig. 7.10 Typical curve of the circumferential

The reason for this is that the maximum value of the adhesion friction in the patch can be achieved only in the transition zone between the adhesion and sliding friction. At very high slip values s the transferred circumferential force F_x decreases to the value $F_{x,G}$, which occurs during pure sliding. That is the case when the wheel, either blocks during braking or spins during acceleration. The force adhesion coefficient corresponds to the sliding friction coefficient

$$\mu_G = \frac{F_{x,G}}{F_Z}. \tag{7.33}$$

Characteristics of the circumferential curve are:

- the initial slope $c_s = \left[\frac{dF_x}{ds}\right]_{s=0}$,
- the slip value s_{max} and the corresponding maximum value of the slip force $F_{x,max}$,
- as well as the sliding force $F_{x,G} = \mu_G F_z$.

When no lateral forces need to be transmitted, then the circumferential slip curve will be influenced by the following variables:

- Constructive tire parameters (material, construction type, measurements). A significant influence is the shear stiffness of the tread strip. With increasing stiffness, the s_{max} moves towards smaller values of slip, whereas, as a rule, the elevation of μ_{max}/μ_G increases, its breath however decreases. The characteristic form of the slip curve is mostly determined through the material, the construction and the measurements of the tire. The two other externally influencing parameters are the tire load F_z and the force adhesion coefficients μ_{max} and μ_G, which distort the fundamental form in different ways.
- Tire normal loadF_z: increasing the tire load F_z leads to the most part to a change of scale of the ordinates of the slip curve. i.e. the position of the maximum is preserved. This can, on the one hand, be explained that with increasing tire load F_z the elastic deflection f and with it the patch length L increases, which in turn results in a bigger shear stiffness of the patch and hence leads to an increase in the initial slope c_s of $F_x(s)$. On the other hand, the force adhesion coefficients μ_{max} and μ_G remain principally the same. Through this the limits of the transferable circumferential forces increases proportionally to F_z. Up to a statistical operating tire load the forces $F_{x,max}$ resp. $F_{x,G}$ change proportionally to F_z. With increasing tire load the values $F_{x,max}$ resp. $F_{x,G}$ only increase slightly degressive. This is due to a deterioration of the friction coefficients resulting from the deformation of the tire structure. Due to this approximate proportionality between F_x and F_z, one often plots the ratio F_x/F_z over the slip s.
- Maximum force adhesion coefficient μ_{max} and the sliding friction coefficient μ_G : if μ_{max} and μ_G were modified proportionally, then the abscissa and the ordinate will be scaled similar to stretching, with respect to the coordinate origin. The initial slope of $F_x(s)$ will be defined primarily through the shear

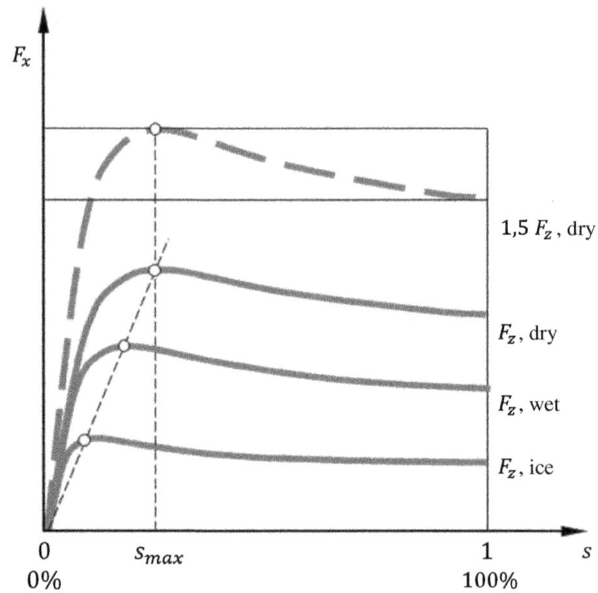

Fig. 7.11 Relationship of the force adhesion with respect to tire load and friction coefficient

stiffness of the tread and hence will remain the same. The limit $F_{x,max}$ however, will move itself in the direction of the larger slip values, see Fig. 7.11.

A fundamental problem in the modeling of the tire behavior is that the tire parameters vary within broad limits even in normal driving situations.

7.3.4 *Tires Subjected to Lateral Forces*

Slip angle and lateral slip A free rolling wheel (no circumferential force) that is acted upon by a lateral force F_y gets a velocity component lateral to the rolling direction. The angle formed between the direction of motion of the tire center point and a vector, that on the one hand lies in the cross-sectional area of the wheel and on the other hand parallel to the road, is called the slip angle α. Hence according to Fig. 7.2:

$$\tan\alpha = \frac{v_y}{v_x} \text{ and } \sin\alpha = \frac{v_y}{v}. \tag{7.34}$$

Lateral force The variable $\tan\alpha$ is also known as the lateral slip or skew slip and the angle α as the slip angle. In normal driving conditions $|\alpha| < 12°$ and the slip angle and the lateral slip can be set to be the same as a good approximation (Heißing and Ersoy 2011). Fundamentally one can assume that the tread exhibits similar deformation characteristics in the longitudinal and the lateral directions,

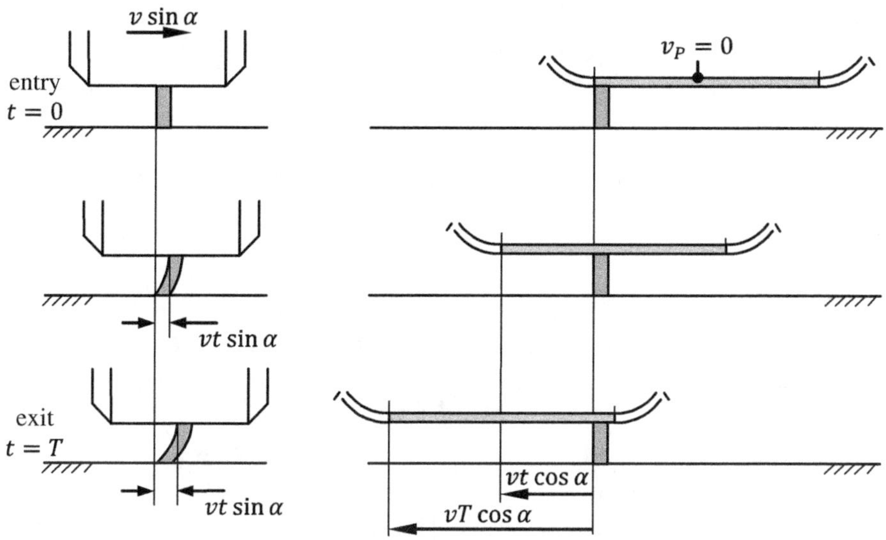

Fig. 7.12 Transmission of side forces

such that the described model descriptions can be used well to model the transmission of the lateral forces. If one were to assume that a profile element running under the patch adheres initially to the surface, then this element will be deformed laterally as it moves through the patch, see Fig. 7.12. At small slip angles the element adheres till the exit out of the patch and the tangential stress in the lateral direction of the tire increases almost linearly till the posterior patch boundary. The lateral force $F_y$is proportional to the surface under the curve of the tangential stress, and hence it follows:

$$F_y = \int\limits_A \tau_y(x)dA = \int\limits_0^{L\cos\alpha} k_\alpha x \tan\alpha b dx = \underbrace{\frac{1}{2} b k_\alpha L^2}_{c_\alpha} \sin\alpha\cos\alpha$$
$$= c_\alpha \sin\alpha\cos\alpha \tag{7.35}$$

with the cornering stiffness of the tire c_α. The constant k_α describes the material characteristic of the tread. For small slip angles α it is then valid

$$F_y = \left[\frac{dF_y}{d\alpha}\right]_{\alpha=0} \alpha = c_\alpha \alpha. \tag{7.36}$$

Analogous to the behavior of the circumferential forces, the lateral force also increases proportional to the slip angle α initially. At larger slip angles the deflection of the patch elements and hence the tangential stress will become so great towards the posterior patch boundary such that the adhesion friction limit will be

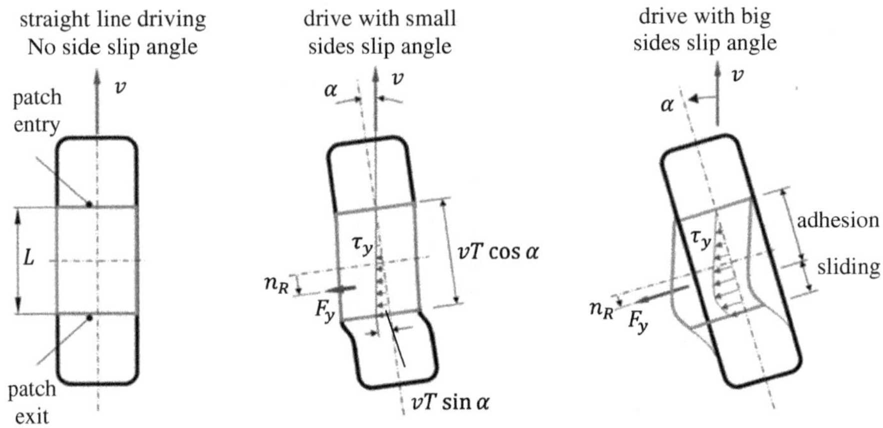

Fig. 7.13 Wheel under lateral forces according to Zomotor (1991)

exceeded and sliding begins. Similar to the circumferential transmission one can differentiate the patch surface into an adhesion region and a sliding region. The lateral force no longer increases linearly; instead it does so digressively with respect to the slip angle. With increasing α the sliding region in the patch moves towards the front, till it (theoretically) encompasses the entire patch surface at $\alpha = 90°$.

Caster and aligning torque As a result of the increasing lateral stress in the patch in the direction of the patch exit, the resulting lateral force acts at a point at a distance of n_R behind the middle of the patch surface, see Fig. 7.13. This leads to a torque, which tries to reduce the slip. In total there results an aligning torque opposing the slip angle α given by

$$M_z = n_R F_y. \tag{7.37}$$

Here n_R describes the so-called tire caster. The aligning torque M_z increases linearly for small slip angles α:

$$M_z = c_{M,\alpha}\alpha \quad \text{with} \quad c_{M,\alpha} = n_R c_\alpha = const. \tag{7.38}$$

With the patch length L one can, as a result of the triangular stress distribution (where the resulting force acts at the center of gravity of the triangle) in the patch in the linear area, estimate the caster for small slip angles as

$$n_R \approx \frac{1}{6} L. \tag{7.39}$$

With increasing sliding region the center of gravity of the surface moves under the tangential stress distribution in the direction of the patch center. With this one gets $n_R \to 0$ and the aligning torque tends to zero for larger slip angles, Ref

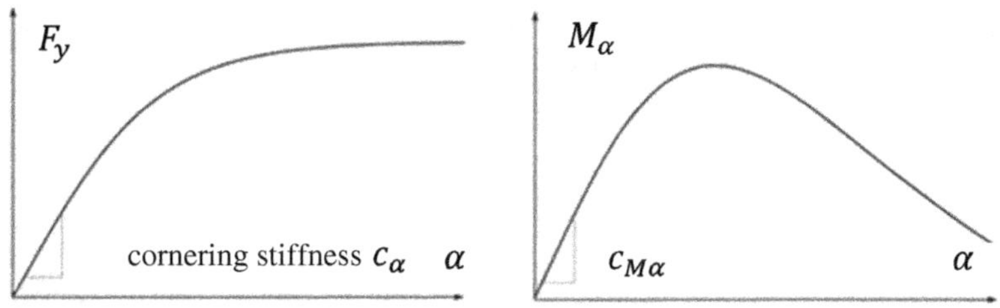

Fig. 7.14 Lateral force and aligning torque

Fig. 7.14. In extreme cases n_R can even become negative and even the aligning torque can change its sign.

7.3.5 *Influence of the Camber on the Tire Lateral Force*

The camber angle γ introduced in Chap. 6 has an influence, however small, on the tire lateral forces. To estimate its contribution to the lateral force one considers a wheel, which rolls with a camber angle γ. Let us assume the lateral force depending on this camber angle be $F_{y,\gamma}$. Its formation is explained in the Fig. 7.15. A free rolling wheel, under the influence of the camber angle γ, will, without any other external influences, move along a circular path about an imaginary apex O. The wheel will however be forced to move along the x-direction due to the wheel suspension. This requires a lateral force $F_{y,\gamma}$ as well as a steering torque $M_{z,\gamma}$, which causes, the wheel to turn in a circular path. For small camber angles $|\gamma| < 5^{\circ}$ the

Fig. 7.15 Formation of the camber—lateral forces

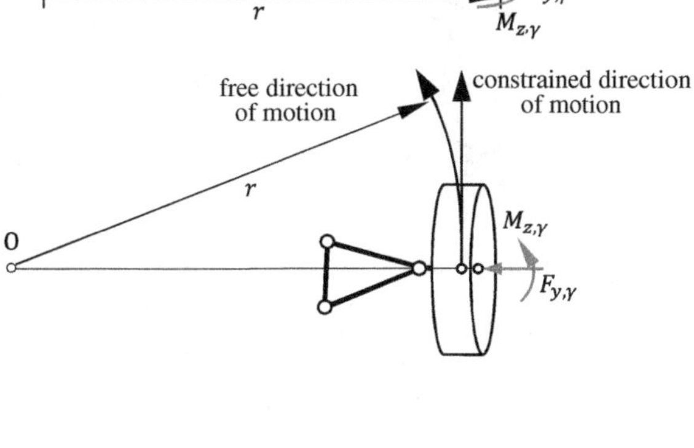

lateral force due to camber and the steering torque due to the camber increases approximately linear and one arrives at the first approximation as follows:

$$F_{y,\gamma} = -c_\gamma \gamma, \tag{7.40}$$

$$M_{z,\gamma} = -c_{M,\gamma} \gamma. \tag{7.41}$$

The forces and torques due to the camber are much smaller than the forces and torques caused by the slip.

7.3.6 *Influence of the Tire Load and the Tire Forces on the Patch Surface*

In normal working conditions there exists an approximately linear relationship between the tire forces F_x and F_y in the patch surface and the current normal force F_z. On exceeding the constructive operating load $F_{z,B}$, the horizontal force potential only increases, slightly degressive, with respect to the tire load, as the friction bond of the rubber in the contact region decreases with increasing contact force. This effect can be considered by for example introducing an effective tire load (see e.g. Ammon 2013)

$$F_{z,eff} = F_z \left(1 - e_z \left(\frac{F_z}{F_{z,B}}\right)^2\right), \tag{7.42}$$

that will be used instead of the current actual tire load F_z to calculate the tire horizontal forces. For this one must still determine the tire dependent positive degression parameter e_z included in the Eq. (7.42).

7.3.7 *Fundamental Structure of the Tire Forces*

Ammon in Ammon (2013) has suggested, organizing the circumferential and lateral forces in the following manner:

$$F_x = \mu_{max,0} \, F_{z,eff} f(\xi), \quad \xi = \frac{s}{s_{max,0} \, \mu_{max,0}}, \tag{7.43}$$

$$F_y = \mu_{max,0} \, F_{z,eff} g(\eta), \quad \eta = \frac{\alpha}{\alpha_{max,0} \, \mu_{max,0}}. \tag{7.44}$$

Here the index 0 represents the friction coefficient and the tire load at a normal operating condition, i.e. the conditions in the vicinity of the normal operating conditions and $F_{z,eff}$ will be used as in Eq. (7.42) with $F_{z,0} = F_{z,B}$. The functions f and g describe typical characteristics as they are shown in the Figs. 7.10 and 7.14. These functions depend only on the characteristics of the tire, such as the shear stiffness, geometry and the constructive structure as a result of the transformation of the slip variables. The first two factors mirror the influence of the surrounding such as the nature of the road surface and the influence of the tire load.

7.3.8 Superposition of Circumferential and Lateral Forces

Pure longitudinal forces or pure lateral forces will be transmitted only during purely straight line driving or during cornering with a constant velocity. In general driving situations the longitudinal and the lateral slip are overlapped, i.e. there will occur along with a circumferential force F_x simultaneously a lateral force F_y. This resulting force must however according to the COULOMB's friction law always fulfill the condition

$$\sqrt{F_x^2 + F_y^2} \leq \mu_{max} F_z. \tag{7.45}$$

Graphically one can explain this through the circle of forces (Kamm's Circle), if the maximally possible force adhesion coefficient μ_{max} in the circumferential and the lateral direction match, see Fig. 7.16. From this one can arrive at the following conclusions:

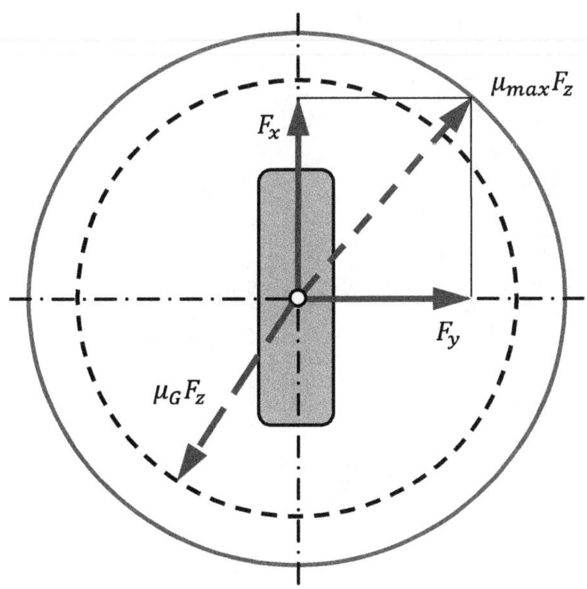

Fig. 7.16 Simultaneous superposition of the lateral and longitudinal forces—Kamm's Circle

- The maximum transmissible lateral force, when a simultaneous occurrence of the longitudinal force, is smaller. This means for real driving situations, that accelerations or braking reduces the maximal possible lateral forces in a curve and vice versa.
- If an additional longitudinal force F_x is applied on a tire with a given lateral force F_y, then the slip angle α and the longitudinal slip s will increase.

In the reality the lateral force—circumferential force characteristics deviate from the theoretical ideal:

- The limit curve of the maximally transmissible horizontal force is an ellipse, as in real tires, normally, the adhesion limit in the longitudinal direction $\mu_{max,x}F_z$ is larger than the adhesion limit in the lateral direction $\mu_{max,y}F_z$.
- The slip force curves for acceleration and braking are normally not exactly symmetrical.
- In many tires due to an asymmetry of the tire construction a lateral force F_y occurs even when the slip angle α vanishes, and respectively for lateral forces $F_y = 0$ a small slip angle $\alpha \neq 0$ is required.

In order to be able to consider the superposition effects during driving situations, in which both longitudinal and lateral slip occur (during braking, acceleration in a curve, etc.), one defines from the longitudinal slip s and the lateral slip α an absolute slip variable

$$s_a = \sqrt{s^2 + \tan^2\alpha} = \frac{\sqrt{v_P^2 + v_y^2}}{|\bar{v}|} \tag{7.46}$$

with the effect direction given by

$$\psi_a = \arctan\frac{\tan\alpha}{s} = \arctan\frac{v_y}{v_P} \tag{7.47}$$

with

$$|\bar{v}| = \max(v, \omega r_{dyn}). \tag{7.48}$$

For the (absolute) tire forces in the direction of the given angle ψ_a one gets:

$$F_{\psi_a}(s_a) = \sqrt{\frac{s^2F_x^2(s_a) + \tan^2\alpha\, F_y^2(s_a)}{s_a^2}}. \tag{7.49}$$

The resulting horizontal components of the tire force are then applied in the coordinate directions of the wheel carrier:

$$F_{a,x} = F_{\psi_a}(s_a)\cos\psi_a = \frac{s}{s_a}F_{\psi_a}(s_a),$$
$$F_{a,y} = F_{\psi_a}(s_a)\sin\psi_a = \frac{\tan\alpha}{s_a}F_{\psi_a}(s_a). \tag{7.50}$$

In Figs. 7.17 and 7.18 the principle characteristics of the longitudinal and the lateral forces of different slip angles are schematically shown.

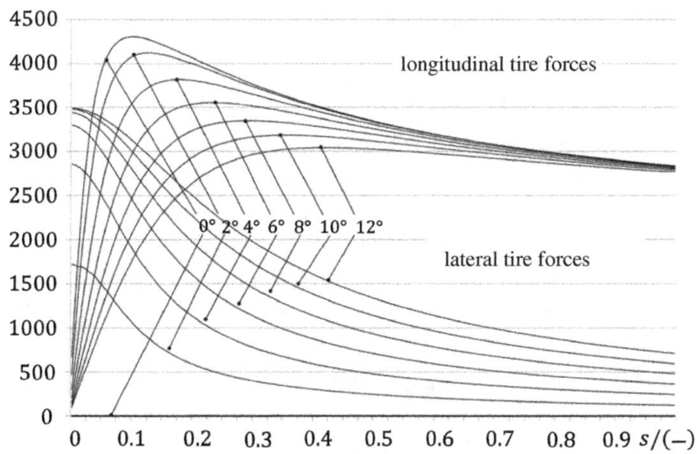

Fig. 7.17 Superposition of the longitudinal and the lateral forces (schematic representation)

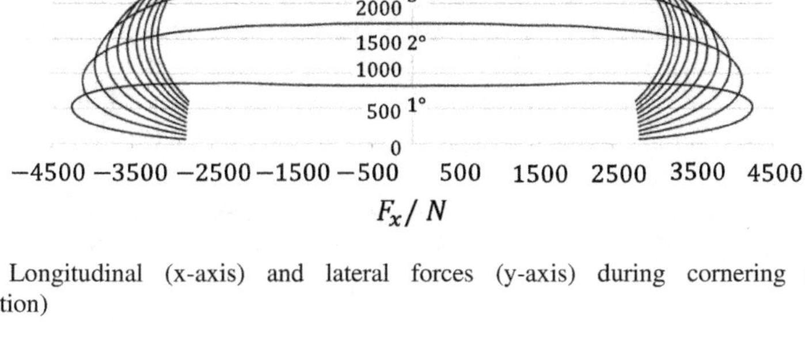

Fig. 7.18 Longitudinal (x-axis) and lateral forces (y-axis) during cornering (schematic representation)

7.4 Tire Models

The modeling of the tire forces requires special attention and care especially when along with the stationary behavior the instationary behavior has to be covered. One can differentiate generally between three types of tire models:

- mathematical models,
- physical models,
- and a combination of the two.

In the following, first models will be considered that can represent the physical characteristics through a purely mathematical description. To this end, one starts at a known, meaning measured, characteristic that has been saved as characteristic curves. For the further processing one can either be accomplished with the help of an approximation through an algebraic function (Magic Formula Tire Model, (Pacejka and Besselink 1997)) or through an interpolation, (Schieschke and Gnadler 1987). This art of modeling is for the simulation of driving maneuvers mostly sufficient as the excitation frequencies in these cases are well below the Eigen frequencies of the belts. Problems can however occur, when a large number of influencing variables have to be considered, as in this case a considerable characteristic curve needs to be saved and evaluated. And it is practically impossible, to change individual parameters, without recreating the entire characteristic curves.

A second major group is the physical models. Here one can differentiate between the following significant types:

- FE (Finite Element)-Models (Gipser 1987),
- models to investigate the membrane vibrations, (Böhm et al. 1985),
- stationary Models (e.g. HSRI model),
- instationary Models, e.g. circular ring model (Schulze 1988) or the patch matrix model, (Schnelle 1990).

The two models mentioned at the beginning are often used to investigate the deformation of the tire in the context of comfort and vibration analysis. Within the context of vehicle dynamics however, one is more often than not only interested in the forces and torques occurring between the tire and the road and not especially in the deformation itself. From the point of view of a reasonable computation time, such models are still not used in the vehicle dynamics simulation in spite of the massive increase in computation power available today. In stationary models only the rim will be considered to have mass and inertia, and the belt will not be modeled as a separate body. These models are similar to the characteristic curve models mostly used in the simulation of stationary driving maneuvers. In order to extend the application area of these models, very often a first order time element (PT_1) is used to represent the delayed building up of the tangential forces. However if excitations in the region of the belts eigenfrequency, between 30 Hz and 50 Hz, were to occur, such as through the use of elastic components in the wheel suspension or through the pulsation of the brake pressure during an ABS controlled braking procedure, and these have an effect on the characteristics of the vehicle behavior, then the eigendynamics of the belt have to be considered as well. To this end the belt will be modeled as a rigid ring with mass. This will however result at low speeds (below around 10 m/s) highly instationary transport processes in the contact surface between the tire and the road, that require a special description.

After this introductory consideration, a kinematic model for the wheel-road contact will be explained at the later part of this chapter, which is based on the following basic idea:

- The contact geometry tire/road are described through a representative mechanism and can be calculated elegantly using simple expressions.
- Simple tire models, in which the longitudinal forces and the lateral forces can be calculated, without considering the eigendynamics of the belts, as a function of the longitudinal slip and the lateral slip, can access these variables.
- For more complex tire models, that need to represent the instationary processes, the patch can be discretized. The required kinematic values are available for these models also.

7.4.1 The Contact Point Geometry

In this section it shall be assumed that the position and orientation of the wheel in space is known. For this one measures the position and velocity of the wheel from the vehicle chassis, see Fig. 7.19. The position of the wheel carrier can be described through the wheel center point R and through the introduced wheel carrier fixed coordinate system introduced in Chap. 6. These values are known through the vehicle kinematics. What is now required is the position and velocity of the wheel-road contact zone, which in this case is given by a contact point A and an orientation of a tangential plane on the road surface in point A.

Figure 7.20 shows a schematic representation, in which the wheel center plane is represented as a flat disc. The road surface is described locally through the tangential plane Σ. This is for the description of a normal road surface without steps or other individual obstacles completely sufficient.

Given are a road surface normal vector $\boldsymbol{n}_s$ and the wheel normal vector $\boldsymbol{n}_R$ as well as the position of the wheel center point R, given by the position vector in the inertial system. The position of the wheel is, in this case, as described in the previous chapter, described through the corresponding geometrical and kinematical relationships originating from the vehicle chassis over the wheel suspension to the wheel carrier. With these values one can now determine the position of the wheel contact point A. This point is given as the contact point between a rigid disc (center point R, radius as yet unknown) and the road surface Σ. The connecting vector from R to A must be orthogonal to both the wheel axis $\boldsymbol{n}_R$ as well as the unit vector $\boldsymbol{n}_L$ in the longitudinal direction of the wheel disc (Fig. 7.21).

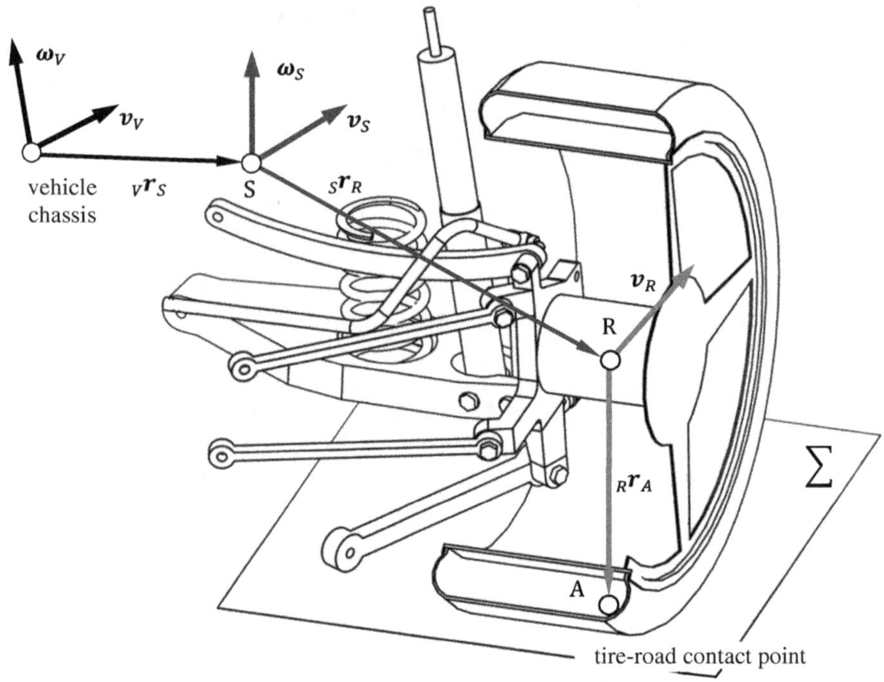

Fig. 7.19 Calculation of the position and velocity of the wheel carrier from the vehicle chassis

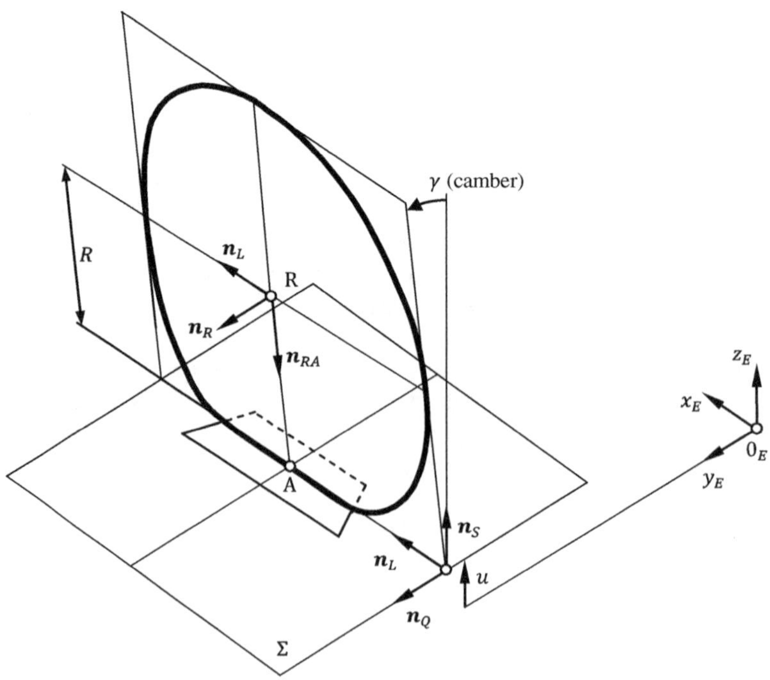

Fig. 7.20 Unit vector at the wheel

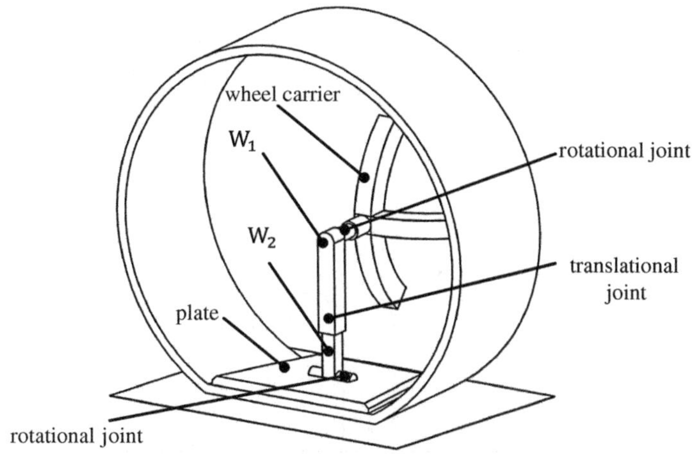

Fig. 7.21 Multibody replacement model of the wheel geometry (Schnelle 1990)

It is advantageous for the process of understanding, to create a mechanical substitute model for the vectors at the wheel (Fig. 7.22). It consists of a wheel carrier R mounted rotating on the angle element W_1, whose lower arm is perpendicular to the rotation axis. This lower arm thus always lies in the wheel plane. Using a translational joint it is connected to a second, angle element W_2 lying vertically on the floor. This is again orthogonal, such that the connecting line from R and A as required not only lies in the wheel plane, but also is perpendicular to the longitudinal direction of $\boldsymbol{n}_L$. At the same time, the plate located on the floor is always pointed in the direction of $\boldsymbol{n}_L$. The average tire radius R, i.e. the length of the distance AR, relates to the displacement of the translational joint. For a given position and orientation of the wheel carrier R, the arms of the angle point in such a way, that between the link of W_2 and the road is a line contact. The plate located on the road surface, which represents the contact patch, can rotate about this link relative to the angle W_2. The presented kinematic structure can also be interpreted as a kinematic chain with six joint DoF, which represents the spatial pose of the wheel carrier relative to the road surface. The calculation of the vectors of this system is now done successively:

Spanning out a tripod $\{\boldsymbol{R}; \boldsymbol{n}_R, \boldsymbol{n}_L, \boldsymbol{n}_{RA}\}$ in the wheel surface The unit vectors in the rolling direction of the wheel is given by

$$\boldsymbol{n}_L = \frac{\boldsymbol{n}_R \times \boldsymbol{n}_S}{\| \boldsymbol{n}_R \times \boldsymbol{n}_S \|}. \tag{7.51}$$

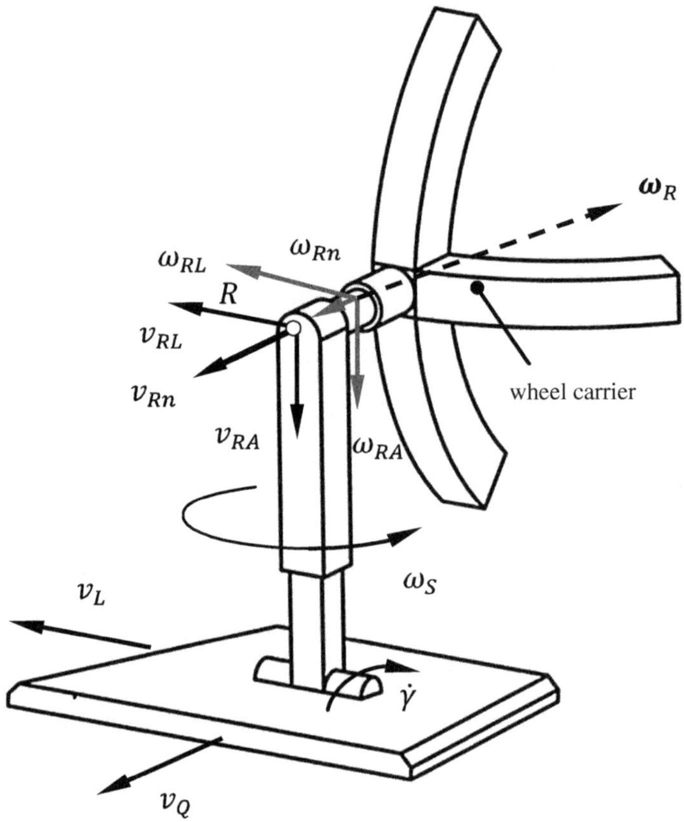

Fig. 7.22 Velocities of the wheel road contact

The unit vector from R to A is given by:

$$\boldsymbol{n}_{RA} = \boldsymbol{n}_R \times \boldsymbol{n}_L. \tag{7.52}$$

For the unit vector at right angles to the rolling direction one arrives at (Fig. 7.21):

$$\boldsymbol{n}_Q = \boldsymbol{n}_S \times \boldsymbol{n}_L. \tag{7.53}$$

Derivation of the current wheel radius *R* The position vector to the point A is given in the inertial system. Its projection on the normal to the road surface $\boldsymbol{n}_S$ gives the height of the road surface (in more detail it gives the height of the point A and hence that of the tangential plane Σ) in the inertial system. This is known and has the value u. Hence one can derive:

$$R = \frac{u - \boldsymbol{r}_A \boldsymbol{n}_S}{\boldsymbol{n}_{RA} \boldsymbol{n}_S}. \tag{7.54}$$

With the position vector $\boldsymbol{r}_R$ of the wheel center point R, the position vector $\boldsymbol{r}_A$ to the point A is given by:

$$\boldsymbol{r}_A = \boldsymbol{r}_R + R\boldsymbol{n}_{RA}. \tag{7.55}$$

Furthermore the position of the two angle elements W_1 and W_2 is known through the unit vectors $\boldsymbol{n}_{RA}, \boldsymbol{n}_L$ and $\boldsymbol{n}_Q$. There now exists two coordinate systems, one wheel carrier fixed through $\{R; \boldsymbol{n}_L, \boldsymbol{n}_R, \boldsymbol{n}_{RA}\}$ as well as a patch fixed system given by $\{A; \boldsymbol{n}_L, \boldsymbol{n}_Q, \boldsymbol{n}_S\}$. In this way one can define the camber angle γ between the wheel plane and the vertical plane. The following simplified relationships are enough to describe the relative rotation of the coordinate systems

$$\cos\gamma = \boldsymbol{n}_R \boldsymbol{n}_Q, \quad \sin\gamma = -\boldsymbol{n}_R \boldsymbol{n}_S. \tag{7.56}$$

Iterative evaluation of the contact point for uneven road surfaces If the road surface is uneven, one can no longer assume a constant road surface normal vector $\boldsymbol{n}_S$. This is more over dependent on the position, i.e. the position vector $\boldsymbol{r}_A$. If however the road surfaces are constrained to be lightly curved, the contact point is clearly definable as described above and also the directions can be specified with respect to the tangential plane with a good approximation in the following manner:

- As a starting value for $\boldsymbol{r}_A$, the projection of the wheel center R on the surface (in the direction of gravity) is chosen.
- The road surface normal vector $\boldsymbol{n}_S$ is formed at this point.
- With this vector $\boldsymbol{n}_S$ and according to (7.51)–(7.55) the contact point geometry is constructed. Even the road height u is part of the iteration, as it is also dependent on the position.
- The value $\boldsymbol{r}_A$ calculated from (7.55) is now a better estimation. The iteration is started again with the new value of $\boldsymbol{r}_A$. The iteration is continued so long till the difference between two successive values of $\boldsymbol{r}_A$ lies within a pre-defined iteration tolerance. At this point it is not appropriate to set too high standards in the precision, as the errors made in this step are to be viewed in relation with the errors already induced due to the simplification during the modeling of the tires.

7.4.2 Contact Velocity

Given are the absolute velocity $\boldsymbol{v}_R$ of the wheel center point R and the absolute angular velocity $\boldsymbol{\omega}_R$ of the wheel carrier. It makes sense to decompose the two vectors into components of the two coordinate systems mentioned earlier, see Fig. 7.22:

$$\boldsymbol{v}_{RL} = (\boldsymbol{v}_R \cdot \boldsymbol{n}_L)\boldsymbol{n}_L = v_{RL}\boldsymbol{n}_L, \tag{7.57}$$

$$\boldsymbol{v}_{Rn} = (\boldsymbol{v}_R \cdot \boldsymbol{n}_R)\boldsymbol{n}_R = v_{Rn}\boldsymbol{n}_R, \tag{7.58}$$

$$\boldsymbol{v}_{RA} = (\boldsymbol{v}_R \cdot \boldsymbol{n}_{RA})\boldsymbol{n}_{RA} = v_{RA}\boldsymbol{n}_{RA}, \tag{7.59}$$

$$\boldsymbol{\omega}_{RL} = (\boldsymbol{\omega}_R \cdot \boldsymbol{n}_L)\boldsymbol{n}_L = \omega_{RL}\boldsymbol{n}_L, \tag{7.60}$$

$$\boldsymbol{\omega}_{Rn} = (\boldsymbol{\omega}_R \cdot \boldsymbol{n}_R)\boldsymbol{n}_R = \omega_{Rn}\boldsymbol{n}_R, \tag{7.61}$$

$$\boldsymbol{\omega}_{RA} = (\boldsymbol{\omega}_R \cdot \boldsymbol{n}_{RA})\boldsymbol{n}_{RA} = \omega_{RA}\boldsymbol{n}_{RA}. \tag{7.62}$$

The contact point A and the patch (defined through the plate P, Fig. 7.22) also exhibit six velocities namely:

- The longitudinal and the lateral velocity of A:

$$\boldsymbol{v}_L = v_L\boldsymbol{n}_L, \tag{7.63}$$

$$\boldsymbol{v}_Q = v_Q\boldsymbol{n}_Q. \tag{7.64}$$

- The angular velocity of the patch about the vertical axis:

$$\boldsymbol{\omega}_S = \omega_S\boldsymbol{n}_S. \tag{7.65}$$

- The velocity of change in camber:

$$\boldsymbol{\omega}_L = -\dot{\gamma}\boldsymbol{n}_L. \tag{7.66}$$

- The change in the radii of the tire:

$$\dot{R}\boldsymbol{n}_{RA}. \tag{7.67}$$

- The relative rotation between the patch and the wheel carrier:

$$\boldsymbol{\omega}_n = \omega_n\boldsymbol{n}_R. \tag{7.68}$$

From the pose and the two known velocities $\boldsymbol{v}_R$ and $\boldsymbol{\omega}_R$, it is now possible to determine the complete motion of the point A and that of the patch, represented through velocities occurring in the Eqs. (7.63)-(7.68). To this end one calculates the velocity of the wheel center point B and the angular velocity of the wheel carrier from the bottom, i.e. beginning from the road surface. Once the velocities (7.63)-(7.68) are known, then through them $\boldsymbol{v}_R$ and $\boldsymbol{\omega}_R$ are known unambiguously. Now, these are projected, as shown above, on the axis of the top tripod leg and arrives at simple expressions that can be collected in a matrix:

$$\begin{bmatrix} \omega_s \\ v_L \\ \omega_n \\ \dot{\gamma} \\ v_Q \\ \dot{R} \end{bmatrix} = \begin{bmatrix} -1/\cos\gamma & 0 & 0 & 0 & 0 & 0 \\ -R\tan\gamma & 1 & 0 & 0 & 0 & 0 \\ \tan\gamma & 0 & -1 & 0 & 0 & 0 \\ 0 & 0 & 0 & -1 & 0 & 0 \\ 0 & 0 & 0 & R/\cos\gamma & 1/\cos\gamma & 0 \\ 0 & 0 & 0 & -R\tan\gamma & -\tan\gamma & -1 \end{bmatrix} \begin{bmatrix} \omega_{RA} \\ v_{RL} \\ \omega_{Rn} \\ \omega_{RL} \\ v_{Rn} \\ v_{RA} \end{bmatrix} \tag{7.69}$$

These relationships contain only the wheel radius R and the camber angle γ, which are both known from the calculation of the position. The kinematic processing makes the analytical differentiation of the position vectors and the rotational tensor completely superfluous. Of particular interest is also the partly decoupled nature of the structure of the Eq. (7.69). It is thus possible, to determine the position and velocity of the wheel-road contact point using a simple method. Which of these variables are finally used in the tire model, or can be used, is dependent on the type and the computational complexity of the model.

7.4.3 *Calculation of the Slip Variables*

From the values calculated in the Sect. 7.4.2, one can now calculate the slip values that are required for calculating the horizontal tire forces. The slip s is given using the following relationship with the derived values from the last section and the current wheel angular velocity $\dot{\rho}$

$$s_{A,B} = \frac{\left|v_L - \dot{\rho} r_{dyn}\right|}{\max(v_L, \dot{\rho} r_{dyn})}, \tag{7.70}$$

$$\tan\alpha = \frac{v_Q}{v_L}. \tag{7.71}$$

In this way, different input variables for the different tire models are available. In this book two such tire models are discussed in the next section.

7.4.4 Magic Formula Model

One of the most widely used tire models is the so-called Magic Formula Model that was developed by Pacejka (2006) and Pacejka and Bakker (1993). The magic formula tire model is a pure mathematical description of the input-output relationship of the tire-road contact under quasi-stationary conditions. The modeling approach allows that the characteristics of the lateral guiding forces, the braking forces and the aligning torque are described mathematically.

The Magic Formula Model was developed, to connect the force variables described in this chapter with the rigid body slip using mathematical functions, i.e.

- the circumferential force F_x with the circumferential slip s,
- the lateral force F_y with the slip angle α and
- the aligning torque M_z with that of the slip angle α.

These relationships are determined during quasi-static rolling or vehicle maneuvers and approximated through mathematical functions. Pacejka und Bakker suggested, that the above discussed typical force functions be approximated through a combination of sine and arctangent functions. The formulae are capable of describing the circumferential force, lateral force and aligning torque as functions of the longitudinal and lateral slip with a high degree of accuracy. The description however is limited to stationary state characteristics. Figure 7.23 shows the fundamental form of these describing functions. The requirements of these describing functions are:

- the description of all the stationary tire state characteristics,
- easy availability of the data,
- possibility of a certain physical interpretation of the relationships,
- high accuracy,
- and easy evaluation.

As a possible describing function one can choose for example, (Pacejka 2006):

$$y(x) = D\sin(C\arctan(Bx - E(Bx - \arctan Bx))), \tag{7.72}$$

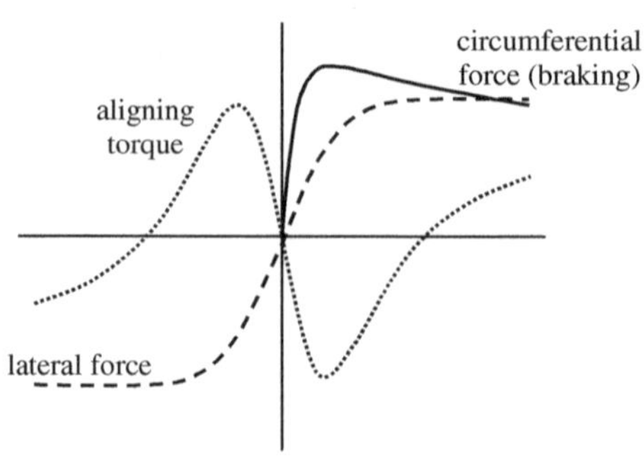

Fig. 7.23 Generic tire characteristic curves from the method of magic formula

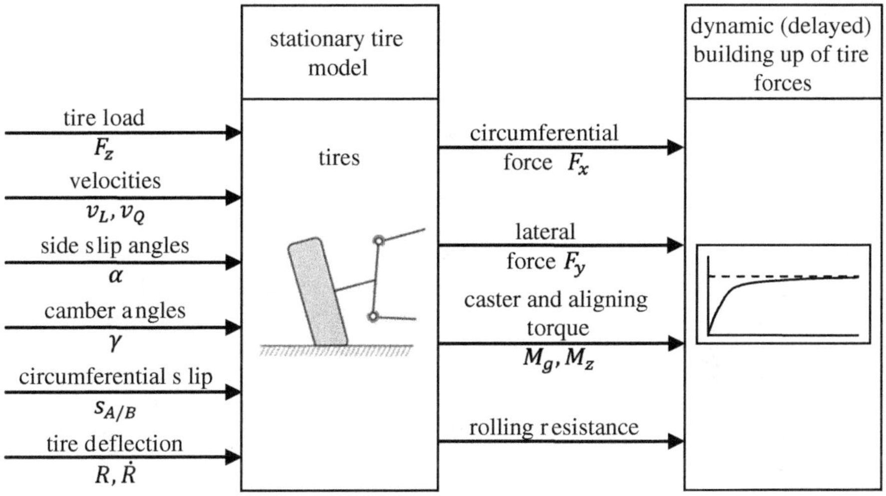

Fig. 7.24 Input- Output values of tire models

Table 7.4 Interpretation of Magic Formula Parameter

Parameter	Interpretation
B	Stiffness factor
D	Maximum force, maximum torque respectively
C	Influences the form of the curve—scaling in the x-direction
E	Additional stretching or compression of the characteristic curve
BCD	Slope of the characteristic curve at the zero slip (stiffness)
S_v	Vertical shifting of the characteristic curves
S_h	Horizontal shifting of the characteristic curves

$$Y(X) = y(x) + S_v, \tag{7.73}$$

$$x = X + S_h. \tag{7.74}$$

In this case $Y(X)$ represents either the circumferential force F_x, the lateral force F_y or the aligning torque M_z. The variable X is either the longitudinal slip s or the slip angle α. The parameters used can be interpreted as given in Table 7.4 and Fig. 7.25.

Between the parameters and the typical characteristic variables of the curves (see Fig 7.23) the following relationships are valid:

$$C = 1 \pm \left(1 - \frac{2}{\pi} \arcsin \frac{y_a}{D}\right), \tag{7.75}$$

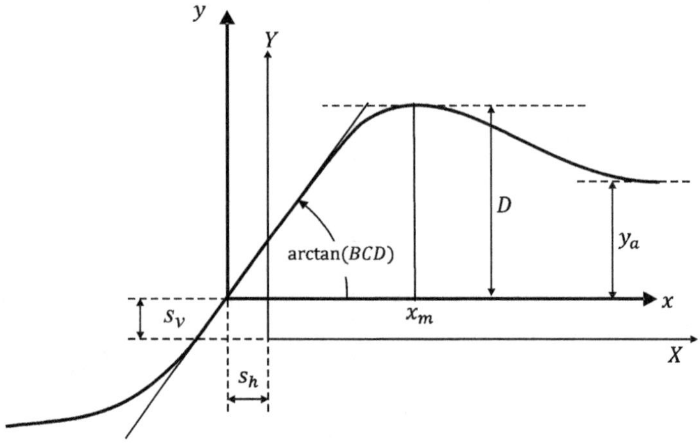

Fig. 7.25 Interpretation of the Magic Formula coefficients

$$E = \frac{Bx_m - \tan(\frac{\pi}{2C})}{Bx_m - \arctan(Bx_m)} \quad \textit{if } C > 1. \tag{7.76}$$

Related data records can be obtained from the literature (e.g. Pacejka and Bakker 1993 and Pacejka 2006) for example. During the generation of realistic vehicle models, the parameters of the used tires derived from the measurements will be made available through approximations, (Pacejka 2006). The fundamental structure of the fundamental curves given by (7.72)–(7.74) can be seen in the Figs. 7.23 and 7.25. It should however be noted that this structure deals with a purely stationary description of the tire forces, the time delay during to generation of the actual tire forces through the finite velocity of the wheels is not yet considered here, Ref. Sect. 7.5 (Fig. 7.24).

7.4.5 Magic Formula Models for Superimposed Slip

Furthermore the circumferential and the slip force characteristics are assumed to be not independent of one another. To represent the mutual dependence, weighting curves are used (Pacejka 2006), which realize the interdependence of the force characteristics. It is concerned with so-called derating curves, that make sure that the forces in the longitudinal and the lateral directions are calculated in such a way that, in total, the friction conditions between the tire and road is represented correctly. A typical approach for the lateral forces is for example:

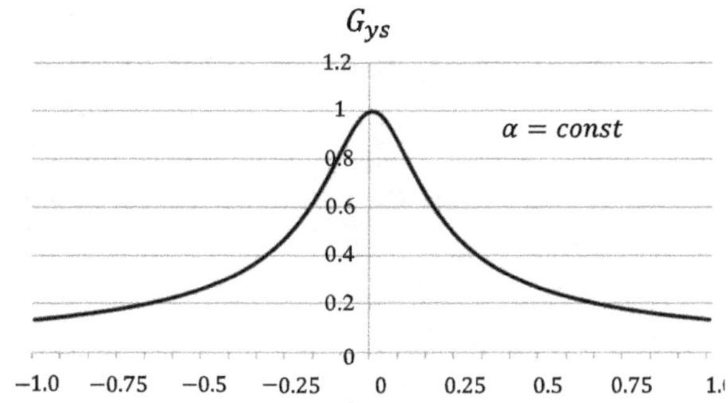

Fig. 7.26 Derating functions according to Pacejka (2006)

$$F_y = G_{ys} F_{y0} + S_{Vys} \tag{7.77}$$

with the lateral force F_{y0} for a longitudinal-free case, the force offset S_{Vys} and the weighting function

$$G_{ys} = \frac{\cos[C_{ys} \arctan(B_{ys}(s + S_{Hys}))]}{\cos[C_{ys} \arctan(B_{ys} S_{Hys})]} > 0. \tag{7.78}$$

The weighting function G_{ys} has the characteristic that for $s = 0$ it holds the value one and with increasing s it tends to zero. For a detailed discussion and the complete list of Eqs. the reader is referred to Pacejka (2006) (Fig. 7.26).

7.4.6 HSRI Tire Model

According to the descriptions in Sects. 7.4.1–7.4.3, the equations of motion of the patch can be determined relative to the camber angle as well as the angular and translational velocity of the wheel carrier and the wheel center point, respectively. This allows the calculation of the tire slip angle and the longitudinal slip. If these values are known, then one could calculate the dynamic tire forces with the help of the HSRI tire model (Highway Safety Research Institute) which in turn has been derived from Dugoff et al. (1969). It was extended by Uffelmann (1980) to also consider the fluctuations in the tire load, while (Wiegner 1974) appended the tire aligning torque. This physical model allows the calculation of the tire longitudinal and tire lateral forces by approximating the deformations taking place in the tire contact patch. These deformations are shown in Fig. 7.13. One can see here that during the overlap of the force adhesion limit a sliding region is formed, while the front end of the patch still adheres to the road; compare also the Sects. 7.3.3 and 7.3.4. The geometrical form of the sliding and the adhesion regions will be approximated as trapezoidal surfaces as shown in Fig. 7.27.

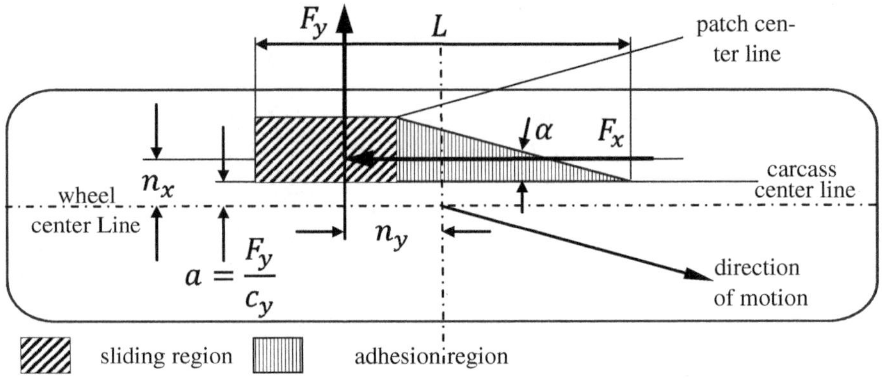

Fig. 7.27 Derivation of the HSRI model through trapezoidal approximation of the contact patch surface with adhesion and sliding regions

The trapezoidal surface results due to the parallel shift of the carcass (belt in contemporary radial tires) from the wheel center line in the contact patch and through the partitioning into the sliding and the adhesion regions (Fig. 7.13). Thereby the deflection of the tire profile results in the adhesion region due to the slip angle and in the sliding region due to the parallel shift compared to the carcass center line. Under the assumption of a constant surface pressure over the entire contact patch, one can deduce that the tire longitudinal force F_x and the tire lateral force F_y act on the surface center of gravity of the approximated trapezoidal surface (comp. Fig. 7.27). To determine the equations of motion of the contact patch (in sliding or in adhesion) one must first calculate the sliding velocity:

$$v_G = v_P\sqrt{s_{A,B}^2 + \tan^2\alpha}, \tag{7.79}$$

where v_P represents the velocity in the contact patch longitudinal direction according to Fig. 7.7 and $s_{A,B}$ the slip according to Table 7.2. With the help of the sliding velocity and the adhesion coefficients μ_0 one can calculate the adhesion coefficient between the road surface and the tires:

$$\mu = \mu_0\left(1 - k_R \tanh(a v_G)^2\right), \tag{7.80}$$

where the force adhesion factors k_R and a depend on the nature of the road surface. Using the circumferential stiffness c_s and the cornering stiffness c_α Eqs. (7.30) and (7.35) one obtains the essential dimensionless value $\bar{s}_R$:

$$\bar{s}_R = \frac{\sqrt{\left(c_s s_{A,B}\right)^2 + \left(c_\alpha \tan\alpha\right)^2}}{\mu F_z\left(1 - s_{A,B}\right)}. \tag{7.81}$$

Is now $\bar{s}_R \leq 0,5$ then there is no sliding but only adhesion in the contact patch. For the longitudinal and the lateral forces the following equations are valid:

$$F_x = c_s\ s_{A,B}\left(\frac{1}{1-s_{A,B}}\right), \tag{7.82}$$

$$F_y = c_\alpha \tan\alpha. \tag{7.83}$$

If however $\bar{s}_R > 0,5$ then both sliding and adhesion are present in the contact patch (see Fig. 7.27), and the tire forces are calculated as follows:

$$F_x = \frac{c_s\ s_{A,B}(\bar{s}_R - 0.25)}{\bar{s}_R^2(1-s_{A,B})}, \tag{7.84}$$

$$F_y = -\frac{c_\alpha \tan\alpha(\bar{s}_R - 0.25)}{\bar{s}_R^2(1-s_{A,B})}. \tag{7.85}$$

For calculating the aligning torque in the tire contact patch one still needs the lever arms n_x and n_y as shown in the Fig. 7.27, which are again dependent on the value of $\bar{s}_R$ previously calculated. During pure adhesion ($\bar{s}_R \leq 0,5$) the lever arms are defined through the position of the center of gravity of the surface of the triangular adhesion region:

$$n_x = \frac{2}{3}L\tan\alpha + \frac{F_y}{c_y}, \tag{7.86}$$

$$n_y = \frac{1}{6}(l + 2\bar{s}_R(0.5 - \bar{s}_R)), \tag{7.87}$$

whereby the tire lateral stiffness c_y can be approximately calculated from the vertical tire stiffness c_z:

$$c_y \approx \frac{c_z}{2}. \tag{7.88}$$

During sliding and adhesion ($\bar{s}_R > 0,5$) one arrives at the lever arms n_x and n_y in the following manner:

$$n_x = \frac{L}{2}\tan\alpha\left(\left(\frac{\bar{s}_R - \frac{1}{3}}{\bar{s}_R\left(\bar{s}_R - \frac{1}{4}\right)}\right) + \frac{F_y}{c_y}\right), \tag{7.89}$$

$$n_y = \frac{L}{2}\left(\frac{12 - \frac{1}{\bar{s}_R^2}}{12 - \frac{3}{\bar{s}_R}} - 1\right)\left(\frac{1 - (\bar{s}_R - 0.5)}{C_{corr}}\right). \tag{7.90}$$

Through this the aligning torque in the contact patch is given through the equilibrium of torque:

$$M_z = F_y n_y - F_x n_x. \tag{7.91}$$

The influence of the camber can be simply added linearly as per Eqs. (7.40) and (7.41). In this way, it is possible to reliably describe the tire forces also up to standstill through the partly empirically derived relationships.

7.5 Instationary Tire Behavior

All the considerations till now were under the assumption that the equations of motion of the tire were stationary or at least quasi stationary. This means in particular that the introduced parameters, such as the values of circumferential slip, slip angle, camber as well as the tire forces and the tire torques remain constant over time or at least only fluctuate very slowly.

During the investigation of a multitude of vehicle dynamic processes, these conditions are either not met or are not met to a sufficient degree to warrant this assumption. This is particularly true for maneuvers such as the steer angle jump, ABS-braking and ESP-intervention. In these cases, the longitudinal slip $s(t)$ and the slip angle $\alpha(t)$ change very fast with time. In these cases, one must consider that the tire forces $F_x(t)$, $F_y(t)$ and torques $M_z(t)$ can only build up with a time delay. The reason for this is that the force generation is only accompanied with the motion of the tire.

In the simplest case, the time delay of the force generation can be considered to be that of a first order system. This way the force generation behaves like a PT_1-system. The time delay in the generation of the force can be approximated as follows as per (Heißing and Ersoy 2011):

$$T_x \frac{dF_x}{dt} + dF_x = F_{x,stat} \tag{7.92}$$

for the tire circumferential forces and

$$T_y \frac{dF_y}{dt} + dF_y = F_{y,stat} \tag{7.93}$$

for the tire lateral forces. The time constants T_x and T_y in (7.92) and (7.93) are given by:

$$T_x = \frac{c_{s,stat}}{c_{x,stat} v_x}, T_y = \frac{c_{\alpha,stat}}{c_{y,stat} v_x}. \tag{7.94}$$

To this end the following parameters were used:

- the circumferential stiffness

$$c_{s,stat} = \left[\frac{dF_x}{ds}\right]_{s=0}, \tag{7.95}$$

- the cornering stiffness

$$c_{\alpha,stat} = \left[\frac{dF_y}{d\alpha}\right]_{\alpha=0} \tag{7.96}$$

- as well as the (static) tire longitudinal stiffness and the lateral stiffness $c_{x,stat}$ resp. $c_{y,stat}$, the longitudinal velocity v_x and the static tire forces $F_{x,stat}$ and $F_{y,stat}$ at s and α respectively.

The inlet lengths

$$\sigma_s = \frac{c_{s,stat}}{c_{x,stat}}, \tag{7.97}$$

$$\sigma_\alpha = \frac{c_{\alpha,stat}}{c_{y,stat}} \tag{7.98}$$

characterize the distance dependence of the determination of the tire forces. The inlet angles describe here the length that the tires have to travel, in order to generate around two thirds of the dynamic tire forces.

References

Ammon D (2013) Modellbildung und Systementwicklung in der Fahrzeugdynamik. Vieweg +Teubner Verlag, Stuttgart. - ISBN 978-3-663-12246-3

Böhm F, Duchow A and Hahn P (1985) Beanspruchung und Verformung des Gürtelreifens unter Innendruck. Automobilindustrie 3. S 317–23

DIN (1986) DIN_70020-5: Automotive engineering; tyres and wheels; concepts and measuring conditions. in Automotive engineering; tyres and wheels; concepts and measuring conditions (ed.), Deutsches Institut für Normung e.V., Berlin.

Dugoff H, Fancher PS and Segel L (1969) Tire Performance Characteristics Affecting Vehicle Response to Steering and Breaking Control Inputs. Highway Safety Research Institute

Gillespie T (1992) Fundamentals of vehicle dynamics. Society of Automotive Engineers, Warrendale, PA. - ISBN 978-1-56091-199-9

Gipser M (1987) DNS-Tire, ein dynamisches, räumliches nichtlineares Reifenmodell. In: Reifen, Fahrwerk, Fahrbahn. In VDI Berichte 650, S 115-35. VDI-Verlag, Düsseldorf

Gipser M (2010) http://www.cosin.eu/prod_ftire. (ed.)

Heißing B and Ersoy M (2011) Chassis Handbook. 1st (ed.), Vieweg+Teubner Verlag, Wiesbaden. - ISBN 978-3-8348-0994-0

Leister G (2009) Fahrzeugreifen und Fahrwerkentwicklung Strategie, Methoden, Tools. Vieweg +Teubner Verlag/GWV Fachverlage GmbH Wiesbaden, Wiesbaden. -ISBN 978-3-8348-0671-0

Pacejka HB (2006) Tyre and vehicle dynamics. 2nd (ed.), Butterworth-Heinemann, Oxford. - ISBN 978-0-7506-6918-4

Pacejka HB and Bakker E (1993) The magic formula tyre model, tyre models for vehicle dynamic analysis. In *1st International Colloquiumon Tyre Models for Vehicle Dynamic Analysis*. Pacejka HB (ed.), S 1–18, Swets&Zeitlinger, Lisse

Pacejka HB and Besselink IJM (1997) Magic formula tyre model with transient properties. In Proceedings of the Berlin Tyre Colloquium, Vehicle System Dynamics Supplement 27, S 145–55, Swets&Zeitlinger, Lisse, Berlin

Reimpell J and Sponagel P (1988) Fahrwerktechnik: Reifen und Räder. Vogel Fachbuch, Würzburg. - ISBN 978-3-8023-0737-9

Rill G (1994) Simulation von Kraftfahrzeugen. Vieweg-Verlag, Braunschweig/Wiesbaden. - ISBN 978-3-528-08931-3

Schieschke R and Gnadler R (1987) Modellbildung und Simulation von Reifeneigenschaften. *VDI-Berichte* 650

Schnelle K-P (1990) Simulationsmodelle für die Fahrdynamik von Personenwagen unter Berücksichtigung der nichtlinearen Fahrwerkskinematik Dissertation, Universität Stuttgart, Stuttgart

Schulze DH (1988) Instationäre Modelle des Luftreifens als Bindungselemente in Mehrkörpersystemen für fahrdynamische Untersuchungen. *TUE TECHNISCHE UEBERWACHUNG* 29, - ISBN 1434-9728

Schuster C (1999) Strukturvariante Modelle zur Simulation der Fahrdynamik bei niedrigen Geschwindigkeiten [Dr.-Ing.]. Dissertation, Universität - Gesamthochschule Duisburg, Duisburg

Uffelmann F (1980) Berechnung des Lenk- und Bremsverhaltens von Kraftfahrzeugzügen auf rutschiger Fahrbahn [Dr.-Ing.]. Dissertation, TU Braunschweig, Braunschweig

Wiegner P (1974) Über den Einfluss von Blockierverhinderern auf das Fahrverhalten von Personenkraftwagen bei Panikbremsungen [Dr.-Ing.]. Dissertation, TU Braunschweig, Braunschweig

Zomotor A (1991) Fahrwerktechnik: Fahrverhalten. Vol. 2. Auflage, Vogel Verlag und Druck KG, Würzburg. - ISBN 978-3-8023-0774-4

Chapter 8
Modeling of the Drivetrain

In the Chaps. 5–7, the subsystems chassis, the wheel suspensions, as well as the tires and wheels—subsystems which are of prior importance for the complete dynamic vehicle model—have been presented and were illustrated by examples. In the Chaps. 5–7, the subsystems chassis, the wheel suspensions, as well as the tires and wheels—subsystems which are of prior importance for the complete dynamic vehicle model—have been presented and were illustrated by examples. A further subsystem of major importance is the drivetrain of a passenger car, as it provides a normally indispensable contribution to the detailed investigation of the drive dynamics of safety features such as antilock braking system (ABS), traction control system (ASR), or electronic stability program (ESP). (In contrary, the dynamics of the drivetrain can only be neglected if only a pure ABS-braking sequence of the vehicle is regarded).

8.1 Drivetrain Concepts

In passenger cars, there are (as shown schematically in Figs. 8.1, 8.2 and 8.3) the three basic drivetrain concepts front-wheel-drive, rear-wheel-drive and four-wheel-drive, presented here with examples of popular vehicle types (Fig. 8.2).

In the following, the general concept of the modeling of a drivetrain will be presented and illustrated using the example of a front-wheel-drive vehicle such as the Volkswagen VW Golf VI.

D. Schramm et al., Vehicle Dynamics,
DOI 10.1007/978-3-662-54483-9_8

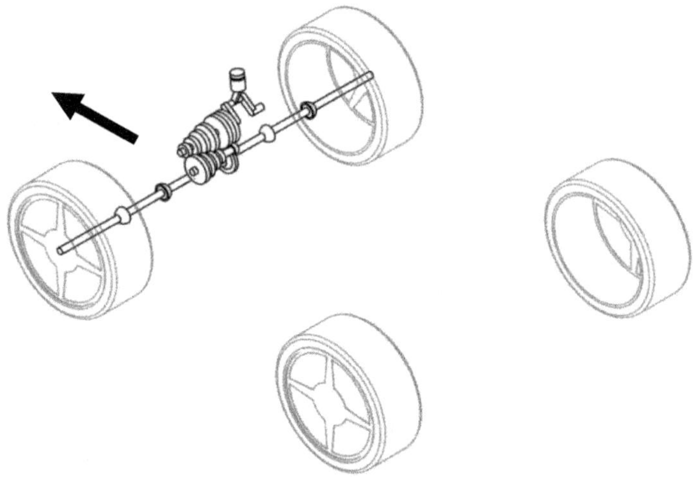

Fig. 8.1 Front-wheel-drive vehicle, e.g. Volkswagen VW Golf VI

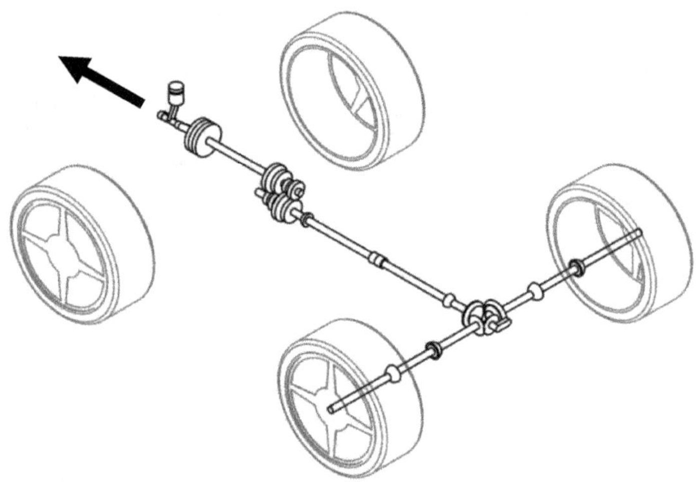

Fig. 8.2 Rear-wheel-drive vehicle, e.g. BMW – 5-series

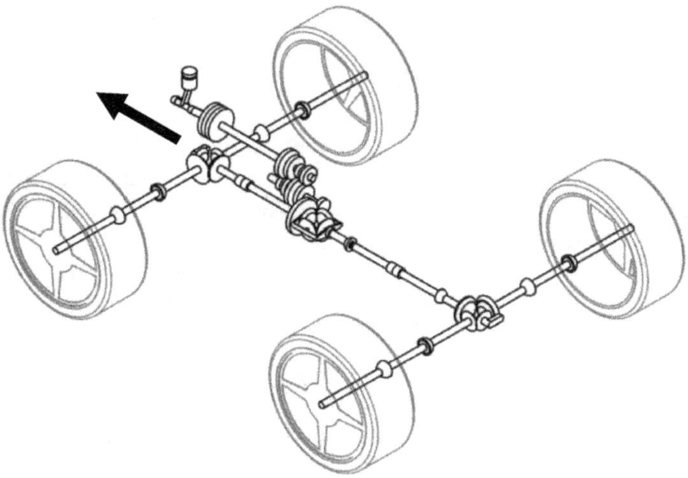

Fig. 8.3 Four-wheel-drive vehicle, e.g. AUDI A5 Coupé Quattro

8.2 Modeling

In the subsequent sections, the major aspects of modeling the influence of the drivetrain to the ride performance of a road vehicle will be discussed. Besides, the analysis of the relative motion of the engine block with respect to the vehicle, and the derivation of a suitable model of the drive train, in particular, the modeling of the homokinetic joints in the drive shaft will be outlined.

8.2.1 Relative Motion of the Engine Block

The motion of the engine block is of great influence to the dynamics of the drivetrain, and it is the aim of this section to analyze the 3-dimensional displacement of the engine block. Here, the description of the motion can be based on an inertial (i.e. space-fixed), as well as, on a relative (i.e. vehicle-fixed) concept.

In case of the inertia-based concept, the displacement of the engine block will be described by three independent translational and three independent rotational DoF, with respect to the inertial reference system "E".

This method has the advantage that the elements of the mass matrix and the gyroscopic terms are decoupled with respect to the motion of the chassis and the engine block, respectively. The coupling of the two motions results solely from the forces of the elastic engine mounts, and is called force coupling. It has to be mentioned that for the description of the engine block displacement, always six DoF are required, with the consequence that the engine block hinges have to be modeled precisely. This requires a high technical effort. A further disadvantage comes from the fact that individual DoF of the overall engine displacement cannot be switched off.

In case of a relative description, the engine block displacement will be described by an arbitrary number (maximum 6) of chosen degrees of freedom (DoF) of the engine relative to the vehicle, based on the technical realization of the engine suspension. The disadvantage here comes from the fact that the elements of the mass matrix, and the gyroscopic terms, respectively, are no longer decoupled. This concept is named mass or inertia coupling.

The investigation of the engine block displacement of a front-wheel-drive vehicle with a laterally mounted engine showed that the engine block displacement can be characterized by two major displacements: A vertical displacement along the vertical vehicle axis and a pitch motion around an axis parallel to the axes of the drive shaft. All other displacements may be neglected in comparison to these two dominant effects. The results of this analysis show a clear advantage of the second method, compared to the inertia-based method (Schmitz 1994).

8.2.2 Modelling of the Drivetrain

In Fig. 8.4, the kinematical model of the drivetrain under consideration is presented. The transmission elements drive shaft and cardan shaft between engine block and wheel are presented only for the left-hand part of the vehicle. The right-hand part of the vehicle is represented by a topologically identical subsystem. From this figure, one recognizes the subsystems engine block, crank shaft, clutch, gearbox, differential gear, drive shaft and drive shaft with homokinetic joints. Directly after the outer homokinetic joint, the subsystems wheel carrier and wheel, which were already discussed in previous chapters, can be detected.

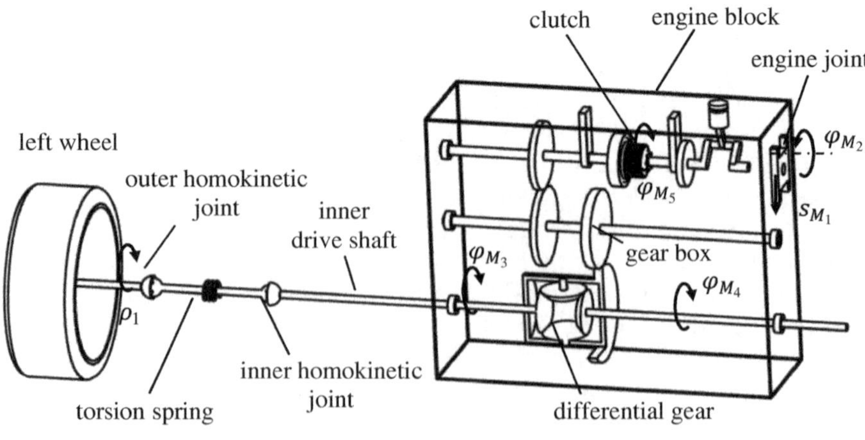

Fig. 8.4 Kinematic model of the drivetrain of a front-wheel-drive vehicle

The drivetrain model is based on the following assumptions:

- In a real system the engine block is supported by two conventional elastomer joints and an additional hydrodynamic joint. The resulting dominant vertical and pitch displacement of the engine block is a consequence of the arrangement of the supporting joints. These effects can be represented in the subsequent kinematic model of the subsystem by a prismatic and a revolute joint, respectively. In the later dynamic model, the resilience and the damping forces in the engine joints can be stated by suitable force elements. Further details of this modeling step will be presented in Sect. 8.2.3.
- The torque, transmitted from the combustion engine to the drive shaft, represents the input for the drive train.
- The clutch transmits the input torque to the gearbox. By the modeling of the clutch as a torsion elastic element, an additional DoF is introduced into the model. The aligning torques can be calculated from the corresponding nonlinear characteristic curve of the clutch.
- Inside the gearbox, engine speed and engine torque are converted. The subsequent differential gear distributes the torque equally onto the right and left output shaft.
- The cardan shafts transmit the rotational motion equally to the two front wheels, and will be, similar to the clutch, modeled for this purpose as torsion-elastic elements (Aucktor et al. 2002; Schütz 1971). As a consequence, two further DoF are introduced into the system. In Sect. 8.2.4, the cardan shafts will be discussed in more detail.
- Details to the gearshift system will not be discussed yet.
- Further details to the gearbox, like gear backlash or friction effects, will also not be taken into account.

As a result of this modeling approach for the drivetrain, one gets $f = 5$ additional DoF for the complete vehicle model with the coordinates:

s_{M_1} lift coordinate of the engine block,
φ_{M_2} pitch angle of the engine block,
φ_{M_3} angle of revolution of the left inner drive shaft,
φ_{M_4} angle of revolution of the right inner drive shaft,
φ_{M_5} angle of revolution of the crankshaft

The complete vehicle model of the VW Golf, used as an example in this section, possesses $f = 21$ DoF, as it will be discussed in more detail in Chap. 12.

8.2.3 Engine Bracket

As a consequence of the increasing need for more driving comfort, the layout of the engine bracket has become an extensive field of engineering development in the automotive industry. A broad field of literature exists about the design, system

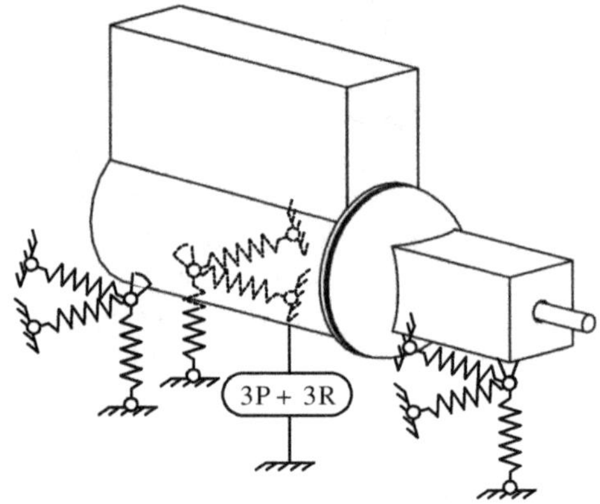

Fig. 8.5 Engine block with elastic bearings

properties, tasks and optimization of engine bearings, e.g. (Johnson and Subhedar 1979; Wallentowitz and Mitschke 2006; Bathelt et al. 1989). The contents of the subsequent explanations are attempts for the modeling of engine bearings within a three-dimensional vehicle model (Schmitz 1994).

Kinematic substitute system While the drive shafts within an engine block are rotating around fixed axes, the engine block can execute a general spatial motion (with in the most general case having six DoF), due to its elastic bearings (Fig. 8.5). Depending on its design concept, the engine block is connected to the vehicle body by three to five bearing positions (Demant et al. 1990).

In this approach, the determination of the parameters of the engine bearings becomes very difficult: Each bearing has to be measured along its axis separately. In addition, the spring constants in lateral and longitudinal direction of the vehicle are very large, and the resulting stiff differential equations have a poor numerical efficiency.

Therefore, an alternative approach will be presented here, based on a method for the modeling of elastic bearings, developed by Schnelle (1990). The main idea is to consider only the relevant engine displacements, which leads to a reduced number of coordinates and DoF, respectively.

From measurements it is well known that some of the displacements of the engine block with respect to the car body are dominant over the others. These are the rotations about axes, which are parallel to the output shafts of the gearbox, transmitting the engine torque, as well as the lift displacement, which is excited by an uneven road. In Fig. 8.6, the kinematic models of the engine mount, consisting of a revolute joint (R) and a prismatic joint (P), respectively, of a vehicle with laterally and longitudinally assembled engines are shown. With these models suitable force laws, as shown in more detail in Chap. 9, can be expressed directly in the corresponding joint coordinates.

Selecting the joint coordinates s_{M_1}, φ_{M_2} of the kinematic substitute system as generalized coordinates $\boldsymbol{q}_M$ for the displacement of the engine block, their spatial displacement with respect to the vehicle can directly be derived by means of the

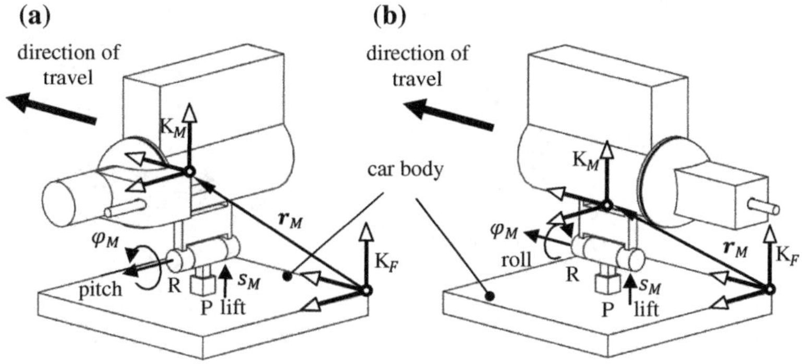

Fig. 8.6 Kinematic substitute model of engine bracket with reduced number of coordinates. **a** laterally assembled engine, **b** longitudinally assembled engine

Fig. 8.7 Engine mount as kinematic transformer

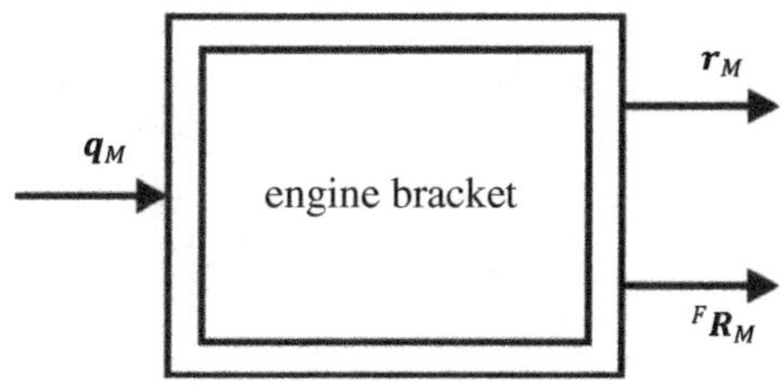

corresponding forward kinematics. The kinematic transformer, shown in Fig. 8.7, can symbolize the kinematic transmission behavior. Here, $\boldsymbol{r}_M$ designates the position vector to the reference point of the engine block and ${}^F\boldsymbol{R}_M$ the force acting on the engine block.

Hydraulic engine bearing The design of the engine mount in the kinematic substitute system influences the mechanical as well as the acoustic vibration behavior of the vehicle. This leads to conflicting goals due to the increasing demand from the customers for more driving smoothness. On the one hand, the bearing should be large and smooth enough to guarantee a good acoustic insulation. On the other hand, a smooth bearing leads, in case of a low-frequency excitation with large amplitudes, to large amplitudes in the lift motion of the engine block, which can be observed especially on an uneven road. The shaking of the relatively large engine mass is transmitted via car body to the passengers. This effect is called "engine shake" (Bürger and Dödelbacher 1988).

For the resolution of this conflict of objectives, the car manufacturers meanwhile have introduced hydraulic bearings which are characterized by a well-defined interdependency of damping behavior and stiffness from the excitation frequency.

Figure 8.8 shows the setup of a hydraulic bearing combined with the frequency-response of the dynamic stiffness and the phase angle.

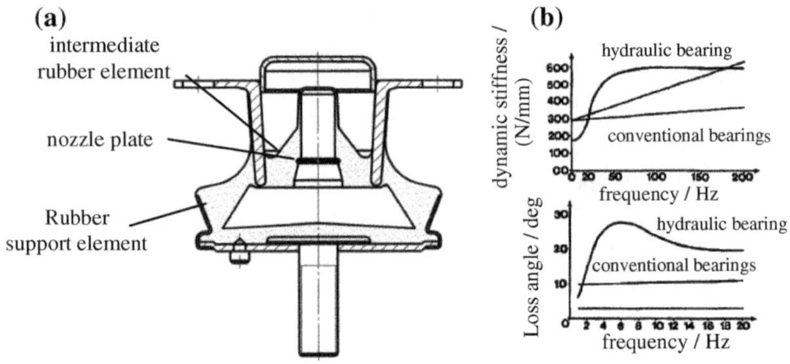

Fig. 8.8 Hydraulic bearing of a Porsche 944 (Freund and von Sivers 1981). **a** Setup, **b** Characteristics

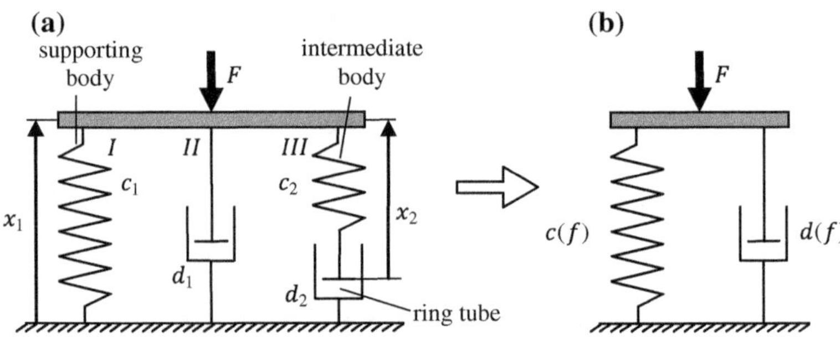

Fig. 8.9 Hydraulic bearing: **a** Mechanical substitute system, **b** Interpretation as parallel circuit of frequency-dependent stiffness and damping

The modeling of the dynamic behavior of hydraulic bearings is analyzed and discussed in various publications (Holzemer 1985; Freudenberg and Freudenberg 1998). The main aspect here is an exact description of the fluid dynamics within the ring-shaped tube and the nozzle-plate, respectively. The mechanical substitute system of Fig. 8.9a represents a good approximation of the dynamic behavior of a hydraulic bearing (Bösenberg and van den Boom 1979; Zamow and Witte 1988).

The force elements in the branches I and III can be directly identified with the help of the bodies in the hydraulic bearings. These subsystems represent the dynamic characteristics of the bearing. In existing hydraulic bearings an undesired increase of the dynamic stiffness in the range of high frequencies can be observed frequently (Göhlich and Köder 1992). An additional damping d_1 in a parallel connection takes this effect into consideration. The dynamic bearing performance in the time-domain can be represented by the subsequent differential equations:

$$F = c_1x_1 + d_1\dot{x}_1 + c_2x_2, \tag{8.1}$$

$$d_2(\dot{x}_1 - \dot{x}_2) = c_2x_2. \tag{8.2}$$

The consideration of the hydraulic bearing in the complete vehicle model leads to an uneven system order, i.e. an uneven number of first order differential equations (Schiehlen and Eberhard 2004). After a transformation of the system into the frequency domain and an identification of the transfer function with the transfer function of Fig. 8.9

$$\frac{F(\mathrm{j}\omega)}{X_1(\mathrm{j}\omega)} \stackrel{!}{=} c(\omega) + \mathrm{j}\omega d(\omega), \tag{8.3}$$

one gets the frequency-dependent stiffness and damping, respectively:

$$c(\omega) = \mathrm{Re}\left\{\frac{F(j\omega)}{X_1(j\omega)}\right\}, \quad d(\omega) = \frac{1}{\omega}\mathrm{Im}\left\{\frac{F(j\omega)}{X_1(j\omega)}\right\}. \tag{8.4}$$

The dynamic stiffness and the dead angle result from

$$c_{dyn} = \left|\frac{F(j\omega)}{X_1(j\omega)}\right|, \quad \phi = \arg\left(\frac{F(j\omega)}{X_1(j\omega)}\right). \tag{8.5}$$

Figure 8.10a shows the dynamic behavior of the model of the hydraulic bearing in comparison with a conventional rubber-metal engine bearing, which is modeled by a parallel circuit of a spring and a damper with constant parameters. One can observe a good alignment with the measurements of (Fig. 8.8b). In addition, in Fig. 8.10b the frequency dependency of stiffness and damping of a hydraulic bearing is shown. The data of the system is shown in Table 8.1.

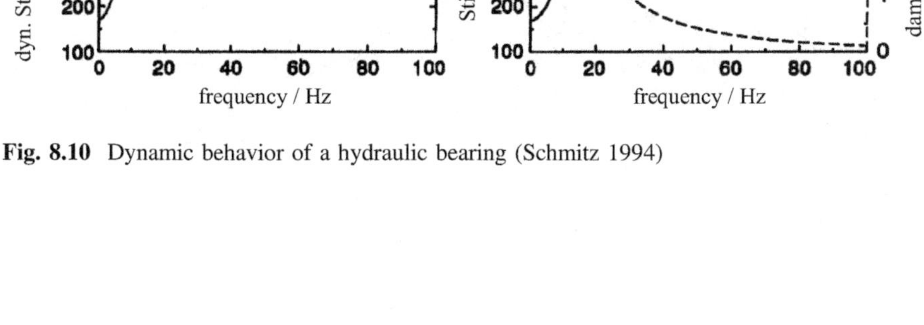

Fig. 8.10 Dynamic behavior of a hydraulic bearing (Schmitz 1994)

Table 8.1 System data of Fig. 8.10

Hydraulic bearing		Rubber bearing
$c_1 = 170$	Ns/mm	$c_1 = 300$ Ns/mm
$c_2 = 430$	Ns/mm	
$d_1 = 0$	Ns/mm	$d_1 = 0.7$ Ns/mm
$d_2 = 4$	Ns/mm	

8.2.4 *Modeling of Homokinetic Joints*

In passenger cars with front-wheel drive a nearly uniform transmission of the rotation from the output shaft onto the wheels is realized by means of constant-velocity joints, also named homokinetic joints. Universal joints on the other hand transmit angular velocity in a much more non-uniform manner. The non-uniformity between input and output angular velocity is increasing with the deflection angle of the two shafts (Aucktor et al. 2002; Schütz 1971).

In this section, the influence of the drivetrain on the drivability as well as on the motion transfer inside the drivetrain will be investigated. For this purpose, an exact modeling of the homokinetic joints is required. The inner homokinetic joint—located in front of the differential gear—can be modeled as a prismatic joint, enabling shifts along the axial direction, whereas the outer homokinetic joint—oriented towards the wheel—is realized as a revolute joint (Fig. 8.11). (In present cars, inner tripod joints are used.)

In Fig. 8.12, the kinematic model of the realized synchronous velocity joint (Matschinsky 2007; Hunt 1990) is presented, while Fig. 8.13 shows the complete drive shaft with the inner (on the engine side), as well as the outer (on the wheel side) synchronous velocity joint. The kinematic topology of the synchronous velocity joints is identical.

According to Fig. 8.12, the two shafts are connected by the spherical joint S_1. The transmission of the angular velocity is used in the mechanism shown in Fig. 8.13 and, while representing a kinematical loop, consists of the two revolute

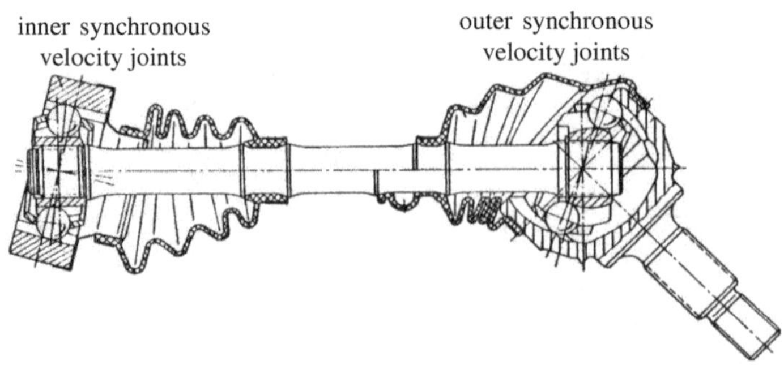

Fig. 8.11 Drive shaft of a front-wheel-drive vehicle

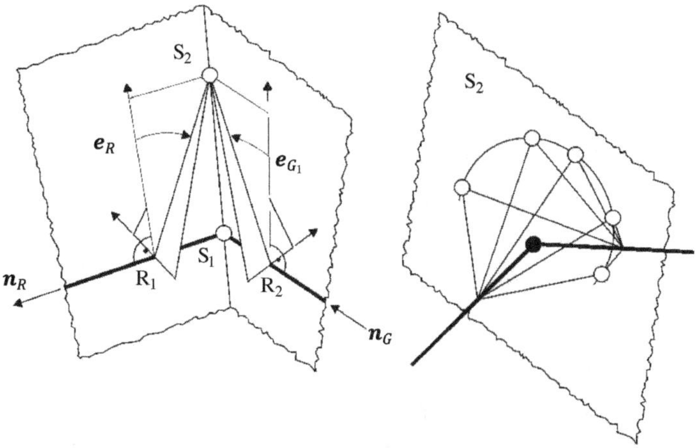

Fig. 8.12 Kinematic model of the synchronous velocity joint (Schmitz 1994)

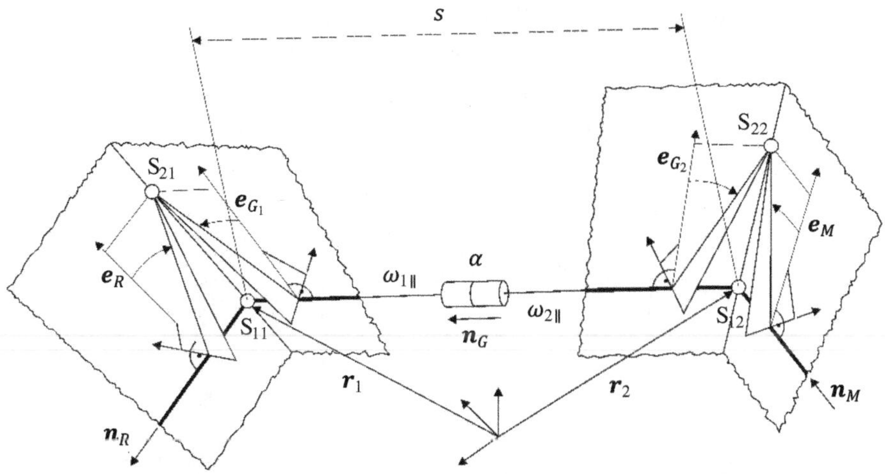

Fig. 8.13 Model of the drive shaft

joints R_1, R_2, and the spherical joint S_2. The joint S_2 is moving in the bisectoring plane of the two shafts. The complete drive shaft, as shown in Fig. 8.13, consists of three individual shafts, which are connected by the two synchronous velocity joints mentioned above.

The torsional DoF is approximated by a concentrated revolute joint (difference angle α, angular velocity $\dot{\alpha}$) (Fig. 8.13), which will be determined below. According to Schmitz (1994) for the two unit vectors $\boldsymbol{e}_{G_1}$ and $\boldsymbol{e}_R$ from the geometry one obtains:

$$\boldsymbol{e}_{G_1} = \boldsymbol{e}_R - \frac{\boldsymbol{e}_R \cdot \boldsymbol{n}_G}{1 + \boldsymbol{n}_R \cdot \boldsymbol{n}_G} (\boldsymbol{n}_R + \boldsymbol{n}_G), \tag{8.6}$$

$$\boldsymbol{e}_{G_2} = \boldsymbol{e}_M - \frac{\boldsymbol{e}_M \cdot \boldsymbol{n}_G}{1 + \boldsymbol{n}_M \cdot \boldsymbol{n}_G} (\boldsymbol{n}_M + \boldsymbol{n}_G). \tag{8.7}$$

For the difference angle α, it follows that:

$$\cos \alpha = \boldsymbol{e}_{G_1} \cdot \boldsymbol{e}_{G_2}, \tag{8.8}$$

$$\sin \alpha = (\boldsymbol{n}_G \times \boldsymbol{e}_{G_2}) \cdot \boldsymbol{e}_{G_1}. \tag{8.9}$$

For the calculation of the difference angular velocity $\dot{\alpha}$, the angular velocity perpendicular to the drive shaft has to be determined:

$$\boldsymbol{\omega}_{\perp} = \frac{(\dot{\boldsymbol{e}}_1 - \dot{\boldsymbol{r}}_1) \times \boldsymbol{n}_G}{s}. \tag{8.10}$$

For the angular velocities parallel to the drive shaft $\omega_{1\|}$ (left side) and $\omega_{2\|}$ (right side), one obtains

$$\omega_{1\|} = (\boldsymbol{\omega}_R - \boldsymbol{\omega}_{\perp}) \cdot \frac{\boldsymbol{n}_R + \boldsymbol{n}_G}{1 + \boldsymbol{n}_R \cdot \boldsymbol{n}_G}, \tag{8.11}$$

$$\omega_{2\|} = (\boldsymbol{\omega}_M - \boldsymbol{\omega}_{\perp}) \cdot \frac{\boldsymbol{n}_M + \boldsymbol{n}_G}{1 + \boldsymbol{n}_M \cdot \boldsymbol{n}_G} \tag{8.12}$$

and the desired angular velocity $\dot{\alpha}$ is:

$$\dot{\alpha} = \omega_{1\|} - \omega_{2\|}. \tag{8.13}$$

8.3 Modeling of the Engine

In Fig. 8.14, the effective interaction of different components for the modeling of a real time-range behavior of the engine-torque is shown. The dominant elements in the engine model are the static characteristics, which provide speed n and throttle-valve position α^*, respectively, the steady-state engine torque M^* for a given crankshaft. The time-delayed throttle-valve position α^* will be determined from an underlying throttle-valve position α_{Gas} (driver command) and a subsequent first-order time-delay PT_1 plus a time-delay element T_t.

The dynamic characteristics (including time-delay effects) of the engine will be considered by a series-connection of the steady-state engine torque M^* with a first-order time-delay term PT_1 in combination with a time-delay element T_t. The PT_1-term includes the inlet manifold dynamics of the engine, whereas the

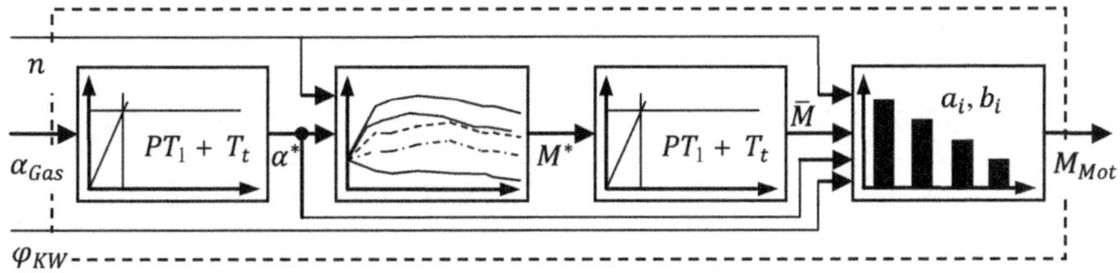

Fig. 8.14 Complete model of the engine torque

time-delay T_t describes the time-delay between cylinder-charge and torque-output. The time-constant of the PT_1-term mainly depends on the length of the inlet manifold. The time-delay T_t corresponds to a half-turn of the crankshaft. From this time-delayed average engine torque $\bar{M}$, one can get—together with the engine rotation n, and the crankshaft angle φ_{KW}, as well as the delayed throttle-valve position α^*, based on Fourier series (following Eq. (8.14))—a good approximation for the periodic input torque M_{Mot} of the combustion engine.

$$M_{Mot}(\alpha, n, \varphi_{KW}) = \bar{M} + \sum_{i=1}^{N} a_i(n, \alpha^*) \cos\left(iz_p\, \varphi_{KW}\right) + b_i(n, \alpha^*) \sin\left(iz_p\, \varphi_{KW}\right). \tag{8.14}$$

Here, the parameters a_i and b_i represent the Fourier coefficients and z_p is an engine-specific parameter (e.g. for a 4-stroke engine $z_p = 0.5$). The Fourier coefficients have to be calculated based on the crankshaft rotation n and the throttle-valve position α^* by a double linear interpolation between the data points at the four corner points, see also Fig. 8.15a. The time-history of the resulting engine torque M_{Mot} is shown in Fig. 8.15b. In this result, the influence of a dual-mass flywheel is not considered.

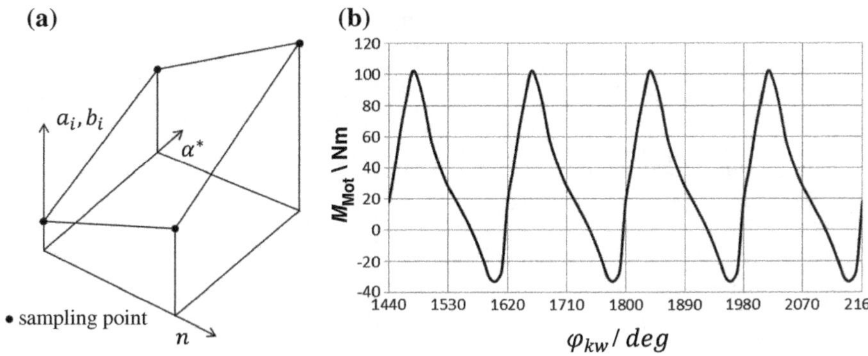

Fig. 8.15 Fourier coefficients and generated oscillation

8.4 Relative Kinematics of the Drivetrain

In the kinematic model of the drivetrain, as presented in Fig. 8.4, one can detect an independent kinematic loop L_1, which is shown in Fig. 8.16. A topologically equivalent kinematic loop L_2 exists for the right part of the vehicle as well. Using the methods for the analysis of kinematic loops (Woernle 1988) presented in Chap. 3, the equations for the relative kinematics of L_1 can be derived in a very simple, straight-forward manner, and can even be solved explicitly in this particular case. In a further step, in addition to the kinematics of L_1 and L_2, a kinematic approach for the gearbox and the differential gear is required.

The differential gear in the drivetrain makes it possible for the two front-wheels, e.g. while cornering, to rotate with different velocity. Differential gears without locking, as usually realized in front-wheel-drive vehicles, distribute the torque equally onto both vehicle sides (Zomotor 1991). The differential gear, shown in Fig. 8.17, consists of a crown wheel, the two differential bevel gears, and the two driven bevel gears.

For the crown wheel, one gets the kinematic relation

$$\dot{\varphi}_T = \frac{1}{2}\left(\dot{\varphi}_{M_3} + \dot{\varphi}_{M_4}\right). \tag{8.15}$$

Between differential gear and clutch we have the manually shifted transmission. Its function is to convert torque and rotation speed. The gearbox of a Volkswagen VW-Golf II is shown in Fig. 8.18. One has to differentiate between two conversion

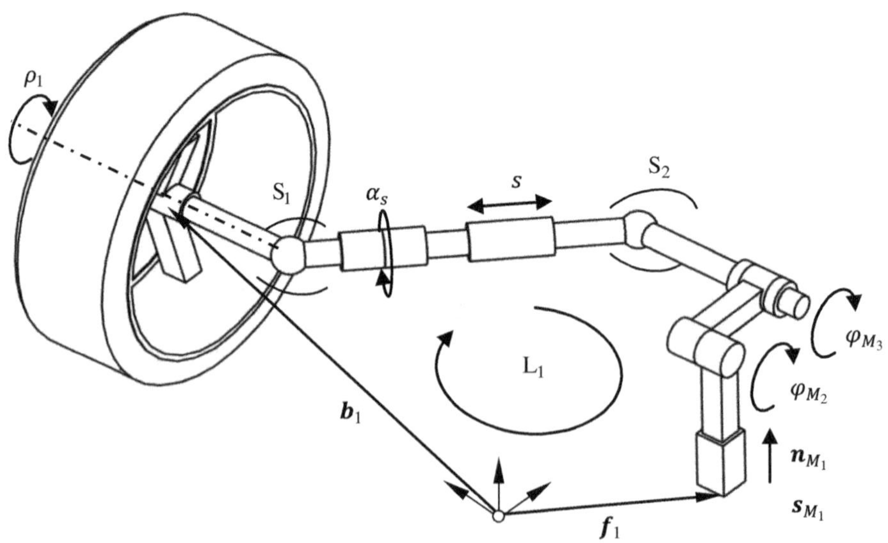

Fig. 8.16 The kinematic loop L_1

$$i_D = -\frac{\dot{\varphi}_{II}}{\dot{\varphi}_T}.$$

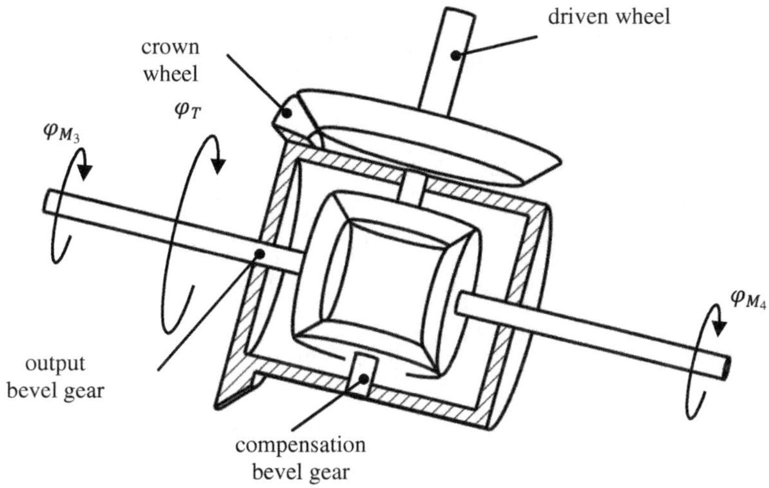

Fig. 8.17 Schematic representation of the differential gear of a front-wheel-drive vehicle

ratios. The conversion ratio i_G represents the conversion between the transmission shafts I and II in the individual gears

$$i_G = -\frac{\dot{\varphi}_I}{\dot{\varphi}_{II}}, \tag{8.16}$$

while the axle conversion ratio i_D represents the constant conversion ratio between transmission shaft I and crown wheel shaft (Bosch 2011)

$$i_D = -\frac{\dot{\varphi}_{II}}{\dot{\varphi}_T}. \tag{8.17}$$

In Eqs. (8.16) and (8.17), the velocities relative to the engine block have to be inserted, resulting in the total conversion ratio

$$i_{Ges} = i_G \cdot i_D \tag{8.18}$$

from which the kinematic correlations between transmission shaft I and shaft II, respectively, and the output bevel gear can be calculated. Finally, the difference angle $\Delta\beta_K$ and its angular velocity $\Delta\dot{\beta}_K$, representing the elastic rotational displacement of the clutch (see also Fig. 8.4), can be expressed as

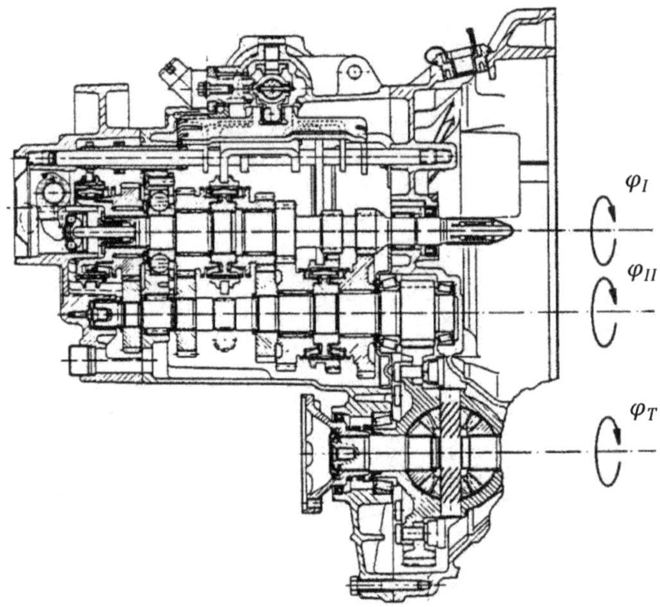

Fig. 8.18 Gearbox VW Golf II

$$\Delta\beta_K = \varphi_I - \varphi_{M_5}, \tag{8.19}$$

$$\Delta\dot{\beta}_K = \dot{\varphi}_I - \dot{\varphi}_{M_5}. \tag{8.20}$$

8.5 Absolute Kinematics of the Drivetrain

The aim of absolute kinematics is to derive the absolute position, velocity, and acceleration for the individual rigid bodies of the drivetrain, modeled here as multibody system with rigid bodies with mass, with the help of the relations from relative kinematics (see Chap. 3). The here considered rigid bodies with mass are:

- drive shafts left and right,
- output shafts left and right,
- engine housing,
- transmission shafts I and II, together with differential crown wheel,
- crank shaft.

Based on the equations of the relative kinematics, the motion of all inherent bodies with respect to vehicle-fixed reference system "*V*" can be determined. In a further step, the absolute velocities and accelerations relative to the inertial system can be calculated (Sect. 5.2).

8.6 Equations of Motion

To state the equations of motion for the multibody system under consideration, in a first step the forces acting on the system have to be considered. The external forces and torques, acting on the drivetrain, are the weights of the individual bodies, spring forces, and torques in the elastic bearings and in the clutch, as well as the engine torque. The spring forces and torques are responsible for a dynamic coupling of the motion of the individual bodies. Reset forces and torques have to be considered in the revolute joints of the two prop shafts, in the translatory and rotatory engine bearing, as well as in the clutch by spring-damper systems with adjusted characteristic curves (see Chap. 9).

Starting with the global kinematics (see Chap. 3.5), consisting of relative kinematics and absolute kinematics, the equations of motion of the (sub-) systems can now be expressed. The equations of motion will be stated using the method of kinematic differentials, as analyzed in detail in Chap. 4.6, and as already demonstrated for other vehicle components and subsystems in previous chapters. It is the aim of this approach to state equations of motion in minimum coordinates. It is a further aim of this method to take advantage of the possibilities in order to reduce the computational amount by using a suitable evaluation of the elements of the equations of motion. A more detailed analysis of the dynamics of the drivetrain and its influence on the dynamics of vehicles (front-wheel driven, rear-wheel driven, and four-wheel drive can be found in Schmitz (1994). The influence of the drivetrain on the structure of the mass-matrix of the complete vehicle is shown in Fig. 8.19, and will be discussed in more detail in Chap. 12.

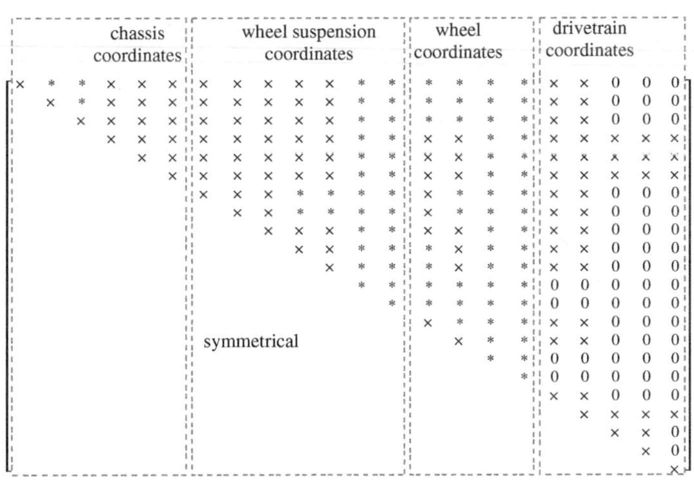

×: Mass matrix element that is influenced by the drivetrain
*: Mass matrix element that is not influenced by the drivetrain
0: Mass matrix element ofthe drivetrain = 0

Fig. 8.19 Influence of the drivetrain on the filling of the mass matrix

8.7 Discussion of Simulation Results

As an example for a simulation with the front-drive-vehicle (VW Golf II) used in this chapter, a sudden steering input in longitudinal direction will be discussed, i.e. the steering wheel will be turned instantaneously by 12° in the right direction (see also Fig. 14.1). The steering wheel angle and engine torque are kept constant. Figure 8.20 shows the time history of the angular velocities of the left and right front wheel.

While the angular velocity of the left front wheel is increasing, the angular velocity of the right front wheel is decreasing due to the smaller curve radius. As a consequence, the angular velocities of the front wheels $\dot{\rho}_1$ and $\dot{\rho}_2$ will decrease due to the cornering resistance. The time-history of the angular velocities $\dot{\varphi}_{M_3}(t)$ and $\dot{\varphi}_{M_4}(t)$ of the left and right drive shaft looks similar to the angular velocities of the wheels (Fig. 8.21). This can be interpreted as a stronger vibration tendency in the shafts due to the smaller moments of inertia of the drivetrain in comparison to the wheels.

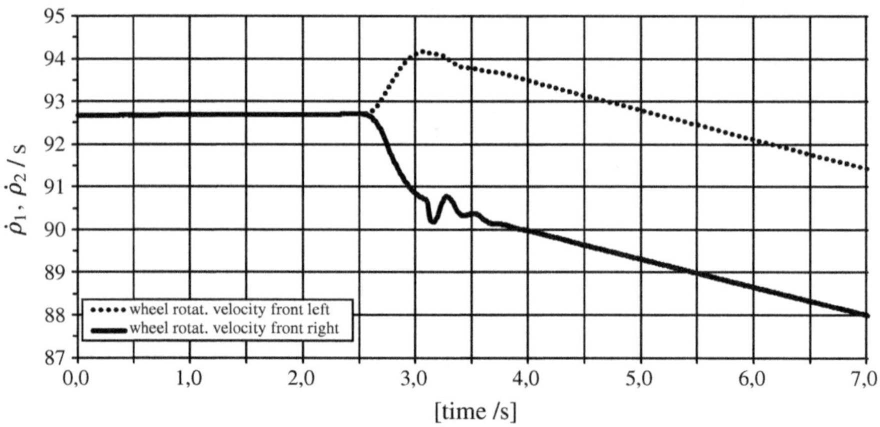

Fig. 8.20 Angular velocities of the right and left front wheel

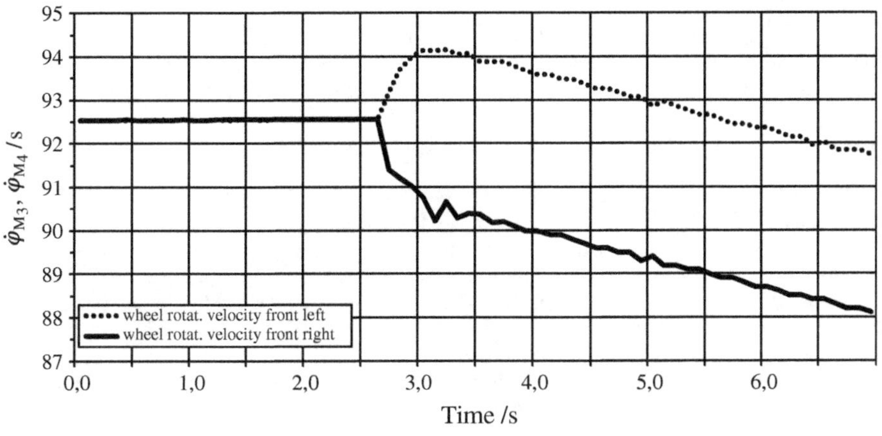

Fig. 8.21 Angular velocities $\dot{\varphi}_{M_3}(t)$ and $\dot{\varphi}_{M_4}(t)$

References

Aucktor E, Schmelz F and Seherr-Thoss H-C (2002) Gelenke und Gelenkwellen: Berechnung, Gestaltung, Anwendungen. 2., erw. Aufl. (ed.), Springer-Verlag, Berlin u.a. - ISBN 978-3-540-41759-0

Bathelt H, Bukovics J and Young D (1989) Die Entwicklung der Aggregatlagerung des Audi V8. *ATZ* 91(2). S 93-102

Bosch (2011) Kraftfahrzeugtechnisches Taschenbuch. 27. Auflage (ed.), Vieweg + Teubner, Plochingen. - ISBN 978-3-8348-1440-1

Bösenberg D and van den Boom J (1979) Motorlagerung im Fahrzeug mit integrierter hydraulischer Dämpfung - ein Weg zur Verbesserung des Fahrkomforts. ATZ 81(10)

Bürger K-H and Dödelbacher G (1988) Verbesserung des Fahrzeugschwingungsverhaltens durch Strukturoptimierung in der Konzeptphase. *VDI-Berichte* 699

Demant H, Roos J and Wolko D (1990) Schwingungstechnische Anforderungen an Triebwerklagerungssysteme. *ATZ. Automobiltechnische Zeitschrift* 92, - ISBN 0001-2785. S 110-8

Freudenberg U and Freudenberg T (1998) Berechnung des dynamischen Verhaltens komplexer Hydrolager. *Automobil-Industrie* 2. S 147-54

Freund J and von Sivers R (1981) Der neue Porsche 944. *ATZ* 83(9). S 415-20

Göhlich D and Köder B (1992) Berechnung der Körperschallübertragung bei Elastomerlagern. VDI Bericht Nr. 1007. S 467-85

Holzemer K (1985) Theorie der Gummilager mit hydraulischer Dämpfung. *ATZ. Automobiltechnische Zeitschrift* 87, - ISBN 0001-2785. S 545-51

Hunt KH (1990) Kinematic Geometry of Mechanisms

Johnson SR and Subhedar JW (1979) Computer optimization of engine mounting systems. *SAE paper* 790974. S 979

Matschinsky W (2007) Radführungen der Straßenfahrzeuge: Kinematik, Elasto-Kinematik und Konstruktion. 3. (ed.), Springer - ISBN 978-3-540-71196-4

Schiehlen W and Eberhard P (2004) Technische Dynamik- Modelle für Regelung und Simulation; mit 4 Tabellen und 44 Beispielen. 2., neubearb. und erg. Aufl. (ed.), Teubner, Stuttgart u.a. - ISBN 978-3-519-12365-1

Schmitz T (1994) Modellbildung und Simulation der Antriebsdynamik von Personenwagen [Dr.-Ing.]. Dissertation, Universität – Gesamthochschule Duisburg, Duisburg
Schnelle K-P (1990) Simulationsmodelle für die Fahrdynamik von Personenwagen unter Berücksichtigung der nichtlinearen Fahrwerkskinematik Dissertation, Universität Stuttgart, Stuttgart.
Schütz KH (1971) Gleichlauf-Kugelgelenke für Kraftfahrzeugantriebe. In *Antriebstechnik*,Vol. 10, S 437–40
Wallentowitz H and Mitschke M (2006) Dynamik der Kraftfahrzeuge (VDI-Buch). Springer-Verlag, Heidelberg. - ISBN 978-3-540-42011-8
Woernle C (1988) Ein systematisches Verfahren zur Aufstellung der geometrischen Schließbedingungen in kinematischen Schleifen mit Anwendung bei der Rückwärtstransformation für Industrieroboter [Dr.-Ing.]. Dissertation, Universität Stuttgart, Düsseldorf: VDI-Verlag
Zamow J and Witte L (1988) Fahrzeugsimulation unter Verwendung des starrkörperprogramms ADAMS. *VDI-Berichte* - ISBN 0083-5560. S 287-309
Zomotor A (1991) Fahrwerktechnik: Fahrverhalten. Vol. 2. Auflage, Vogel Verlag und Druck KG, Würzburg. - ISBN 978-3-8023-0774-4

Chapter 9
Force Components

Force components can occur as active or passive elements in vehicles. Some examples for passive force components are shown on a wheel suspension in Fig. 9.1. The requirements for the force components of a chassis keep increasing steadily and are at times contradictory, Ref. Fig. 9.2. This is why the implementation of the safety and comfort functions requires the substitution of the passive components with active ones. This way, the driving behavior can for example be influenced sustainably by the actuation of the usually passive anti-roll bars, Ref. Chap. 14.

9.1 Forces and Torques in Multibody Systems

The simulation and analysis of vehicle models is influenced significantly by the type and the temporal behavior of the applied forces. In mechanics, one differentiates between different fundamental types of forces (Schiehlen and Eberhard 2004):

- external and internal forces and torques,
- applied and reaction forces and torques,
- surface and volume forces.

Category 1 depends on the boundaries of the system in question. The outside forces and torques acting on said system are referred to as external forces. The forces acting within a system are thus referred to as the internal forces. Through a change in the system boundaries (such as cutting to arrive at a free body diagram), the internal forces become external forces of the new system boundaries. At the boundary where the cut is implemented, two external forces, based on the principle of Newton's Third Law (action equals reaction) occur whose amplitudes are equal but whose directions are opposite to each other.

D. Schramm et al., Vehicle Dynamics,
DOI 10.1007/978-3-662-54483-9_9

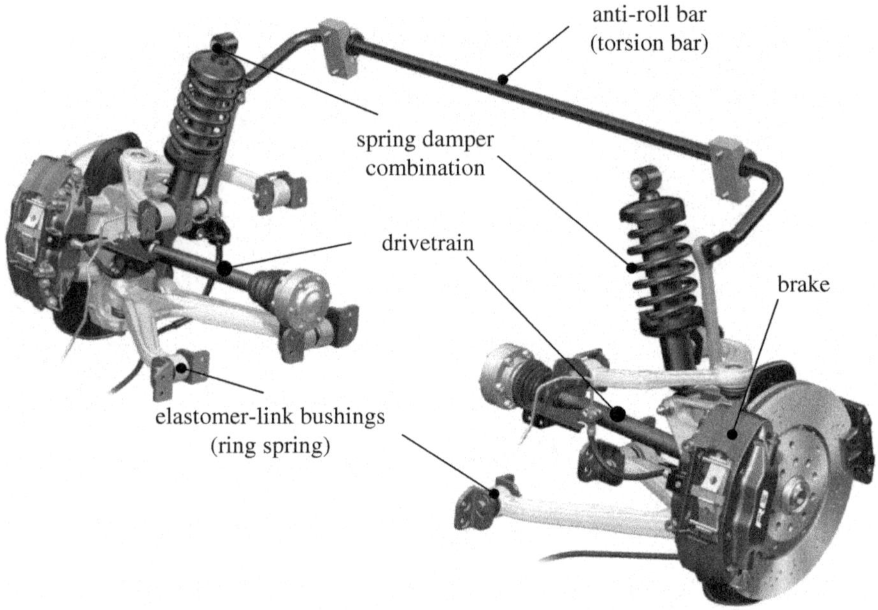

Fig. 9.1 Force elements in a wheel suspension (by courtesy of Audi AG)

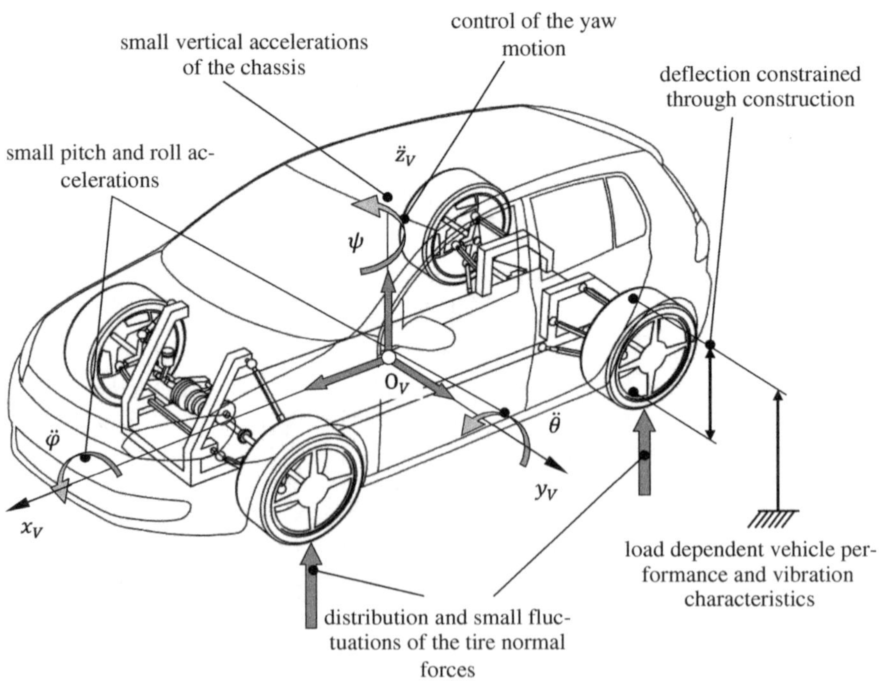

Fig. 9.2 Requirements on force elements

Category 2 refers to the source of the forces. Applied forces can be described using the physical force laws with respect to position and velocity. In systems with COULOMB's friction however, further considerations are necessary, as the applied forces in this case are generally dependent on the reaction forces (normal forces) (Schramm 1986a). Examples of applied forces in vehicle systems are the forces due to gravity, the spring and damper forces, but also the gas forces in the cylinders of the combustion engine, the air forces and especially the tire forces.

Examples of reaction forces are the forces occurring inside the joints. If modeling the wheel and the road surface as rigid bodies, the normal forces from the road to the wheels are reaction forces. In the case of the modeling of the wheel with an elastic tire however, they are considered to be applied forces.

The third category refers to the distribution of the forces. Surface forces are constrained to a surface, and the volume forces are distributed spatially. Examples of surface forces in vehicles are the wind forces and, strictly speaking, also the forces between the road surface and the thread of the tire. The volume forces in vehicles consist of all the forces that are due to gravity, but also for example, the electrical and magnetic field forces.

Another differentiation results from the temporal characteristic of the forces. One differentiates between forces with deterministic, and those with stochastic temporal characteristics. Deterministic forces are defined explicitly through their time functions. Stochastic forces do not possess a rule based temporal characteristic and need to be described using statistical methods (Smith 2003).

9.1.1 Reaction Forces

For further analysis of the reaction forces, one first considers the Newton-Euler equations for the components of the considered multibody system (see Chap. 4):

$$m_i \ddot{\boldsymbol{s}}_i = \boldsymbol{F}_i^e + \boldsymbol{F}_i^z, \quad i = 1, \ldots, n_B, \tag{9.1}$$

$$\boldsymbol{\Theta}_{S_i} \dot{\boldsymbol{\omega}}_i + \boldsymbol{\omega}_i \times \boldsymbol{\Theta}_{S_i} \boldsymbol{\omega}_i = \boldsymbol{T}_i^e + \boldsymbol{T}_i^z, \quad i = 1, \ldots, n_B. \tag{9.2}$$

On the right hand side of the Eqs. (9.1) and (9.2), the resulting forces on each of the bodies K_i are partitioned into the resulting applied forces $\boldsymbol{F}_i^e$ and torques $\boldsymbol{T}_i^e$ as well as the reaction forces $\boldsymbol{F}_i^z$ and torques $\boldsymbol{T}_i^z$. Similar to how the local velocities can be represented by the generalized velocities

$$\dot{\boldsymbol{s}}_i = \boldsymbol{J}_{s_i} \dot{\boldsymbol{q}}, \quad \boldsymbol{\omega}_i = \boldsymbol{J}_{\varphi_i} \dot{\boldsymbol{q}}, \quad i = 1, \ldots, n_B, \tag{9.3}$$

the local reaction forces can be represented by dependencies of the generalized reaction forces

$$\boldsymbol{Z} = \begin{bmatrix} Z_1 & \ldots & Z_r \end{bmatrix}^T, \; r = 6n_B - f, \tag{9.4}$$

and

$$\boldsymbol{F}_i^z = \boldsymbol{F}_i^z(\boldsymbol{Z}), \; \boldsymbol{T}_i^z = \boldsymbol{T}_i^z(\boldsymbol{Z}), i = 1, \ldots n_B. \tag{9.5}$$

The relationships derived from Eq. (9.5) are generally obtained from the joint forces of the individual bodies. To this end it is most convenient to apply the natural bearing forces, i.e. the bearing forces pointing in the locked bearing directions, in order to obtain the generalized reaction forces; Ref. Chaps. 3 and 4 as well as (Schramm 1985, 1986a, b) and (Schiehlen and Eberhard 2004). This is however only required if the joint forces are necessary, for example to determine the inner loads on the system. However, as is usually the case, if the behavior of motion of the system is under investigation, the consideration of Eq. (9.5) is not required, as the reaction forces vanish in both the principle of virtual work as well as in Jourdain's principle, Ref. Chap. 4.

9.1.2 Applied Forces

As already mentioned, the applied forces in the vehicle systems depend on the position and velocity values, as well as other parameters of the systems and, if required, also explicitly on time:

$$\boldsymbol{F}_i^e = \boldsymbol{F}_i^e(\boldsymbol{q}, \dot{\boldsymbol{q}}, t), \quad \boldsymbol{T}_i^e = \boldsymbol{T}_i^e(\boldsymbol{q}, \dot{\boldsymbol{q}}, t), \quad i = 1, \ldots n_B. \tag{9.6}$$

For each applied force, information regarding the action direction of the force as well as information about the corresponding force law is required.

9.2 Operating Brake System

An indispensable vehicle component is the brake system, which is nowadays constructed mainly as a disc brake (Burckhardt 1991; Breuer and Bill 2012b). Accurate models of the disc brakes were developed mainly to reproduce noise disturbances during braking such as brake squealing (squeal, $f = 1 - 15$ kHz) or the brake groan (groan, $f = 200 - 500$ Hz), Ref. (Gouya and Nishiwaki 1990; Samie and Sheridan 1990) or (Shi 1996). As the influence of the structure dynamics dominates in this kind of problems, mostly finite elements analysis is used. For vehicle dynamics investigations however, these effects do not normally play a role, which significantly simplifies the modeling in this case.

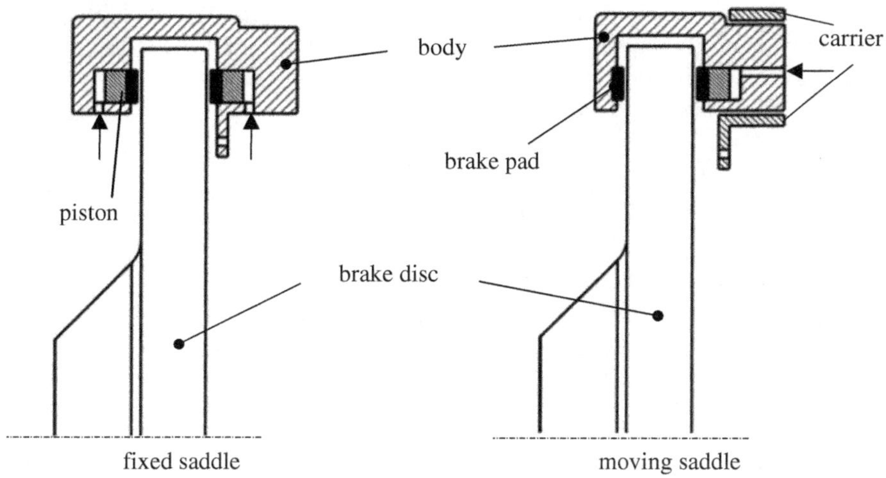

Fig. 9.3 Construction of a disc brake

In the following, the constructive design and the modeling of a disc brake will be explained, Ref. Fig. 9.3. The exact construction and references relating to the design can be found in (Burckhardt 1991) or (Breuer and Bill 2012a). A common type of construction is the gray cast brake disc with externally embracing brake pads. This can be differentiated further into fixed and moving (swimming) brake saddles, Ref. Fig. 9.3.

The modeling of a fixed-brake-saddle brake will be considered exemplarily. The focus of this modeling is the conversion of the braking pressure into a braking torque on the brake disc. For a given braking pressure p_B, the braking torque M_B on a rotating brake disc is given by

$$M_B = p_B A_K r_B \mu_G(p_B, T_B). \tag{9.7}$$

Here, A_K refers to the effective contact surface between the brake pad and the brake disc while r_B describes the distance between the axis center and the brake pad center. The sliding friction coefficient μ_G depends on the effective acting brake pressure p_B and the pad temperature T_B. As per (Burckhardt 1991) and (Schuster 1999), these relationships can be described through approximations. If one were to neglect the sliding velocity, the friction value would increase to a maximum at the optimum temperature $T_{B,opt}$ with increasing temperature T_B, after which it would start to drop. Mathematically, this relationship can be approximated using a parabola:

$$\mu_G(T_B) = \mu_{G,max}\left[1 - c_P\left(\frac{T_B - T_{B,opt}}{T_{B,opt}}\right)^2\right]. \tag{9.8}$$

In this case the value c_P is a dimensionless material constant, which represents the temperature dependency of the material pair between the brake disc and the brake pad. Very similarly, one can also approximate the relationship between the friction and the brake pressure:

$$\mu_G(p_B) = \mu_{G,p_B=0}\left[1 - c_T\left(\frac{p_B}{p_B^*}\right)^2\right]. \tag{9.9}$$

Here, the value c_T is another dimensionless material constant, $\mu_{G,p_B=0}$ is the friction coefficient at $p_B = 0$, and p_B^* represents the maximum brake pressure. The equation for the friction coefficient then results from Eqs. (9.8) and (9.9) as a first approximation:

$$\mu_G(p_B, T_B) = \mu_{G,p_B=0}\left[1 - c_T\left(\frac{p_B}{p_B^*}\right)^2\right] \cdot \mu_{G,max}\left[1 - c_P\left(\frac{T_B - T_{B,opt}}{T_{B,opt}}\right)^2\right]. \tag{9.10}$$

9.3 Aerodynamic Forces

At average or higher velocities the aerodynamic forces and torques do not only have a significant influence on the vehicle performance and fuel consumption, but also on its dynamic behavior. The air resistance can be fundamentally attributed to three physical effects, (Popp and Schiehlen 1993):

- The form resistance results from the turbulence of the air stream at the vehicle rear. This effect accounts for around 85 % of the complete resistance.
- The friction resistance (surface resistance) results from the shear flow on the chassis and depends on its surface area (accounts for approx. 10 %).
- The inner resistance is due to the fact that part of the air stream flows through the chassis (accounts for approx. 5 %). This stream serves to cool the motor and the ventilation of the passenger space.

The wind forces are, at large, caused by turbulent streams. They are proportional to the ram pressure:

$$p_L = \frac{1}{2}\rho_L v_L^2, \tag{9.11}$$

with the air density ρ_L and the flow velocity v_L. With an effective cross-sectional area A of the vehicle and a dimensionless coefficient c_W (Table 9.1) one attains the force due to wind resistance

Table 9.1 Examples of air resistance coefficients

Vehicle type	Wind resistance coefficients c_W
Passenger vehicle	0.2–0.4
Truck	0.6–1.0
Bus	0.6–0.7
Motorcycle	0.6

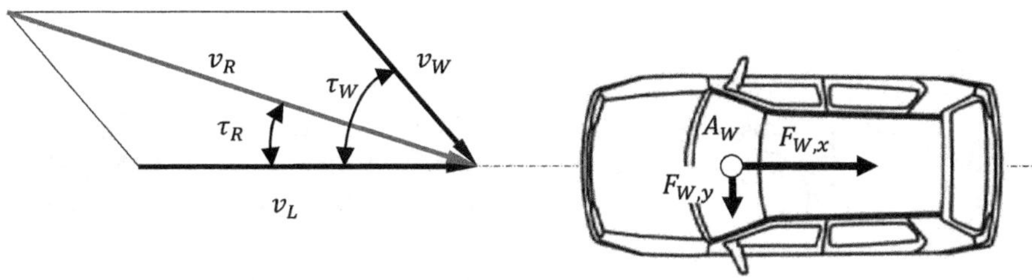

Fig. 9.4 Determination of the relative velocity

$$F_W = c_W A p_L. \tag{9.12}$$

If one now also wants to consider the environmental (surrounding) wind speed along with the inflow velocity due to airflow, one must then calculate using the actual relative velocity. In Fig. 9.4, the relative velocity and the flow angle is represented. Here, the direction of the environmental velocity is given as the angle τ_W. The relative velocity is obtained by the application of the cosine law:

$$v_R = \sqrt{v_L^2 + v_W^2 + 2 v_L v_W \cos \tau_W}. \tag{9.13}$$

For the air flow angle one then gets:

$$\tau_R = \arcsin\left(\frac{v_W}{v_R} \sin \tau_W\right). \tag{9.14}$$

Through the slanted inflow, along with a lateral force, there are also changes in the direction of the longitudinal force as well as aerodynamic lift forces. In total, the following force components are bound to occur (Haken 2008):

Air resistance (longitudinal force):

$$F_{W,x} = c_x A_W \frac{\rho}{2} v_R^2, \tag{9.15}$$

lateral force:

$$F_{W,y} = c_y A_W \frac{\rho}{2} v_R^2, \tag{9.16}$$

lift (buoyant force):

$$F_{W,z} = c_z A_W \frac{\rho}{2} v_R^2. \tag{9.17}$$

Notes

- For an airflow angle with $\tau_R = 0$ in Eq. (9.15) again becomes $c_x = c_W$.
- For reasons of simplicity, the cross-sectional area is assumed to be A_W in Eqs. (9.16) and (9.17), even though this cannot be explained physically in these cases.
- The coefficients c_x, c_y and c_z depend on the angle of inflow τ_R.

9.4 Spring and Damper Components

9.4.1 Spring Elements

The Fig. 9.5 shows two bodies K_i and K_j, which are connected by a spring element at points $\mathrm{P}_{i,l}$ and $\mathrm{P}_{j,l}$. If one were to name the position vector from one reference system to the point of action as $\boldsymbol{r}_{P_{i,l}}$ and $\boldsymbol{r}_{P_{j,l}}$ and the corresponding position vectors from the reference point of the bodies with ${}_i\boldsymbol{r}_{P_{i,l}}$ and ${}_j\boldsymbol{r}_{P_{j,l}}$ respectively, and the distance vector with

$${}_{i,l}\boldsymbol{r}_{j,l} = \boldsymbol{r}_{P_{j,l}} - \boldsymbol{r}_{P_{i,l}}, \tag{9.18}$$

then one arrives at the spring force of the body K_i:

$$\boldsymbol{F}_l = \frac{{}_{i,l}\boldsymbol{r}_{j,l}}{\left|{}_{i,l}\boldsymbol{r}_{j,l}\right|} f\left(\left|{}_{i,l}\boldsymbol{r}_{j,l}\right|\right) = \boldsymbol{u}_l f(u) = \boldsymbol{u}_l(f(u_{l_0}) + f(\Delta u_l)). \tag{9.19}$$

Here, the unit vector in the direction of the force $\boldsymbol{F}_l$ on the body K_i and $f(u)$ with $u_l = \left|{}_{i,l}\boldsymbol{r}_{j,l}\right|$ describes the force law. Equation (9.19) is valid for various force laws, as long as these force laws can be represented as dependent on displacement u only. The expression $f(u_{l_0})$ in Eq. (9.19) refers to the so-called uncompressed spring length, whereas $f(\Delta u_l)$ with $\Delta u_l = u_l - u_{l_0}$ refers to the variable part of the spring force. One obtains:

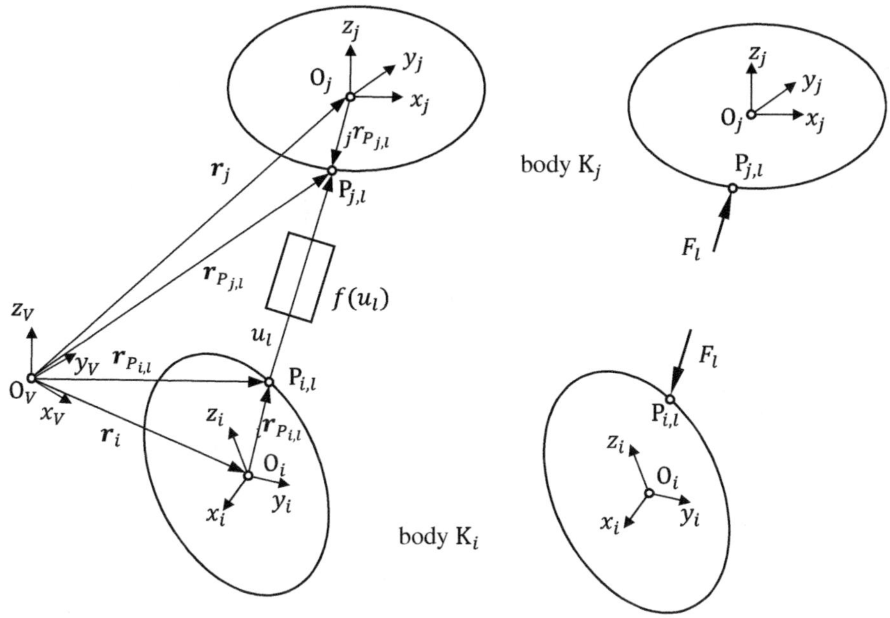

Fig. 9.5 Force elements between two bodies

$$\boldsymbol{F}_l = \boldsymbol{u}_l(f(u_{l_0}) + f(\Delta u_l)) = \boldsymbol{F}_{l_0} + \boldsymbol{F}_l(\Delta u_l), \tag{9.20}$$

with the spring preload F_{l_0} and the variable force part $F_l(\Delta u)$. A force acts on the body K_j, with the same magnitude but in opposite direction. In addition to the spring force, a torsional torque has to be considered, which for example with respect to the reference point of the body K_i is:

$$\boldsymbol{T}_{i,l} = {}_i\boldsymbol{r}_{P_{i,l}} \times \boldsymbol{F}_{i,l}. \tag{9.21}$$

The torsional torque on the body K_j is obtained in a similar way, by replacing i with j. A typical force law $f(\Delta u_l)$ is shown in Fig. 9.6. Here a force law is shown which has slightly progressive characteristics in the central region and a significantly progressive characteristic in the boundary regions, modeling end stops.

9.4.2 Damper Elements

The force of a damper element is described with respect to the displacement velocity v_l. This is calculated from the temporal change of the distances between the points $P_{i,l}$ and $P_{j,l}$ and finally the projection on the direction vector $\boldsymbol{u}_l$ to:

Fig. 9.6 Typical characteristic of a chassis spring

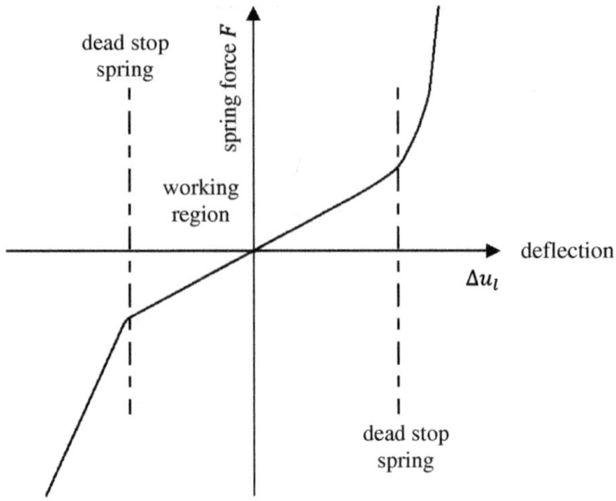

$$v_l = \boldsymbol{u}_l^T{}_{i,l}\dot{\boldsymbol{r}}_{j,l} = \boldsymbol{u}_l^T\left({}_j\dot{\boldsymbol{r}}_{P_{j,l}} - {}_i\dot{\boldsymbol{r}}_{P_{i,l}}\right)$$
$$= \boldsymbol{u}_l^T\left(\dot{\boldsymbol{r}}_j + \boldsymbol{\omega}_j \times_j \boldsymbol{r}_{P_{j,l}} - \dot{\boldsymbol{r}}_i - \boldsymbol{\omega}_i \times_i \boldsymbol{r}_{P_{i,l}}\right). \tag{9.22}$$

For a damper force, one arrives at

$$\boldsymbol{F}_l = \boldsymbol{u}_l g(v_l). \tag{9.23}$$

A typical force law for a chassis damper is shown in Fig. 9.7.

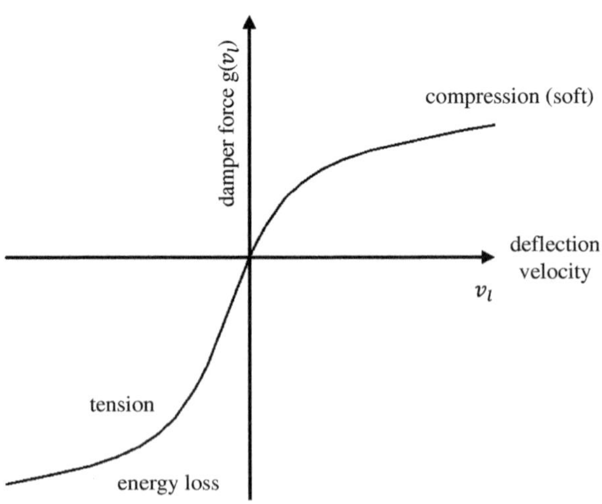

Fig. 9.7 Typical characteristic of a chassis damper

9.4.3 Force Elements Connected in Parallel

Force elements connected in parallel can simply be added.

9.4.4 Force Elements in Series

In vehicle technology, several force components will often be linked in series, for example to achieve an elastic beading of dampers and hence to increase the driving comfort of the chassis or to attain a suitable bearing for the motor. The corresponding substitute model is shown in Fig. 9.8. The total displacement of the end points of the force elements shall be given by w, and that of the spring shall be described as u. Thus, the force equilibrium at the connecting point Q of the spring and the damper leads to:

$$f(u) = g(\dot{w} - \dot{u}). \tag{9.24}$$

This relationship represents a differential equation of the first order for a spring deflection u:

$$\dot{u} = \dot{w} - g^{-1}(f(u)). \tag{9.25}$$

The input of these differential equations consists of the displacement velocities $\dot{w}$ of the combined elements. These can easily be obtained by using the kinematic values of the end points. The output of the equation is the length u of the spring element. The dynamics of the force elements is described by the characteristic curves derived from the function $g^{-1}(f(u))$. This combination of spring and

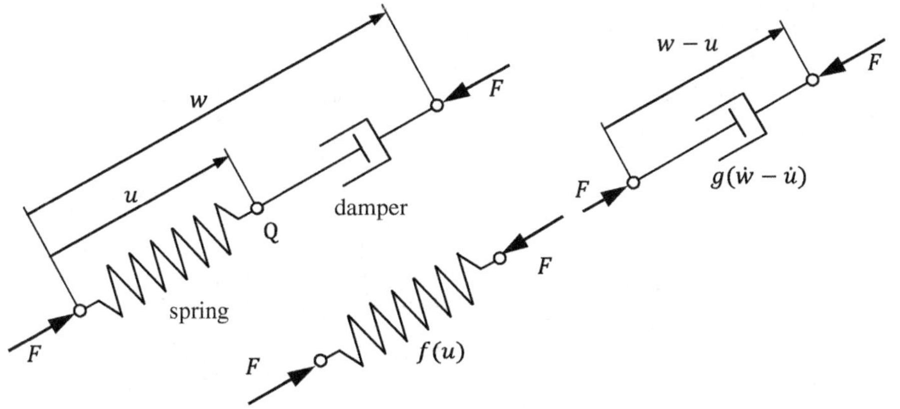

Fig. 9.8 Series connection of spring and damper elements

damper elements possesses its own dynamics. Through the additional spring, the force peaks which would normally be present if only a damper was used, are be filtered out.

9.5 Anti-Roll Bars

Anti-roll bars are used to increase the driving comfort by reducing the roll angle during cornering and to improve the driving behavior by a positively influencing the self-steering behavior of the vehicle chassis (Ammon 2013). Their working principle is based on increasing the rolling stiffness of the vehicle by coupling the forces between the two tires of a single axle during different compression travels. This results in a reduction of the normal force difference between the two tires on the relevant axle.

9.5.1 *Passive Anti-Roll Bars*

An anti-roll bar is usually a bent, U-shaped torsion bar spring. It is mounted on the vehicle chassis using rubber sleeves at A, A′, allowing a rotational degree of freedom. Their ends B, B′ are connected to the two sides of the wheel suspension of the concerned axle, Ref. Fig. 9.9. At different deflections of the spring of the two tires, the levers are subject to different displacements resulting in a twist of the torsion beam, and thus a corresponding torsion torque on the chassis. The connection with the wheel suspension is realized either by the use of rubber sleeves or

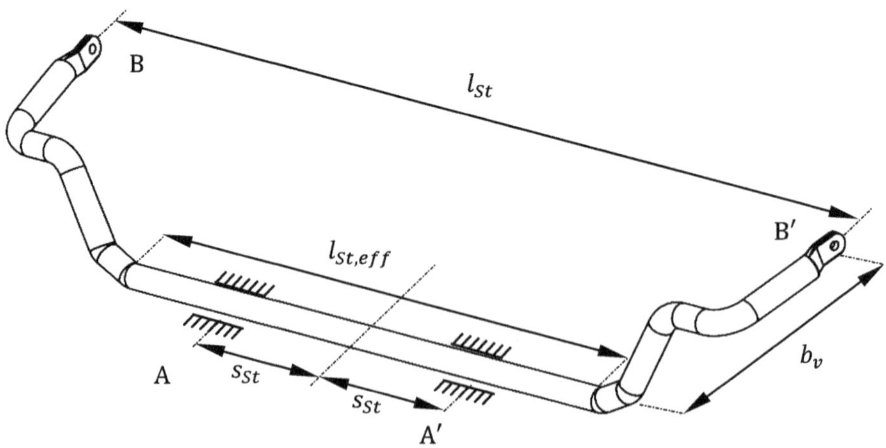

Fig. 9.9 Passive anti-roll bar

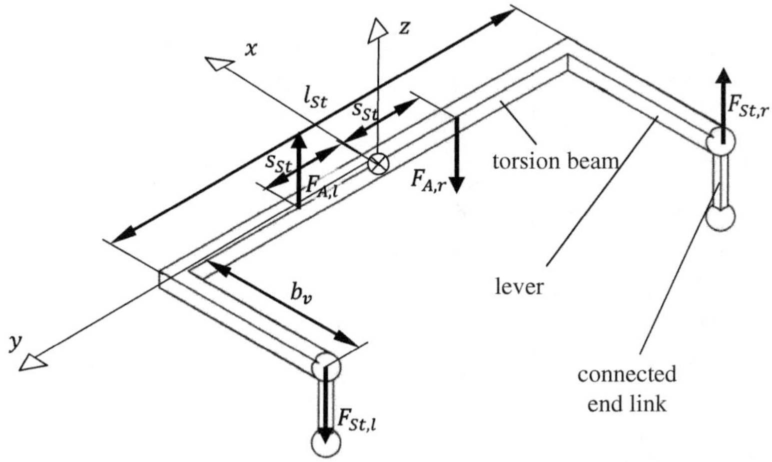

Fig. 9.10 Modeling of a passive anti-roll bar

through an anti-roll-bar connecting link, ref. Fig. 9.10. For modeling the anti-roll bar as a force element of a multibody system it will be assumed that the torsion elasticity of the torsion spring is concentrated in a rotating joint in the direction of the torsion axis of the anti-roll bar. It shall be assumed that the bending of the levers is negligible when compared to the torsion.

Since the mass of the anti-roll bar is small in comparison to the rest of the bodies of the suspension, it shall also be neglected, i.e. the anti-roll bar is modeled as a force element without mass. Otherwise, the modeling would have to contain a subsystem including the mass of the anti-roll bar (Frik 1994). Other than that, only the forces generated by the anti-roll bar will be a part of the equation of motion as applied forces. This assumption leads to the modeling in Fig. 9.10. It shall also be assumed that each anti-roll bar half is connected to the vehicle using a rotational joint. This way, the torsion angle required for the torsion element can be calculated using the difference between the deflection angles of the two halves of the anti-roll bar.

For the torque of the torsion beams with the torsion angle γ, the following formula is obtained:

$$M_{St} = c_{St}\gamma. \tag{9.26}$$

Here, the torsion spring constant is provided as:

$$c_{St} = \frac{GJ_T}{l_{St,eff}} \tag{9.27}$$

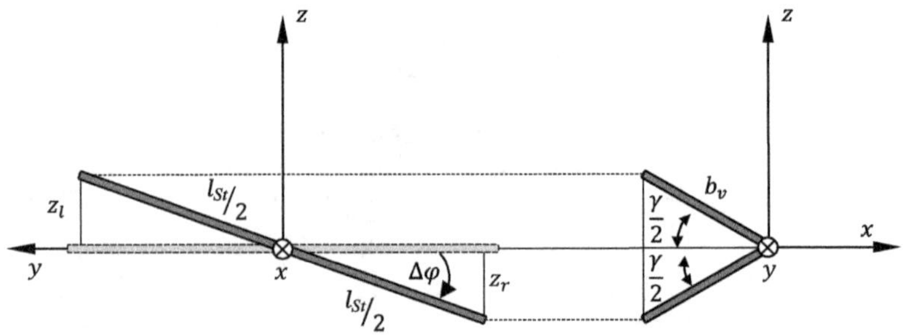

Fig. 9.11 Modeling of a passive anti-roll bar (rear view)

In Eq. (9.27), G represents the modulus of shear of the spring material, J_T is the polar surface moment of inertia of the spring's cross-sectional area and $l_{St,eff}$ is the effective (torsion) length of the rotating beam. For a mostly vertical displacement of the points of action of the force on the road suspension, the following relationships are obtained (ref Fig. 9.11):

$$z_r = -z_l = -b_v \sin\frac{\gamma}{2},\; z_r = -z_l = -\frac{l_{St}}{2}\sin\Delta\varphi. \tag{9.28}$$

From Eq. (9.28) results the relationship between the torsion of the torsion beam and the lateral tilt angle of the vehicle

$$\gamma = 2\arcsin\left(\frac{l_{St}}{2b_v}\sin\Delta\varphi\right) \approx \frac{l_{St}}{b_v}\Delta\varphi \tag{9.29}$$

and with it the torsion torque of the anti-roll bar dependent on the lateral tilt of the vehicle relative to the road surface:

$$M_{St} = c_{St}\gamma \approx \frac{l_{St}}{b_v}c_{St}\Delta\varphi. \tag{9.30}$$

The forces on the wheel suspension and the forces on the chassis are calculated by evaluating the force and torque equilibrium of the anti-roll bar to:

$$F_{St,l} = F_{St,r} = \frac{l_{St}}{b_v^2}c_{St}\Delta\varphi, \tag{9.31}$$

$$F_{A,l} = F_{A,r} = \frac{l_{St}}{2s_{St}}F_{St,l} = \frac{l_{St}^2}{2s_{St}b_v^2}c_{St}\Delta\varphi. \tag{9.32}$$

This leads to the anti-roll bar torque on the chassis

$$M_{St,A} = -s_{St}(F_{A,l} + F_{A,r}) = -\frac{l_{St}^2}{b_v^2} c_{St} \Delta\varphi. \tag{9.33}$$

9.5.2 *Active Anti-Roll Bars*

Over the past few years, active anti-roll bars have been implemented as actuators in order to achieve rolling stabilization and to influence the vehicle's dynamics (Öttgen 2005). To this end, a passive anti-roll bar is basically spilt into two and the free ends are then usually re-linked using an electrical actuator Ref. Fig. 9.12. Instead of the passive anti-roll bar, which is dependent on the torsion of the anti-roll bar, an active torque $M_{St,A}$ is created, which is supported by the wheel carrier over the anti-roll bar connecting link, and through the reaction forces, a torque is applied onto the vehicle chassis over the vehicle's longitudinal axis. Thus, roll stabilization is realizable and the self-steering characteristics of a vehicle can be influenced through the partitioning of the applied torques in the front and rear axle. In models the actuator can usually be assumed to be ideal, even though the applied torques are naturally constrained as control variables.

An example of the implementation of an active anti-roll bar in a production vehicle is shown in Fig. 9.13.

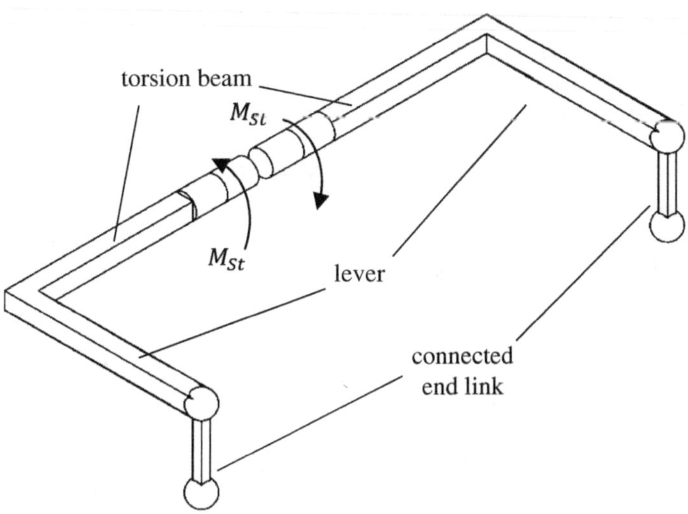

Fig. 9.12 Modeling of an active anti-roll bar

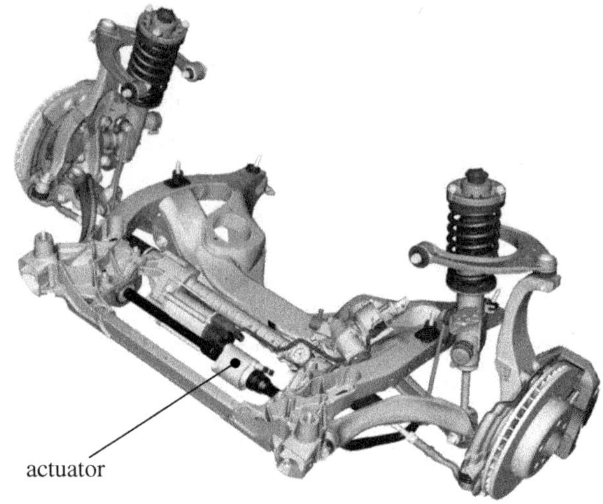

Fig. 9.13 Double wishbone front axle with an integrated anti-roll bar, 5-series (by courtesy of BMW Group)

9.6 Rubber Composite Elements

Rubber composite elements play an important role in designing modern wheel suspension systems. Along with the decoupling of the vehicle's chassis and the wheel suspension with respect to the transfer of body sound and vibrations and the damping of the resonance appearance, their specific influence on the wheel suspension kinematics under load is relevant for vehicle dynamics. As a result of the great forces which occur for example during braking, the consideration of these elements for a multitude of applications is indispensable. Generally, however, the complex geometry of the rubber composites makes modeling such elements very difficult. Hence, usually the method of finite elements is used to describe such elements (Mohr-Matuschek 1992; Gaul and Zastrau 1989). Additionally, even under static loading, the effects of hysteresis are prominent, (Göbel 1974).

The simulation is also made difficult by the complex material behavior of the rubber composite material. For example, the dynamic behavior of the rubber elements is extremely frequency-dependent, especially in the regions of higher frequencies (Nicolin and Dellmann 1985). In order to still include such elements in the multibody simulations, one generally uses simple multi-dimensional characteristic models. In such a way for example, (Schuster 1999) uses a six-dimensional element to represent the three translational and three rotational stiffness and damping characteristics, Fig. 9.14.

The model used here has three degrees of rotation and three translation degrees of freedom, which can be blocked or released, depending on the application, to represent only the relevant elasticities. The stiffness and damping characteristics

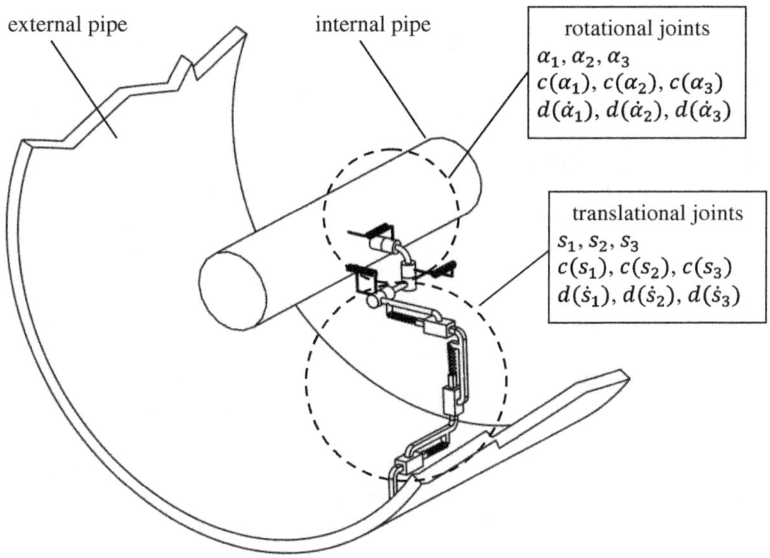

Fig. 9.14 Model of an elastic bearing

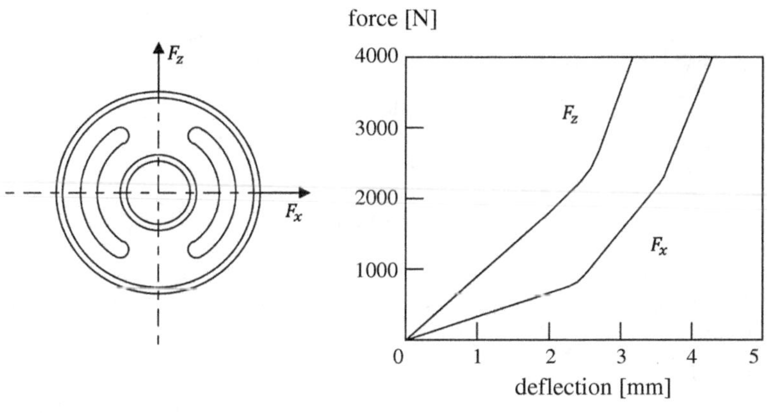

Fig. 9.15 Elastic bearing and bearing forces (as per (Reimpell 1988))

$c(\alpha_i), c(s_i), d(\dot{\alpha}_i)$ and $d(\dot{s}_i)$ need to be obtained via measurements, Fig. 9.15. In Fig. 9.15, one can also recognize, with the help of the forces in x- and y-direction, how the form of the bearing is related to the resulting stiffness. At high frequency excitations, these simple models are however, only applicable within certain limitations. The stiffness effects occurring in rubber bearings at high frequency excitation is of less significance for the problems considered in this book.

References

Ammon D (2013) Modellbildung und Systementwicklung in der Fahrzeugdynamik. Vieweg +Teubner Verlag, Stuttgart—ISBN 978-3-663-12246-3

Breuer B and Bill KH (2012a) Bremsenhandbuch. 4. überarbeitete und erweiterte Auflage (ed.), Springer Vieweg—ISBN 978-3-8348-1796-9

Breuer B and Bill KH (2012b) Bremsenhandbuch-Grundlagen, Komponenten, Systeme, Fahrdynamik. 4., überar. u. erw. Aufl. 2012 (ed.), Vieweg+Teubner Verlag+GWV-Fachverlage GmbH, Wiesbaden.—ISBN 978-3-8348-2225-3

Burckhardt M (1991) Fahrwerktechnik: Bremsdynamik und PKW-Bremsanlagen. 1. Aufl. (ed.), Vogel-Verlag, Würzburg.—ISBN 978-3-8023-0184-1

Frik S (1994) Untersuchungen zur erforderlichen Modellkomplexität bei der Fahrdynamiksimulation [Dr.-Ing.]. Dissertation, Universität - Gesamthochschule Duisburg

Gaul L and Zastrau B (1989) Nichtlineare und viskoelastische Elemente in Mehrkörpersystemen. In *Sonderdruck zum Kolloquium des DFG-Schwerpunktes Dynamik von Mehrkörpersystemen*

Göbel EF (1974) Rubber Springs Design

Gouya M and Nishiwaki M (1990) Study on disc brake groan. *SAE 90007*

Haken K-L (2008) Grundlagen der Kraftfahrzeugtechnik mit 36 Tabellen. Hanser, München.—ISBN 978-3-446-22812-2

Mohr-Matuschek U (1992) Auslegung von Kunststoff- und Elastomerformteilen mittels Finite-Element-Simulationen. *Fortschrittberichte VDI* 20

Nicolin J and Dellmann T (1985) Über die modellhafte Nachbildung der dynamischen Eigenschaften einer Gummifeder. *ZEV-Glasers Annalen* 109

Öttgen O (2005) Zur modellgestützten Entwicklung eines mechatronischen Fahrwerkregelungssystems für Personenkraftwagen Dissertation, Universität Duisburg-Essen, Düsseldorf: VDI-Verlag

Popp.K. and Schiehlen.W. (1993) Fahrzeugdynamik: eine Einführung in die Dynamik des Systems Fahrzeug - Fahrweg; mit 27 Beispielen. Teubner, Stuttgart.—ISBN 978-3-519-02373-9

Reimpell J (1988) Fahrwerktechnik: Radaufhängungen. Vol. Auflage: 2 überarb. u. erw. Aufl., Vogel Fachbuch, Würzburg.—ISBN 978-3-8023-0738-6

Samie F and Sheridan D (1990) Contact analysis for a passenger car disc brake. *SAE 900005*

Schiehlen W and Eberhard P (2004) Technische Dynamik- Modelle für Regelung und Simulation ; mit 4 Tabellen und 44 Beispielen. 2., neubearb. und erg. Aufl. (ed.), Teubner, Stuttgart u.a.—ISBN 978-3-519-12365-1

Schramm D (1986a) Ein Beitrag zur Dynamik reibungsbehafteter Mehrkörpersysteme, Düsseldorf

Schramm D (1986b) Ein effizienter Algorithmus zur numerischen Berechnung der Zwangskräfte mechanischer Systeme. *Z. Angew. Math. U. Mecha.* 66

Schramm D (1985) Eine Methode zur Lösung der Zwangskraftgleichungen. *Z. Angew. Math. U. Mecha* 65

Schuster C (1999) Strukturvariante Modelle zur Simulation der Fahrdynamik bei niedrigen Geschwindigkeiten [Dr.-Ing.]. Dissertation, Universität - Gesamthochschule Duisburg, Duisburg

Shi X (1996) Entstehung des Bremsenquitschens [Dr.-Ing.]. Dissertation, Technische Universität Braunschweig

Smith SW (2003) Digital Signal Processing

Chapter 10
Single Track Models

Single track models allow a physically plausible description of the driving behavior of vehicles without major modeling and parameterization effort. Hence, in this chapter a number of linear and nonlinear single track models will be described.

10.1 Linear Single Track Model

The classic linear single track model (Riekert and Schunk 1940) allows the approximate, yet physically plausible description of the lateral dynamics of a vehicle, Ref. Fig. 10.1.

The modeling is based on a series of simplifications:

- The velocity of the vehicle's center of gravity is considered to be constant along the longitude of its trajectory.
- All lifting, rolling and pitching motion will be neglected.
- The vehicle's mass is assumed to be concentrated at the center of gravity S.
- The front and the rear tires will be represented as one single tire on each axle. The imaginary tire contact points V and H, which the tire forces are to act upon, lie along the center of the axle.
- The pneumatic trail and the aligning torque resulting from the slip angle of the tire will be neglected.
- The wheel-load distribution between front and rear axle is assumed to be constant.
- The longitudinal forces on the tires, resulting from the assumption of a constant longitudinal velocity, will be neglected.

The first two assumptions lead to four constraints for the six degrees of freedom of rigid bodies in the model. As a result, the only possible motion left is the heading angle (yaw angle) ψ_V, which only occurs in the form of the yaw rate $\dot{\psi}_V$, and the side slip angle β. The side slip angle represents the direction of the deviation of the

D. Schramm et al., Vehicle Dynamics,
DOI 10.1007/978-3-662-54483-9_10

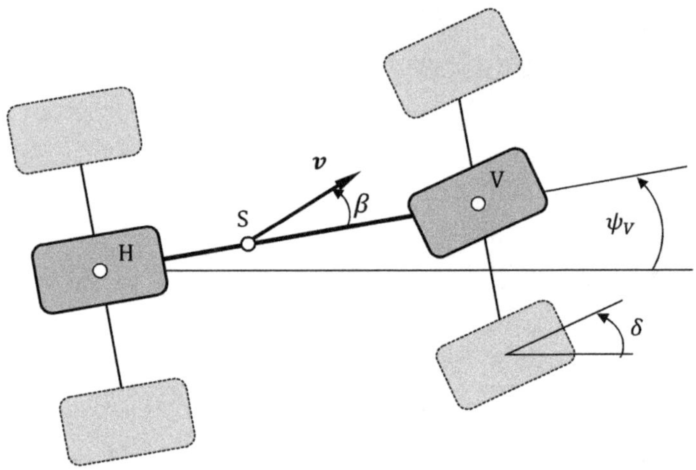

Fig. 10.1 Linear single track model

center of gravity from the vehicle's steering axis. The steering angle δ of the front axle serves as the input parameter. This greatly idealized vehicle model allows the investigation of the fundamental driving dynamic relationships within the lateral acceleration region of

$$a_y \leq 0,4g \approx 4\frac{\mathrm{m}}{\mathrm{s}^2} \tag{10.1}$$

on dry roads (Ammon 2013).

10.1.1 *Equations of Motion of the Linear Single Track Model*

To generate the equations of motion the rigid body kinematics of the vehicle has to be reviewed. To this end, the kinematics of the vehicle in the x_E, y_E-plane of the inertial system can be described as $\mathbf{K}_E = \{\mathrm{O}_E; x_E, y_E, z_E\}$, Ref. Fig. 10.2. Especially in the quasi-stationary situation, i.e. for very small velocities $\boldsymbol{v}$ of the center of gravity S, all points of the vehicle move along a circle with the center of the curvature being K_A. In this case, this coincides with the instantaneous center of rotation M of the motion. The steering angle required to execute this motion is, under the assumption of small steering motion and large radii of curvature relative to the measurements of the vehicle, given as:

$$\tan \delta_A = \frac{l}{\sqrt{\rho_M^2 - l_h^2}} \overset{|\delta_A| \ll 1, l_h \ll \rho_M}{\longrightarrow} \delta_A \approx \frac{l}{\rho_M}. \tag{10.2}$$

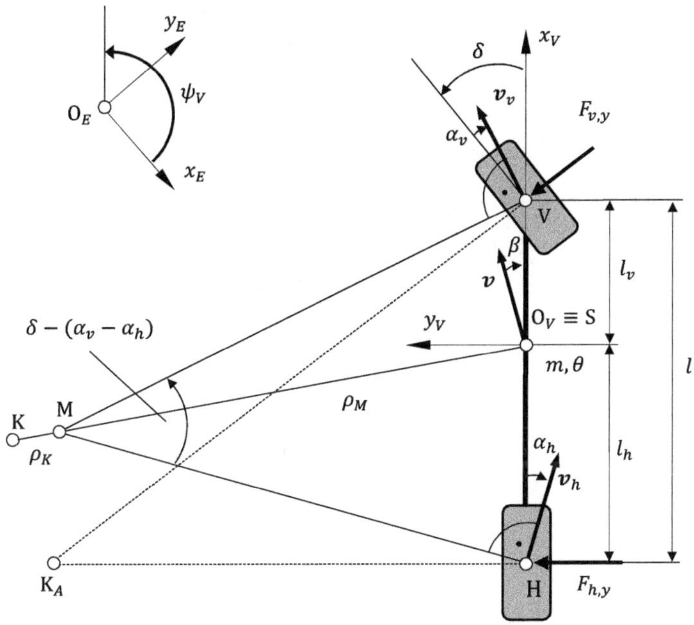

Fig. 10.2 Mathematical description of the linear single track model

The steering angle δ_A of the front wheels resulting from Eq. (10.2) is called the Ackermann steering angle.

In general, the vehicle velocity is given according to (Fig. 10.2) in the vehicle fixed coordinate system $\mathbf{K}_V = \{O_V; x_V, y_V, z_V\}$:

$$ {}^V\boldsymbol{v} = \begin{bmatrix} v\cos\beta \\ v\sin\beta \\ 0 \end{bmatrix}. \tag{10.3}$$

The acceleration of the vehicle's center of gravity S, represented in the coordinate system $\mathbf{K}_V$, is therefore:

$$\begin{aligned} {}^V\boldsymbol{a} &= \frac{d^V\boldsymbol{v}}{dt} + {}^V\boldsymbol{\omega} \times {}^V\boldsymbol{v} = \begin{bmatrix} -v\sin\beta\dot{\beta} \\ v\cos\beta\dot{\beta} \\ 0 \end{bmatrix} + \begin{bmatrix} 0 \\ 0 \\ \dot{\psi}_V \end{bmatrix} \times \begin{bmatrix} v\cos\beta \\ v\sin\beta \\ 0 \end{bmatrix} \\ &= \begin{bmatrix} -v\left(\dot{\psi}_V + \dot{\beta}\right)\sin\beta \\ v\left(\dot{\psi}_V + \dot{\beta}\right)\cos\beta \\ 0 \end{bmatrix}. \end{aligned} \tag{10.4}$$

Due to the assumption of a constant longitudinal velocity $v = const$, the acceleration $\boldsymbol{a}$ is only a purely normal acceleration $\boldsymbol{a}_n$, perpendicular to the vehicle ($\boldsymbol{a}^T\boldsymbol{v} = 0$). Its magnitude is given by:

$$a_n = |\boldsymbol{a}_n| = v\left(\dot{\psi}_V + \dot{\beta}\right). \tag{10.5}$$

From Fig. 10.2 it can be inferred that the radius of curvature ρ_K of the path curve of the center of gravity is described by

$$\rho_K = \frac{v}{\left(\dot{\psi}_V + \dot{\beta}\right)}. \tag{10.6}$$

For the following observations, the acceleration of the center of gravity perpendicular to the longitudinal axis of the vehicle is required. For small side slip angles β according to Eq. (10.6), this results in:

$$a_y = v\left(\dot{\psi}_V + \dot{\beta}\right)\cos\beta \approx v\left(\dot{\psi}_V + \dot{\beta}\right) = \frac{v^2}{\rho_K}. \tag{10.7}$$

The calculation of the horizontal tire forces still requires the velocities of the tire contact point. These are calculated according to Fig. 10.2:

$$\begin{aligned} {}^V\boldsymbol{v}_v &= {}^V\boldsymbol{v} + {}^V\boldsymbol{\omega} \times {}^V_S\boldsymbol{r}_V \\ &= \begin{bmatrix} v\cos\beta \\ v\sin\beta \\ 0 \end{bmatrix} + \begin{bmatrix} 0 \\ 0 \\ \dot{\psi} \end{bmatrix} \times \begin{bmatrix} l_v \\ 0 \\ 0 \end{bmatrix} = \begin{bmatrix} v\cos\beta \\ v\sin\beta + l_v\dot{\psi}_V \\ 0 \end{bmatrix}, \end{aligned} \tag{10.8}$$

at the front wheels and

$$\begin{aligned} {}^V\boldsymbol{v}_h &= {}^V\boldsymbol{v} + \boldsymbol{\omega} \times {}^V_S\boldsymbol{r}_h \\ &= \begin{bmatrix} v\cos\beta \\ v\sin\beta \\ 0 \end{bmatrix} + \begin{bmatrix} 0 \\ 0 \\ \dot{\psi}_V \end{bmatrix} \times \begin{bmatrix} -l_h \\ 0 \\ 0 \end{bmatrix} = \begin{bmatrix} v\cos\beta \\ v\sin\beta - l_h\dot{\psi}_V \\ 0 \end{bmatrix} \end{aligned} \tag{10.9}$$

at the rear wheels. Here, ${}_S\boldsymbol{r}_v$ and ${}_S\boldsymbol{r}_h$ are the position vectors based on the center of gravity S of the vehicle to the front tire contact point V and to the rear contact point H respectively. The current velocity $\boldsymbol{v}_v$ of the front wheels can now be expressed as the side slip angle β and the steering angle δ in the vehicle fixed coordinate system $\mathbf{K}_V$:

$$ {}^V\boldsymbol{v}_v = \begin{bmatrix} v\cos\beta \\ v\sin\beta + l_v\dot{\psi}_V \\ 0 \end{bmatrix} = \begin{bmatrix} v_v\cos(\delta - \alpha_v) \\ v_v\sin(\delta - \alpha_v) \\ 0 \end{bmatrix}. \tag{10.10} $$

The first two components in Eq. (10.10) offer the relationship to the front slip angle α_v for small steering angles δ:

$$ \begin{aligned} \tan(\delta - \alpha_v) &= \frac{v\sin\beta + l_v\dot{\psi}_V}{v\cos\beta} \approx \beta + l_v\frac{\dot{\psi}_V}{v} \\ &\Rightarrow \alpha_v = \delta - \beta - l_v\frac{\dot{\psi}_V}{v}. \end{aligned} \tag{10.11} $$

This procedure can be applied similarly to the rear axle:

$$ {}^V\boldsymbol{v}_h = \begin{bmatrix} v\cos\beta \\ v\sin\beta - l_h\dot{\psi}_V \\ 0 \end{bmatrix} = \begin{bmatrix} v_h\cos\alpha_h \\ -v_h\sin\alpha_h \\ 0 \end{bmatrix}. \tag{10.12} $$

From the first and the second components of this vector equation, one then gets:

$$ \begin{aligned} -\tan\alpha_h &= \frac{v\sin\beta - l_h\dot{\psi}}{v\cos\beta} \approx \beta - l_h\frac{\dot{\psi}_V}{v}, \\ &\Rightarrow \alpha_h \approx -\beta + l_h\frac{\dot{\psi}_V}{v}. \end{aligned} \tag{10.13} $$

To set up the equations of motion, the values of the forces acting on the vehicle along with the kinematic descriptions are still required. While still considering the position of the center of gravity of the vehicle (Fig. 10.3), the normal forces on the tires are given by:

$$ F_{v,z} = mg\frac{l_h}{l} \quad \text{and} \quad F_{h,z} = mg\frac{l_v}{l}. \tag{10.14} $$

The tire forces may then be calculated assuming a linear relationship between the lateral forces and the tire slip angles as discussed in Chap. 7:

$$ F_{v,y} = c_{\alpha,v}\,\alpha_v \quad \text{and} \quad F_{h,y} = c_{\alpha,h}\,\alpha_h, \tag{10.15} $$

with the cornering stiffnesses $c_{\alpha,v}$ and $c_{\alpha,h}$ (Ref. Fig. 10.4).

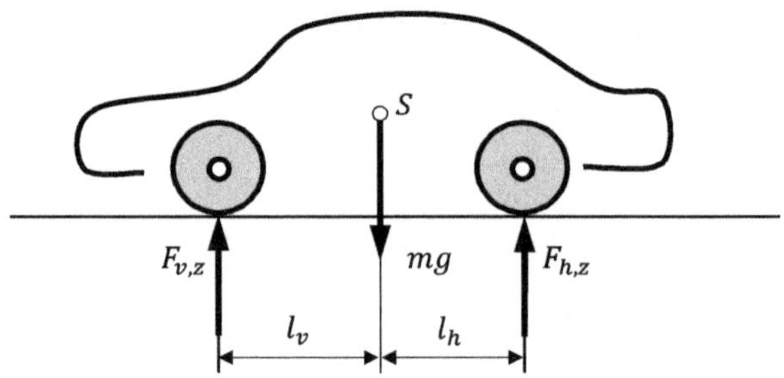

Fig. 10.3 Tire loads in a linear single track model

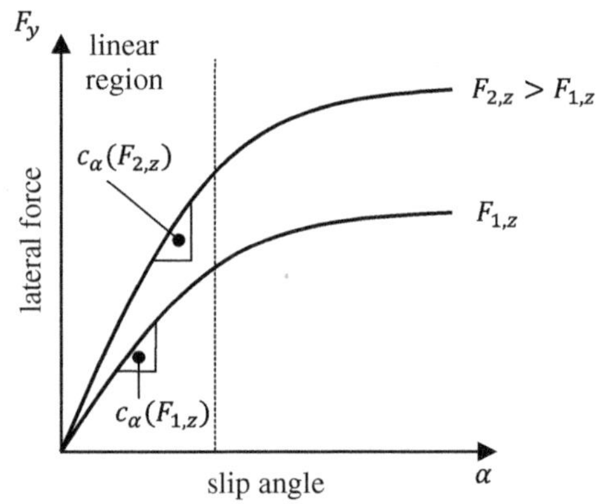

Fig. 10.4 Relationship between the tire lateral forces and the slip angle

In Fig. 10.4 it is noteworthy that the cornering stiffness is a function of the tire loads. This will be dealt with in more detail in Sect. 10.3. With help of the accelerations (10.4) and Fig. 10.4, the principle of linear momentum in the lateral direction yields

$$mv\left(\dot{\psi}_V + \dot{\beta}\right)\cos\beta = \cos\delta\, F_{v,y} + F_{h,y}. \tag{10.16}$$

The corresponding principle of angular momentum about the vertical axis is

$$\theta\ddot{\psi}_V = F_{v,y}\cos\delta\, l_v - F_{h,y}\, l_h. \tag{10.17}$$

If one were to substitute the expressions for the tire's lateral forces with the relationships given in Eq. (10.15) as well as in Eqs. (10.11) and (10.13) and considering $\cos\beta \approx 1, \cos\delta \approx 1$ for $|\beta|, |\delta| \ll 1$, one finally arrives at the two equations of motion of the linear single track model:

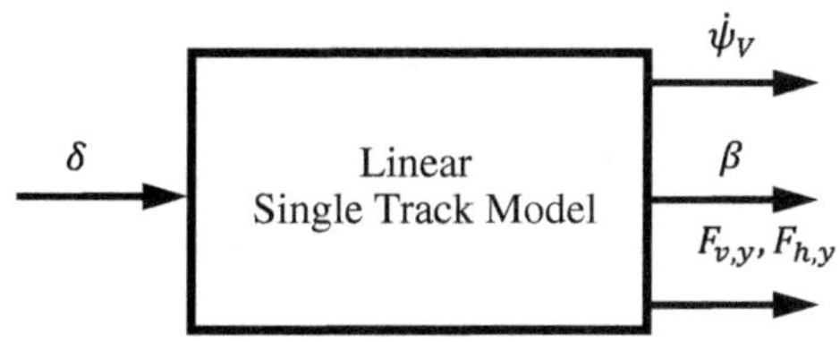

Fig. 10.5 Linear single track model as a dynamic system

$$mv\dot{\beta} + \left(mv^2 + c_{\alpha,v}l_v - c_{\alpha,h}l_h\right)\frac{\dot{\psi}_V}{v} + \left(c_{\alpha,v} + c_{\alpha,h}\right)\beta = c_{\alpha,v}\delta, \tag{10.18}$$

$$\theta\ddot{\psi}_V + \left(c_{\alpha,v}l_v^2 + c_{\alpha,h}l_h^2\right)\frac{\dot{\psi}_V}{v} + \left(c_{\alpha,v}l_v - c_{\alpha,h}l_h\right)\beta = c_{\alpha,v}l_v\delta. \tag{10.19}$$

With the substitution

$$\boldsymbol{x} = \begin{bmatrix} x_1 \\ x_2 \end{bmatrix} = \begin{bmatrix} \dot{\psi}_V \\ \beta \end{bmatrix}, \tag{10.20}$$

one arrives at the state space representation

$$\underbrace{\begin{bmatrix} \dot{x}_1 \\ \dot{x}_2 \end{bmatrix}}_{\dot{x}} = \underbrace{\begin{bmatrix} -\frac{1}{v}\frac{c_{\alpha,v}l_v^2 + c_{\alpha,h}l_h^2}{\theta} & -\frac{c_{\alpha,v}l_v - c_{\alpha,h}l_h}{\theta} \\ -1 - \frac{1}{v^2}\frac{c_{\alpha,v}l_v - c_{\alpha,h}l_h}{m} & -\frac{1}{v}\frac{c_{\alpha,v} + c_{\alpha,h}}{m} \end{bmatrix}}_{\boldsymbol{A}} \underbrace{\begin{bmatrix} x_1 \\ x_2 \end{bmatrix}}_{x} + \underbrace{\begin{bmatrix} \frac{c_{\alpha,v}l_v}{\theta} \\ \frac{1}{v}\frac{c_{\alpha,v}}{m} \end{bmatrix}}_{\boldsymbol{B}} \underbrace{[\delta]}_{\boldsymbol{u}} \tag{10.21}$$

with the $[2 \times 1]$-state vector $\boldsymbol{x}$, the $[2 \times 2]$-system matrix $\boldsymbol{A}$, the $[2 \times 1]$-control matrix $\boldsymbol{B}$ and the $[1 \times 1]$-input vector $\boldsymbol{u}$. This leads to the representation of the linear single track model as a dynamic system with a corresponding transfer function (Fig. 10.5). The representation (10.21) is a suitable basis for fundamental analysis of vehicle dynamics. This will be exemplified in the following sections. For a more detailed analysis, the interested reader is referred to (Willumeit 1998).

10.1.2 Stationary Steering Behavior and Cornering

For cornering along a circle with a constant radius ρ, the steering angle δ as well as the yaw rate $\dot{\psi}_V$ and the side slip angle β are all constant, i.e. it follows:

$$\delta = const, \dot{\delta} = 0, \tag{10.22}$$

$$\dot{\psi}_V = const, \ddot{\psi}_V = 0, \tag{10.23}$$

$$\beta = const, \dot{\beta} = 0, \tag{10.24}$$

$$\rho_K = \frac{v}{\dot{\psi}_V + \dot{\beta}} = \frac{v}{\dot{\psi}_V} = \rho. \tag{10.25}$$

With the additional constraints (10.22)–(10.25) and by implementing the Eqs. (10.16) and (10.17) as well as considering the Eq. (10.15), in a single step one obtains the relationship:

$$\alpha_v - \alpha_h = \frac{mv^2}{\rho l}\left(\frac{l_h}{c_{\alpha,v}} - \frac{l_v}{c_{\alpha,h}}\right) = \underbrace{\frac{m}{l}\left(\frac{l_h c_{\alpha,h} - l_v c_{\alpha,v}}{c_{\alpha,v} c_{\alpha,h}}\right)}_{EG}\frac{v^2}{\rho}. \tag{10.26}$$

The expression *EG* in Eq. (10.26) is called the self-steering gradient which characterizes the typical driving behavior of a given vehicle for a given steering motion. As a result, one can for example solve the following practical problems:

- Which steering angle $\delta_H = i_L\delta$ with the steering transmission ratio i_L is necessary for a vehicle with a velocity v to follow a circle with a radius ρ?
- Which parameters become stationary when a steering angle δ_H is applied to a vehicle travelling in a straight line?
- What happens in the transition region (instationary steering behavior)?

To answer the first question, one first calculates the slip angle and the steering angle for a given circle with the radius ρ. From the side slip angle of the rear wheels (10.13), the side slip angle of the vehicle can be calculated using the transformation:

$$\beta = l_h\frac{\dot{\psi}_V}{v} - \alpha_h = \frac{l_h}{\rho} - \frac{m}{c_{\alpha,h}}\frac{l_v}{l}\frac{v^2}{\rho}. \tag{10.27}$$

The steering angle required for the circular path is then given with the slip angle at the front wheels (10.11):

$$\begin{aligned}\delta &= l_v \frac{\dot{\psi}_V}{v} + \alpha_v + \beta = \frac{l}{\rho} + \alpha_v - \alpha_h \\ &= \underbrace{\frac{l}{\rho}}_{\delta_A} + \underbrace{\frac{m}{l}\left(\frac{l_h c_{\alpha,h} - l_v c_{\alpha,v}}{c_{\alpha,v} c_{\alpha,h}}\right)}_{EG} \underbrace{\frac{v^2}{\rho}}_{a_y} = \delta_A + EG \cdot a_y.\end{aligned} \tag{10.28}$$

The first summand occurring in Eq. (10.28) is the Ackermann steering angle (compare Eq. (10.2)). With increasing velocity v, the required steering angle increases or decreases for a given circular path, depending on the sign of the self-steering gradient EG, Ref. (Fig. 10.6). If the required steering angle is greater than the Ackermann steering angle ($EG > 0$), this is called under steering driving behavior, in the case of ($EG < 0$), it is called over steering driving behavior. In case of $EG = 0$, it is characterized as a neutral driving behavior.

From Eq. (10.28), one can also derive the relationship

$$EG \cdot a_y = \alpha_v - \alpha_h \tag{10.29}$$

i.e. the self-steering behavior depends on the difference of the slip angles between the front and rear wheels. Now, the remaining parameters can also be calculated. The yaw rate is defined as

$$\dot{\psi}_V = \frac{v}{\rho} = const, \tag{10.30}$$

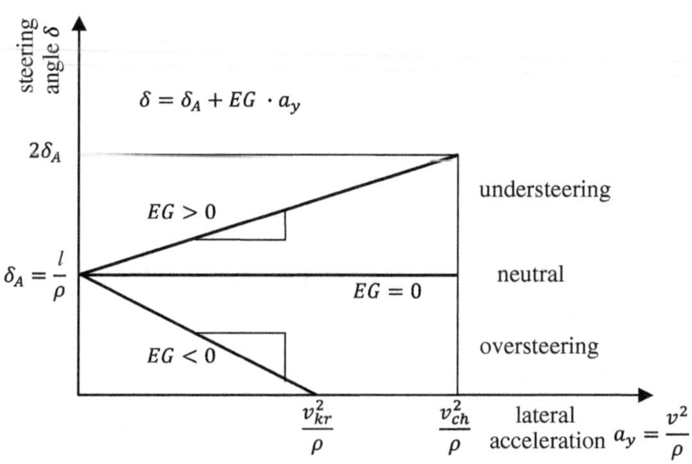

Fig. 10.6 Self steering gradient in the linear region

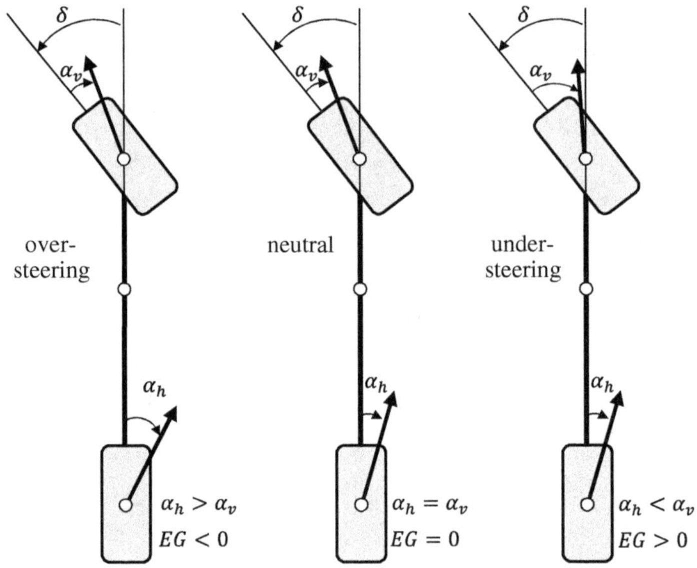

Fig. 10.7 Self-steering behavior of a linear single track model

the tire loads are

$$F_{v,y} = m\frac{l_h}{l}\frac{v^2}{\rho}, \quad F_{h,y} = m\frac{l_v}{l}\frac{v^2}{\rho}, \tag{10.31}$$

and the tire slip angle is (Fig. 10.7)

$$\alpha_v = \frac{F_{v,y}}{c_v} = \frac{m}{c_{\alpha,v}}\frac{l_h}{l}\frac{v^2}{\rho}, \quad \alpha_h = \frac{F_{h,y}}{c_h} = \frac{m}{c_{\alpha,h}}\frac{l_v}{l}\frac{v^2}{\rho}. \tag{10.32}$$

If a steering angle δ is applied to a vehicle driving in a straight line, then the yaw rate

$$\dot{\psi}_{V,stat} = \frac{v}{\rho} = \frac{v}{l + EG \cdot v^2}\delta_{stat} \tag{10.33}$$

has to be considered. This means that the yaw rate takes on different values depending on the self-steering gradient. One denotes

$$\frac{\dot{\psi}_V}{\delta} = \frac{v}{l + EG \cdot v^2} \tag{10.34}$$

as the yaw amplification factor for a velocity v. This factor is small for big self-steering gradients (understeering vehicle) and large for small (negative) self-steering gradients (oversteering vehicle). For

$$EG = -\frac{l}{v^2} < 0 \tag{10.35}$$

the numerator in Eq. (10.34) becomes zero and the yaw amplification factor strives toward an infinite value. In reality, this means that the vehicle will tend to become instable (more precise: it leaves the linear region), as even very small steering inputs would lead to infinite yaw rates. The velocity

$$v_{kr} = \sqrt{-\frac{l}{EG}}, \tag{10.36}$$

necessary for this to occur (consider $EG < 0$) is defined as the critical velocity. Vice versa, one can calculate the maximum yaw amplification for a given positive self-steering gradient. Through differentiation of Eq. (10.34) with respect to the velocity v one obtains:

$$\frac{d}{dv}\left(\frac{\dot{\psi}_V}{\delta}\right) = \frac{l - EG \cdot v^2}{(l + EG \cdot v^2)^2} = 0 \Rightarrow v_{ch}^2 = \frac{l}{EG}. \tag{10.37}$$

The velocity v_{ch}, at which the yaw amplification factor reaches its maximum, is called the characteristic velocity. It is interpreted as the vehicle velocity at which the vehicle reacts most sensitively to steering inputs. Typical values of v_{ch} are between 65 and 100 km/h.

10.1.3 Instationary Steering Behavior: Vehicle Stability

In order to investigate the driving stability during straight line driving, one assumes the steering angle to be equal to zero in Eq. (10.21). This way, one arrives at the linear homogenous state space equation:

$$\underbrace{\begin{bmatrix} \dot{x}_1 \\ \dot{x}_2 \end{bmatrix}}_{\dot{x}} = \underbrace{\begin{bmatrix} -\frac{1}{v}\frac{c_{\alpha,v}l_v^2 + c_{\alpha,h}l_h^2}{\theta} & -\frac{c_{\alpha,v}l_v - c_{\alpha,h}l_h}{\theta} \\ -1 - \frac{1}{v^2}\frac{c_{\alpha,v}l_v - c_{\alpha,h}l_h}{m} & -\frac{1}{v}\frac{c_{\alpha,v} + c_{\alpha,h}}{m} \end{bmatrix}}_{A} \underbrace{\begin{bmatrix} x_1 \\ x_2 \end{bmatrix}}_{x}, \tag{10.38}$$

or in short

$$\underbrace{\begin{bmatrix} \dot{x}_1 \\ \dot{x}_2 \end{bmatrix}}_{\dot{x}} = \underbrace{\begin{bmatrix} -\frac{1}{v}a_{11} & -a_{12} \\ -1-\frac{1}{v^2}a_{21} & -\frac{1}{v}a_{22} \end{bmatrix}}_{A} \underbrace{\begin{bmatrix} x_1 \\ x_2 \end{bmatrix}}_{x}, \tag{10.39}$$

with the coefficients

$$\begin{aligned} a_{11} &= \frac{c_{\alpha,v}l_v^2 + c_{\alpha,h}l_h^2}{\theta}, \quad a_{12} = \frac{c_{\alpha,v}l_v - c_{\alpha,h}l_h}{\theta}, \\ a_{21} &= \frac{c_{\alpha,v}l_v - c_{\alpha,h}l_h}{m}, \quad a_{22} = \frac{c_{\alpha,v} + c_{\alpha,h}}{m}. \end{aligned} \tag{10.40}$$

As a result, one derives the polynomial for the characteristic equation of the system matrix $\boldsymbol{A}$:

$$\begin{aligned} \det(\lambda \boldsymbol{E} - \boldsymbol{A}) &= \lambda^2 + \frac{1}{v}(a_{11} + a_{22})\lambda - a_{12} + \frac{1}{v^2}(a_{11}a_{22} - a_{12}a_{21}) \\ &= \lambda^2 + a_1\lambda + a_2. \end{aligned} \tag{10.41}$$

The linear system (10.38) is known to be stable when both coefficients of the characteristic polynomial are positive. This is obviously always the case for a_1. From the constraint for a_2 it follows that:

$$\begin{aligned} a_2 &= -a_{12} + \frac{1}{v^2}(a_{11}a_{22} - a_{12}a_{21}) \\ &= \frac{c_{\alpha,v}c_{\alpha,h}l^2}{m\theta v^2}\left(1 + \frac{c_{\alpha,h}l_h - c_{\alpha,v}l_v}{c_{\alpha,v}c_{\alpha,h}l^2}mv^2\right) > 0. \end{aligned} \tag{10.42}$$

This condition is valid for any velocity v, if:

$$c_{\alpha,h}l_h > c_{\alpha,v}l_v. \tag{10.43}$$

In any other case, the velocity is limited by following equation:

$$v^2 < \frac{1}{m}\frac{c_{\alpha,v}c_{\alpha,h}l^2}{c_{\alpha,v}l_v - c_{\alpha,h}l_h}. \tag{10.44}$$

The right side of the inequality (10.44) corresponds exactly to the critical velocity v_{kr} calculated earlier. This means that an oversteering vehicle can become unstable after a certain velocity, whereas this is not the case in understeering vehicles.

10.2 Nonlinear Single Track Model

The single track model covered in the previous section already allows for a conclusive insight into the typical self-steering (eigen-) behavior of a vehicle. However, it neither includes a description of the drivetrain, nor does it allow the representation of the vehicle behavior at larger steering angles or with higher lateral accelerations. As a result, an extended model will be introduced in the following section, allowing the description of the nonlinear vehicle behavior in spite of a few restrictions.

10.2.1 Kinetics of the Nonlinear Single Track Model

The nonlinear single track model (Figs. 10.8 and 10.9) consists of

- The vehicle chassis as a rigid body, with the translational degrees of freedom x_V, y_V and the rotation ψ_V about the vertical axis.
- One imaginary front and rear wheel respectively (indices v and h), characterized by the wheel speed and tire forces.
- A given steering angle (toe angle) δ at the front axle and the steering transmission ratio i_L which are derived from the steering wheel angle δ_H as $\delta = \frac{1}{i_L}\delta_H$
- The description of the wheel driving torques $M_{A,v}$ and $M_{A,h}$ as functions of the engine torque M_M and the current, depending on the chosen gear transmission ratio i_G(G). The engine torque is a function of the motor rpm ω_M and the acceleration pedal position $p_F(t)$. The entire driving torque $M_A = M_{A,v} + M_{A,h}$ can be distributed arbitrarily by a factor ξ_a (also temporal) onto the front and rear axles, which allows for the simulation of a four-wheel drive configuration.
- The description of the air resistance.
- Specifying the brake torques $M_{B,v}$ and $M_{B,h}$ on the wheels as functions of the brake pedal position $p_B(t)$. The brake force distribution ξ_b can similar to ξ_a also be specified arbitrarily.

To generate the necessary equations of motion, one first applies the principle of linear momentum on the chassis, which is considered to be a rigid body:

$$m\ddot{\boldsymbol{r}}_V = \boldsymbol{F}_v + \boldsymbol{F}_h + \boldsymbol{F}_W + \boldsymbol{F}_G. \tag{10.45}$$

The acceleration of the vehicle chassis is obtained by twofold differentiation with respect to time of the position vector $\boldsymbol{r}_V$ to the center of gravity S of the vehicle given in the inertial system $\mathbf{K}_E$:

$$\boldsymbol{r}_V = \begin{bmatrix} x_V \\ y_V \\ h_S \end{bmatrix}, \dot{\boldsymbol{r}}_V = \begin{bmatrix} \dot{x}_V \\ \dot{y}_V \\ 0 \end{bmatrix}, \ddot{\boldsymbol{r}}_V = \begin{bmatrix} \ddot{x}_V \\ \ddot{y}_V \\ 0 \end{bmatrix}. \tag{10.46}$$

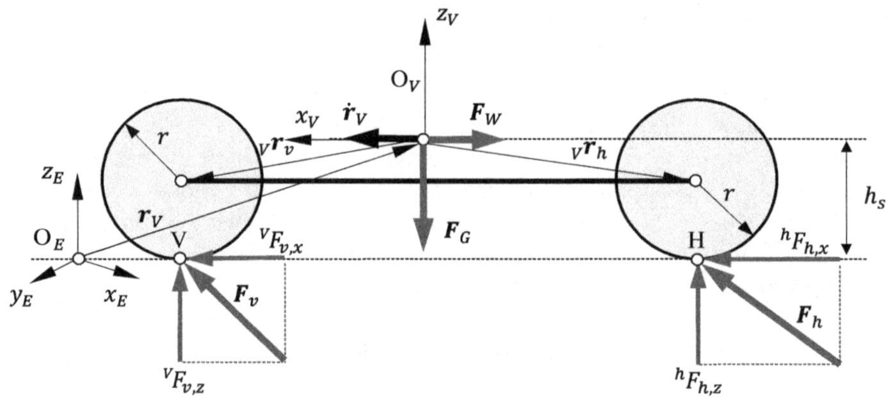

Fig. 10.8 Nonlinear single track model—*side view*

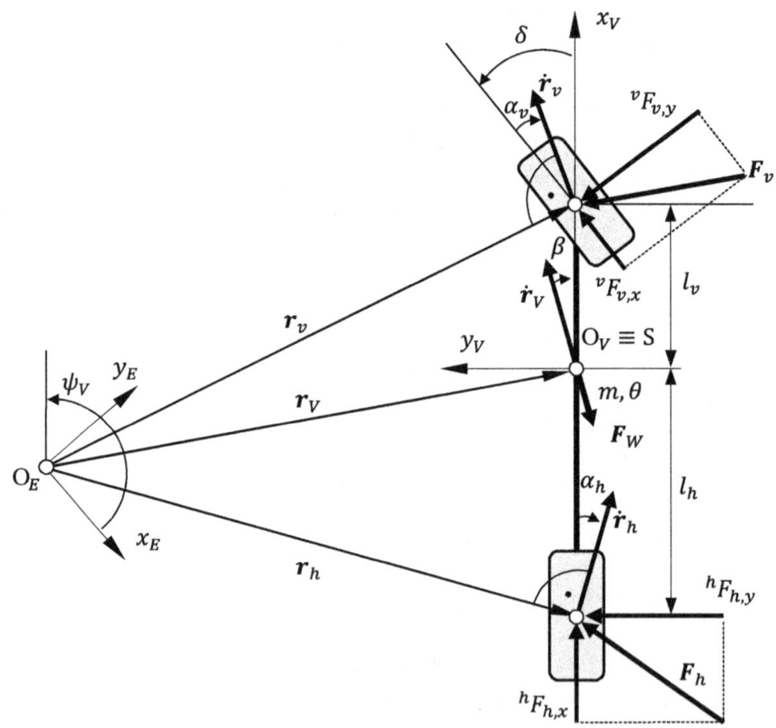

Fig. 10.9 Nonlinear single track model–*top view*

The forces on the front and rear wheel as well as the weights are given in coordinates of the inertial system as:

$$\boldsymbol{F}_v = \begin{bmatrix} F_{v,x} \\ F_{v,y} \\ F_{v,z} \end{bmatrix}, \quad \boldsymbol{F}_h = \begin{bmatrix} F_{h,x} \\ F_{h,y} \\ F_{h,z} \end{bmatrix}, \quad \boldsymbol{F}_G = \begin{bmatrix} 0 \\ 0 \\ -mg \end{bmatrix}. \tag{10.47}$$

As a single external force, the air resistance is given by:

$$\boldsymbol{F}_W = \frac{1}{2} c_w \rho_L A \dot{\boldsymbol{r}}_V |\dot{\boldsymbol{r}}_V| = \begin{bmatrix} F_{W,x} \\ F_{W,y} \\ 0 \end{bmatrix} = \begin{bmatrix} \frac{1}{2} c_w \rho_L A \dot{x}_V \sqrt{\dot{x}_V^2 + \dot{y}_V^2} \\ \frac{1}{2} c_w \rho_L A \dot{y}_V \sqrt{\dot{x}_V^2 + \dot{y}_V^2} \\ 0 \end{bmatrix}, \tag{10.48}$$

with an air resistance coefficient c_W, the air density ρ_L and the front surface area A of the vehicle. In this case, only the air flow opposite to the trajectory of the center of gravity of the vehicle is considered. If the influences of both the side stream and the vertical forces need to be considered, then Eq. (10.48) needs to be extended correspondingly as discussed in Chap. 9. From Eq. (10.45), and combined with the Eqs. (10.46)-(10.48), one can finally derive the complete principle of linear momentum in coordinates of the inertial system:

$$\begin{bmatrix} m\ddot{x}_V \\ m\ddot{y}_V \\ 0 \end{bmatrix} = \begin{bmatrix} F_{v,x} + F_{h,x} - F_{W,x} \\ F_{v,y} + F_{h,y} - F_{W,y} \\ F_{v,z} + F_{h,z} - mg \end{bmatrix}. \tag{10.49}$$

In a similar manner, one can arrive at the principle of angular momentum of the vehicle with respect to its center of gravity in the general form as:

$$\boldsymbol{\Theta}_V \dot{\boldsymbol{\omega}}_V + \boldsymbol{\omega}_V \times (\boldsymbol{\Theta}_V \cdot \boldsymbol{\omega}_V) = {}_V\boldsymbol{r}_v \times \boldsymbol{F}_v + {}_V\boldsymbol{r}_h \times \boldsymbol{F}_h. \tag{10.50}$$

With the moment of inertia matrix in the vehicle fixed coordinate system

$$\boldsymbol{\Theta}_V = \begin{bmatrix} 0 & 0 & 0 \\ 0 & 0 & 0 \\ 0 & 0 & \theta_{zz} \end{bmatrix}, \tag{10.51}$$

the vectors for the angular velocity and acceleration

$${}^V\boldsymbol{\omega}_V = \begin{bmatrix} 0 \\ 0 \\ \dot{\psi}_V \end{bmatrix}, \quad {}^V\dot{\boldsymbol{\omega}}_V = \begin{bmatrix} 0 \\ 0 \\ \ddot{\psi}_V \end{bmatrix}, \tag{10.52}$$

and the vectors for the points of application of the tire forces

$$ {}^V_V\boldsymbol{r}_v = \begin{bmatrix} l_v \\ 0 \\ -h_S \end{bmatrix}, {}^V_V\boldsymbol{r}_h = \begin{bmatrix} -l_h \\ 0 \\ -h_S \end{bmatrix}, \tag{10.53} $$

which can finally be simplified as a final vector Eq. (10.50) of the complete vehicle as a scalar equation:

$$ \theta_{zz}\ddot{\psi}_V = l_v^V F_{v,y} - l_h^V F_{h,y}. \tag{10.54} $$

10.2.2 Tire Forces

In order to evaluate the Eqs. (10.49) and (10.54), the tire forces in the wheel fixed coordinate system are required which are initially only stationary and linear with respect to cornering stiffness $c_{s,v/h}$ and $c_{\alpha,v/h}$ and the slip variables $s_{v/h}$ and $\alpha_{v/h}$:

$$ \begin{bmatrix} {}^vF_{v,x,stat} \\ {}^vF_{v,y,stat} \end{bmatrix} = \begin{bmatrix} c_{s,v}s_v \\ c_{\alpha,v}\alpha_v \end{bmatrix}, \quad \begin{bmatrix} {}^hF_{h,x,stat} \\ {}^hF_{h,y,stat} \end{bmatrix} = \begin{bmatrix} c_{s,h}s_h \\ c_{\alpha,h}\alpha_h \end{bmatrix}. \tag{10.55} $$

The force components in Eq. (10.55) later need to be included in the inertial system or the vehicle fixed system respectively. This approach is applied in Sect. 10.3, in which it shall however be considered that the four coefficients $c_{s,v}$, $c_{s,h}$, $c_{\alpha,v}$ and $c_{\alpha,h}$ are generally nonlinear and dependent on the tire loads. Alternatively to Eq. (10.55) it is also possible to use a more detailed description with a (simplified) Magic Formula (Ref. Chap. 7), especially for higher lateral accelerations, which will be dealt with below, Ref. for example (Gipser 1999; Orend 2007).

For the sake of clarity, only the definitions for the front axle are below. The corresponding definitions for the rear axle are derived simply by replacing the index "*v*" with "*h*". A simplified Magic Formula approach for the tire forces yields:

$$ \begin{bmatrix} F_{v,x,stat} \\ F_{v,y,stat} \end{bmatrix} = F_{v,z,eff} \begin{bmatrix} \mu_{v,x} \sin\left(c_{v,x} \arctan\left(b_{v,x} \frac{s_{v,a}}{\mu_{v,x}}\right)\right) \\ \mu_{v,y} \sin\left(c_{v,y} \arctan\left(b_{v,y} \frac{s_{v,a}}{\mu_{v,y}}\right)\right) \end{bmatrix}, \tag{10.56} $$

with the effective tire load

$$ F_{v,z,eff} = F_{v,z}\left(1 - e_z\left(\frac{F_{v,z}}{F_{v,z,0}}\right)^2\right). \tag{10.57} $$

Equations (10.56) and (10.57) contain the friction coefficients $\mu_{v,x}$ and $\mu_{v,y}$, the tire parameters $c_{v,x}$, $c_{v,y}$, $c_{h,x}$, $c_{h,y}$ and $b_{v,x}$, $b_{v,y}$, $b_{h,x}$, $b_{h,y}$ as well as the variable tire loads ${}^{v}F_{v,z}$ and ${}^{h}F_{h,z}$. The degressive dependency of the horizontal tire forces on the tire loads is considered using the degressive parameter e_z. Furthermore, the slip variables s_v, α_v and s_h, α_h are included as the input variables for whose calculation the velocity of the wheel center point is required. This is represented by the kinematic relationships in the inertial system:

$$\begin{aligned}\dot{\boldsymbol{r}}_v &= \dot{\boldsymbol{r}}_V + \boldsymbol{\omega}_V \times \boldsymbol{T}_V(\psi_V){}^V_V\boldsymbol{r}_v \\ &= \begin{bmatrix} \dot{x}_V \\ \dot{y}_V \\ 0 \end{bmatrix} + \begin{bmatrix} 0 \\ 0 \\ \dot{\psi}_V \end{bmatrix} \times \begin{bmatrix} \cos\psi_V & -\sin\psi_V & 0 \\ \sin\psi_V & \cos\psi_V & 0 \\ 0 & 0 & 1 \end{bmatrix} \begin{bmatrix} l_v \\ 0 \\ -(h_s - r) \end{bmatrix} \\ \Rightarrow \begin{bmatrix} \dot{x}_v \\ \dot{y}_v \\ \dot{z}_v \end{bmatrix} &= \begin{bmatrix} \dot{x}_V - l_v\dot{\psi}_V \sin\psi_V \\ \dot{y}_V + l_v\dot{\psi}_V \cos\psi_V \\ 0 \end{bmatrix},\end{aligned} \tag{10.58}$$

and

$$\begin{aligned}\dot{\boldsymbol{r}}_h &= \dot{\boldsymbol{r}}_V + \boldsymbol{\omega}_V \times \boldsymbol{T}_V(\psi_V){}^V_V\boldsymbol{r}_h \\ &= \begin{bmatrix} \dot{x}_V \\ \dot{y}_V \\ 0 \end{bmatrix} + \begin{bmatrix} 0 \\ 0 \\ \dot{\psi}_V \end{bmatrix} \times \begin{bmatrix} \cos\psi_V & -\sin\psi_V & 0 \\ \sin\psi_V & \cos\psi_V & 0 \\ 0 & 0 & 1 \end{bmatrix} \begin{bmatrix} -l_h \\ 0 \\ -(h_s - r) \end{bmatrix} \\ \Rightarrow \begin{bmatrix} \dot{x}_h \\ \dot{y}_h \\ \dot{z}_h \end{bmatrix} &= \begin{bmatrix} \dot{x}_V + l_h\dot{\psi}_V \sin\psi_V \\ \dot{y}_V - l_h\dot{\psi}_V \cos\psi_V \\ 0 \end{bmatrix}.\end{aligned} \tag{10.59}$$

For calculating the slip values however, the velocities are required in the wheel fixed coordinate system. To this end, if one were to consider the rotation of the wheels with respect to the vehicle fixed coordinate system it would yield:

$${}^v\dot{\boldsymbol{r}}_v = \begin{bmatrix} {}^v\dot{x}_v \\ {}^v\dot{y}_v \\ {}^v\dot{z}_v \end{bmatrix} = {}^v\boldsymbol{T}_E\dot{\boldsymbol{r}}_v = \begin{bmatrix} \mathrm{c}(\psi_V + \delta) & \mathrm{s}(\psi_V + \delta) & 0 \\ -\mathrm{s}(\psi_V + \delta) & \mathrm{c}(\psi_V + \delta) & 0 \\ 0 & 0 & 1 \end{bmatrix} \begin{bmatrix} \dot{x}_v \\ \dot{y}_v \\ \dot{z}_v \end{bmatrix}, \tag{10.60}$$

$${}^h\dot{\boldsymbol{r}}_h = \begin{bmatrix} {}^h\dot{x}_h \\ {}^h\dot{y}_h \\ {}^h\dot{z}_h \end{bmatrix} = {}^h\boldsymbol{T}_E\dot{\boldsymbol{r}}_h = \begin{bmatrix} \mathrm{c}\psi_V & \mathrm{s}\psi_V & 0 \\ -\mathrm{s}\psi_V & \mathrm{c}\psi_V & 0 \\ 0 & 0 & 1 \end{bmatrix} \begin{bmatrix} \dot{x}_h \\ \dot{y}_h \\ \dot{z}_h \end{bmatrix}. \tag{10.61}$$

Note: In the following Eqs. (10.62)–(10.67) the tire forces are only given for the front axle. For the rear axle, the index "*v*" is to be replaced with the index "*h*". With the components of the velocity vector (10.60) and (10.61), after subtraction of the rolling velocity $r\dot{\rho}_v$ and normalizing, the longitudinal and lateral slips at the front one obtains:

$$s_v = \frac{r\dot{\rho}_v - {}^v\dot{x}_v}{\max(|r\dot{\rho}_v|, |{}^v\dot{x}_v|)}, \tag{10.62}$$

$$\alpha_v = -\frac{{}^v\dot{y}_v}{|r\dot{\rho}_v|}. \tag{10.63}$$

Now, as described in Chap. 7, the normalized total slip

$$s_{v,a} = \sqrt{s_v^2 + \tan^2 \alpha_v} \tag{10.64}$$

is calculated at the front and rear axle. From Eq. (10.56) and the direction of action of the slip

$$\psi_v = \operatorname{atan} \frac{\alpha_v}{s_v}, \tag{10.65}$$

the resulting tire forces are given by the Eqs. (10.56)–(10.65), at first the magnitude:

$$F_{\psi_v}(s_{v,a}) = \sqrt{\frac{s_v^2 F_{v,x,stat}^2 + \alpha_v^2 F_{v,y,stat}^2}{s_{v,a}^2}} \tag{10.66}$$

and from it, the tire forces in the wheel fixed coordinate system:

$$\begin{bmatrix} {}^vF_{v,x,stat} \\ {}^vF_{v,y,stat} \end{bmatrix} = F_{\psi_v}(s_{v,\alpha}) \begin{bmatrix} \cos \psi_v \\ \sin \psi_v \end{bmatrix} = \frac{1}{s_{v,a}} F_{\psi_v}(s_{v,\alpha}) \begin{bmatrix} s_v \\ \alpha_v \end{bmatrix}. \tag{10.67}$$

To consider the settling time of the tires during fast changes of course or velocity according to Chap. 7, an addition to Eq. (10.56) is necessary. Suitable time delay constants $T_{v,x}$ and $T_{v,y}$ are chosen to represent the first order response of the system. As the conditional equation for the dynamic tire forces $\boldsymbol{F}_v$ and $\boldsymbol{F}_h$ based on the already known quasi-stationary forces $\boldsymbol{F}_{v,stat}$ and $\boldsymbol{F}_{h,stat}$ a first order differential equation is used. Exemplarily for the front axle they read:

$$\begin{aligned} \begin{bmatrix} {}^v\dot{F}_{v,x} \\ {}^v\dot{F}_{v,y} \end{bmatrix} &= \begin{bmatrix} \frac{1}{T_{v,x}} & 0 \\ 0 & \frac{1}{T_{v,y}} \end{bmatrix} \left(\begin{bmatrix} {}^vF_{v,x,stat} \\ {}^vF_{v,y,stat} \end{bmatrix} - \begin{bmatrix} {}^vF_{v,x} \\ {}^vF_{v,y} \end{bmatrix} \right) \\ &= \begin{bmatrix} \frac{c_{v,x,stat}|r\dot{\rho}_v|}{c_{v,s,stat}} & 0 \\ 0 & \frac{c_{v,y,stat}|r\dot{\rho}_v|}{c_{v,\alpha,stat}} \end{bmatrix} \left(\begin{bmatrix} {}^vF_{v,x,stat} \\ {}^vF_{v,y,stat} \end{bmatrix} - \begin{bmatrix} {}^vF_{v,x} \\ {}^vF_{v,y} \end{bmatrix} \right). \end{aligned} \tag{10.68}$$

To this end, the time constants for the x- and y-directions are calculated according to Chap. 7 as follows:

$$\frac{1}{T_{v,x}} = \frac{c_{v,x,stat}|r\dot{\rho}_v|}{c_{v,s,stat}}, \quad \frac{1}{T_{v,y}} = \frac{c_{v,y,stat}|r\dot{\rho}_v|}{c_{v,\alpha,stat}}. \tag{10.69}$$

With the equilibrium of momentum about the vehicle center of gravity S, and the force equilibrium in the z-direction of the inertial system, the tire loads are determined. Hence, the tire normal forces at the front and the rear are:

$$^VF_{v,z} = \frac{l_h}{l}mg - \frac{h_s}{l}\left(^VF_{v,x} + {}^VF_{h,x}\right), \tag{10.70}$$

$$^VF_{h,z} = \frac{l_v}{l}mg + \frac{h_s}{l}\left(^VF_{v,x} + {}^VF_{h,x}\right). \tag{10.71}$$

Finally, the principle of momentum conservation at the front and the rear wheels with respect to the wheel center is required:

$$\theta_v\ddot{\rho}_v = M_{A,v} - \text{sign}(\dot{\rho}_v)M_{B,v} - r^vF_{v,x} \tag{10.72}$$

$$\theta_h\ddot{\rho}_h = M_{A,h} - \text{sign}(\dot{\rho}_h)M_{B,h} - r^hF_{h,x}. \tag{10.73}$$

10.2.3 Drive and Brake Torques

A random distribution of the driving torques between the front and the rear axle $M_{A,v}$ and $M_{A,h}$ is modeled in this example. The eigen-dynamics of the drivetrain will not be considered. For a dimensionless factor $0 \leq \xi_a \leq 1$, it follows:

$$M_{A,v} = (1 - \xi_a)M_A = M_A - M_{A,h}, \tag{10.74}$$

$$M_{A,h} = \xi_a M_A. \tag{10.75}$$

With this, $\xi_a = 0$ represents a front wheel drive, while $\xi_a = 1$ represents a rear wheel drive. For all other values $0 < \xi_a < 1$, a four wheel drive configuration with variable distribution on the axles is obtained. For the calculation of the driving torque M_A one first needs an approximation of the engine speed of rotation:

$$\omega_M = i_D i_G(G)((1 - \xi_a)\dot{\rho}_v + \xi_a\dot{\rho}_h), 0 \leq \xi_a \leq 1. \tag{10.76}$$

The drivetrain parameters i_D and i_G represent the transmission of the central differential and that of the gearbox respectively. In this manner, the total driving

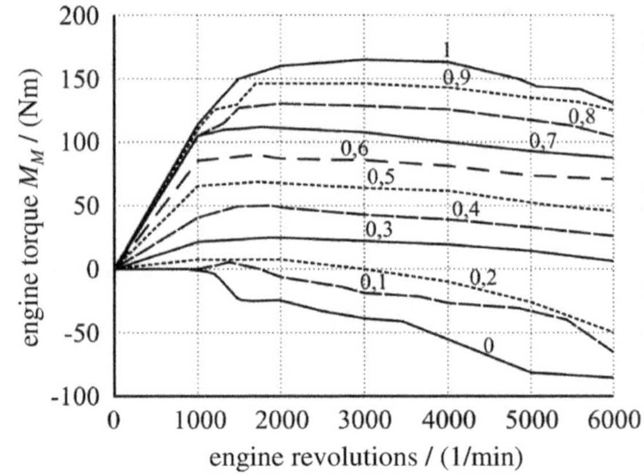

Fig. 10.10 Schematic representation of a engine characteristic curve dependent on the rpm and acceleration pedal position

torque M_A, required for the evaluation of the Eqs. (10.72) and (10.73), based on the engine torque M_M is given by:

$$M_A = i_D i_G(G) M_M(\omega_M, p_F). \tag{10.77}$$

The engine torque is interpolated from a two dimensional engine torque characteristic curve (Fig. 10.10). Along with the engine speed ω_M, another dimensionless input parameter, the acceleration pedal position $0 \leq p_F \leq 1$ is also required. The pedal position is normally interpreted as an excitation function $p_F(t)$.

Analogous to this the brake torques are calculated as follows:

$$M_{B,v} = (1 - \xi_b) M_B(p_B) = M_B(p_B) - M_{B,h}, \tag{10.78}$$

$$M_{B,h} = \xi_b M_B(p_B). \tag{10.79}$$

Here, $0 \leq \xi_b \leq 1$ is a dimensionless distribution parameter again and $p_B(t)$ is the brake pedal travel dependent on time, Ref. Fig. 10.11.

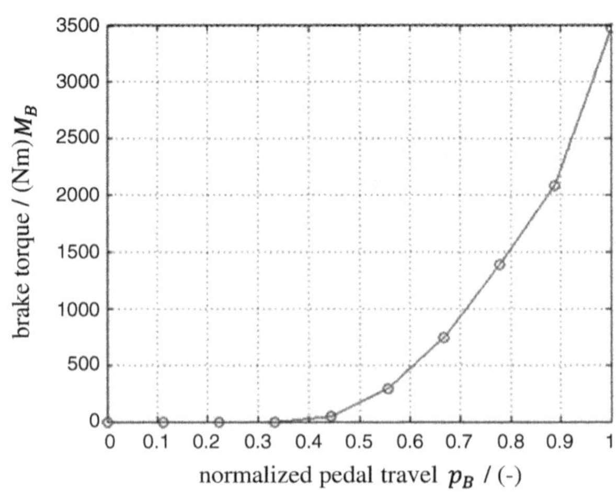

Fig. 10.11 Brake torque dependent on the brake pedal positions

10.2.4 Equations of Motion

After the initial preliminary work in the past sections it is now possible to list the complete set of equations of motion of the nonlinear single track model.

- Principle of conservation of linear momentum:

$$\begin{bmatrix} m\ddot{x}_V \\ m\ddot{y}_V \end{bmatrix} = \begin{bmatrix} F_{v,x} + F_{h,x} - F_{W,x} \\ F_{v,y} + F_{h,y} - F_{W,y} \end{bmatrix}, \tag{10.80}$$

with

$$\begin{aligned} F_{W,x} &= \frac{1}{2} c_W \rho_L A \dot{x}_V \sqrt{\dot{x}_V^2 + \dot{y}_V^2}, \\ F_{W,y} &= \frac{1}{2} c_W \rho_L A \dot{y}_V \sqrt{\dot{x}_V^2 + \dot{y}_V^2}, \end{aligned} \tag{10.81}$$

$$\begin{aligned} F_{v,x} &= \cos(\psi_V + \delta)^v F_{v,x} - \sin(\psi_V + \delta)^v F_{v,y}, \\ F_{v,y} &= \sin(\psi_V + \delta)^v F_{v,x} + \cos(\psi_V + \delta)^v F_{v,y}, \end{aligned} \tag{10.82}$$

$$\begin{aligned} F_{h,x} &= \cos\psi_V\, {}^h F_{h,x} - \sin\psi_V\, {}^h F_{h,y}, \\ F_{h,y} &= \sin\psi_V\, {}^h F_{h,x} + \cos\psi_V\, {}^h F_{h,y}. \end{aligned} \tag{10.83}$$

- Principle of conservation of the angular momentum for the chassis in the vehicle fixed coordinate system:

$$\theta_{zz}\ddot{\psi}_V = l_v\, {}^V F_{v,y} - l_h\, {}^V F_{h,y}. \tag{10.84}$$

with

$$\begin{aligned} {}^V F_{v,y} &= \sin\delta\, {}^v F_{v,x} + \cos\delta\, {}^v F_{v,y}, \\ {}^V F_{h,y} &= {}^h F_{h,y}. \end{aligned} \tag{10.85}$$

- Principle of conservation of the angular momentum for the front and rear axle:

$$\theta_v \ddot{\rho}_v = M_{A,v} - M_{B,v}\mathrm{sign}(\dot{\rho}_v) - r^v F_{v,x}, \tag{10.86}$$

$$\theta_h \ddot{\rho}_h = M_{A,h} - M_{B,h}\mathrm{sign}(\dot{\rho}_h) - r^h F_{h,x}. \tag{10.87}$$

- Dynamic tire forces:

$$\begin{bmatrix} {}^v\dot{F}_{v,x} \\ {}^v\dot{F}_{v,y} \end{bmatrix} = \begin{bmatrix} \frac{c_{v,x,stat}|r\dot{\rho}_v|}{c_{s,v,stat}} & 0 \\ 0 & \frac{c_{v,y,stat}|r\dot{\rho}_v|}{c_{v,\alpha,stat}} \end{bmatrix} \left(\begin{bmatrix} {}^vF_{v,x,stat} \\ {}^vF_{v,y,stat} \end{bmatrix} - \begin{bmatrix} {}^vF_{v,x} \\ {}^vF_{v,y} \end{bmatrix} \right), \tag{10.88}$$

$$\begin{bmatrix} {}^h\dot{F}_{h,x} \\ {}^h\dot{F}_{h,x} \end{bmatrix} = \begin{bmatrix} \frac{c_{h,x,stat}|r\dot{\rho}_h|}{c_{h,s,stat}} & 0 \\ 0 & \frac{c_{h,y,stat}|r\dot{\rho}_h|}{c_{h,\alpha,stat}} \end{bmatrix} \left(\begin{bmatrix} {}^hF_{h,x,stat} \\ {}^hF_{h,y,stat} \end{bmatrix} - \begin{bmatrix} {}^hF_{h,x} \\ {}^hF_{h,y} \end{bmatrix} \right). \tag{10.89}$$

10.2.5 Equations of State

One can now transfer the equations of motion into the state space form:

$$\dot{\boldsymbol{x}} = \boldsymbol{f}(\boldsymbol{x}, t, \boldsymbol{u}), \tag{10.90}$$

with the state vector

$$\boldsymbol{x} = \left[x_V, y_V, \psi_V, \dot{x}_V, \dot{y}_V, \dot{\psi}_V, \dot{\rho}_v, \dot{\rho}_h, {}^vF_{v,x}, {}^vF_{v,y}, {}^hF_{h,x}, {}^hF_{h,y}\right]^T \tag{10.91}$$

and the excitation vector

$$\boldsymbol{u} = [\delta_H, p_F, p_B, G]^T. \tag{10.92}$$

Along with the acceleration and brake pedal position p_F and p_B, the steering wheel angle δ_H and the gear parameter G (defining the gear engaged) also appear. As a result, the nonlinear single track model can be represented as a dynamic system as shown in Fig. 10.12.

As a whole, the Eqs. (10.90)–(10.92) read:

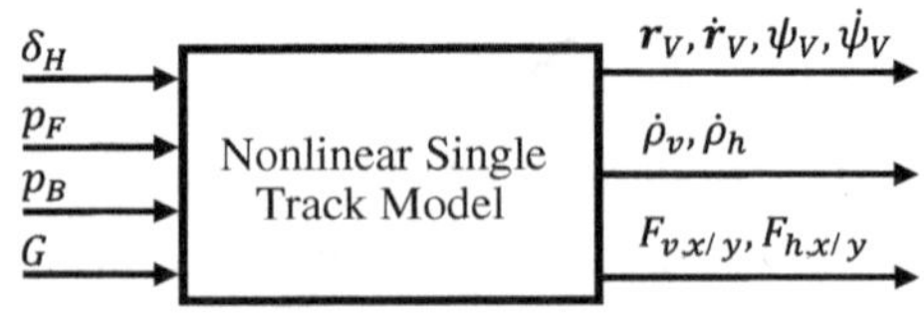

Fig. 10.12 Nonlinear single track model of a dynamic system

$$\underbrace{\begin{bmatrix} \dot{x}_V \\ \dot{y}_V \\ \dot{\psi}_V \\ \ddot{x}_V \\ \ddot{y}_V \\ \ddot{\psi}_V \\ \ddot{\rho}_v \\ \ddot{\rho}_h \\ {}^V\dot{F}_{v,x} \\ {}^V\dot{F}_{v,y} \\ {}^V\dot{F}_{h,x} \\ {}^V\dot{F}_{h,y} \end{bmatrix}}_{\dot{\boldsymbol{x}}} = \underbrace{\begin{bmatrix} \dot{x}_V \\ \dot{y}_V \\ \dot{\psi}_V \\ \frac{1}{m}(F_{v,x} + F_{h,x} - F_{W,x}) \\ \frac{1}{m}(F_{v,y} + F_{h,y} - F_{W,y}) \\ \frac{1}{\theta_{zz}}(l_v{}^V F_{v,y} - l_h{}^V F_{h,y}) \\ \frac{1}{\theta_v}(M_{A,v} - M_{B,v}\mathrm{sign}(\dot{\rho}_v) - r\,{}^vF_{v,x}) \\ \frac{1}{\theta_h}(M_{A,h} - M_{B,h}\mathrm{sign}(\dot{\rho}_h) - r\,{}^hF_{h,x}) \\ \frac{c_{v,x,stat}|r\dot{\rho}_v|}{c_{v,s,stat}}\left({}^vF_{v,x,stat} - {}^vF_{v,x}\right) \\ \frac{c_{v,y,stat}|r\dot{\rho}_v|}{c_{v,\alpha,stat}}\left({}^vF_{v,y,stat} - {}^vF_{v,y}\right) \\ \frac{c_{h,x,stat}|r\dot{\rho}_h|}{c_{h,s,stat}}\left({}^hF_{h,x,stat} - {}^hF_{h,x}\right) \\ \frac{c_{h,y,stat}|r\dot{\rho}_h|}{c_{h,\alpha,stat}}\left({}^hF_{h,y,stat} - {}^hF_{h,y}\right) \end{bmatrix}}_{\boldsymbol{f}(\boldsymbol{x},t,\boldsymbol{u})} \tag{10.93}$$

10.3 Linear Roll Model

Due to their modeling constraints, the single track models discussed in this chapter so far do not allow the description and investigation of effects resulting from different tire loading, for example during cornering. These effects will obviously be included in the spatial modeling, which is the focus of Chaps. 11 and 12. It is however also possible for real time simulations or for simple fundamental investigations in example, to model and consider such effects in the single track models discussed until now. However, the following constraints still hold:

- Changes in the chassis geometry as a result of the forces will not be considered. This means that all equilibrium conditions need to be formulated from the output geometry.
- The roll angle and the steering angle of the front wheels will be assumed to be small.
- The inertial force ma_y, perpendicular to the direction of motion, will be assumed to be an external force acting on the center of gravity of the chassis. Centrifugal forces of the tires or of the components of the wheel suspension will not be considered.
- The chassis weight m_A is equal to the weight m of the entire vehicle.

To this end, so-called linear roll models for small roll angles φ are introduced. In a linear roll model the front and the rear axles will be combined to form one axle, thus creating a roll center W, which will be assumed to be at the height of the axle center, (Ref. Fig. 10.15), (Ammon 2013; Öttgen 2005), see Figs. 10.13, 10.14, 10.15 and 10.16.

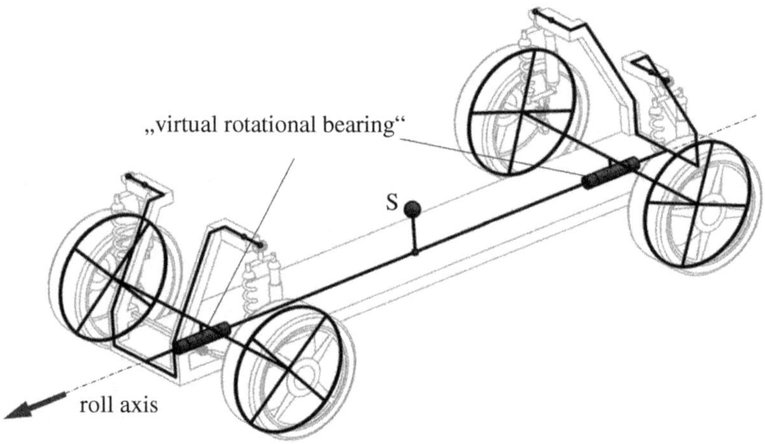

Fig. 10.13 Roll model–schematic representation

As a reaction to the roll torque, which results from the lateral acceleration a_y of the chassis, forces $A_{l,v}, A_{r,v}, A_{l,h}, A_{r,h}$ are created due to the springs, dampers and the anti-roll bars acting opposite to the roll motion.

The stationary forces $W_{v,y}$, $W_{v,z}$, $W_{h,y}$, $W_{h,z}$ of the chassis concerning the rolling motion are transferred to the axles via virtual rotational joints, and result from the quasi-stationary equilibrium conditions.

As the center of gravity is generally not on the same level as that of the rolling center, the roll angle, which is set during cornering, is dependent on the lateral acceleration.

10.3.1 Equation of Motion for the Rolling of the Chassis

The equilibrium of momentums for the entire vehicle about the roll center W with the corresponding sum of vertical and lateral tire forces left and right $F_{l,z}$, $F_{r,z}$ and $F_{l,y}$, $F_{r,y}$ of the vehicle total mass m, the moment of inertia θ_{xx} about the roll axis, the height of the center of gravity h_s and the track width $2s_R$, lead to the relationship:

$$0 = (h_S - h_W) \cdot ma_y + h_W\left(F_{l,y} + F_{r,y}\right) + s_R(F_{l,z} - F_{r,z}). \tag{10.94}$$

If one now also considers the (quasi-stationary) force balance in lateral direction

$$F_{l,y} + F_{r,y} = ma_y, \tag{10.95}$$

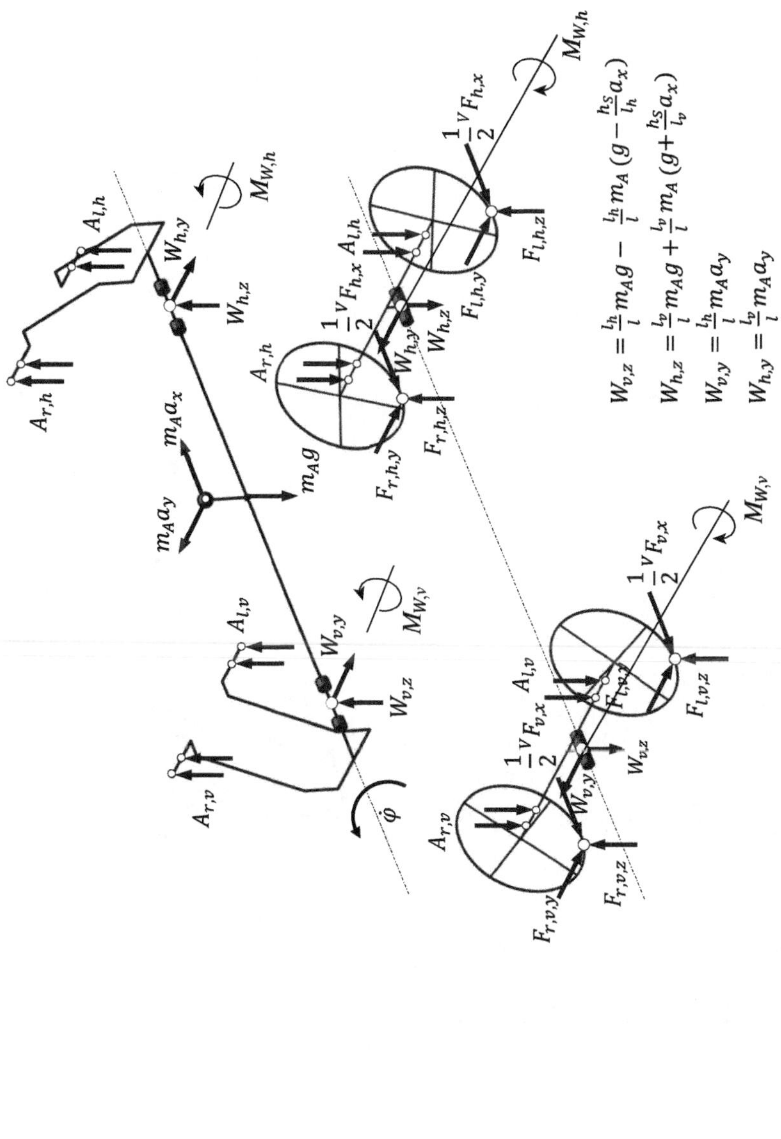

Fig. 10.14 Roll model–free body diagram

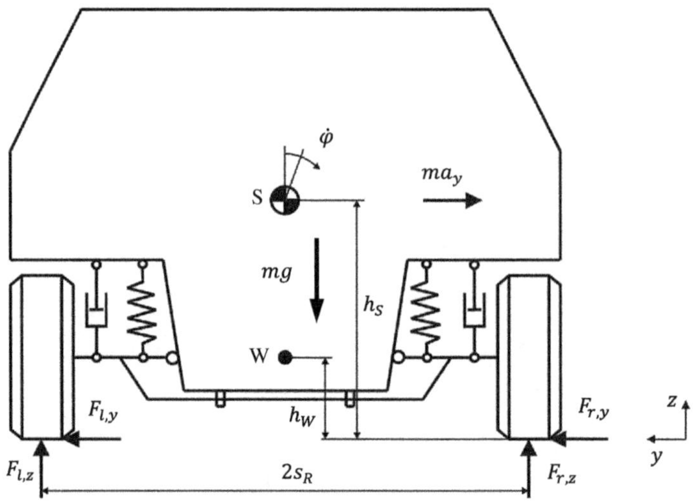

Fig. 10.15 Linear roll model—forces front and rear combined

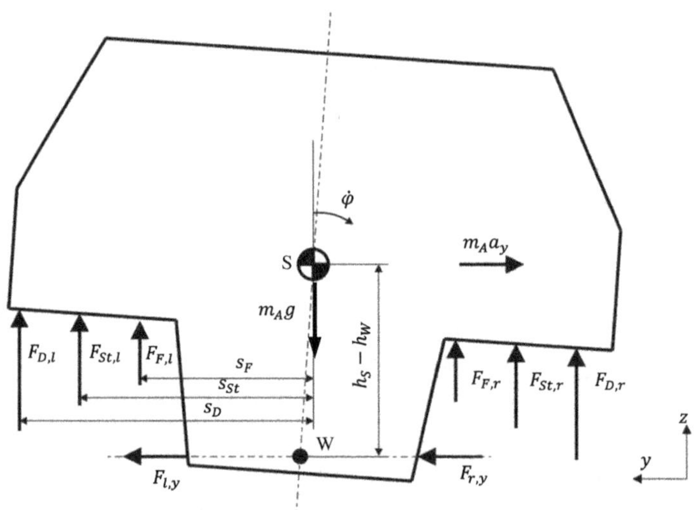

Fig. 10.16 Description of the construction of the rolling dynamics

thus Eq. (10.94) can be simplified as

$$0 = h_S \cdot ma_y + s_R(F_{l,z} - F_{r,z}). \tag{10.96}$$

Because the solidification principle was applied during the derivation of Eq. (10.96), the tire normal forces for the individual tires cannot be determined due to the resulting static indefiniteness. Additionally, one should consider the elasticity due to the wheel suspension. To this end, the vehicle chassis will be displayed as a free body diagram (Fig. 10.16). The principle of moment conservation for the chassis with respect to the rolling center leads to the following equations of motion for the chassis:

$$\begin{aligned}\theta_A\ddot{\varphi} =& (h_S - h_W)\cos\varphi \cdot m_A a_y \\ &+ s_{F,v}\left(F_{F,l,v} - F_{F,r,v}\right) + s_{F,h}\left(F_{F,l,h} - F_{F,r,h}\right) \\ &+ s_{D,v}\left(F_{D,l,v} - F_{D,r,v}\right) + s_{D,h}\left(F_{D,l,h} - F_{D,r,h}\right) \\ &+ s_{St,v}\left(F_{St,l,v} - F_{St,r,v}\right) + s_{St,h}\left(F_{St,l,h} - F_{St,r,h}\right) \\ &+ (h_S - h_W)\sin\varphi m_A g\end{aligned} \tag{10.97}$$

Here, h_w is the height of the rolling center, and $s_{F,(v/h)}$, $s_{D,(v/h)}$ and $s_{St,(v/h)}$ are the distances of the spring, damper and anti-roll bars action points from the central plane of the symmetrically assumed vehicle chassis.

Due to the assumption of small roll angles φ, the shift of the center of gravity and that of the action points of the axle lateral forces will be neglected.

Considering the resulting spring and damper forces on the left and the right side, under the assumption of linear force elements, one arrives at the following relationships for the front axle:

$$F_{F,l,v} = -c_{F,v}s_{F,v}\sin\varphi \approx -c_{F,v}s_{F,v}\varphi, \tag{10.98}$$

$$F_{F,r,v} = c_{F,v}s_{F,v}\sin\varphi \approx c_{F,v}s_{F,v}\varphi, \tag{10.99}$$

$$F_{D,l,v} = -d_v s_{D,v}\cos\varphi\dot{\varphi} \approx -d_v s_{D,v}\dot{\varphi}, \tag{10.100}$$

$$F_{D,r,v} = d_v s_{D,v}\cos\varphi\dot{\varphi} \approx d_v s_{D,v}\dot{\varphi}. \tag{10.101}$$

The corresponding relationships for the rear axle are obtained by substituting the indices "v" through "h" in Eqs. (10.98)–(10.101). The forces of the anti-roll bars are due to the torsion of the torsion beam with the torsion stiffness $c_{St,v}$ and $c_{St,h}$, the beam lengths $l_{St,v}$ and $l_{St,h}$ as well as the anti-roll bar lever arms $b_{St,v}$ and $b_{St,h}$ can be approximated as (compare also Chap. 9)

$$F_{St,l,v} = -\frac{c_{St,v}}{b_{St,v}}\arcsin\left(\frac{l_{St,v}}{2b_{St,v}}\sin\varphi\right) \approx -\frac{1}{2}\frac{c_{St,v}l_{St,v}}{b_{St,v}^2}\varphi, \tag{10.102}$$

$$F_{St,r,v} = \frac{c_{St,v}}{b_{St,v}}\arcsin\left(\frac{l_{St,v}}{2b_{St,v}}\sin\varphi\right) \approx \frac{1}{2}\frac{c_{St,v}l_{St,v}}{b_{St,v}^2}\varphi. \tag{10.103}$$

Here too, the formula for the rear axle is obtained by substituting the index "v" by "h". With the help of the Eqs. (10.98)–(10.103), the equation of motion (10.97) can now be written as

$$
\begin{aligned}
&\theta_A\ddot{\varphi} + 2\left(s_{D,v}^2 d_v + s_{D,h}^2 d_h\right)\dot{\varphi} \\
&\quad + 2\left(s_{F,v}^2 c_{F,v} + \frac{c_{St,v} l_{St,v} s_{St,v}}{2b_{St,v}^2} + s_{F,h}^2 c_{F,h} + \frac{c_{St,h} l_{St,h} s_{St,h}}{2b_{St,h}^2}\right)\varphi \\
&\quad = m_A a_y (h_S - h_W) + m_A g \varphi (h_S - h_W).
\end{aligned}
\tag{10.104}
$$

The Eq. (10.104) represents the roll angle equation of the chassis about a roll center assumed to be constant. The inertial force (centrifugal force) of the chassis is the only excitation. This additional equation can be used to calculate the dynamic tire normal forces of all the four tires in both the linear as well as, allowing certain constraints, the nonlinear single track model. This is particularly helpful during the investigation of reactions to load fluctuations during cornering, including the influence of the cornering forces on the over and under steering behavior of vehicles. Furthermore, such models allow the investigation of the influence of the passive and active anti-roll bars on the driving characteristics of vehicles, Ref. Chap. 14.

The inertias and masses as well as the length measurements in Eq. (10.104) are generally available for a given vehicle. However, the stiffness and damping need to be identified, for example with different driving maneuvers and the parameter identification algorithm. This can either be done experimentally or with virtual tests using a more complex vehicle model, Ref. for example (Öttgen 2005).

Finally, the dynamic model as given by Fig. 10.17 is derived, although one should note that in the Block Tire Model, the tire forces are first calculated in the horizontal direction for all the four tires and they should then be added axle-wise.

10.3.2 Dynamic Tire Loads

With the help of the dynamic tire loads in a suitable tire model, which includes the degressive nature of the influence of the tire load, the cornering stiffness of the tires

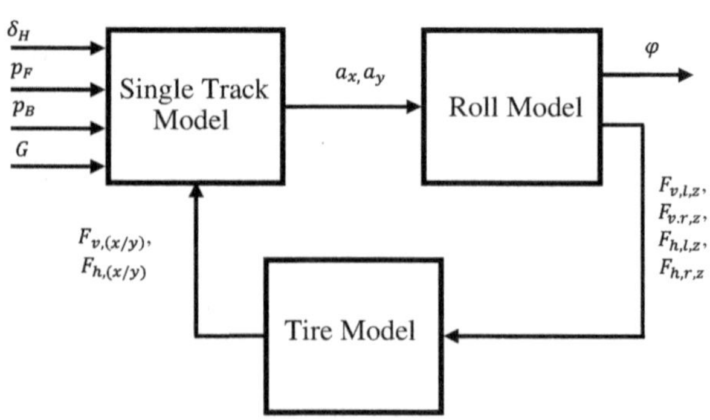

Fig. 10.17 Nonlinear single track model with linear roll dynamics as a dynamic system

can be calculated in order to calculate the lateral tire forces for a single track model. For calculating the tire loads, the centrifugal forces are initially distributed on the front and rear axle depending on the position of the center of gravity, and finally the equilibrium condition under the influence of the inertial forces and the wheel suspension, Ref. Figs. 10.14 and 10.16 is considered. In order to make the computations a bit clearer, the forces and torques are collected. One should note that in the case of the torques, the force components are acting upon different points:

$$A_{l,v} = -F_{F,l,v} - F_{D,l,v} - F_{St,l,v}, \tag{10.105}$$

$$A_{r,v} = -F_{F,r,v} - F_{D,r,v} - F_{St,r,v}, \tag{10.106}$$

$$A_{l,h} = -F_{F,l,h} - F_{D,l,h} - F_{St,l,h}, \tag{10.107}$$

$$A_{r,h} = -F_{F,r,h} - F_{D,r,h} - F_{St,r,h}, \tag{10.108}$$

$$\begin{aligned} M_{A,v} =& - s_{F,v}\left(F_{F,l,v} - F_{F,r,v}\right) - s_{D,v}\left(F_{D,l,v} - F_{D,r,v}\right) \\ & - s_{St,v}\left(F_{St,l,v} - F_{St,r,v}\right), \end{aligned} \tag{10.109}$$

$$\begin{aligned} M_{A,h} =& - s_{F,h}\left(F_{F,l,h} - F_{F,r,h}\right) - s_{D,h}\left(F_{D,l,h} - F_{D,r,h}\right) \\ & - s_{St,h}\left(F_{St,l,h} - F_{St,r,h}\right), \end{aligned} \tag{10.110}$$

From the equilibrium conditions one obtains:

- Force equilibrium in the vertical direction on the front axle:

$$F_{l,v,z} + F_{r,v,z} = \frac{l_h}{l} mg - \frac{h_s}{l} ma_x + A_{l,v} + A_{r,v}. \tag{10.111}$$

- Force equilibrium in the vertical direction on the rear axle:

$$F_{l,h,z} + F_{r,h,z} = \frac{l_v}{l} mg + \frac{h_s}{l} ma_x + A_{l,h} + A_{r,h}. \tag{10.112}$$

- Force equilibrium in the lateral direction on the front axle:

$$F_{l,v,y} + F_{r,v,y} = \frac{l_h}{l} ma_y. \tag{10.113}$$

- Force equilibrium in the lateral direction on the rear axle:

$$F_{l,h,y} + F_{r,h,y} = \frac{l_v}{l} m a_y. \tag{10.114}$$

- Torque equilibrium about the rolling center on the front axle:

$$\left(F_{r,v,z} - F_{l,v,z}\right) s_R - \left(F_{r,v,y} + F_{l,v,y}\right) h_W = M_{A,v}. \tag{10.115}$$

- Torque equilibrium about the rolling center on the rear axle:

$$\left(F_{r,h,z} - F_{l,h,z}\right) s_R - \left(F_{r,h,y} + F_{l,h,y}\right) h_W = M_{A,h}. \tag{10.116}$$

The Eqs. (10.111)-(10.116) can be solved easily for the tire loads. One obtains:

$$F_{l,v,z} = -\frac{1}{2s_R} M_{A,v} + \frac{l_h}{2l} m\left(g - \frac{h_W}{s_R} a_y\right) - \frac{h_s}{2l} m a_x \tag{10.117}$$

$$F_{r,v,z} = \frac{1}{2s_R} M_{A,v} + \frac{l_h}{2l} m\left(g + \frac{h_W}{s_R} a_y\right) - \frac{h_s}{2l} m a_x \tag{10.118}$$

$$F_{l,h,z} = -\frac{1}{2s_R} M_{A,h} + \frac{l_v}{2l} m\left(g - \frac{h_W}{s_R} a_y\right) + \frac{h_s}{2l} m a_x \tag{10.119}$$

$$F_{r,h,z} = \frac{1}{2s_R} M_{A,h} + \frac{l_v}{2l} m\left(g + \frac{h_W}{s_R} a_y\right) + \frac{h_s}{2l} m a_x \tag{10.120}$$

And correspondingly, the sum of the resulting lateral forces on the front and rear axles:

$$F_{l,v,y} + F_{r,v,y} = \frac{l_h}{l} m a_y, \tag{10.121}$$

$$F_{l,h,y} + F_{r,h,y} = \frac{l_v}{l} m a_y. \tag{10.122}$$

10.3.3 Influence of the Self-steering Behavior

A single track model with extended linear roll dynamics can, for example, be applied in order to explain the influence of the anti-roll bar on the vehicle dynamics. To this end, the influence of the calculated vertical force distributions on the tire horizontal forces is investigated

$$\begin{bmatrix} {}^{V}F_{v,x,stat} \\ {}^{V}F_{v,y,stat} \end{bmatrix} = \begin{bmatrix} c_{v,x}\left({}^{V}F_{v,z}\right)s_v \\ c_{v,y}\left({}^{V}F_{v,z}\right)\alpha_v \end{bmatrix}. \tag{10.123}$$

The corresponding relationship to the rear axle is given when replacing the index "v" with "h" in Eq. (10.123).

In order to continue with the calculations, one still requires a description of the nonlinear relationship between the tire normal forces at the tires and the cornering stiffness, Ref. Fig. 10.18. This can for example, as an approximate description of the relationship, be considered to be a polynomial (Öttgen 2005):

$$c_{(v/h),y} = \sum_{k=1}^{n} c_{k,(v/h)} F_{(v/h),z}^{k}. \tag{10.124}$$

The (equally nonlinear) relationship of the longitudinal stiffness of the tire normal forces shall not be considered here.

For the following observations, the lateral guiding forces in the quasi-static driving conditions shall be distributed between the front and the rear axles. From Eq. (10.95) and the torque equilibrium about the center of gravity S of the vehicle, one can deduce the lateral guiding forces on the front and rear axle:

$$F_{(v/h),y} = F_{l,(v/h),y} + F_{r,(v/h),y} = \frac{m a_y l_{(v/h)}}{l}. \tag{10.125}$$

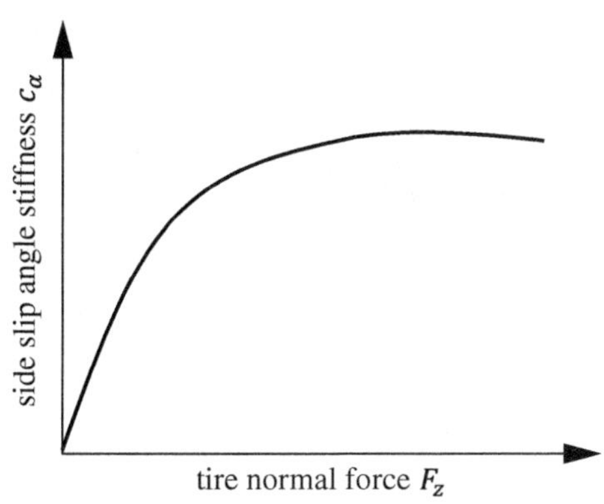

Fig. 10.18 Dependence of the cornering stiffness depending with the tire normal forces (qualitative representation)

The following calculation is shown exemplarily for the rear axle. In order to generate a force $F_{h,y}$ on the two rear wheels, the rear axle, under the assumption that the two sides are equal, must assume a slip angle of α_h. The slip angle that needs to be applied depends on (comp. Eq. (10.15))

$$\alpha_h = \frac{F_{h,y}}{c_{\alpha,h}}, \tag{10.126}$$

the cornering stiffness $c_{\alpha,h}$, which in turn (comp. Fig. 10.4) is a degressive function of the tire load. The tire load consists of the static tire load and the tire load change or fluctuation $\Delta F_{h,z}$ during cornering. For the inside tire during cornering (index "*i*") and for the outside tire during cornering (index "*a*") the tire loads are:

$$F_{h,z,i} = \frac{l_v}{2l} mg - \Delta F_{h,z}, \quad F_{h,z,a} = \frac{l_v}{2l} mg + \Delta F_{h,z}. \tag{10.127}$$

From this, the lateral guiding forces can be determined:

$$F_{h,y,i} = c_{\alpha,h}\left(F_{h,z,i}\right)\alpha_h, F_{h,y,a} = c_{\alpha,h}\left(F_{h,z,a}\right)\alpha_h. \tag{10.128}$$

Due to the degressive characteristics of the cornering stiffness as a function of the tire load, the loss of the lateral guiding force on the inside tire during cornering is greater than the gain in the lateral guiding force on the outside tire, i.e. it holds:

$$F_{h,y,i} + F_{h,y,a} < F_{h,y}. \tag{10.129}$$

As a consequence, a new slip angle $\bar{\alpha}_h > \alpha_h$ has to be applied so that the following holds:

$$F_{h,y,i} + F_{h,y,a} = \left(c_{\alpha,h}\left(F_{h,z,i}\right) + c_{\alpha,h}\left(F_{h,z,a}\right)\right)\bar{\alpha}_h = F_{h,y}. \tag{10.130}$$

This physical effect facilitates the possibility of influencing the eigen-dynamic of the vehicle by changing the tire load distribution between the front and rear axle. Additionally, the force elements in the anti-roll bar given in Eqs. (10.117)–(10.120) can be used either passively through predefined and predetermined anti-roll bar stiffness, or even changed actively during operation. A change in the eigen-dynamics of the vehicle can, according to Eq. (10.29), for example lead to an increase in the anti-roll bar stiffness at the front axle and to a decrease in the anti-roll bar stiffness at the rear axle. Alternatively, one can for example increase the chassis elasticity of the front axle or even decrease the track width at the front axle. An active intervention of vehicle dynamics through an active anti-roll bar is outlined in Chap. 14.

References

Ammon D (2013) Modellbildung und Systementwicklung in der Fahrzeugdynamik. Vieweg+ Teubner Verlag, Stuttgart.—ISBN 978-3-663-12246-3

Gipser M (1999) Systemdynamik und Simulation. Vieweg+Teubner, Stuttgart u.a.—ISBN 978-3-519-02743-0

Orend R (2007) Integrierte Fahrdynamikregelung mit Einzelradaktorik. S 159, Shaker Verlag, Aachen. - ISBN 978-3-8322-5800-9

Öttgen O (2005) Zur modellgestützten Entwicklung eines mechatronischen Fahrwerkregelungssystems für Personenkraftwagen Dissertation, Universität Duisburg-Essen, Düsseldorf: VDI-Verlag

Riekert P and Schunk TE (1940) Zur Fahrmechanik des gummibereiften Kraftfahrzeugs. Vol. 11, S 210–224—ISBN ISSN: 0020-1154

Willumeit H-P (1998) Modelle und Modellierungsverfahren in der Fahrzeugdynamik. B.G. Teubner, Stuttgart/Leipzig.—ISBN 978-3-663-12248-7

Chapter 11
Twin Track Models

11.1 Twin Track Model Without Suspension Kinematics

In this section, the equations of motion for a basic spatial twin track model without any suspension kinematics will be set up. The foundation for this will be the substitute model represented in Fig. 11.1.

This vehicle model is characterized by the fact that no kinematic wheel suspension is modeled at all, i.e. the wheels are simply connected to the vehicle chassis by the spring and damper forces, which are applied forces. Additionally, it shall be assumed that the center of the wheel can only move relative to the chassis and perpendicular to the road. This means that with this model it is not possible to investigate the influence of the camber or the influence of other spatial motions of the wheels. This model is however sufficient for initial principle investigations or even for a basic vehicle simulator model. In detail, this model incorporates the DoF of the vehicle as described in Table 11.1.

A general vehicle model that can be adapted to include more complex wheel suspensions to a certain extent is dealt with in Sect. 11.2.

The vehicle model has $f = 14$ DoF. In order to describe the spatial motion of the chassis, one defines a global coordinate system (inertial system)

$$\mathbf{K}_E = \{\mathrm{O}_E; x_E, y_E, z_E\}, \tag{11.1}$$

as well as vehicle fixed coordinate system

$$\mathbf{K}_V = \{\mathrm{O}_V; x_V, y_V, z_V\}. \tag{11.2}$$

The rotations about the three CARDAN angles $\theta_V, \psi_V, \varphi_V$ are chosen in order to describe the motion of the chassis relative to the inertial system. The mathematical description can be achieved using the transformation matrix (rotational matrix) ${}^E\boldsymbol{T}_V$, which describes the rotation of the vehicle fixed coordinate system into the inertial system:

D. Schramm et al., Vehicle Dynamics,
DOI 10.1007/978-3-662-54483-9_11

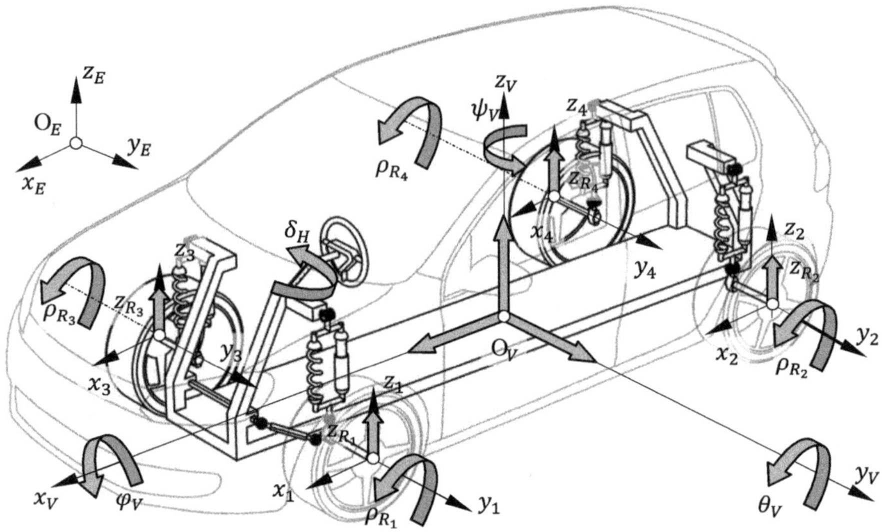

Fig. 11.1 Degrees of freedom (DoF) of a spatial twin track model without suspension kinematics

Table 11.1 DoF of a basic spatial vehicle model

Expression	Description
x_V, y_V, z_V	Coordinates of the center of gravity of the vehicle chassis in the inertial system
$\psi_V, \theta_V, \varphi_V$	CARDAN angles of the vehicle chassis
$\rho_{R_i},\ i = 1, \ldots, 4$	Rotation angle of the ith wheel about its own rotational axis
$z_{R_i}, i = 1, \ldots, 4$	Vertical motion of the ith wheel in the vehicle fixed coordinate system

$$
\begin{gathered}
{}^{E}\boldsymbol{T}_V = \boldsymbol{T}_z(\psi_V) \cdot \boldsymbol{T}_y(\theta_V) \cdot \boldsymbol{T}_x(\varphi_V) = \\
\begin{bmatrix} c\theta_V c\psi_V & s\varphi_V s\theta_V c\psi_V - c\varphi_V s\psi_V & c\varphi_V s\theta_V c\psi_V + s\varphi_V s\psi_V \\ c\theta_V s\psi_V & s\varphi_V s\theta_V s\psi_V + c\varphi_V c\psi_V & c\varphi_V s\theta_V s\psi_V - s\varphi_V c\psi_V \\ -s\theta_V & s\varphi_V c\theta_V & c\varphi_V c\theta_V \end{bmatrix}
\end{gathered}
\tag{11.3}
$$

For the angular velocity of the chassis relative to the inertial system and represented in the coordinates of the vehicle fixed coordinate system one obtains (the kinematic CARDAN equation, Sect. 2.5.5):

$$
{}_{E}^{V}\boldsymbol{\omega}_V = \begin{bmatrix} \omega_{V,x} \\ \omega_{V,y} \\ \omega_{V,z} \end{bmatrix} = \begin{bmatrix} \dot{\varphi}_V - \dot{\psi}_V s\theta_V \\ \dot{\theta}_V c\varphi_V + \dot{\psi}_V c\theta_V s\varphi_V \\ -\dot{\theta}_V s\varphi_V + \dot{\psi}_V c\theta_V c\varphi_V \end{bmatrix} = \boldsymbol{T}_\omega \begin{bmatrix} \dot{\psi}_V \\ \dot{\theta}_V \\ \dot{\varphi}_V \end{bmatrix},
\tag{11.4}
$$

with the [3 × 3]-rotation matrix

$$\boldsymbol{T}_\omega = \begin{bmatrix} -s\theta_V & 0 & 1 \\ c\theta_V s\varphi_V & c\varphi_V & 0 \\ c\theta_V c\varphi_V & -s\varphi_V & 0 \end{bmatrix}. \tag{11.5}$$

Correspondingly, the spatial motion of the individual wheels, not including their own rotation about the wheel axis, can be described using the rotation matrix ${}^E\boldsymbol{T}_{R_i}, i = 1, \ldots, 4$ relative to the inertial coordinate system $\mathbf{K}_E$. For the (steered) front wheels, this results in:

$${}^E\boldsymbol{T}_{R_1} = {}^E\boldsymbol{T}_{R_3} = \boldsymbol{T}_z(\psi_V + \delta) = \begin{bmatrix} c(\psi_V + \delta) & -s(\psi_V + \delta) & 0 \\ s(\psi_V + \delta) & c(\psi_V + \delta) & 0 \\ 0 & 0 & 1 \end{bmatrix}. \tag{11.6}$$

For reasons of simplicity it can be assumed here that the steering action of both wheels is always equal and is linearly dependent on the steering wheel angle δ_H as a function of time:

$$\delta_1 = \delta_3 = \delta = i_L \cdot \delta_H(t). \tag{11.7}$$

For the rear wheel (without steering, i.e. $\delta_2 = \delta_4 = 0$) it correspondingly holds:

$${}^E\boldsymbol{T}_{R_2} = {}^E\boldsymbol{T}_{R_4} = \boldsymbol{T}_z(\psi_V) = \begin{bmatrix} c\psi_V & -s\psi_V & 0 \\ s\psi_V & c\psi_V & 0 \\ 0 & 0 & 1 \end{bmatrix}. \tag{11.8}$$

Furthermore, the rotation of the wheel fixed coordinate system $\mathbf{K}_{R_i}$ relative to the vehicle fixed coordinate system $\mathbf{K}_V$ is required. For the front axle this results in:

$$\begin{aligned} {}^V\boldsymbol{T}_{R_1} &= {}^V\boldsymbol{T}_{R_3} = \boldsymbol{T}_x^T(\varphi_V) \cdot \boldsymbol{T}_y^T(\theta_V) \cdot \boldsymbol{T}_z(\delta) \\ &= \begin{bmatrix} c\delta c\theta_V & -s\delta c\theta_V & -s\theta_V \\ s\varphi_V s\theta_V c\delta + c\varphi_V s\delta & -s\varphi_V s\theta_V s\delta + c\varphi_V c\delta & s\varphi_V c\theta_V \\ c\varphi_V s\theta_V c\delta - s\varphi_V s\delta & -c\varphi_V s\theta_V s\delta - s\varphi_V c\delta & c\varphi_V c\theta_V \end{bmatrix} \end{aligned} \tag{11.9}$$

and for the non-steered rear axle

$$\begin{aligned} {}^V\boldsymbol{T}_{R_2} &= {}^V\boldsymbol{T}_{R_4} = \boldsymbol{T}_x^T(\varphi_V) \cdot \boldsymbol{T}_y^T(\theta_V) \\ &= \begin{bmatrix} c\theta_V & 0 & -s\theta_V \\ s\varphi_V s\theta_V & c\varphi_V & s\varphi_V c\theta_V \\ c\varphi_V s\theta_V & -s\varphi_V & c\varphi_V c\theta_V \end{bmatrix} \end{aligned} \tag{11.10}$$

11.1.1 NEWTON's and EULER's Equations for a Basic Spatial Twin Track Model

The Newton equations for the vehicle chassis in vector form are:

$$m_V \ddot{\boldsymbol{r}}_V = \sum_{i=1}^{4} \boldsymbol{F}_i + \boldsymbol{G} + \boldsymbol{F}_W = \sum_{i=1}^{4} \boldsymbol{F}_i - m_V g \boldsymbol{e}_z + \boldsymbol{F}_W. \tag{11.11}$$

Here, $\boldsymbol{F}_i$ represents the vector of transmitted forces between the wheel suspensions and the chassis. These are on the one hand the horizontal reaction forces and on the other hand the applied forces of the chassis springs and dampers at the pivot point A_i, Fig. 11.2. The vector $\boldsymbol{e}_z$ represents the unit vector in positive direction of the z-axis of the inertial coordinate system, thus assuming that the force due to gravity acts in the direction opposite to that of $\boldsymbol{e}_z$. The vector $\boldsymbol{F}_W$ is a collection of all the air resistance and aerodynamic forces.

Written out in components, represented in the vehicle fixed coordinate system, this results in:

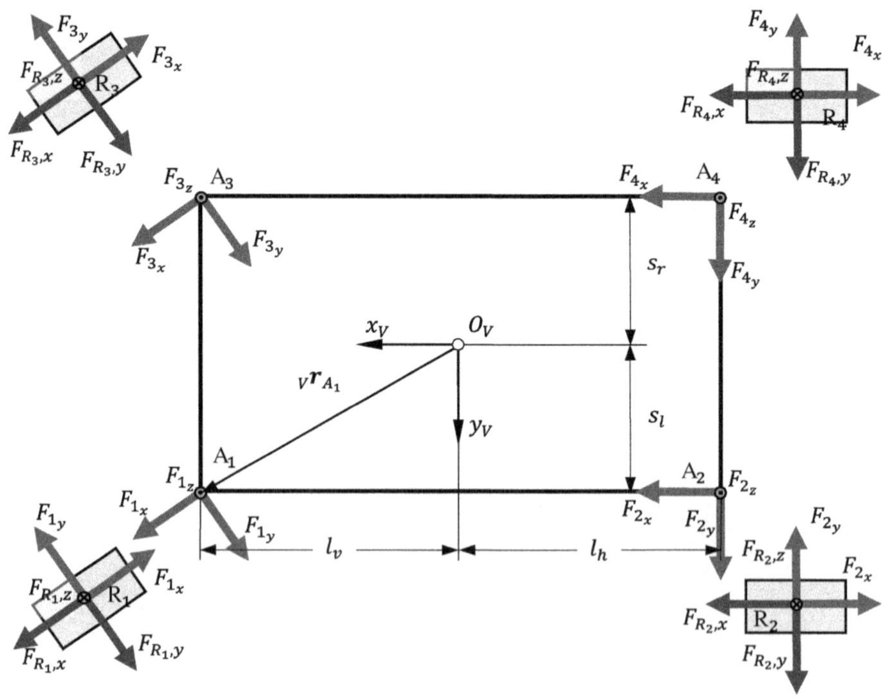

Fig. 11.2 Free body diagram of a twin track model—*top view*

$$m_V(\dot{v}_x + \omega_y v_z - \omega_z v_y) = \sum_{i=1}^{4} F_{i_x} - m_V g(\boldsymbol{e}_z)_x + F_{W_x}, \tag{11.12}$$

$$m_V(\dot{v}_y + \omega_z v_x - \omega_x v_z) = \sum_{i=1}^{4} F_{i_y} - m_V g(\boldsymbol{e}_z)_y + F_{W_y}, \tag{11.13}$$

$$m_V(\dot{v}_z + \omega_x v_y - \omega_y v_x) = \sum_{i=1}^{4} F_{i_z} - m_V g(\boldsymbol{e}_z)_z + F_{W_z}. \tag{11.14}$$

Here, $(\boldsymbol{e}_z)_{x/y/z}$ represent the coordinates of the unit vector transformed into the vehicle fixed coordinate system. The forces on the vehicle chassis at the pivot points A_i of the four wheel suspensions is given by:

$$\boldsymbol{F}_i = \boldsymbol{F}_{F_i} + \boldsymbol{F}_{D_i} + \boldsymbol{F}_{H_i}. \tag{11.15}$$

The influence of the anti-roll bar will later be considered as an applied torque. The meaning of the individual force components can be seen in Table 11.2.

EULER's equations of the chassis with respect to the center of gravity, represented in the chassis fixed (vehicle fixed) coordinate system, are given as (Fig. 11.3):

$$\boldsymbol{\Theta}_V \dot{\boldsymbol{\omega}}_V + \boldsymbol{\omega}_V \times (\boldsymbol{\Theta}_V \boldsymbol{\omega}_V) = \sum_{i=1}^{4} {}_V\boldsymbol{r}_{A_i} \times \boldsymbol{F}_i + \boldsymbol{r}_W \times \boldsymbol{F}_W + \boldsymbol{T}_{St,v} + \boldsymbol{T}_{St,h}, \tag{11.16}$$

where

$$\boldsymbol{\Theta}_V = \begin{bmatrix} \theta_{V,xx} & 0 & 0 \\ 0 & \theta_{V,yy} & 0 \\ 0 & 0 & \theta_{V,zz} \end{bmatrix} = const \tag{11.17}$$

represents the inertia matrix of the chassis with respect to the chassis fixed coordinate system. For reasons of simplicity, the non-diagonal moments of deviation are neglected. The other values of Eq. (11.16) are explained in Table 11.3 and Fig. 11.4.

Table 11.2 Forces between the wheels and the chassis

Force	Relevance for the wheel "*i*"
$\boldsymbol{F}_{H_i}$	Horizontal forces on the chassis (reaction forces in the x-, y-plane)
$\boldsymbol{F}_{F_i}$	Forces of the suspension spring
$\boldsymbol{F}_{D_i}$	Forces of the suspension dampers

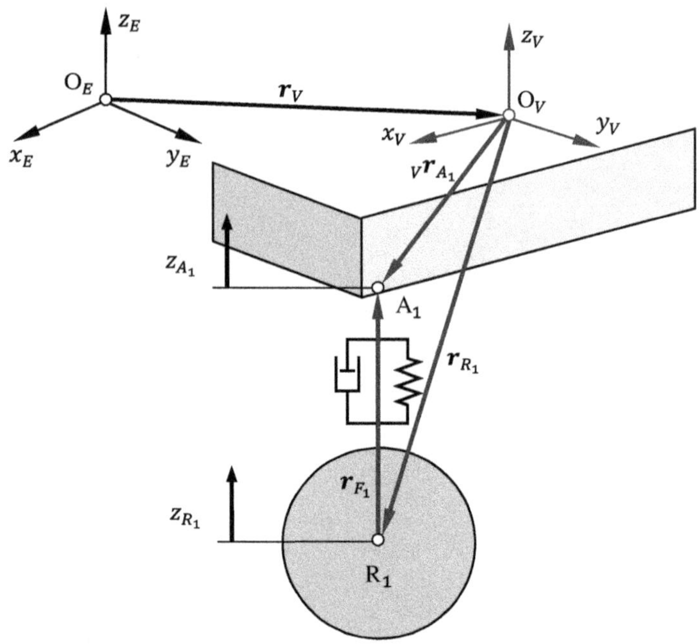

Fig. 11.3 Position vector of the wheel suspension

Table 11.3 Physical values for the EULER equation

Variable	Meaning
ω_V	Absolute angular velocity of the chassis
$_V\boldsymbol{r}_{A_i}$	Position vector of the center of gravity of the chassis O_V from the pivot point A_i
$\boldsymbol{T}_{St,v}, \boldsymbol{T}_{St,h}$	Torque due to anti-roll bars front (v) and rear (h)
$\boldsymbol{F}_i$	Forces from the tires acting on the chassis
$\boldsymbol{r}_W$	Position vector from the center of gravity of the chassis to the resulting aerodynamic forces

11.1.2 Spring and Damper Forces

In order to complete setting up the equations of motion of the system, all that remains to be done is the calculation of the applied forces (also Ref. to Chap. 9).

Spring and damper forces For the spring forces between the wheels and the chassis, the following relationship holds as given in Chap. 9

$$\boldsymbol{F}_{F_i} = \underbrace{f_i(l_{F_i})}_{\substack{\text{force}\\ \text{law}}} \underbrace{\frac{\boldsymbol{l}_{F_i}}{l_{F_i}}}_{\substack{\text{force}\\ \text{direction}}} = f_i(l_{F_i})\boldsymbol{u}_{F_i}, \tag{11.18}$$

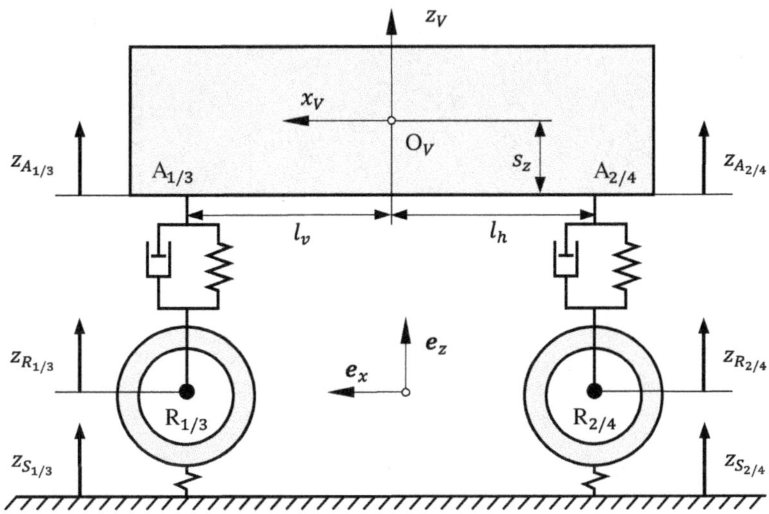

Fig. 11.4 Spatial twin track model—*side view*

taking into consideration the spring force law $f_i(l_{F_i})$. Furthermore,

$$ {}^V\boldsymbol{l}_{F_i} = {}^V_V\boldsymbol{r}_{A_i} - {}^V_V\boldsymbol{r}_{R_i} = {}^V\boldsymbol{T}_E \begin{bmatrix} 0 \\ 0 \\ 1 \end{bmatrix} (z_{A_i} - z_{R_i} + l_{stat}), \tag{11.19} $$

with the current spring length $l_{F_i} = |\boldsymbol{l}_{F_i}|$ and the undeflected spring length l_{stat}. For subsequent calculation of the torque, one still needs the position vector of the point of action of the spring and damper forces on the chassis:

$$ {}^V\boldsymbol{r}_{A_1} = \begin{bmatrix} l_v \\ s_l \\ -s_z \end{bmatrix}, {}^V\boldsymbol{r}_{A_2} = \begin{bmatrix} -l_h \\ s_l \\ -s_z \end{bmatrix}, {}^V\boldsymbol{r}_{A_3} = \begin{bmatrix} l_v \\ -s_r \\ -s_z \end{bmatrix}, {}^V\boldsymbol{r}_{A_4} = \begin{bmatrix} -l_h \\ -s_r \\ -s_z \end{bmatrix}. \tag{11.20} $$

With this, assuming linear spring characteristics, the spring forces at the front left can be exemplarily calculated as

$$ \begin{aligned} {}^V\boldsymbol{F}_{F_1} &= -\left(c_{a_1} \underbrace{(z_{A_1} - z_{R_1})}_{\Delta l_{F_1}} + c_{a_1}(l_{stat} - l_0) \right) {}^V\boldsymbol{T}_E \begin{bmatrix} 0 \\ 0 \\ 1 \end{bmatrix} \\ &= -(c_{a_1}\Delta l_{F_1} + F_{Fstat_1}) {}^V\boldsymbol{T}_E \begin{bmatrix} 0 \\ 0 \\ 1 \end{bmatrix}, \end{aligned} \tag{11.21} $$

with the undeflected spring length l_0. Correspondingly, the damper forces can be calculated as

$$ {}^V\boldsymbol{F}_{D_1} = -d_{a_1} \Delta \dot{l}_{F_1} {}^V\boldsymbol{T}_E \begin{bmatrix} 0 \\ 0 \\ 1 \end{bmatrix}, \tag{11.22} $$

with

$$ \Delta \dot{l}_{F1} = \dot{z}_{A_1} - \dot{z}_{R_1}. \tag{11.23} $$

Anti-roll bar forces The anti-roll bar is to be modeled here as a linear force element connected to the axles (Chap. 9). With the torsional spring constant $c_{St,h}$, one arrives for example at the torque on the chassis at the rear axle as Fig. 11.5:

$$ M_{St,h} = (s_l + s_r) F_{St,h} = c_{St,h}((z_{A_2} - z_{R_2}) - (z_{A_4} - z_{R_4})). \tag{11.24} $$

11.1.3 NEWTON's and EULER's Equations of the Wheels

Newton's equations for the wheels, given in the vertical direction of the inertial coordinate system, are:

$$ m_{R_i} \ddot{z}_{R_i} = F_{R_i,z} - F_{i,z} - m_R g. \qquad i = 1, \ldots, 4 \tag{11.25} $$

Correspondingly, EULER's equations for the wheels are

$$ \Theta_{R_i} \ddot{\rho}_{R_i} = M_{A,R_i} - M_{B,R_i} - r_{R_i} F_{R_i,x}. \qquad i = 1, \ldots, 4 \tag{11.26} $$

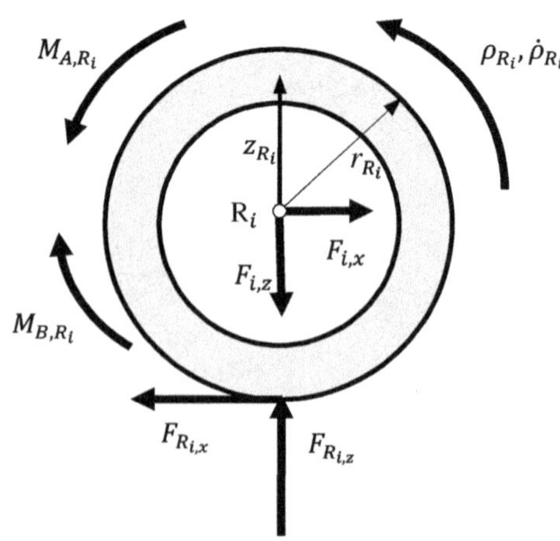

Fig. 11.5 Forces and torques on the wheels

11.1.4 Tire-Road Contact

To calculate the tire forces $\boldsymbol{F}_{R_i}$, the coordinates of the tire-road contact point as well as the absolute velocities of the wheel center point are required:

$$ {}^V\boldsymbol{r}_{R_i} = {}^V\boldsymbol{r}_{A_i} + {}^V\boldsymbol{T}_E \begin{bmatrix} 0 \\ 0 \\ -(z_{A_i} - z_{R_i} + l_{stat_i}) \end{bmatrix}, i = 1, \ldots, 4, \tag{11.27} $$

$$ {}^V\boldsymbol{v}_{R_i} = \begin{bmatrix} {}^V\dot{x}_{R_i} \\ {}^V\dot{y}_{R_i} \\ {}^V\dot{z}_{R_i} \end{bmatrix} = {}^V\boldsymbol{v}_V + {}^V\boldsymbol{\omega}_V \times {}^V\boldsymbol{r}_{A_i} + {}^V\boldsymbol{T}_E \begin{bmatrix} 0 \\ 0 \\ -(\dot{z}_{A_i} - \dot{z}_{R_i}) \end{bmatrix}, i = 1, \ldots, 4 \tag{11.28} $$

In order to calculate the circumferential slip however, the knowledge of the velocity of the wheel in the wheel fixed coordinate system is required:

$$ {}^{R_i}\boldsymbol{v}_{R_i} = {}^{R_i}\boldsymbol{T}_V \, {}^V\boldsymbol{v}_{R_i}. \tag{11.29} $$

According to Chap. 7, this results in the wheel circumferential slip (longitudinal slip) as:

$$ s_i = \frac{r_{R_i}\dot{\rho}_{R_i} - {}^{R_i}\dot{x}_{R_i}}{\max\left(\left|r_{R_i}\dot{\rho}_{R_i}\right|, \left|{}^{R_i}\dot{x}_{R_i}\right|\right)} \tag{11.30} $$

as well as the tire slip angle

$$ \alpha_i = -\arctan\left(\frac{{}^{R_i}\dot{y}_{R_i}}{\max\left(\left|r_{R_i}\dot{\rho}_{R_i}\right|, \left|{}^{R_i}\dot{x}_{R_i}\right|\right)}\right). \tag{11.31} $$

Finally, the tire forces

$$ {}^V\boldsymbol{F}_{R_i} = {}^V\boldsymbol{T}_{R_i} \begin{bmatrix} F_{R_i,x} \\ F_{R_i,y} \\ F_{R_i,z} \end{bmatrix} \tag{11.32} $$

acting on the wheels $R_i, i = 1, \ldots, 4$ need to be calculated. To evaluate Eq. (11.32), the force components $F_{R_i,x}$ and $F_{R_i,y}$ are calculated from the corresponding tire models and transformed into the vehicle fixed coordinate system. The components $F_{R_i,z}$ on the other hand are derived from the wheel spring forces (wheel elasticity) as

$$\begin{aligned} F_{R_1,z} &= -\max\left(c_R\left(z_{R,1} - z_{S,1} + l_{R,stat}\right) + F_{St,v}, 0\right), \\ F_{R_2,z} &= -\max\left(c_R\left(z_{R,2} - z_{S,2} + l_{R,stat}\right) + F_{St,h}, 0\right), \\ F_{R_3,z} &= -\max\left(c_R\left(z_{R,3} - z_{S,3} + l_{R,stat}\right) - F_{St,v}, 0\right), \\ F_{R_4,z} &= -\max\left(c_R\left(z_{R,4} - z_{S,4} + l_{R,stat}\right) - F_{St,h}, 0\right). \end{aligned} \tag{11.33}$$

For the stationary tire forces, represented in the wheel fixed coordinate system with

$$\begin{bmatrix} F_{R_i,x,stat} \\ F_{R_i,y,stat} \end{bmatrix} = \begin{bmatrix} \mu_{x_i} \sin\left(c_x \arctan\left(b_x \frac{s_i}{\mu_{x_i}}\right)\right) \\ \mu_{y_i} \sin\left(c_y \arctan\left(b_y \frac{\alpha_i}{\mu_{y_i}}\right)\right) \end{bmatrix} F_{R_i,z,eff}, \tag{11.34}$$

and

$$F_{R_i,z,eff} = F_{R_i,z}\left(1 - e_z\left(\frac{F_{R_i,z}}{F_{R_i,z,0}}\right)^2\right), \tag{11.35}$$

a simplified version of the Magic Formula (Chap. 7) is used. There, the input values used are again the slip values (11.30) and (11.31), the wheel loads $F_{R_i,z}$, and the friction coefficients μ_{x_i/y_i}. The adaptation of the characteristics of the curve for a specific tire is achieved by using the constant parameters $c_{x/y}, b_{x/y}$ for the lateral and longitudinal stiffness as well as c_R for the vertical stiffness of the tires. Furthermore, the constant parameter e_z is required for the approximate representation of the degressive dependency of the tire forces on the wheel load.[1]

The curves derived from the simplified Magic Formula Model are exemplarily represented in Fig. 11.6.

The dynamic tire forces are then calculated from the formula:

$$\begin{bmatrix} \dot{F}_{R_i,x} \\ \dot{F}_{R_i,y} \end{bmatrix} = \begin{bmatrix} \frac{1}{T_{x_i}} & 0 \\ 0 & \frac{1}{T_{y_i}} \end{bmatrix} \left(\begin{bmatrix} F_{R_i,x,stat} \\ F_{R_i,y,stat} \end{bmatrix} - \begin{bmatrix} F_{R_i,x} \\ F_{R_i,y} \end{bmatrix}\right), \tag{11.36}$$

with the reciprocal time constants:

$$\frac{1}{T_{x_i/y_i}} = \frac{\left|r_{R_i}\dot{\rho}_{R_i}\right|}{\sigma_{x_i/y_i}}. \tag{11.37}$$

[1]Here "i" stands for x and y respectively.

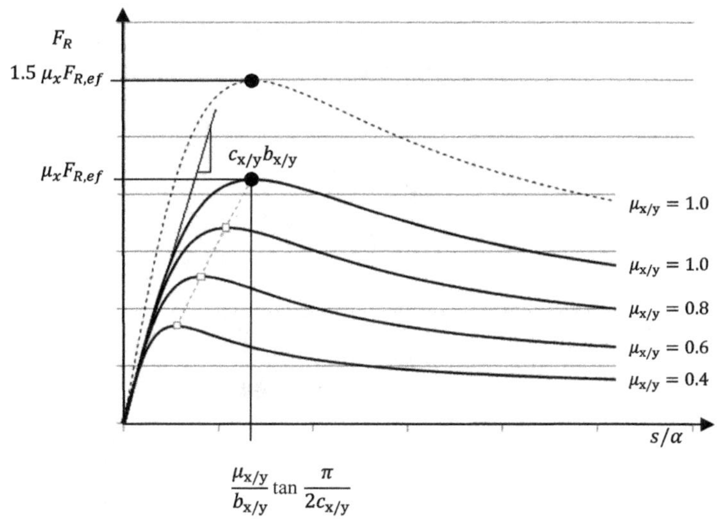

Fig. 11.6 Simplified magic formula model–c, b = *const*, μ, $F_{R,eff}$ variable

For the aerodynamic forces, one obtains the relationship

$$F_W = \frac{1}{2} c_W \varrho A \dot{r}_V \|\dot{r}_V\|_2 = \frac{1}{2} c_W \varrho_L A \sqrt{v_x^2 + v_y^2} \begin{bmatrix} v_x \\ v_y \\ 0 \end{bmatrix}, \tag{11.38}$$

in which again only the influence of the opposing airstream (i.e. opposite and parallel to the trajectory of the center of gravity of the chassis) is considered. If for example the influence of the side wind also needs to be included, then the velocity $\dot{r}_V$ has to be adjusted accordingly (Chap. 9).

11.1.5 Drivetrain

The modeling of the drivetrain can be derived from Fig. 11.7. In this case, unlike in Chap. 8, a one-dimensional modeling is chosen. Similarly, for the sake of simplicity, no elastic elements are used.

In order to specify the driving torques on the front and the rear axles, one first needs to obtain the engine speed (rpm). This is available through a basic kinematic observation while considering the following terms

- i_D gear transmission ration of the central differential,
- $i_G(G)$ gear-dependent gear ratio,
- i_v, i_h differential transmission front/back

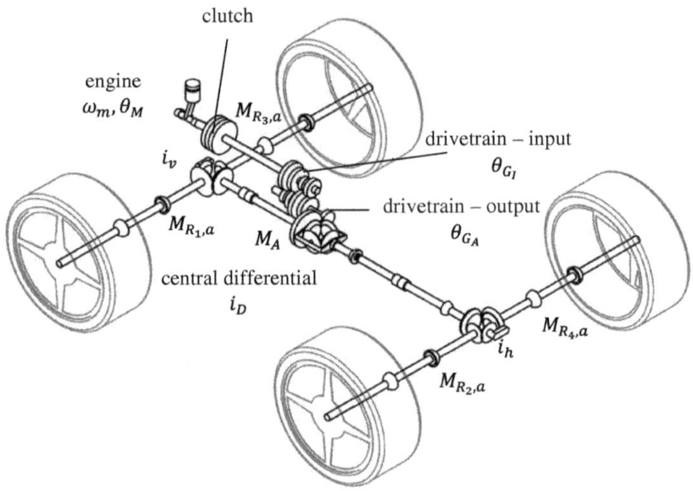

Fig. 11.7 Drivetrain — example four-wheel drive

- $\xi_a, \bar{\xi}_a$ distribution of the driving torque on the front- and rear axle with $0 \leq \xi_a \leq 1,\ \bar{\xi}_a = 1 - \xi_a$,
- θ_M moment of inertia of the engine,
- θ_{G_I} moment of inertia of the engine-sided driveshaft,
- θ_{G_A} moment of inertia of the drivetrain-sided driveshaft

as

$$\omega_M = i_D i_G(G)\left(\bar{\xi}_a i_v\left(\frac{\dot{\rho}_{R_1} + \dot{\rho}_{R_3}}{2}\right) + \xi_a i_h\left(\frac{\dot{\rho}_{R_2} + \dot{\rho}_{R_4}}{2}\right)\right). \tag{11.39}$$

From this, through the evaluation of the engine torque characteristic map $M_M(\omega_m, p_f)$ and the angular acceleration $\alpha_M = \dot{\omega}_M$ of the engine, one obtains, the driving torque applied by the engine on the clutch

$$M_K = M_M - \theta_M \alpha_M, \tag{11.40}$$

and from this the calculation of the entire drive torque on the wheels

$$M_A = (i_v \bar{\xi}_a + i_h \xi_a) i_D \left[i_G M_K - \frac{1}{i_G}(\theta_{G_I} i_G^2 + \theta_{G_A}) \alpha_M \right]. \tag{11.41}$$

The drive torque on each of the wheels is therefore:

$$
\begin{aligned}
M_{R_1,a} = M_{R_3,a} &= \frac{1}{2} i_v \bar{\xi}_a i_D \left[i_G M_K - \frac{1}{i_G} \left[\theta_{G_I} i_G^2 + \theta_{G_A} \right] \alpha_M \right], \\
M_{R_2,a} = M_{R_4,a} &= \frac{1}{2} i_h \xi_a i_D \left[i_G M_K - \frac{1}{i_G} \left[\theta_{G_I} i_G^2 + \theta_{G_A} \right] \alpha_M \right].
\end{aligned}
\tag{11.42}
$$

The driving torque consists of two parts, one of which is simply the transmitted engine torque, the other being the moment of inertia of the drivetrain, including the central differential. The moments of inertia of the four driveshafts are completely transferred onto the tires. The moment of inertia of the CARDAN shaft however is only transferred proportionally to the tires.

Clutch Equation (11.42) is initially only valid for a closed clutch. If the clutch is open, no torque will be transferred and the numbers of revolutions of the two clutch plates are different from each other. During the clutching and declutching procedures, a torque is transmitted which is dependent on the applied clutch pressure and other factors while the number of revolutions of the clutch plates are assimilated. As soon as the number of revolutions of the clutch plates is no longer tightly coupled, the number of DoF increases. As the gear changing sequence is very short and a detailed investigation and inclusion of these effects is not part of the simplified model required here, it shall be neglected. Instead, it shall be assumed in a very simplified manner that the speeds of rotation of the two clutch plates are always the same and that the transferred torque $M_M - \theta_M \alpha_M$ only depends on one parameter representing the activation of the clutch:

$$
M_{K\xi} = (M_M - \theta_M \alpha_M)(1 - \xi_K), \qquad 0 \leq \xi_K \leq 1. \tag{11.43}
$$

The clutch is closed for $\xi_K = 0$ (clutch pedal not used) and open for $\xi_K = 1$ (pedal fully extended). If M_K is now replaced with $M_{K\xi}$, in the Eqs. (11.41) and (11.42) a basic gear change procedure becomes possible by defining ξ_K as a function of time.

11.1.6 Brake System

The brake torque can also be distributed randomly on the front and the rear axles:

$$
\begin{aligned}
M_{B,h} &= \xi_b M_B, \\
M_{B,v} &= (1 - \xi_b) M_B = M_B - M_{B,h}.
\end{aligned}
\tag{11.44}
$$

For the braking torque on each wheel one gets:

$$\begin{aligned} M_{B,R_1} &= M_{B,R_3} = \frac{1}{2} M_{B,v}, \\ M_{B,R_2} &= M_{B,R_4} = \frac{1}{2} M_{B,h}. \end{aligned} \tag{11.45}$$

11.1.7 Equations of Motion

Using the preliminary work of Sects. 11.1.1–11.1.6 it is now possible to list the complete equations of motion of a twin track model:

$$\begin{aligned} m_V \left(\begin{bmatrix} \dot{v}_x \\ \dot{v}_y \\ \dot{v}_z \end{bmatrix} + \boldsymbol{\omega}_V \times \begin{bmatrix} v_x \\ v_y \\ v_z \end{bmatrix} \right) &= \sum_{i=1}^{4} \begin{bmatrix} F_{R_i,x} \\ F_{R_i,y} \\ F_{R_i,z} \end{bmatrix} \\ &\quad - \frac{1}{2} c_W \rho A \sqrt{v_x^2 + v_y^2} \begin{bmatrix} v_x \\ v_y \\ 0 \end{bmatrix} + \begin{bmatrix} 0 \\ 0 \\ -m_V g \end{bmatrix}, \end{aligned} \tag{11.46}$$

$$\boldsymbol{\Theta}_V \dot{\boldsymbol{\omega}}_V + \boldsymbol{\omega}_V \times (\boldsymbol{\Theta}_V \boldsymbol{\omega}_V) = \sum_{i=1}^{4} \boldsymbol{r}_i \times \boldsymbol{F}_i + \boldsymbol{r}_W \times \boldsymbol{F}_W, \tag{11.47}$$

$$\theta_{R_i} \ddot{\rho}_{R_i} = M_{a,R_i} - M_{b,R_i} \operatorname{sign}(\dot{\rho}_{R_i}) - r F_{R_i,x}, \qquad i = 1, \ldots, 4, \tag{11.48}$$

$$m_{R_i} \ddot{z}_{Ri} = F_{R_i,z} - F_{i,z} - m_{R_i} g, \qquad i = 1, \ldots, 4, \tag{11.49}$$

$$\begin{bmatrix} \dot{F}_{R_i,x} \\ \dot{F}_{R_i,y} \end{bmatrix} = \begin{bmatrix} \frac{|r \dot{\rho}_{R_i}|}{\sigma_{x_i}} & 0 \\ 0 & \frac{|r \dot{\rho}_{R_i}|}{\sigma_{y_i}} \end{bmatrix} \left(\begin{bmatrix} F_{R_i x,stat} \\ F_{R_i y,stat} \end{bmatrix} - \begin{bmatrix} F_{R_i,x} \\ F_{R_i,y} \end{bmatrix} \right), \qquad i = 1, \ldots, 4 \tag{11.50}$$

$$\boldsymbol{x}_A = \left[x y z \psi_V \theta_V \varphi_V | v_x v_y v_z \omega_x \omega_y \omega_z \right]^T, \tag{11.51}$$

$$\boldsymbol{x}_{RR} = \left[\rho_{R_1} \rho_{R_2} \rho_{R_3} \rho_{R_4} | \dot{\rho}_{R_1} \dot{\rho}_{R_2} \dot{\rho}_{R_3} \dot{\rho}_{R_4} \right]^T, \tag{11.52}$$

$$\boldsymbol{x}_{Rz} = \left[z_{R_1} z_{R_2} z_{R_3} z_{R_4} | \dot{z}_{R_1} \dot{z}_{R_2} \dot{z}_{R_3} \dot{z}_{R_4} \right]^T, \tag{11.53}$$

$$\boldsymbol{x}_{RF} = \left[F_{x_1} F_{y_1} F_{x_2} F_{y_2} F_{x_3} F_{y_3} F_{x_4} F_{y_4} \right]^T, \tag{11.54}$$

$$\boldsymbol{x} = \left[x_A \quad x_{RR} \quad x_{Rz} \quad x_{RF} \right]^T. \tag{11.55}$$

The excitation vector $\boldsymbol{u}$ is again, as in Chap. 10:

$$\boldsymbol{u} = \begin{bmatrix} \delta_H & p_F & p_B & G \end{bmatrix}^T. \tag{11.56}$$

Here, δ_H is the steering wheel angle. The position of the acceleration pedal p_F and that of the brake pedal p_B is normalized between the values 0 and 1. The input value G defines the gear chosen with the forward gears 1…5, the reverse gear –1 and neutral equal to 0. Hence, the state space equation of the spatial twin track model is:

$$\dot{\boldsymbol{x}} = \boldsymbol{f}(\boldsymbol{x}, t, \boldsymbol{u}), \tag{11.57}$$

with the [36 $\times$ 1]-vector $\boldsymbol{x}$, the [4 $\times$ 1]-vector $\boldsymbol{u}$ and the time t.

11.2 Twin Track Models with Kinematic Wheel Suspensions

The vehicle model introduced below is an extension of the model described in the previous section but now including the wheel suspension kinematics. To this end, the front and the rear axle suspensions will be modeled, which can perform spatial motion, Ref. Fig. 11.8.

11.2.1 Degrees of Freedom of the Twin Track Model

For the wheel suspension, it shall only be assumed that its motion can be described through a generalized coordinate $\varphi_i, i = 1, \ldots, 4$. This coordinate typically represents the (spatial) motion of the wheel. The constructive design of the wheel suspension is left undefined at this time, i.e. the kinematics of the wheel carriers and the wheels have to be represented as a function of φ_i later. Additionally, the steering behavior $\delta_i, i = 1, \ldots, 4$ is available as a function of time. In this case, the corresponding steering angle at the front axle is given by the steering angle of the steering wheel and the steering ratio as:

$$\delta_1 = \delta_3 = \delta = \frac{1}{i_L} \delta_H(t). \tag{11.58}$$

Instead of Eq. (11.58) one can naturally also consider a non-proportional relationship with the steering input as well as a more realistic steering system with $\delta_1 \neq \delta_3$, without influencing the procedure below, Ref. Sect. 11.2.9. At the rear axle, the steering angles can either be assumed to be constant or can be used to describe an active rear wheel steering.

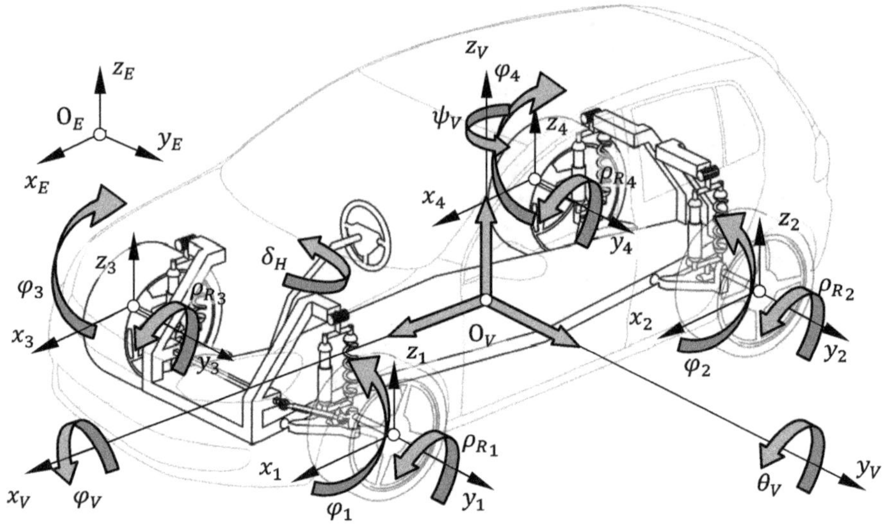

Fig. 11.8 DoF of a basic spatial twin track model

The description of the wheel rotations about its own axis is realized through the unit vectors $\boldsymbol{u}_{R_i}, i = 1,\ldots,4$, Ref. Fig. 11.9, which defines the wheel rotation axis relative to the wheel carrier, as well as the rotational angles $\rho_{R_i}, i = 1,\ldots,4$ and the corresponding rotational velocities of the wheel $\dot{\rho}_{R_i}, i = 1,\ldots,4$. Notably through this definition is that the wheel carrier and the wheel with respect to their spatial orientation and the corresponding angular velocities only differ by the wheel's own rotation about its axis. It is for this reason that, not including the moments of inertia about the wheel axis, all the moments of inertia of the wheels are added to the corresponding wheel carriers. The model now consists of 9 bodies, Ref. Table 11.4 neglecting the steering rod and the other components of the wheel suspension due to the dominating masses and moments of inertia of the wheel carriers and the wheels.

The model considers the summarized generalized coordinates and generalized velocities as given in Tables 11.5 and 11.6.

The vehicle model described here therefore has a total of $f = 14$ DoF

Notes

- The steering motion is not considered to be a DoF but is treated as an input $\delta_H(t)$.
- The input variables are actually those introduced during the modeling of the nonlinear single track model and the basic spatial two track model, Ref. Table 11.7.

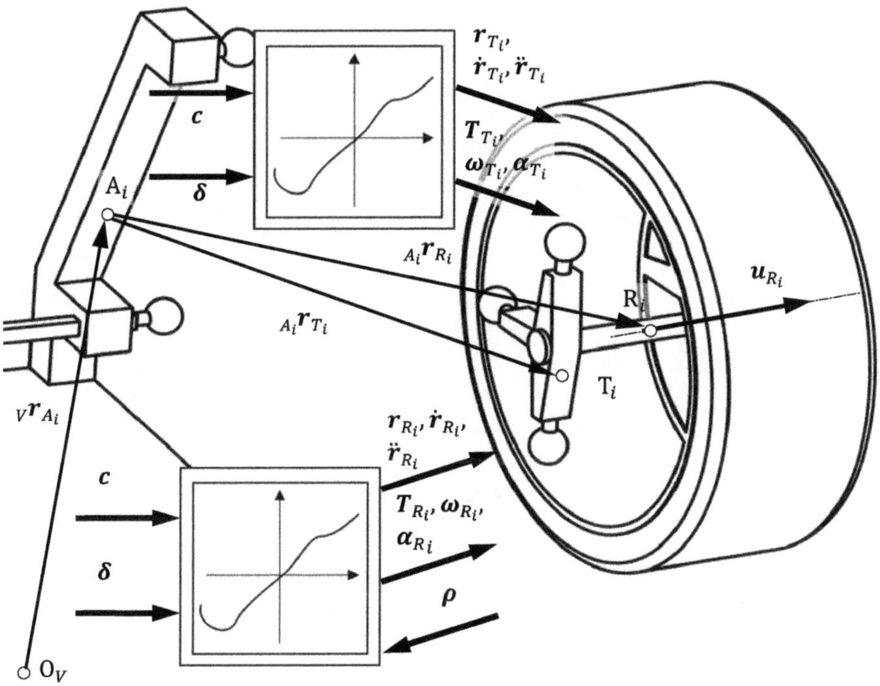

Fig. 11.9 Twin track model with a generalized wheel suspension

Table 11.4 Bodies $K_i, i = 1, \ldots, 9$

Body	Description	Mass	Inertia tensor in body fixed coordinate system with respect to the corresponding center of gravity
K_1	Chassis	m_V	$\boldsymbol{\Theta}_V$
$K_2, \ldots, K_5$	Wheel carriers	m_{T_i}	$\boldsymbol{\Theta}_{T_i}$
$K_6, \ldots, K_9$	Wheels	m_{R_i}	$\boldsymbol{\Theta}_{R_i}$

Table 11.5 Generalized coordinates $q_i, i = 1, \ldots, 14$

Variable	Description
x_V, y_V, Z_V	Coordinates of the center of gravity of the vehicle chassis in the inertial system
$\psi_V, \theta_V, \varphi_V$	CARDAN angles to describe the vehicle chassis relative to the inertial system
$\varphi_i, i = 1, \ldots, 4$	Deflection motion of the wheel carrier relative to the vehicle chassis
$\rho_{Ri}, i = 1, \ldots, 4$	Angle of rotation of the wheels about their own wheel carrier fixed rotational axis $\boldsymbol{u}_{R_i}$

Table 11.6 Generalized velocities $z_i, i = 1, \ldots, 14$

Variable	Description
v_x, v_y, v_z	Coordinates of the translational velocity in the vehicle fixed coordinate system
$\omega_x, \omega_y, \omega_z$	Coordinates of the angular velocity in the vehicle fixed coordinate system
$\dot{\varphi}_i, i = 1, \ldots, 4$	Deflection velocity of the wheel carrier
$\dot{\rho}_{Ri}, i = 1, \ldots, 4$	Rotational velocity of the wheels about their own wheel carrier fixed rotational axis $\boldsymbol{u}_{R_i}$

Table 11.7 Input variables

Variables	Description
p_F	Normalized acceleration pedal position $(0 \leq p_F \leq 1)$
p_B	Normalized brake pedal position $(0 \leq p_B \leq 1)$
G	Gear chosen
$\delta_H(t)$	Steering wheel angle
$\delta_2(t), \delta_4(t)$	Steering angle at the rear tires

11.2.2 *Kinematics of the Vehicle Chassis*

Translation The position vector from the origin of the inertial system O_E to the center of gravity O_V of the vehicle chassis is:

$$^E\boldsymbol{r}_1 = {}^E\boldsymbol{r}_V = \begin{bmatrix} x_V & y_V & z_V \end{bmatrix}^T. \tag{11.59}$$

With this, the absolute translational velocity of the chassis represented in the coordinates of the inertial system can be obtained:

$$^E\boldsymbol{v}_1 = {}^E\boldsymbol{v}_V = \begin{bmatrix} \dot{x}_V \\ \dot{y}_V \\ \dot{z}_V \end{bmatrix} = {}^E\boldsymbol{T}_V \begin{bmatrix} v_x \\ v_y \\ v_z \end{bmatrix}. \tag{11.60}$$

The Matrix $^E\boldsymbol{T}_V$ (Ref. Eq. (11.63)) represents the transformation from the vehicle fixed coordinate system into that of the inertial system. It is to be noted here that the generalized velocities are the velocities in the directions of the coordinates of the chassis fixed coordinate system. Through an additional differentiation of the Eq. (11.60), one arrives at the translational accelerations of the chassis relative to the inertial system:

$$^E\boldsymbol{a}_1 = {}^E\boldsymbol{a}_V = \begin{bmatrix} \ddot{x}_V \\ \ddot{y}_V \\ \ddot{z}_V \end{bmatrix} = {}^E\boldsymbol{T}_V \left(\begin{bmatrix} \dot{v}_x \\ \dot{v}_y \\ \dot{v}_z \end{bmatrix} + {}^V\boldsymbol{\omega}_V \times \begin{bmatrix} v_x \\ v_y \\ v_z \end{bmatrix} \right). \tag{11.61}$$

The Jacobian matrix of the translation of the chassis is normally represented in the chassis-fixed coordinate system as a [3 × 14]-block matrix, in order to characterize the assignment of the generalized velocities represented by the regions of the matrix to the individual component groups of the vehicle:

$$\boldsymbol{J}_{Tra_1} = \boldsymbol{J}_{T_V} = \left[\underbrace{\left(\boldsymbol{I}_{[3\times3]} | 0_{[3\times3]} \right)}_{\text{chassis}} | \underbrace{0_{[3\times4]}}_{\text{wheelcarrier}} | \underbrace{0_{[3\times4]}}_{\text{wheel}} \right]. \tag{11.62}$$

Rotation As in Sect. 11.1, the rotation matrix is required first, describing the orientation of the chassis-fixed coordinate system relative to the inertial system (Ref. also Eq. (11.3)):

$${}^{E}\boldsymbol{T}_V = \boldsymbol{T}_z(\psi_V) \cdot \boldsymbol{T}_y(\theta_V) \cdot \boldsymbol{T}_x(\varphi_V). \tag{11.63}$$

From this, as in Sect. 11.1, the angular velocities of the chassis relative to the inertial system, represented in the chassis coordinate system (Ref. Eqs. (11.4) and (11.5)), are:

$${}^{V}_{E}\boldsymbol{\omega}_1 = {}^{V}_{E}\boldsymbol{\omega}_V = \begin{bmatrix} \omega_x \\ \omega_y \\ \omega_z \end{bmatrix} = \begin{bmatrix} \dot{\varphi}_V - \dot{\psi}_V s\theta_V \\ \dot{\theta}_V c\varphi_V + \dot{\psi}_V c\theta_V s\varphi_V \\ -\dot{\theta}_V s\varphi_V + \dot{\psi}_V c\theta_V c\varphi_V \end{bmatrix} = \boldsymbol{T}_\omega \begin{bmatrix} \dot{\psi}_V \\ \dot{\theta}_V \\ \dot{\varphi}_V \end{bmatrix}. \tag{11.64}$$

For the acceleration of the chassis relative to the inertial system, represented in the chassis fixed coordinate system, one obtains, the generalized accelerations of the vehicle chassis dependent on the generalized coordinates from Eq. (11.64):

$${}^{V}_{E}\boldsymbol{\alpha}_1 = {}^{V}_{E}\boldsymbol{\alpha}_V = \left[\dot{\omega}_x \quad \dot{\omega}_y \quad \dot{\omega}_z \right]^T. \tag{11.65}$$

From Eq. (11.64), one can also obtain the Jacobian matrix of the rotation of the vehicle chassis with the same block distribution as that in the translation case:

$$\boldsymbol{J}_{Rot_1} = \boldsymbol{J}_{Rot_V} = \left[\underbrace{\left(0_{[3\times3]} | \boldsymbol{I}_{[3\times3]} \right)}_{\text{chassis}} | \underbrace{0_{[3\times4]}}_{\text{wheelcarrier}} | \underbrace{0_{[3\times4]}}_{\text{wheel}} \right]. \tag{11.66}$$

Relationship between generalized velocities and generalized coordinates For the following investigation, all generalized coordinates and generalized velocities are collected as two [14 × 1]-Vectors respectively:

$$\boldsymbol{q} = \left[x\, y\, z\, \psi_V\, \theta_V\, \varphi_V \,|\, \varphi_1\, \varphi_2\, \varphi_3\, \varphi_4 \,|\, \rho_{R_1}\, \rho_{R_2}\, \rho_{R_3}\, \rho_{R_4} \right]^T, \tag{11.67}$$

$$\boldsymbol{z} = \left[v_x\, v_y\, v_z\, \omega_x\, \omega_y\, \omega_z \,|\, \dot{\varphi}_1\, \dot{\varphi}_2\, \dot{\varphi}_3\, \dot{\varphi}_4 \,|\, \dot{\rho}_{R_1}\, \dot{\rho}_{R_2}\, \dot{\rho}_{R_3}\, \dot{\rho}_{R_4} \right]^T. \tag{11.68}$$

Now, the relationship between the generalized coordinates and the generalized velocities is needed. It is obvious, that the relationship between the generalized velocities and the generalized coordinates is trivial, except in the case of the vehicle chassis. However, the Eqs. (11.60) and (11.64) are to be considered for the vehicle chassis. Summarizing, one arrives at the block diagonal matrix $\boldsymbol{K}(\boldsymbol{q})$ with the relationship:

$$z = \boldsymbol{K}(\boldsymbol{q})\dot{\boldsymbol{q}} = \begin{bmatrix} {}^V\boldsymbol{T}_E & 0 & 0 & 0 \\ 0 & \boldsymbol{T}_\omega & 0 & 0 \\ 0 & 0 & \boldsymbol{I}_{[4\times4]} & 0 \\ 0 & 0 & 0 & \boldsymbol{I}_{[4\times4]} \end{bmatrix} \dot{\boldsymbol{q}}. \tag{11.69}$$

For the solution of the equations of motion below, one still needs the inverse of this relationship, which can be calculated easily due to the special makeup of $\boldsymbol{K}(\boldsymbol{q})$

$$\dot{\boldsymbol{q}} = \boldsymbol{K}^{-1}(\boldsymbol{q})z, \tag{11.70}$$

as long as matrix $\boldsymbol{T}_\omega$ is invertable. This will only be the case if the kinematic CARDAN equation (11.64) is invertible, i.e. for $\theta_V \neq 90^\circ$, and one derives the inverse matrix as:

$$\boldsymbol{T}_\omega^{-1} = \frac{1}{c\theta_V} \begin{bmatrix} 0 & s\varphi_V & c\varphi_V \\ 0 & c\theta_V c\varphi_V & -c\theta_V s\varphi_V \\ c\theta_V & s\theta_V s\varphi_V & s\theta_V c\varphi_V \end{bmatrix}. \tag{11.71}$$

11.2.3 Generalized Kinematics of the Wheel Suspension

Translation First, a vehicle fixed reference point $A_i, i = 1\ldots4$ for each of the four independent wheel suspensions is defined via the position vector from the center of gravity of the chassis O_V to the point A_i, Ref. Fig. 11.9:

$${}^V_V\boldsymbol{r}_{A_i} = \begin{bmatrix} x_{A_i} & y_{A_i} & z_{A_i} \end{bmatrix}^T, \qquad i = 1\ldots4. \tag{11.72}$$

Each of these vectors is constant in the vehicle fixed coordinate system and hence moves along with it.

The connection of the wheel suspension is set up according to Fig. 11.9, i.e. each wheel carrier is allotted a generalized coordinate φ_i, which for example describes the generalized deflection of the wheel or the turning angle of the link. As an additional input variable, the angle δ_i is used, which is a time dependent function. Through this, the vectors of the generalized coordinates for the four suspensions are

$$\boldsymbol{c} = \begin{bmatrix} \varphi_1 & \varphi_2 & \varphi_3 & \varphi_4 \end{bmatrix}^T, \tag{11.73}$$

and those of the four wheels of the vehicle are defined as:

$$\boldsymbol{\rho} = \begin{bmatrix} \rho_{R_1} & \rho_{R_2} & \rho_{R_3} & \rho_{R_4} \end{bmatrix}^T. \tag{11.74}$$

There are also the steering angles as time dependent functions

$$\boldsymbol{\delta} = \begin{bmatrix} \delta_1 & \delta_2 & \delta_3 & \delta_4 \end{bmatrix}^T. \tag{11.75}$$

Now, the velocity and acceleration values of the wheel carrier and the wheels, relative to the inertial system dependent on the vectors $\boldsymbol{c}$, $\boldsymbol{\rho}$ and $\boldsymbol{\delta}$ are required. It should be considered that the components of the vectors $\boldsymbol{c}$ and $\boldsymbol{\rho}$ are generalized coordinates, while those in the vector $\boldsymbol{\delta}$ contain the prescribed function of time:

- Translational variables for the wheel carriers mass centers T_i relative to the inertial system

$$\begin{aligned} \boldsymbol{r}_{T_i} &= \boldsymbol{r}_{T_i}(\boldsymbol{c}, \boldsymbol{\delta}), \\ \boldsymbol{v}_{T_i} &= \dot{\boldsymbol{r}}_{T_i}\left(\boldsymbol{c}, \boldsymbol{\delta}, \dot{\boldsymbol{c}}, \dot{\boldsymbol{\delta}}\right), \\ \boldsymbol{a}_{T_i} &= \ddot{\boldsymbol{r}}_{T_i}\left(\boldsymbol{c}, \boldsymbol{\delta}, \dot{\boldsymbol{c}}, \dot{\boldsymbol{\delta}}, \ddot{\boldsymbol{c}}, \ddot{\boldsymbol{\delta}}\right). \end{aligned} \tag{11.76}$$

- Rotational variables for the wheel carriers relative to the inertial system

$$\begin{aligned} {}^{E}\boldsymbol{T}_{T_i} &= {}^{E}\boldsymbol{T}_{T_i}(\boldsymbol{c}, \boldsymbol{\delta}), \\ \boldsymbol{\omega}_{T_i} &= \boldsymbol{\omega}_{T_i}\left(\boldsymbol{c}, \boldsymbol{\delta}, \dot{\boldsymbol{c}}, \dot{\boldsymbol{\delta}}\right), \\ \boldsymbol{\alpha}_{T_i} &= \dot{\boldsymbol{\omega}}_{T_i} = \dot{\boldsymbol{\omega}}_{T_i}\left(\boldsymbol{c}, \boldsymbol{\delta}, \dot{\boldsymbol{c}}, \dot{\boldsymbol{\delta}}, \ddot{\boldsymbol{c}}, \ddot{\boldsymbol{\delta}}\right). \end{aligned} \tag{11.77}$$

Below, these relationships will be used generically using the kinematic tranformators, Ref. Chap. 3, and then derived explicitly in Sect. 11.2.4 for the simple case of a semi trailing arm. For the position vector $\boldsymbol{r}_{T_i}$ to the center of gravity T_i of the wheel carrier relative to the inertial system, one has:

$$\boldsymbol{r}_{T_i} = \boldsymbol{r}_V + {}_V\boldsymbol{r}_{T_i}, \tag{11.78}$$

with the position vector of the wheel carrier ${}_V\boldsymbol{r}_{T_i}$ relative to the chassis and the position vector $\boldsymbol{r}_V$ of the chassis center of gravity. The position vector ${}_V\boldsymbol{r}_{T_i}$ is made up of the vehicle fixed position vector ${}_V\boldsymbol{r}_{A_i}$ from the chassis center of gravity O_V to the reference point A_i of the wheel suspension as well as the position vector ${}_{A_i}\boldsymbol{r}_{T_i}$ from A_i to the center of gravity T_i of the wheel carrier:

$$ {}_V\boldsymbol{r}_{T_i}(\boldsymbol{c},\boldsymbol{\delta}) = \underbrace{{}_V\boldsymbol{r}_{A_i}}_{\text{const.}} + {}_{A_i}\boldsymbol{r}_{T_i}(\boldsymbol{c},\boldsymbol{\delta}). \tag{11.79} $$

Through this, the relative velocity of the wheel carrier with respect to the chassis is defined as:

$$ {}_V\boldsymbol{v}_{T_i} = {}_V\dot{\boldsymbol{r}}_{T_i} = \frac{\partial\, {}_{A_i}\boldsymbol{r}_{T_i}}{\partial \boldsymbol{c}}\dot{\boldsymbol{c}} + \frac{\partial\, {}_{A_i}\boldsymbol{r}_{T_i}}{\partial \boldsymbol{\delta}}\dot{\boldsymbol{\delta}} = \boldsymbol{H}_{Tra_c,T_i}\dot{\boldsymbol{c}} + \boldsymbol{H}_{Tra_\delta,T_i}\dot{\boldsymbol{\delta}} \tag{11.80} $$

and the relative acceleration is

$$ {}_V\boldsymbol{a}_{T_i} = {}_V\ddot{\boldsymbol{r}}_{T_i} = \boldsymbol{H}_{Tra_c,T_i}\ddot{\boldsymbol{c}} + \dot{\boldsymbol{H}}_{Tra_c,T_i}\dot{\boldsymbol{c}} + \boldsymbol{H}_{Tra_\delta,T_i}\ddot{\boldsymbol{\delta}} + \dot{\boldsymbol{H}}_{T_{ra\delta},T_i}\dot{\boldsymbol{\delta}}, \tag{11.81} $$

with the translational [3 × 4]-JACOBI matrices $\boldsymbol{H}_{Tra_c,T_i}$ and $\boldsymbol{H}_{Tra_\delta,T_i}$ as well as their derivatives with respect to time. The partial differentiation or the Jacobi matrices depend on the wheel suspension used and can for example be described using the methods (kinematic transformers) described in Chap. 6; they can also be calculated using direct kinematic considerations (Sect. 11.2.4).

For the relative kinematics of the rotation, the rotational matrix ${}^V\boldsymbol{T}_{T_i}(\boldsymbol{c},\boldsymbol{\delta})$ is required first, which describes the spatial orientation of the wheel carrier with respect to the chassis. From this, one arrives at the angular velocity

$$ {}_V\boldsymbol{\omega}_{T_i} = \boldsymbol{H}_{Rot_c,T_i}\dot{\boldsymbol{c}} + \boldsymbol{H}_{Rot_\delta,T_i}\dot{\boldsymbol{\delta}} \tag{11.82} $$

and the angular acceleration

$$ {}_V\boldsymbol{\alpha}_{T_i} = {}_V\,\dot{\boldsymbol{\omega}}_{T_i} = \boldsymbol{H}_{Rot_c,T_i}\ddot{\boldsymbol{c}} + \dot{\boldsymbol{H}}_{Rot_c,T_i}\dot{\boldsymbol{c}} + \boldsymbol{H}_{Rot_\delta,T_i}\ddot{\boldsymbol{\delta}} + \dot{\boldsymbol{H}}_{Rot_\delta,T_i}\dot{\boldsymbol{\delta}}, \tag{11.83} $$

with the rotational [3 × 4]-JACOBI matrices $\boldsymbol{H}_{Rot_c,T_i}$ and $\boldsymbol{H}_{Rot_\delta,T_i}$ as well as their derivatives with respect to time.

From Eqs. (11.80) to (11.83), the absolute velocities and accelerations can be derived as:

$$
\begin{aligned}
\boldsymbol{v}_{T_i} &= \dot{\boldsymbol{r}}_{T_i} = \underbrace{\dot{\boldsymbol{r}}_V + \boldsymbol{\omega}_V \times {}_V\boldsymbol{r}_{T_i}}_{\text{guiding motion}} + \underbrace{{}_V\dot{\boldsymbol{r}}_{T_i}}_{\text{relative motion}} \\
&= \boldsymbol{v}_V + \boldsymbol{\omega}_V \times {}_V\boldsymbol{r}_{T_i} + \boldsymbol{H}_{Tra_c,T_i}\dot{\boldsymbol{c}} + \boldsymbol{H}_{Tra_\delta,T_i}\dot{\boldsymbol{\delta}} \\
&= \underbrace{\left[\boldsymbol{I}_{[3\times3]} \middle| -{}_V\tilde{\boldsymbol{r}}_{T_i} \middle| \boldsymbol{H}_{Tra_c,T_i} \middle| \boldsymbol{0}_{[3\times4]}\right]}_{\boldsymbol{J}_{Tra,T_i}} \underbrace{\begin{bmatrix} \boldsymbol{v}_V \\ \boldsymbol{\omega}_V \\ \dot{\boldsymbol{c}} \\ \dot{\boldsymbol{\rho}} \end{bmatrix}}_{z} + \underbrace{\boldsymbol{H}_{Tra_\delta,T_i}\dot{\boldsymbol{\delta}}}_{\bar{\boldsymbol{v}}_{T_i}} \\
&= \boldsymbol{J}_{Tra,T_i}\boldsymbol{z} + \bar{\boldsymbol{v}}_{T_i}
\end{aligned}
\tag{11.84}
$$

and

$$
\begin{aligned}
\boldsymbol{a}_{T_i} =& \ddot{\boldsymbol{r}}_{T_i} = \boldsymbol{a}_V + \boldsymbol{\alpha}_V \times_V \boldsymbol{r}_{T_i} + \boldsymbol{\omega}_V \times_V \dot{\boldsymbol{r}}_{T_i} + \boldsymbol{H}_{Tra_c,T_i}\ddot{\boldsymbol{c}} + \dot{\boldsymbol{H}}_{Tra_c,T_i}\dot{\boldsymbol{c}} \\
& + \boldsymbol{H}_{Tra_\delta,T_i}\ddot{\boldsymbol{\delta}} + \dot{\boldsymbol{H}}_{Tra_\delta,T_i}\dot{\boldsymbol{\delta}} \\
& + \boldsymbol{\omega}_V \times (\boldsymbol{\omega}_V \times_V \boldsymbol{r}_{T_i} + \underbrace{\boldsymbol{H}_{Tra_c,T_i}\dot{\boldsymbol{c}} + \boldsymbol{H}_{Tra_\delta,T_i}\dot{\boldsymbol{\delta}}}_{{}_V\boldsymbol{v}_{T_i}}) \\
=& \boldsymbol{J}_{Tra,T_i}\dot{\boldsymbol{z}} \\
& + \underbrace{2\left(\omega_V \times_V \rho_c H_{T_i}\right) + \dot{\boldsymbol{H}}_{Tra_c,T_i}\dot{\boldsymbol{c}} + \boldsymbol{H}_{Tra_\delta,T_i}\boldsymbol{\delta} + \dot{\boldsymbol{H}}_{Tra_\delta,T_i}\dot{\boldsymbol{\delta}} + \boldsymbol{\omega}_V \times (\boldsymbol{\omega}_V \times_V \boldsymbol{r}_{T_i})}_{\bar{\boldsymbol{a}}_{T_i}} \\
=& \boldsymbol{J}_{Tra,T_i}\dot{\boldsymbol{z}} + \bar{\boldsymbol{a}}_{T_i}.
\end{aligned}
\tag{11.85}
$$

To this end, the translational [3 × 14]-JACOBI matrices

$$
\boldsymbol{J}_{Tra,T_i} = \boldsymbol{J}_{Tra,T_i}(\boldsymbol{q}) \tag{11.86}
$$

and the residual acceleration vector $\bar{\boldsymbol{a}}_{T_i}$ for the wheel carrier is introduced. Analogous, for the absolute rotational velocities and absolute rotational accelerations the following relationships are derived respectively:

$$
\begin{aligned}
\boldsymbol{\omega}_{T_i} &= \boldsymbol{\omega}_V + {}_V\boldsymbol{\omega}_{T_i} = \boldsymbol{\omega}_V + \boldsymbol{H}_{Rot_c,T_i}\dot{\boldsymbol{c}} + \boldsymbol{H}_{Rot_\delta,T_i}\dot{\boldsymbol{\delta}} \\
&= \underbrace{\left[\boldsymbol{0}_{[3\times3]} \middle| \boldsymbol{I}_{[3\times3]} \middle| \boldsymbol{H}_{Rot_c,T_i} \middle| \boldsymbol{0}_{[3\times4]}\right]}_{\boldsymbol{J}_{Rot,T_i}} \boldsymbol{z} + \underbrace{\boldsymbol{H}_{Rot_\delta,T_i}\dot{\boldsymbol{\delta}}}_{\bar{\boldsymbol{\omega}}_{T_i}} \\
&= \boldsymbol{J}_{Rot,T_i}\boldsymbol{z} + \bar{\boldsymbol{\omega}}_{T_i}
\end{aligned}
\tag{11.87}
$$

and

$$\begin{aligned}\boldsymbol{\alpha}_{T_i} &= \boldsymbol{\alpha}_V + {}_V\boldsymbol{\alpha}_{T_i} = \boldsymbol{\alpha}_V + \boldsymbol{\omega}_V \times {}_V\boldsymbol{\omega}_{T_i} + {}_V\dot{\boldsymbol{\omega}}_{T_i} \\ &= \boldsymbol{\alpha}_V + \boldsymbol{\omega}_V \times \left(\boldsymbol{H}_{Rot_c,T_i}\dot{\boldsymbol{c}} + \boldsymbol{H}_{Rot_\delta,T_i}\dot{\boldsymbol{\delta}}\right) + \boldsymbol{H}_{Rot_c,T_i}\ddot{\boldsymbol{c}} + \dot{\boldsymbol{H}}_{Rot_c,T_i}\dot{\boldsymbol{c}} + \boldsymbol{H}_{Rot_\delta,T_i}\ddot{\boldsymbol{\delta}} + \dot{\boldsymbol{H}}_{Rot_\delta,T_i}\dot{\boldsymbol{\delta}} \\ &= \boldsymbol{J}_{Rot,T_i}\dot{\boldsymbol{z}} + \underbrace{\boldsymbol{\omega}_V \times_V \boldsymbol{\omega}_{T_i} + \dot{\boldsymbol{H}}_{Rot_c,T_i}\dot{\boldsymbol{c}} + \boldsymbol{H}_{Rot_\delta,T_i}\ddot{\boldsymbol{\delta}} + \dot{\boldsymbol{H}}_{Rot_\delta,T_i}\dot{\boldsymbol{\delta}}}_{\bar{\boldsymbol{\alpha}}_{T_i}} \\ &= \boldsymbol{J}_{Rot,T_i}\dot{\boldsymbol{z}} + \bar{\boldsymbol{\alpha}}_{T_i}.\end{aligned} \tag{11.88}$$

The corresponding translational kinematic values for the wheel are simply derived by replacing the index $\cdot_{T_i}$ with $\cdot_{R_i}$ in Eqs. (11.78) to (11.81) and Eqs. (11.84) to (11.85). For rotations however, along with an adjustment of the indices, the rotation of the wheels about their own rotation axis needs to be appended. In this case on arrives at:

$$\begin{aligned}\boldsymbol{\omega}_{R_i} &= \boldsymbol{\omega}_V + {}_V\boldsymbol{\omega}_{R_i} = \boldsymbol{\omega}_V + \boldsymbol{H}_{Rot_c,T_i}\dot{\boldsymbol{c}} + \boldsymbol{H}_{Rot_\rho,R_i}\dot{\boldsymbol{\rho}} + \boldsymbol{H}_{Rot_\delta,T_i}\dot{\boldsymbol{\delta}} \\ &= \underbrace{\left[0_{[3\times3]}|\boldsymbol{I}_{[3\times3]}|\boldsymbol{H}_{Rot_c,T_i}|\boldsymbol{H}_{Rot_\rho,R_i}\right]}_{\boldsymbol{J}_{Rot,R_i}} z + \underbrace{\boldsymbol{H}_{Rot_\delta,T_i}\dot{\boldsymbol{\delta}}}_{\bar{\boldsymbol{\omega}}_{R_i}} \\ &= \boldsymbol{J}_{Rot,R_i} z + \bar{\boldsymbol{\omega}}_{R_i}\end{aligned} \tag{11.89}$$

and

$$\begin{aligned}\boldsymbol{\alpha}_{R_i} &= \boldsymbol{\alpha}_V + {}_V\boldsymbol{\alpha}_{R_i} = \boldsymbol{\alpha}_V + \boldsymbol{\omega}_V \times \boldsymbol{\omega}_{R_i} + {}_V\dot{\boldsymbol{\omega}}_{R_i} \\ &= \boldsymbol{\alpha}_V + \boldsymbol{\omega}_V \times \left(\boldsymbol{H}_{Rot_c,T_i}\dot{\boldsymbol{c}} + \boldsymbol{H}_{Rot_\rho,R_i}\dot{\boldsymbol{\rho}} + \boldsymbol{H}_{Rot_\delta,T_i}\dot{\boldsymbol{\delta}}\right) + \boldsymbol{H}_{Rot_c,T_i}\ddot{\boldsymbol{c}} \\ &\quad + \dot{\boldsymbol{H}}_{Rot_c,T_i}\dot{\boldsymbol{c}} + \boldsymbol{H}_{Rot_\rho,R_i}\ddot{\boldsymbol{\rho}} + \dot{\boldsymbol{H}}_{Rot_\rho,R_i}\dot{\boldsymbol{\rho}} + \boldsymbol{H}_{Rot_\delta,T_i}\ddot{\boldsymbol{\delta}} + \dot{\boldsymbol{H}}_{Rot_\delta,T_i}\dot{\boldsymbol{\delta}} \\ &= \boldsymbol{J}_{Rot,R_i}\dot{\boldsymbol{z}} \\ &\quad + \underbrace{\boldsymbol{\omega}_V \times_V \boldsymbol{\omega}_{R_i} + \dot{\boldsymbol{H}}_{Rot_c,T_i}\dot{\boldsymbol{c}} + \dot{\boldsymbol{H}}_{Rot_\rho,R_i}\dot{\boldsymbol{\rho}} + \; + \dot{\boldsymbol{H}}_{Rot_\delta,T_i}\dot{\boldsymbol{\delta}} + \boldsymbol{H}_{Rot_\delta,T_i}\ddot{\boldsymbol{\delta}}}_{\bar{\boldsymbol{\alpha}}_{R_i}} \\ &= \boldsymbol{J}_{Rot,R_i}\dot{\boldsymbol{z}} + \bar{\boldsymbol{\alpha}}_{R_i}.\end{aligned} \tag{11.90}$$

To facilitate a simplified notation for the equations of motion, the Jacobi matrices for the translation and rotations will be numbered as follows:

$$\text{chassis} \quad \boldsymbol{J}_{Tra_1} = \boldsymbol{J}_{Tra_V}, \; \boldsymbol{J}_{Rot_1} = \boldsymbol{J}_{Rot_V}, \tag{11.91}$$

$$\text{wheel carrier} \quad \boldsymbol{J}_{Tra_{1+i}} = \boldsymbol{J}_{Tra,T_i}, \boldsymbol{J}_{Rot_{1+i}} = \boldsymbol{J}_{Rot,T_i}, \quad i = 1, \ldots, 4, \tag{11.92}$$

$$\text{wheels} \quad \boldsymbol{J}_{Tra_{5+i}} = \boldsymbol{J}_{Tra,R_i}, \boldsymbol{J}_{Rot_{5+i}} = \boldsymbol{J}_{Rot,R_i}, \quad i = 1, \ldots, 4. \tag{11.93}$$

11.2.4 Wheel Suspension with a Trailing Arm Suspension

The kinematic relationships derived in Sect. 11.2.3 are independent of the type of the wheel suspension considered. They are simply based on the methods developed in the previous chapters, which were used to calculate the transfer functions for the input and output variables described in Fig. 11.9. This will be demonstrated below using a basic trailing arm suspension, Ref Fig. 11.10. The unit vectors $\boldsymbol{u}_{S_i}, i = 1,\ldots,4$ are used to describe this suspension, and they define the vehicle fixed rotational axis of the semi-trailing arm:

$$^{V}\boldsymbol{u}_{S_i} = \begin{bmatrix} u_{S_i,x} & u_{S_i,y} & u_{S_i,z} \end{bmatrix}^T, \qquad i = 1,\ldots,4. \tag{11.94}$$

The orientation of the trailing arm is then defined by one rotation matrix each, which describes the orientation of the semi trailing arm fixed coordinate system relative to the chassis fixed coordinate system. The suspension considered here, can be described through independent rotations about two spatial axes. Therefore it makes sense to describe the kinematics using the rotational-displacement pair, (Chap. 3). Hence, for the rotation of the semi trailing arm about the vehicle fixed rotational axis one obtains:

$$^{V}\boldsymbol{T}_{S_i} = \boldsymbol{u}_{S_i}\boldsymbol{u}_{S_i}^T + \left(\boldsymbol{I} - \boldsymbol{u}_{S_i}\boldsymbol{u}_{S_i}^T\right)\cos\varphi_{S_i} + \tilde{\boldsymbol{u}}_{S_i}\sin\varphi_{S_i}. \tag{11.95}$$

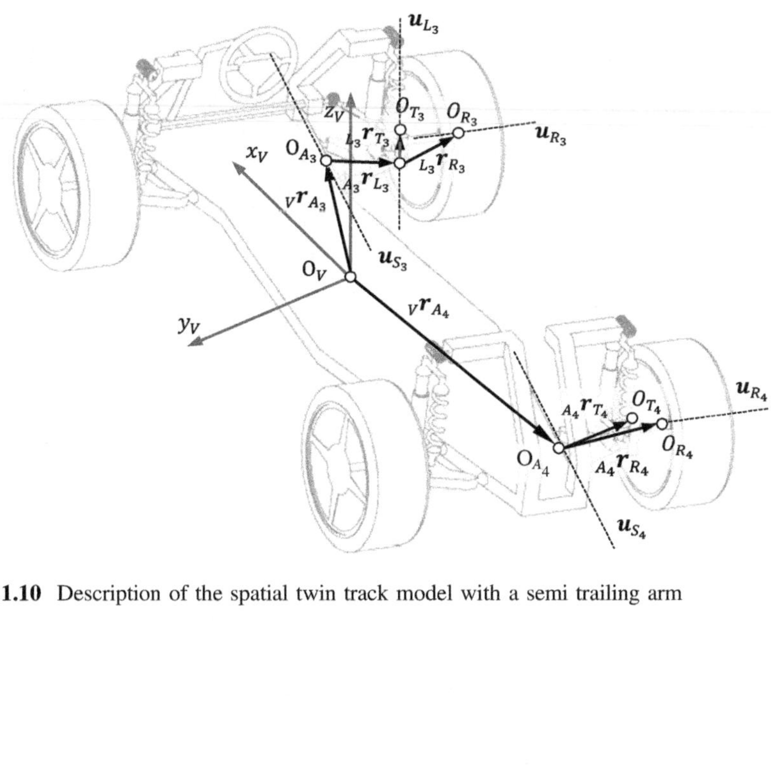

Fig. 11.10 Description of the spatial twin track model with a semi trailing arm

Here, $\varphi_{S_i}, i = 1, \ldots, 4$ describes the tilting angle of the ith semi trailing arm about its own rotation axis. With the relationship in Sect. 11.2.3 it then holds that:

$$\varphi_i := \varphi_{S_i}, \quad i = 1, \ldots, 4 \tag{11.96}$$

and hence

$$\boldsymbol{c} := \begin{bmatrix} \varphi_{S_1} & \varphi_{S_2} & \varphi_{S_3} & \varphi_{S_4} \end{bmatrix}^T. \tag{11.97}$$

Furthermore, the trailing arm fixed position vector from each of the reference points $A_i, i = 1, \ldots, 4$ of the trailing arm rotational axis to the reference point L_i of the corresponding steering axis

$${}^{S_i}_{A_i}\boldsymbol{r}_{L_i} = \begin{bmatrix} x_{L_i} & y_{L_i} & z_{L_i} \end{bmatrix}^T \qquad i = 1, \ldots, 4 \tag{11.98}$$

as well as the semi trailing arm fixed unit vector is needed

$${}^{S_i}\boldsymbol{u}_{L_i} = \begin{bmatrix} u_{L_i,x} & u_{L_i,y} & u_{L_i,z} \end{bmatrix}^T \qquad i = 1, \ldots, 4 \tag{11.99}$$

in order to describe the steering axis of the wheels. The required rotational matrices to determine the rotations $\delta_i, i = 1, \ldots, 4$ of the tires with respect to the semitrailing arms are given as explained in Chap. 3:

$${}_{S_i}\boldsymbol{T}_{T_i} = \boldsymbol{u}_{L_i}\boldsymbol{u}_{L_i}^T + \left(\boldsymbol{I} - \boldsymbol{u}_{L_i}\boldsymbol{u}_{L_i}^T\right)\cos\delta_i + \tilde{\boldsymbol{u}}_{L_i}\sin\delta_i. \tag{11.100}$$

What still remains is the description of the center of gravity T_i and $R_i, i = 1, \ldots, 4$ of the wheel carriers and wheels, respectively. The points T_i are described, relative to the reference points of the steering axis, by the wheel carrier fixed position vector from the reference point L_i of the steering axis to the reference point T_i of the wheel carrier, Ref. Fig. 11.11:

$${}^{T_i}_{L_i}\boldsymbol{r}_{T_i} = \begin{bmatrix} x_{T_i} & y_{T_i} & z_{T_i} \end{bmatrix}^T, \quad i = 1, \ldots, 4. \tag{11.101}$$

The translation of the wheel is defined via the wheel carrier fixed position vector from the reference point L_i of the corresponding steering axis to the reference point R_i of the wheel:

$${}^{T_i}_{L_i}\boldsymbol{r}_{R_i} = \begin{bmatrix} x_{R_i} & y_{R_i} & z_{R_i} \end{bmatrix}^T, \quad i = 1, \ldots, 4. \tag{11.102}$$

The wheel rotational axis is fixed to the wheel carrier and can again be described using a unit vector:

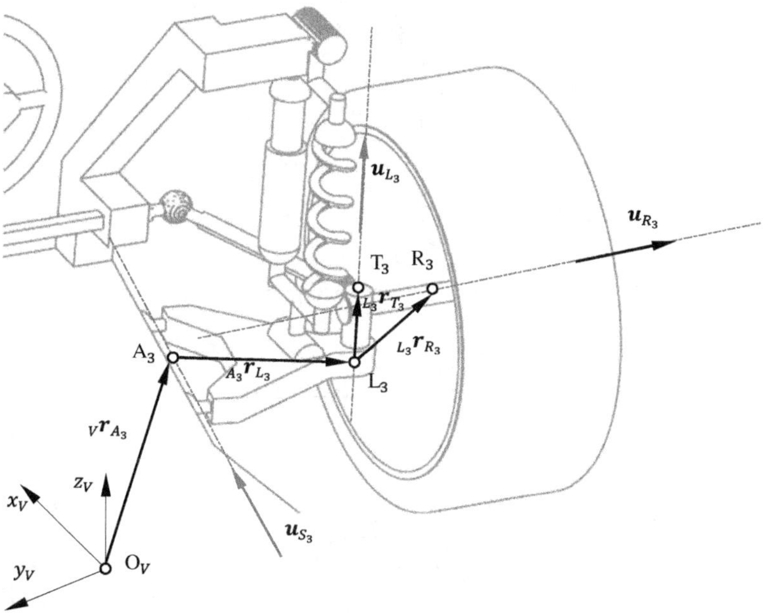

Fig. 11.11 Wheel suspension twin track model with a trailing arm

$$ {}^{T_i}\boldsymbol{u}_{R_i} = \left[u_{R_i,x} \quad u_{R_i,y} \quad u_{R_i,z} \right]^T, \quad i = 1, \ldots, 4. \tag{11.103} $$

Translation After this preliminary work, the position vector from the center of gravity of the chassis to the center of gravity of the wheel carrier can be given as, Ref. Fig. 11.11:

$$ {}_V\boldsymbol{r}_{T_i} = {}_V\boldsymbol{r}_{A_i} + {}_{A_i}\boldsymbol{r}_{L_i} + {}_{L_i}\boldsymbol{r}_{T_i}, \quad i = 1, \ldots, 4. \tag{11.104} $$

With this, the velocities and accelerations of the wheel carrier relative to the chassis can be calculated as

$$ \begin{aligned} {}_V\boldsymbol{v}_{T_i} &= \boldsymbol{\omega}_{S_i} \times \left({}_{A_i}\boldsymbol{r}_{L_i} + {}_{L_i}\boldsymbol{r}_{T_i} \right) + \boldsymbol{\omega}_{L_i} \times {}_{L_i}\boldsymbol{r}_{T_i} \\ &= \boldsymbol{u}_{S_i}\dot{\varphi}_{S_i} \times \left({}_{A_i}\boldsymbol{r}_{L_i} + {}_{L_i}\boldsymbol{r}_{T_i} \right) + \boldsymbol{u}_{L_i}\dot{\delta}_i \times {}_{L_i}\boldsymbol{r}_{T_i} \\ &= \underbrace{\left[\tilde{\boldsymbol{u}}_{S_i A_i}\boldsymbol{r}_{T_i} \right]}_{\boldsymbol{H}_{Trac,T_i}} \dot{\varphi}_{S_i} + \underbrace{\left[\tilde{\boldsymbol{u}}_{L_i L_i}\boldsymbol{r}_{T_i} \right]}_{\boldsymbol{H}_{Tra_\delta,T_i}} \dot{\delta}_i \qquad i = 1, \ldots, 4, \end{aligned} \tag{11.105} $$

and

$$
\begin{aligned}
{}_{V}\boldsymbol{a}_{T_i} &= \dot{\boldsymbol{\omega}}_{S_i} \times {}_{A_i}\boldsymbol{r}_{T_i} + \boldsymbol{\omega}_{S_i} \times {}_{A_i}\dot{\boldsymbol{r}}_{T_i} + \dot{\boldsymbol{\omega}}_{L_i} \times {}_{L_i}\boldsymbol{r}_{T_i} + \boldsymbol{\omega}_{L_i} \times {}_{L_i}\dot{\boldsymbol{r}}_{T_i} \\
&= \dot{\boldsymbol{\omega}}_{S_i} \times {}_{A_i}\boldsymbol{r}_{T_i} + \boldsymbol{\omega}_{S_i} \times (\boldsymbol{\omega}_{S_i} \times {}_{A_i}\boldsymbol{r}_{T_i} + \boldsymbol{\omega}_{L_i} \times {}_{L_i}\boldsymbol{r}_{T_i}) \\
&\quad + \dot{\boldsymbol{\omega}}_{L_i} \times {}_{L_i}\boldsymbol{r}_{T_i} + \boldsymbol{\omega}_{L_i} \times ((\boldsymbol{\omega}_{S_i} + \boldsymbol{\omega}_{L_i}) \times {}_{L_i}\boldsymbol{r}_{T_i}) \\
&= [\tilde{\boldsymbol{u}}_{S_i A_i}\boldsymbol{r}_{T_i}]\ddot{\varphi}_{S_i} + [\tilde{\boldsymbol{u}}_{L_i L_i}\boldsymbol{r}_{T_i}]\ddot{\delta}_i \\
&\quad + \underbrace{\left[\tilde{\boldsymbol{u}}_{S_i}\left(\boldsymbol{u}_{S_i}\dot{\varphi}_{S_i} \times {}_{A_i}\boldsymbol{r}_{T_i} + \boldsymbol{u}_{L_i}\dot{\delta}_i \times {}_{L_i}\boldsymbol{r}_{T_i}\right)\right]}_{\dot{\boldsymbol{H}}_{Tra_c,T_i}} \dot{\varphi}_{S_i} \\
&\quad + \underbrace{\left[\boldsymbol{u}_{L_i} \times \left(\left(\boldsymbol{u}_{S_i}\dot{\varphi}_{S_i} + \boldsymbol{u}_{L_i}\dot{\delta}_i\right) \times {}_{L_i}\boldsymbol{r}_{T_i}\right)\right]}_{\dot{\boldsymbol{H}}_{Tra_\delta,T_i}} \dot{\delta}_i \\
&= [\tilde{\boldsymbol{u}}_{S_i A_i}\boldsymbol{r}_{T_i}]\ddot{\varphi}_{S_i} + [\tilde{\boldsymbol{u}}_{L_i L_i}\boldsymbol{r}_{T_i}]\ddot{\delta}_i \\
&\quad + [\tilde{\boldsymbol{u}}_{S_i V}\boldsymbol{v}_{T_i}]\dot{\varphi}_{S_i} + \tilde{\boldsymbol{u}}_{L_i}\left(\left[\tilde{\boldsymbol{u}}_{S_i}\dot{\varphi}_{S_i} + \tilde{\boldsymbol{u}}_{L_i}\dot{\delta}_i\right]_{L_i}\boldsymbol{r}_{T_i}\right)\dot{\delta}_i.
\end{aligned}
\tag{11.106}
$$

With the help of Eq. (11.106), the relationship with the Jacobian matrices in Sect. 11.2.3 becomes obvious as

$$
\boldsymbol{H}_{Tra_c,T_i} = \begin{bmatrix} \boldsymbol{0}_{[3\times\ldots]} & | \underbrace{\tilde{\boldsymbol{u}}_{S_i}\ {}_{A_i}\boldsymbol{r}_{T_i}}_{i^{th}\text{column}} | & \boldsymbol{0}_{[3\times\ldots]} \end{bmatrix}_{[3\times 4]}, \tag{11.107}
$$

and

$$
\boldsymbol{H}_{Tra_\delta,T_i} = \begin{bmatrix} \boldsymbol{0}_{[3\times\ldots]} & | \underbrace{\tilde{\boldsymbol{u}}_{L_i}\ {}_{L_i}\boldsymbol{r}_{T_i}}_{i^{th}\text{column}} | & \boldsymbol{0}_{[3\times\ldots]} \end{bmatrix}_{[3\times 4]}. \tag{11.108}
$$

Correspondingly, the derivatives are calculated as

$$
\dot{\boldsymbol{H}}_{Tra_c,T_i} = \begin{bmatrix} \boldsymbol{0}_{[3\times\ldots]} & | \underbrace{\tilde{\boldsymbol{u}}_{S_i}\left(\tilde{\boldsymbol{u}}_{S_i}\dot{\varphi}_{S_i}\ {}_{A_i}\boldsymbol{r}_{T_i} + \tilde{\boldsymbol{u}}_{L_i}\dot{\delta}_i\ {}_{L_i}\boldsymbol{r}_{T_i}\right)}_{i^{th}\text{column}} | & \boldsymbol{0}_{[3\times\ldots]} \end{bmatrix}_{[3\times 4]}, \tag{11.109}
$$

and

$$
\dot{\boldsymbol{H}}_{Tra_\delta,T_i} = \begin{bmatrix} \boldsymbol{0}_{[3\times\ldots]} & \left| \underbrace{\tilde{\boldsymbol{u}}_{L_i}\left(\left(\tilde{\boldsymbol{u}}_{S_i}\dot{\varphi}_{S_i} + \tilde{\boldsymbol{u}}_{L_i}\dot{\delta}_i\right)\ {}_{L_i}\boldsymbol{r}_{T_i}\right)}_{i^{th}\text{column}} \right| & \boldsymbol{0}_{[3\times\ldots]} \end{bmatrix}_{[3\times 4]}. \tag{11.110}
$$

Considering the motion of the vehicle chassis (guiding motion), one can derive the velocities and accelerations with respect to the inertial system as:

$$_E\boldsymbol{v}_{T_i} = {}_E\,\boldsymbol{v}_V + \boldsymbol{\omega}_V \times {}_V\,\boldsymbol{r}_{T_i} + {}_V\boldsymbol{v}_{T_i}, \tag{11.111}$$

and

$$\begin{aligned} _E\boldsymbol{a}_{T_i} &= {}_E\boldsymbol{a}_V + \boldsymbol{\omega}_V \times {}_V\,\boldsymbol{v}_{T_i} + \boldsymbol{\alpha}_V \times {}_V\,\boldsymbol{r}_{T_i} + {}_V\boldsymbol{a}_{T_i} \\ &+ \boldsymbol{\omega}_V \times (\boldsymbol{\omega}_V \times {}_V\,\boldsymbol{r}_{T_i} + {}_V\boldsymbol{v}_{T_i}) \end{aligned} \tag{11.112}$$

The Jacobians $\boldsymbol{J}_{Tra,T_i}$ are then to be calculated as provided in Eq. (11.85):

$$\boldsymbol{J}_{Tra,T_i} = \left[\boldsymbol{I}_{[3\times3]}\middle| -_V\,\tilde{\boldsymbol{r}}_{T_i}\middle|\boldsymbol{H}_{Tra_c,T_i}\middle|\boldsymbol{0}_{[3\times4]}\right], i = 2, \ldots, 5. \tag{11.113}$$

Using this, the translational acceleration of the wheel carrier can also be described as

$$_E\boldsymbol{a}_{T_i} = \boldsymbol{J}_{Tra,T_i}\dot{\boldsymbol{z}} + {}_E\bar{\boldsymbol{a}}_{T_i}, \tag{11.114}$$

with the translational residual accelerations of the wheel carrier, which are obtained by substituting Eq. (11.106) in (11.112) to:

$$\begin{aligned} _E\bar{\boldsymbol{a}}_{T_i} &= 2\boldsymbol{\omega}_V \times {}_V\,\boldsymbol{v}_{T_i} + \dot{\boldsymbol{H}}_{Tra_c,T_i}\dot{\boldsymbol{c}} + \boldsymbol{H}_{Tra_\delta,T_i}\ddot{\boldsymbol{\delta}} + \dot{\boldsymbol{H}}_{Tra_\delta,T_i}\dot{\boldsymbol{\delta}} + \boldsymbol{\omega}_V \\ &\times\ \boldsymbol{\omega}_V \times {}_V\,\boldsymbol{r}_{T_i} \\ &= 2_E\boldsymbol{\omega}_V \times {}_V\,\boldsymbol{v}_{T_i} + [\tilde{\boldsymbol{u}}_{S_i}\ {}_V\boldsymbol{v}_{T_i}]\dot{\varphi}_{S_i} + [\tilde{\boldsymbol{u}}_{L_i}\ {}_{L_i}\boldsymbol{r}_{T_i}]\ddot{\delta}_i \\ &+ \tilde{\boldsymbol{u}}_{L_i}\left(\left[\tilde{\boldsymbol{u}}_{S_i}\dot{\varphi}_{S_i} + \tilde{\boldsymbol{u}}_{L_i}\dot{\delta}_i\right]\ {}_{L_i}\boldsymbol{r}_{T_i}\right)\dot{\delta}_i + \boldsymbol{\omega}_V \times (\boldsymbol{\omega}_V \times {}_V\,\boldsymbol{r}_{T_i}). \end{aligned} \tag{11.115}$$

Rotation For the absolute angular velocity of the wheel carriers one obtains:

$$_E\boldsymbol{\omega}_{T_i} = {}_E\,\boldsymbol{\omega}_V + {}_V\boldsymbol{\omega}_{T_i} = {}_E\,\boldsymbol{\omega}_V + \underbrace{\boldsymbol{u}_{S_i}}_{\boldsymbol{H}_{Rot_c,T_i}}\ \dot{\varphi}_{S_i} + \underbrace{\boldsymbol{u}_{L_i}}_{\boldsymbol{H}_{Rot_\delta,T_i}}\ \dot{\delta}_i. \tag{11.116}$$

From Eq. (11.116) the angular acceleration of the wheel carrier is obtained through differentiation again:

$$\begin{aligned} _E\boldsymbol{\alpha}_{T_i} &= {}_E\boldsymbol{\alpha}_V + {}_V\boldsymbol{\alpha}_{T_i} \\ &= {}_E\boldsymbol{\alpha}_V + {}_E\boldsymbol{\omega}_V \times {}_V\,\boldsymbol{\omega}_{T_i} + {}_V\dot{\boldsymbol{\omega}}_{T_i} \\ &= {}_E\boldsymbol{\alpha}_V + {}_E\boldsymbol{\omega}_V \times {}_V\,\boldsymbol{\omega}_{T_i} + \boldsymbol{u}_{S_i}\ddot{\varphi}_{S_i} + \boldsymbol{u}_{L_i}\ddot{\delta}_i \\ &+ \underbrace{[\tilde{\boldsymbol{u}}_{S_i}\boldsymbol{u}_{L_i}]\dot{\varphi}_{S_i}}_{\dot{\boldsymbol{H}}_{Rot_\delta,T_i\,[3\times4]}}\ \dot{\delta}_i. \end{aligned} \tag{11.117}$$

Here too, the direct relationship to the Jacobian matrices is noticeable:

$$\boldsymbol{H}_{Rot_c,T_i} = \Big[\boldsymbol{0}_{[3\times\ldots]} \; \Big| \; \underbrace{\boldsymbol{u}_{S_i}}_{i^{th}column} \; \Big| \; \boldsymbol{0}_{[3\times\ldots]}\Big]_{[3\times 4]}, \tag{11.118}$$

$$\boldsymbol{H}_{Rot_\delta,T_i} = \Big[\boldsymbol{0}_{[3\times\ldots]} \; \Big| \; \underbrace{\boldsymbol{u}_{L_i}}_{i^{th}column} \; \Big| \; \boldsymbol{0}_{[3\times\ldots]}\Big]_{[3\times 4]}, \tag{11.119}$$

$$\boldsymbol{J}_{Rot,T_i} = \left[\boldsymbol{0}_{[3\times 3]} | \boldsymbol{I}_{[3\times 3]} | \boldsymbol{H}_{Rot_c,T_i} | \boldsymbol{0}_{[3\times 4]}\right], i = 2, \ldots, 5. \tag{11.120}$$

Correspondingly, the derivatives can be calculated as:

$$\dot{\boldsymbol{H}}_{Rot_c,T_i} = \Bigg[\boldsymbol{0}_{[3\times\ldots]} \; \Bigg| \; \underbrace{\boldsymbol{0}}_{i^{th}column} \; \Bigg| \; \boldsymbol{0}_{[3\times\ldots]}\Bigg]_{[3\times 4]}, \tag{11.121}$$

and

$$\dot{\boldsymbol{H}}_{Rot_\delta,T_i} = \Bigg[\boldsymbol{0}_{[3\times\ldots]} \; \Bigg| \; \underbrace{[\tilde{\boldsymbol{u}}_{S_i}\boldsymbol{u}_{L_i}]\dot{\varphi}_{S_i}}_{i^{th}column} \; \Bigg| \; \boldsymbol{0}_{[3\times\ldots]}\Bigg]_{[3\times 4]}. \tag{11.122}$$

Hence, the angular accelerations of the wheel carrier can also be written as

$${}^{V}_{E}\boldsymbol{\alpha}_{T_i} = \boldsymbol{J}_{R,T_i}\dot{\boldsymbol{z}} + {}_{E}\bar{\boldsymbol{\alpha}}_{T_i}, \tag{11.123}$$

with the rotational residual acceleration of the wheel carrier:

$${}_{E}\bar{\alpha}_{T_i} = {}_{E}\,\omega_V \times {}_{V}\,\omega_{T_i} + \boldsymbol{u}_{L_i}\ddot{\delta}_i + [\tilde{\boldsymbol{u}}_{S_i}\boldsymbol{u}_{L_i}]\dot{\varphi}_{S_i}\dot{\delta}_i. \tag{11.124}$$

11.2.5 Kinematics of the Wheels While Using a Trailing Arm Suspension

Translation The position and motion of each wheel only differ with respect to the wheel rotation about its own axis as well as through another wheel carrier fixed reference point. Therefore the Jacobian matrices for translation are very similar in construction to those of the wheel carrier. The vector ${}_{L_i}\boldsymbol{r}_{T_i}$ can simply be replaced with the vector ${}_{L_i}\boldsymbol{r}_{R_i}$. With this, one obtains, similar to Eqs. (11.107), (11.108) and (11.113):

$$\boldsymbol{H}_{Tra_c,R_i} = \begin{bmatrix} \mathbf{0}_{[3\times\ldots]} & \left|\underbrace{\tilde{\boldsymbol{u}}_{S_i}\ {}_{A_i}\boldsymbol{r}_{R_i}}_{i^{th}column}\right| & \mathbf{0}_{[3\times\ldots]} \end{bmatrix}_{[3\times4]}, \tag{11.125}$$

and

$$\boldsymbol{H}_{Tra_\delta,R_i} = \begin{bmatrix} \mathbf{0}_{[3\times\ldots]} & \left|\underbrace{\tilde{\boldsymbol{u}}_{L_i}\ {}_{L_i}\boldsymbol{r}_{R_i}}_{i^{th}column}\right| & \mathbf{0}_{[\ldots\times4]} \end{bmatrix}_{[3\times4]}, \tag{11.126}$$

as well as

$$\boldsymbol{J}_{Tra,R_i} = \left[\boldsymbol{I}_{[3\times3]}\middle|-\ {}_V\tilde{\boldsymbol{r}}_{R_i}\middle|\boldsymbol{H}_{Tra_c,R_i}\middle|0_{[3\times4]}\right], i = 6, \ldots, 9. \tag{11.127}$$

Correspondingly, the derivatives can be calculated as follows:

$$\dot{\boldsymbol{H}}_{Tra_c,R_i} = \begin{bmatrix} \mathbf{0}_{[3\times\ldots]} & \left|\underbrace{\tilde{\boldsymbol{u}}_{S_i}\left(\tilde{\boldsymbol{u}}_{S_i}\dot{\varphi}_{S_i}\ {}_{A_i}\boldsymbol{r}_{R_i} + \tilde{\boldsymbol{u}}_{L_i}\dot{\delta}_i\ {}_{L_i}\boldsymbol{r}_{R_i}\right)}_{i^{th}column}\right| & \mathbf{0}_{[3\times\ldots]} \end{bmatrix}_{[3\times4]}, \tag{11.128}$$

$$\dot{\boldsymbol{H}}_{Tra_\delta,R_i} = \begin{bmatrix} \mathbf{0}_{[3\times\ldots]} & \left|\underbrace{\tilde{\boldsymbol{u}}_{L_i}\left(\left(\tilde{\boldsymbol{u}}_{S_i}\dot{\varphi}_{S_i} + \tilde{\boldsymbol{u}}_{L_i}\dot{\delta}_i\right)\ {}_{L_i}\boldsymbol{r}_{R_i}\right)}_{i^{th}column}\right| & \mathbf{0}_{[3\times\ldots]} \end{bmatrix}_{[3\times4]}. \tag{11.129}$$

With this, one obtains the translational velocities of the wheel center as:

$$_E\boldsymbol{v}_{R_i} = {}_E\,\boldsymbol{v}_V + \boldsymbol{\omega}_V \times {}_V\,\boldsymbol{r}_{R_i} + {}_V\boldsymbol{v}_{R_i}. \tag{11.130}$$

The same applies for the accelerations:

$$\begin{aligned} {}_V\boldsymbol{a}_{R_i} = {} & [\tilde{\boldsymbol{u}}_{S_i}\ {}_{A_i}\boldsymbol{r}_{R_i}]\ddot{\varphi}_{S_i} + [\tilde{\boldsymbol{u}}_{L_i}\ {}_{L_i}\boldsymbol{r}_{R_i}]\ddot{\delta}_i + [\tilde{\boldsymbol{u}}_{Si}\ {}_V\boldsymbol{v}_{R_i}]\dot{\varphi}_{S_i} + \tilde{\boldsymbol{u}}_{L_i}\dot{\delta}_i \cdot \\ & \left(\left[\tilde{\boldsymbol{u}}_{S_i}\dot{\varphi}_{S_i} + \tilde{\boldsymbol{u}}_{L_i}\dot{\delta}_i\right] \cdot {}_{L_i}\,\boldsymbol{r}_{R_i}\right) \end{aligned}. \tag{11.131}$$

For the translational acceleration of the wheel relative to the inertial system, one obtains:

$$_E\boldsymbol{a}_{R_i} = {}_E\,\boldsymbol{a}_V + \boldsymbol{\omega}_V \times {}_V\,\boldsymbol{v}_{R_i} + {}_V\boldsymbol{a}_{R_i}. \tag{11.132}$$

Hence, the accelerations can again be written as:

$$_{E}\boldsymbol{a}_{R_i} = \boldsymbol{J}_{Tra,R_i}\dot{\boldsymbol{z}} + {}_{E}\bar{\boldsymbol{a}}_{R_i}, \tag{11.133}$$

with the residual accelerations of the tires as

$$\begin{aligned} {}_{E}\bar{\boldsymbol{a}}_{R_i} =& 2{}_{E}\boldsymbol{\omega}_V \times {}_{V}\boldsymbol{v}_{R_i} + [\tilde{\boldsymbol{u}}_{S_i} \cdot {}_{V}\boldsymbol{v}_{R_i}]\dot{\varphi}_{S_i} + [\tilde{\boldsymbol{u}}_{L_i} \cdot {}_{L_i}\boldsymbol{r}_{R_i}]\ddot{\delta}_i \\ &+ \tilde{\boldsymbol{u}}_{L_i} \cdot \left(\left[\tilde{\boldsymbol{u}}_{S_i}\dot{\varphi}_{S_i} + \tilde{\boldsymbol{u}}_{L_i}\dot{\delta}_i \right] \cdot {}_{L_i}\boldsymbol{r}_{R_i} \right)\dot{\delta}_i + \boldsymbol{\omega}_V \times \boldsymbol{\omega}_V \times {}_{V}\boldsymbol{r}_{R_i}. \end{aligned} \tag{11.134}$$

Rotation A bigger difference is found between the wheel carriers and the wheels in the case of rotation values, as the wheels also rotate about the wheel rotational axis. The angular velocity and acceleration of the wheels is derived from Eq. (11.116) and by appending the wheel rotation about its own axis $\dot{\rho}_{R_i}$:

$$\begin{aligned} {}_{E}\boldsymbol{\omega}_{R_i} &= {}_{E}\boldsymbol{\omega}_V + {}_{V}\boldsymbol{\omega}_{T_i} + {}_{T_i}\boldsymbol{\omega}_{R_i} \\ &= {}_{E}\boldsymbol{\omega}_V + \underbrace{\boldsymbol{u}_{S_i}}_{\boldsymbol{H}_{Rot_c,T_i}} \dot{\varphi}_{S_i} + \underbrace{\boldsymbol{u}_{L_i}}_{\boldsymbol{H}_{Rot_\delta,T_i}} \dot{\delta}_i + \underbrace{\boldsymbol{u}_{R_i}}_{\boldsymbol{H}_{Rot_\rho,R_i}} \dot{\rho}_{R_i}, \end{aligned} \tag{11.135}$$

and

$$\begin{aligned} {}_{E}\boldsymbol{\alpha}_{R_i} =& {}_{E}\boldsymbol{\alpha}_V + {}_{V}\boldsymbol{\alpha}_{T_i} + {}_{T_i}\boldsymbol{\alpha}_{R_i} \\ =& {}_{E}\boldsymbol{\alpha}_V + {}_{E}\boldsymbol{\omega}_V \times {}_{V}\boldsymbol{\omega}_{T_i} + \boldsymbol{u}_{S_i}\ddot{\varphi}_{S_i} + \boldsymbol{u}_{L_i}\ddot{\delta}_i + \boldsymbol{u}_{R_i}\ddot{\rho}_{R_i} \\ &+ \underbrace{[\tilde{\boldsymbol{u}}_{S_i}\boldsymbol{u}_{L_i}]\dot{\varphi}_{S_i}}_{\dot{\boldsymbol{H}}_{Rot_\delta,T_i}} \dot{\delta}_i + \underbrace{\left([\tilde{\boldsymbol{u}}_{S_i}\boldsymbol{u}_{R_i}]\dot{\varphi}_{S_i} + [\tilde{\boldsymbol{u}}_{L_i}\boldsymbol{u}_{R_i}]\dot{\delta}_i \right)}_{\dot{\boldsymbol{H}}_{Rot_\rho,R_i}} \dot{\rho}_{R_i} \end{aligned} \tag{11.136}$$

respectively. Here, the relationship between the rotational axis of the trailing arm suspension and the Jacobian matrices from Sect. 11.2.3 can be observed as well:

$$\boldsymbol{H}_{Rot_c,T_i} = \left[\begin{array}{c|c|c} \boldsymbol{0}_{[3\times\ldots]} & \underbrace{\boldsymbol{u}_{S_i}}_{i^{th}column} & \boldsymbol{0}_{[3\times\ldots]} \end{array} \right]_{[3\times4]}, \tag{11.137}$$

$$\boldsymbol{H}_{Rot_\delta,T_i} = \left[\begin{array}{c|c|c} \boldsymbol{0}_{[3\times\ldots]} & \underbrace{\boldsymbol{u}_{L_i}}_{i^{th}column} & \boldsymbol{0}_{[3\times\ldots]} \end{array} \right]_{[3\times4]}, \tag{11.138}$$

$$\boldsymbol{H}_{Rot_\rho,R_i} = \left[\begin{array}{c|c|c} \boldsymbol{0}_{[3\times\ldots]} & \underbrace{\boldsymbol{u}_{R_i}}_{i^{th}column} & \boldsymbol{0}_{[3\times\ldots]} \end{array} \right]_{[3\times4]}, \tag{11.139}$$

$$\boldsymbol{J}_{Rot,R_i} = \left[0_{[3\times3]}|\boldsymbol{I}_{[3\times3]}|\boldsymbol{H}_{Rot_c,T_i}|\boldsymbol{H}_{Rot_\rho,R_i}\right], i = 2,\ldots,5. \tag{11.140}$$

Correspondingly, the derivatives can be calculated as:

$$\dot{\boldsymbol{H}}_{Rot_c,T_i} = \left[\begin{array}{c|c|c} \mathbf{0}_{[3\times\ldots]} & \underbrace{\mathbf{0}}_{i^{th} column} & \mathbf{0}_{[3\times\ldots]} \end{array}\right], \tag{11.141}$$

$$\dot{\boldsymbol{H}}_{Rot_\delta,T_i} = \left[\begin{array}{c|c|c} \mathbf{0}_{[3\times\ldots]} & \underbrace{[\tilde{\boldsymbol{u}}_{S_i}\cdot \boldsymbol{u}_{L_i}]\dot{\varphi}_{S_i}}_{i^{th} column} & \mathbf{0}_{[3\times\ldots]} \end{array}\right] \tag{11.142}$$

and

$$\dot{\boldsymbol{H}}_{Rot_\rho,R_i} = \left[\begin{array}{c|c|c} \mathbf{0}_{[3\times\ldots]} & \underbrace{[\tilde{\boldsymbol{u}}_{S_i}\boldsymbol{u}_{R_i}]\dot{\varphi}_{S_i} + [\tilde{\boldsymbol{u}}_{L_i}\boldsymbol{u}_{R_i}]\dot{\delta}_i}_{i^{th} column} & \mathbf{0}_{[3\times\ldots]} \end{array}\right]. \tag{11.143}$$

With this, the angular acceleration of the wheel can be expressed as:

$${}_E^V\boldsymbol{\alpha}_{R_i} = \boldsymbol{J}_{Rot,R_i}\dot{\boldsymbol{z}} + {}_E\bar{\boldsymbol{\alpha}}_{R_i}, \tag{11.144}$$

and the rotational residual accelerations of the wheel are

$$\begin{aligned} {}_E\bar{\boldsymbol{\alpha}}_{R_i} &= {}_E\boldsymbol{\omega}_V \times {}_V\boldsymbol{\omega}_{R_i} + \boldsymbol{u}_{L_i}\ddot{\delta}_i + [\tilde{\boldsymbol{u}}_{S_i}\boldsymbol{u}_{L_i}]\dot{\varphi}_{S_i}\dot{\delta}_i \\ &+ \left([\tilde{\boldsymbol{u}}_{S_i}\boldsymbol{u}_{R_i}]\dot{\varphi}_{S_i} + [\tilde{\boldsymbol{u}}_{L_i}\boldsymbol{u}_{R_i}]\dot{\delta}_i\right)\dot{\rho}_{R_i}. \end{aligned} \tag{11.145}$$

11.2.6 Tire Forces and Torques

Calculation of the tire-road contact point To calculate the contact values between the tire and the road, the relationships derived in Chap. 7 are used. To this end, one first calculates the kinematic variables of the wheels dependent on the generalized variables and the generalized velocities described in the previous section. For the tire normal vector $\boldsymbol{n}_{B_i}$ described in Chap. 7 on the cross-sectional area of the tire R_i, it holds:

$$\boldsymbol{n}_{B_i} = \boldsymbol{u}_{R_i}. \tag{11.146}$$

The position vector to the center of the ith wheel (Chap. 7) is known as

$$\boldsymbol{b}_i = {}_E\boldsymbol{r}_V + {}_V\boldsymbol{r}_{A_i} + {}_{A_i}\boldsymbol{r}_{R_i} \tag{11.147}$$

as well as the translational and rotational velocities of the wheel center point $\boldsymbol{v}_{R_i}$ and $\boldsymbol{\omega}_{R_i}$ from Sect. 11.2.3.

Calculation of the tire-road contact forces—tire model With this and the results from Chap. 7, the relative velocities at the contact points can be calculated. These can now be used with any tire model, for example the Magic Formula (Chap. 7) or its simplified version respectively, to calculate the tire forces $\boldsymbol{F}_{R_i}, i = 1, \ldots 4$.

11.2.7 Suspension Springs and Dampers

Force pivot points The force pivot points of the suspension spring and the suspension damper should be still freely definable with the help of parameter estimation. It is for this reason that the springs and the dampers are assigned different pivot points for the force elements. Let the chassis fixed pivot points be ${}_V\boldsymbol{r}_{F,A_i}$ for the suspension springs and ${}_V\boldsymbol{r}_{D,A_i}$ for the suspension damper. The corresponding points on the wheel carriers are:

$${}_V\boldsymbol{r}_{F,T_i} = {}_V\boldsymbol{r}_{T_i} + {}_{T_i}\boldsymbol{r}_{F,T_i}, \tag{11.148}$$

$${}_V\boldsymbol{r}_{D,T_i} = {}_V\boldsymbol{r}_{T_i} + {}_{T_i}\boldsymbol{r}_{D,T_i}. \tag{11.149}$$

With this, the difference vector for the spring force in the vehicle fixed coordinate system are obtained as

$$\Delta\boldsymbol{r}_{F_i} = {}_V\boldsymbol{r}_{A_i,F} - {}_V\boldsymbol{r}_{T_i,F}. \tag{11.150}$$

For the calculation of the damping force, the difference vector is still needed

$$\Delta\boldsymbol{r}_{D_i} = {}_V\boldsymbol{r}_{A_i,D} - {}_V\boldsymbol{r}_{T_i,D}, \tag{11.151}$$

as well as the velocity vector

$$\Delta^V_{S_i}\boldsymbol{v}_{D_i} = \left[\tilde{\boldsymbol{u}}_{S_i}\ {}^V\boldsymbol{T}_{S_i}\ {}^{S_i}_{A_i}\boldsymbol{r}_{T_i,D}\right]\dot{\varphi}_{S_i}, \tag{11.152}$$

with the velocity

$$\dot{l}_{D_i} = \operatorname{sign}(\dot{\varphi}_{S_i})\left\|\Delta^V\boldsymbol{v}_{D_i}\right\|_2. \tag{11.153}$$

Force laws For the spring forces, this results in:

$$\boldsymbol{F}_{F_i} = \underbrace{f_i(l_{F_i})}_{force-law} \underbrace{\frac{\Delta \boldsymbol{r}_{F_i}}{l_{F_i}}}_{force-direction} = f_i(l_{F_i})\boldsymbol{u}_{F_i}. \tag{11.154}$$

With the change in length $\Delta l_{F_i} = \left\|\Delta^V \boldsymbol{r}_{F_i}\right\|_2$, one finally arrives at the spring force

$$\boldsymbol{F}_{F_i} = c_{F_i}(l_{0F_i} - l_{F_i})\frac{\Delta \boldsymbol{r}_{F_i}}{l_{F_i}}, \tag{11.155}$$

and the damper force

$$^V\boldsymbol{F}_{D_i} = d_i \dot{l}_{D_i}\frac{\Delta \boldsymbol{r}_{D_i}}{l_{D_i}}. \tag{11.156}$$

11.2.8 Aerodynamic Forces

Now the aerodynamic forces as described in Chap. 9 are included

$$\boldsymbol{F}_W = -\frac{1}{2}c_W \varrho A \dot{\boldsymbol{r}}_V \|\dot{\boldsymbol{r}}_V\|_2 = -\frac{1}{2}c_W \varrho A \sqrt{v_x^2 + v_y^2}\begin{bmatrix} v_x \\ v_y \\ 0 \end{bmatrix}, \tag{11.157}$$

with the position vector ${}_V^V\boldsymbol{r}_W$ describing the imaginary point of action of the force.

11.2.9 Steering

In this chapter, only a simplified steering model is used which does not include any steering elasticities. First it shall be assumed that the steering angle, prescribed by the driver (here as a function of time), is applied after a constant transmission ratio i_L as a displacement s_H to the steering rack, Fig. 11.12:

$$\delta_H = i_L s_H. \tag{11.158}$$

The relationship between the displacement of the steering rack and the steering angle of the tires is generally nonlinear, i.e. the steering angles δ_1 and δ_3 at the front axle are nonlinear functions of the steering rack displacement and are hence indirectly also using Eq. (11.158) of the steering angle $\delta_H(t)$:

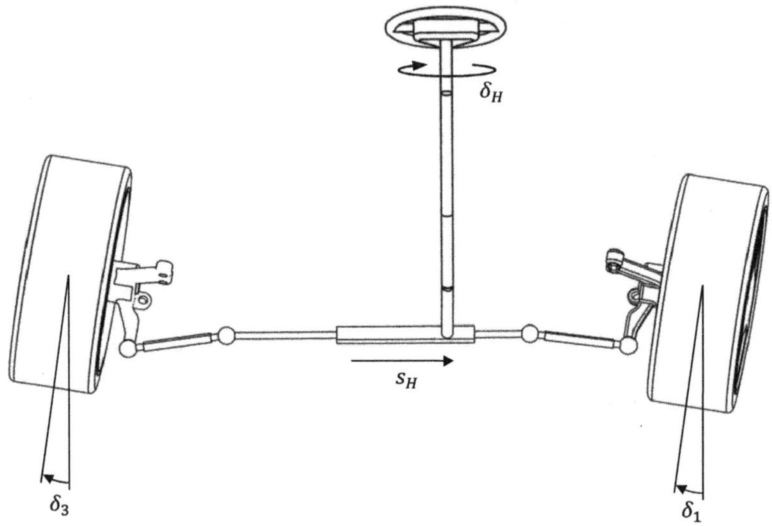

Fig. 11.12 Basic steering model

$$\delta_1 = \bar{\delta}_1(s_H) = \delta_1(\delta_H), \quad \delta_3 = \bar{\delta}_3(s_H) = \delta_3(\delta_H)\,. \tag{11.159}$$

The derivatives with respect to time of the steering angles at the tires are obtained using differentiation with respect to time (11.159):

$$\dot{\delta}_1 = \frac{\partial \bar{\delta}_1}{\partial s_H}\dot{s}_H = \frac{\partial \bar{\delta}_1}{\partial s_H}\frac{1}{i_L}\dot{\delta}_H \quad \ddot{\delta}_1 = \frac{\partial \bar{\delta}_1}{\partial s_H}\frac{1}{i_L}\ddot{\delta}_H + \frac{\partial^2 \bar{\delta}_1}{\partial^2 s_H}\frac{1}{i_l^2}\dot{\delta}_H^2. \tag{11.160}$$

Hereafter the steering angles δ_2 and δ_4 at the rear axle are assumed to be constant, i.e. the rear axle will not be steered.

11.2.10 Anti-roll Bar

For the forces resulting from the anti-roll bars, the following relationships hold according to Chap. 9:

$$\boldsymbol{F}_{St_3} = -\boldsymbol{F}_{St_1} = -\frac{l_{St,v}^2}{2 s_{St,v} b_{St,v}^2} c_{St,v}\varphi_V \boldsymbol{e}_z, \tag{11.161}$$

$$\boldsymbol{F}_{St_4} = -\boldsymbol{F}_{St_2} = -\frac{l_{St,h}^2}{2 s_{St,h} b_{St,h}^2} c_{St,h}\varphi_V \boldsymbol{e}_z. \tag{11.162}$$

Hence the torque

$$\begin{aligned} \boldsymbol{M}_{St} &= \left(l_{St,v}F_{St,v} + l_{St,h}F_{St,h}\right)\boldsymbol{e}_x \\ &= \left(\frac{l_{St,v}^2}{b_{St,v}^2}c_{St,v} + \frac{l_{St,h}^2}{b_{St,h}^2}c_{St,h}\right)\varphi_V\boldsymbol{e}_x \end{aligned} \tag{11.163}$$

acts on the chassis about the chassis fixed x-axis.

11.2.11 Applied Forces and Torques

To set up the equations of motion, the applied forces and torques on the chassis are still required. The reaction forces occurring between the chassis and the wheel carrier as a result of the wheel suspension need not be calculated, as they are cancelled out during the calculation of the Newton–Euler equations of motion through the application of Jourdains principle of virtual work, Chap. 4.

Applied forces and torques on the chassis

$$\boldsymbol{F}_V^e = \boldsymbol{F}_W + \sum_{i=1}^{4}(\boldsymbol{F}_{F_i} + \boldsymbol{F}_{D_i}) - m_V g\boldsymbol{e}_z, \tag{11.164}$$

$$\begin{aligned} \boldsymbol{T}_V^e &= {}_V\boldsymbol{r}_W \times \boldsymbol{F}_W \\ &+ \sum_{i=1}^{4}\left({}_V\boldsymbol{r}_{A_i,F} \times \boldsymbol{F}_{F_i} + {}_V\boldsymbol{r}_{A_i,D} \times \boldsymbol{F}_{D_i}\right) + \boldsymbol{M}_{St}. \end{aligned} \tag{11.165}$$

Applied forces and torques on the wheel carrier $(\boldsymbol{i} = 1, \ldots, 4)$

$$\boldsymbol{F}_{T_i}^e = -(\boldsymbol{F}_{F_i} + \boldsymbol{F}_{D_i} + \boldsymbol{F}_{St_i}) + m_{T_i} g\boldsymbol{e}_z, \tag{11.166}$$

$$\boldsymbol{T}_{T_i}^e = -\left({}_{T_i}\boldsymbol{r}_{F,T_i} \times \boldsymbol{F}_{F_i} + {}_{T_i}\boldsymbol{r}_{D,Ti} \times \boldsymbol{F}_{D_i}\right) + {}_{T_i}\boldsymbol{r}_{T_i,St} \times \boldsymbol{F}_{St_i}. \tag{11.167}$$

Forces and torques on the tires $(\boldsymbol{i} = 1, \ldots, 4)$

$$\boldsymbol{F}_{R_i}^e = \boldsymbol{F}_{R_i} - m_{R_i} g\boldsymbol{e}_z, \tag{11.168}$$

$$\boldsymbol{T}_{R_i}^e = {}_{R_i}\boldsymbol{r}_{R_i,A} \times \boldsymbol{F}_{R_i} + \left(M_{R_i,A} - M_{R_i,B}\mathrm{sign}(\dot{\rho}_{R_i})\right)\boldsymbol{u}_{R_i}. \tag{11.169}$$

The driving torques $M_{R_i,A}$ and the braking torque $M_{R_i,B}$ are again calculated according to Eqs. (11.42) and (11.45).

11.2.12 NEWTON's and EULER's Equations

With the preliminary work of the past sections, it is now possible to set up the Newton and Euler equations for all bodies inside the vehicle fixed coordinate system. The general form of the Newton equation according to Chap. 4 is:

$$m_i \boldsymbol{J}_{Tra_i}\dot{\boldsymbol{z}} + m_i \bar{\boldsymbol{a}}_i(\boldsymbol{y}, \boldsymbol{z}, t) = \boldsymbol{F}_i^e + \boldsymbol{F}_i^r, \qquad i = 1, \ldots, 9, \tag{11.170}$$

with the vector of applied forces $\boldsymbol{F}_i^e$ and the vector of reaction forces $\boldsymbol{F}_i^r$. Similarly, the Euler equations with respect to the center of gravity can be obtained as

$$\boldsymbol{\Theta}_i \boldsymbol{J}_{Rot_i}\dot{\boldsymbol{z}} + \boldsymbol{\Theta}_i \bar{\boldsymbol{\alpha}}_i(\boldsymbol{y}, \boldsymbol{z}, t) + \tilde{\boldsymbol{\omega}}_i(\boldsymbol{\Theta}_i \boldsymbol{\omega}_i) = \boldsymbol{T}_i^e + \boldsymbol{T}_i^r, \qquad i = 1, \ldots, 9, \tag{11.171}$$

with the vector of applied torques $\boldsymbol{T}_i^e$ and the vector of reaction torques $\boldsymbol{T}_i^r$. It is notable that all values in Eqs. (11.170) and (11.171) are absolute values, always with respect to the corresponding centers of gravity and represented in the chassis fixed coordinate system. This most notably means that each of the moment of inertia matrices need to be transformed into the chassis fixed coordinate system:

$${}^V\boldsymbol{\Theta}_{T_i} = {}^V\boldsymbol{T}_{T_i} \underbrace{{}^{T_i}\boldsymbol{\Theta}_{T_i}}_{const} {}^V\boldsymbol{T}_{T_i}^T, \qquad i = 1, \ldots, 4, \tag{11.172}$$

$${}^V\boldsymbol{\Theta}_{R_i} = {}^V\boldsymbol{T}_{R_i} \underbrace{{}^{R_i}\boldsymbol{\Theta}_{R_i}}_{const} {}^V\boldsymbol{T}_{R_i}^T, \qquad i = 1, \ldots, 4. \tag{11.173}$$

Form Eqs. (11.170) and (11.171), and multiplying with the transposed Jacobi matrices on the left side and finally summing all the bodies as given in Chap. 4 the equations of motion of the system result as:

$$\begin{aligned} &\underbrace{\sum_{i=1}^{9} \left[\boldsymbol{J}_{Tra_i}^T m_i \boldsymbol{J}_{Tra_i} + \boldsymbol{J}_{Rot_i}^T \, {}^V\boldsymbol{\Theta}_i \boldsymbol{J}_{Roti} \right]}_{massmatrix\, \boldsymbol{M}(\boldsymbol{q},t)} \dot{\boldsymbol{z}} \\ &+ \underbrace{\sum_{i=1}^{9} m_i \boldsymbol{J}_{Tra_i}^T \bar{\boldsymbol{a}}_i + \boldsymbol{J}_{Rot_i}^T \, {}^V\boldsymbol{\Theta}_i \bar{\boldsymbol{\alpha}}_i + \boldsymbol{J}_{Rot_i}^T \boldsymbol{\omega}_i \times \left({}^V\boldsymbol{\Theta}_i \boldsymbol{\omega}_i \times \boldsymbol{\omega}_i \right)}_{\text{centrifugal, gyro and coriolis forces } \boldsymbol{b}(\boldsymbol{z},\boldsymbol{q},t)} \\ &= \underbrace{\sum_{i=1}^{9} \boldsymbol{J}_{Tra_i}^T \boldsymbol{F}_i^e + \boldsymbol{J}_{Rot_i}^T \boldsymbol{T}_i^e}_{generalized\, forces\, \boldsymbol{Q}(\boldsymbol{z},\boldsymbol{q},t)} + \underbrace{\sum_{i=1}^{9} \boldsymbol{J}_{Tra_i}^T \boldsymbol{F}_i^r + \boldsymbol{J}_{Rot_i}^T \boldsymbol{T}_i^r}_{\boldsymbol{0}} \end{aligned} \tag{11.174}$$

The equations of motion consist of three parts as defined in Chap. 4:

- the $[f \times f]$-mass matrix $\boldsymbol{M}(\boldsymbol{q}, t)$,
- the $[f \times 1]$-vector of the generalized centrifugal-, gyro- and coriolis- forces $\boldsymbol{b}(z, \boldsymbol{q}, t)$,
- and the $[f \times 1]$-Vector of the generalized forces $\boldsymbol{Q}(z, \boldsymbol{q}, t)$.

The reaction forces cancel each other out due to the principle of Jourdain.

Mass matrix For a simple and clear representation, the symmetric $[14 \times 14]$-mass matrix $\boldsymbol{M}(\boldsymbol{q}, t)$ will be calculated separately for the individual component groups:

$$\boldsymbol{M}(\boldsymbol{q},t) = \begin{bmatrix} \boldsymbol{M}_{VV} & \boldsymbol{M}_{V\omega} & \boldsymbol{M}_{VT} & \boldsymbol{M}_{VR} \\ & \boldsymbol{M}_{\omega\omega} & \boldsymbol{M}_{\omega T} & \boldsymbol{M}_{\omega R} \\ & & \boldsymbol{M}_{TT} & \boldsymbol{M}_{TR} \\ & & & \boldsymbol{M}_{RR} \end{bmatrix}, \tag{11.175}$$

with

$$\boldsymbol{M}_{VV} = \boldsymbol{I}\left(m_v + \sum_{i=1}^{4} m_{T_i} + m_{R_i}\right), \tag{11.176}$$

$$\boldsymbol{M}_{V\omega} = -\sum_{i=1}^{4}({}_V\tilde{\boldsymbol{r}}_{T_i} m_{T_i} + {}_V\tilde{\boldsymbol{r}}_{R_i} m_{R_i}), \tag{11.177}$$

$$\boldsymbol{M}_{VT} = \sum_{i=1}^{4} \boldsymbol{H}_{Tra_c,T_i} m_{T_i} + \boldsymbol{H}_{Tra_c,R_i} m_{R_i}, \tag{11.178}$$

$$\boldsymbol{M}_{VR} = 0, \tag{11.179}$$

$$\begin{aligned} \boldsymbol{M}_{\omega\omega} &= {}^V\boldsymbol{\Theta}_V \\ &+ \sum_{i=1}^{4}\left({}^V\boldsymbol{\Theta}_{T_i} + {}^V\boldsymbol{\Theta}_{R_i} + {}_V\tilde{\boldsymbol{r}}_{T_i}^T\, {}_V\tilde{\boldsymbol{r}}_{T_i} m_{T_i} + {}_V\tilde{\boldsymbol{r}}_{R_i}^T\, {}_V\tilde{\boldsymbol{r}}_{R_i} m_{R_i}\right), \end{aligned} \tag{11.180}$$

$$\boldsymbol{M}_{\omega T} = \sum_{i=1}^{4}\left(-\, {}_V\tilde{\boldsymbol{r}}_{T_i}^T \boldsymbol{H}_{Tra_c,T_i} m_{T_i} - {}_V\tilde{\boldsymbol{r}}_{R_i}^T \boldsymbol{H}_{Tra_c,R_i} m_{R_i} + \left({}^V\boldsymbol{\Theta}_{T_i} + {}^V\boldsymbol{\Theta}_{R_i}\right)\boldsymbol{H}_{Rot_c,T_i}\right), \tag{11.181}$$

$$\boldsymbol{M}_{\omega R} = \sum_{i=1}^{4} {}^V\boldsymbol{\Theta}_{R_i} \boldsymbol{H}_{Rot_p,R_i}, \tag{11.182}$$

$$\boldsymbol{M}_{TT} = \sum_{i=1}^{4} \left(\boldsymbol{H}^{T}_{Tra_c,T_i} \boldsymbol{H}_{tra_c,T_i} m_{T_i} + \boldsymbol{H}^{T}_{Tra_c,R_i} \boldsymbol{H}_{Tra_c,R_i} m_{R_i} + \boldsymbol{H}^{T}_{Rot_c,T_i} ({}^{V}\boldsymbol{\Theta}_{T_i} + {}^{V}\boldsymbol{\Theta}_{R_i}) \boldsymbol{H}_{Rot_c,T_i} \right), \tag{11.183}$$

$$\boldsymbol{M}_{TR} = \sum_{i=1}^{4} \boldsymbol{H}^{T}_{Rot_c,T_i} \; {}^{V}\boldsymbol{\theta}_{R_i} \boldsymbol{H}_{Rot_\rho,R_i}, \tag{11.184}$$

$$\boldsymbol{M}_{RR} = \sum_{i=1}^{4} \boldsymbol{H}^{T}_{Rot_\rho,R_i} \; {}^{V}\boldsymbol{\Theta}_{R_i} \boldsymbol{H}_{Rot_\rho,R_i}. \tag{11.185}$$

It is notable that the matrix product

$$\boldsymbol{H}^{T}_{Tra_c,T_i} \boldsymbol{H}_{Tra_c,T_i} \tag{11.186}$$

deals with diagonal matrices. Considering the Eqs. (11.176) to (11.186), the mass matrix with the following structure is obtained:

$$\begin{bmatrix}
\blacksquare & 0 & 0 & 0 & \blacksquare & \blacksquare & \blacksquare & \blacksquare & \blacksquare & \blacksquare & 0 & 0 & 0 & 0 \\
 & \blacksquare & 0 & \blacksquare & 0 & \blacksquare & \blacksquare & \blacksquare & \blacksquare & \blacksquare & 0 & 0 & 0 & 0 \\
 & & \blacksquare & \blacksquare & \blacksquare & 0 & \blacksquare & \blacksquare & \blacksquare & \blacksquare & 0 & 0 & 0 & 0 \\
 & & & \blacksquare & \blacksquare & \blacksquare & \blacksquare & \blacksquare & \blacksquare & \blacksquare & \blacksquare & \blacksquare & \blacksquare & \blacksquare \\
 & & & & \blacksquare & \blacksquare & \blacksquare & \blacksquare & \blacksquare & \blacksquare & \blacksquare & \blacksquare & \blacksquare & \blacksquare \\
 & & & & & \blacksquare & \blacksquare & \blacksquare & \blacksquare & \blacksquare & \blacksquare & \blacksquare & \blacksquare & \blacksquare \\
 & & & & & & \blacksquare & 0 & 0 & 0 & \blacksquare & 0 & 0 & 0 \\
 & & & & & & & \blacksquare & 0 & 0 & 0 & \blacksquare & 0 & 0 \\
 & & & & & & & & \blacksquare & 0 & 0 & 0 & \blacksquare & 0 \\
 & & & & & & & & & \blacksquare & 0 & 0 & 0 & \blacksquare \\
 & & & & & & & & & & \blacksquare & 0 & 0 & 0 \\
 & & & & & & & & & & & \blacksquare & 0 & 0 \\
 & & & & & & & & & & & & \blacksquare & 0 \\
 & & & & & & & & & & & & & \blacksquare
\end{bmatrix}_{[14\mathrm{x}14]}. \tag{11.187}$$

Vector of generalized centrifugal, gyroscopic and coriolis forces

$$
\begin{aligned}
\boldsymbol{b}(\boldsymbol{q},z,t) &= \sum_{i=1}^{9} m_i \boldsymbol{J}_{Tra_i}^T \bar{\boldsymbol{a}}_i + \boldsymbol{J}_{Rot_i}^T \left({}^V\boldsymbol{\Theta}_i \bar{\boldsymbol{\alpha}}_i + \tilde{\boldsymbol{\omega}}_i (\boldsymbol{\Theta}_i \boldsymbol{\omega}_i) \right) \\
&= \begin{bmatrix} m_V \tilde{\boldsymbol{\omega}}_V \boldsymbol{v}_V \\ \tilde{\boldsymbol{\omega}}_V ({}^V\boldsymbol{\Theta}_V \boldsymbol{\omega}_V) \\ \boldsymbol{0} \\ \boldsymbol{0} \end{bmatrix} + \sum_{i=1}^{4} \begin{bmatrix} m_{T_i} \bar{\boldsymbol{a}}_{T_i} + m_{R_i} \bar{\boldsymbol{a}}_{R_i} \\ -(m_{T_i}\, {}_V\tilde{\boldsymbol{r}}_{T_i}^T \bar{\boldsymbol{a}}_{T_i} + m_{R_i}\, {}_V\tilde{\boldsymbol{r}}_{R_i}^T \bar{\boldsymbol{a}}_{R_i}) \\ m_{T_i} \boldsymbol{H}_{Tra_c,T_i}^T \bar{\boldsymbol{a}}_{T_i} + m_{R_i} \boldsymbol{H}_{Tra_c,R_i}^T \bar{\boldsymbol{a}}_{R_i} \\ \boldsymbol{0} \end{bmatrix} \\
&+ \begin{bmatrix} \boldsymbol{0} \\ {}^V\boldsymbol{\Theta}_{T_i} \bar{\boldsymbol{\alpha}}_{T_i} + {}^V\boldsymbol{\Theta}_{R_i} \bar{\boldsymbol{\alpha}}_{R_i} \\ \boldsymbol{H}_{Rot_c,T_i}^T\, {}^V\boldsymbol{\Theta}_{T_i} \bar{\boldsymbol{\alpha}}_{T_i} + \boldsymbol{H}_{Rot_c,T_i}^T\, {}^V\boldsymbol{\Theta}_{R_i} \bar{\boldsymbol{\alpha}}_{R_i} \\ \boldsymbol{H}_{Rot_\rho,T_i}^T\, {}^V\boldsymbol{\Theta}_{R_i} \bar{\boldsymbol{\alpha}}_{R_i} \end{bmatrix} \\
&+ \begin{bmatrix} \boldsymbol{0} \\ \tilde{\boldsymbol{\omega}}_{T_i} ({}^V\boldsymbol{\Theta}_{T_i} \boldsymbol{\omega}_{T_i}) + \tilde{\boldsymbol{\omega}}_{R_i} ({}^V\boldsymbol{\Theta}_{R_i} \boldsymbol{\omega}_{R_i}) \\ \boldsymbol{H}_{Rot_c,T_i}^T \tilde{\boldsymbol{\omega}}_{T_i} ({}^V\boldsymbol{\Theta}_{T_i} \boldsymbol{\omega}_{T_i}) + \boldsymbol{H}_{Rot_c,T_i}^T \tilde{\boldsymbol{\omega}}_{R_i} ({}^V\boldsymbol{\Theta}_{R_i} \boldsymbol{\omega}_{R_i}) \\ \boldsymbol{H}_{Rot_\rho,R_i}^T \tilde{\boldsymbol{\omega}}_{R_i} ({}^V\boldsymbol{\Theta}_{R_i} \boldsymbol{\omega}_{R_i}) \end{bmatrix}
\end{aligned}
\tag{11.188}
$$

Generalized forces

$$
\begin{aligned}
\boldsymbol{Q}(\boldsymbol{q},z,t) &= \sum_{i=1}^{9} \boldsymbol{J}_{Tra_i}^T \boldsymbol{F}_i^e + \boldsymbol{J}_{Rot_i}^T \boldsymbol{T}_i^e \\
&= \begin{bmatrix} \boldsymbol{F}_V^e \\ \boldsymbol{T}_V^e \\ \boldsymbol{0}_{[4\times1]} \\ \boldsymbol{0}_{[4\times1]} \end{bmatrix} + \sum_{i=1}^{4} \begin{bmatrix} \boldsymbol{F}_{T_i}^e \\ -{}_V\tilde{\boldsymbol{r}}_{T_i}^T \boldsymbol{F}_{T_i}^e \\ \boldsymbol{H}_{Tra_c,T_i}^T \boldsymbol{F}_{T_i}^e \\ \boldsymbol{0}_{[4\times1]} \end{bmatrix} + \begin{bmatrix} \boldsymbol{0}_{[3\times1]} \\ \boldsymbol{T}_{T_i}^e \\ \boldsymbol{H}_{Rot_c,T_i}^T \boldsymbol{T}_{T_i}^e \\ \boldsymbol{0}_{[4\times1]} \end{bmatrix} \\
&+ \sum_{i=1}^{4} \begin{bmatrix} \boldsymbol{F}_{R_i}^e \\ -\,{}_V\tilde{\boldsymbol{r}}_{R_i}^T \boldsymbol{F}_{R_i}^e \\ \boldsymbol{H}_{Tra_c,R_i}^T \boldsymbol{F}_{R_i}^e \\ \boldsymbol{0}_{[4\times1]} \end{bmatrix} + \begin{bmatrix} \boldsymbol{0}_{[3\times1]} \\ \boldsymbol{T}_{R_i}^e \\ \boldsymbol{H}_{Rot_c,T_i}^T \boldsymbol{T}_{R_i}^e \\ \boldsymbol{H}_{Rot_\rho,R_i}^T \boldsymbol{T}_{R_i}^e \end{bmatrix}.
\end{aligned}
\tag{11.189}
$$

11.2.13 Motion and State Space Equations

With the [28 × 1]–state vector

$$x = \begin{bmatrix} q \\ z \end{bmatrix} \tag{11.190}$$

and the relationship

$$\dot{q} = K^{-1}(q,t)z = \begin{bmatrix} {}^{V}T_E^T & 0 & 0 & 0 \\ 0 & T_\omega^{-1} & 0 & 0 \\ 0 & 0 & I_{4\times4} & 0 \\ 0 & 0 & 0 & I_{4\times4} \end{bmatrix} z, \tag{11.191}$$

$$\dot{z} = M^{-1}(q,t)(Q(z,q,t) - b(z,q,t)) \tag{11.192}$$

the final state space representation of the system results as follows:

$$\dot{x} = \begin{bmatrix} \dot{q} \\ \dot{z} \end{bmatrix} = \begin{bmatrix} K^{-1}(q,t)z \\ M^{-1}(q,t)(Q(z,q,t) - b(z,q,t)) \end{bmatrix}. \tag{11.193}$$

11.3 Simplified Driver Model

The focus of this chapter is placed on the description of simulation models for vehicles. However, in order to actually carry out complex driving maneuvers, such as for example lane change, slalom etc. using the simulation models created, models for vehicle guidance or driver models are also required. Hence, a basic driver model is to be discussed in this section. For a more complex treatment of this subject, the reader is requested to refer to the literature available in this field. Here, a concept for a single driver model based on a linear controller is described. This model was initially developed by Adamski (2001) and is implemented in the vehicle simulation environment FASIM C++ (Opgen-Rhein 1998).

11.3.1 Controller Concept

For FASIM C++, Ref. Chap. 12, a basic driver model was developed which contains a course controller and a velocity controller. The model can not only control the vehicle velocity but also follow a given trajectory (Opgen-Rhein 1998). For simplicity, the velocity controller is realized as a PID controller (Fig. 11.13).

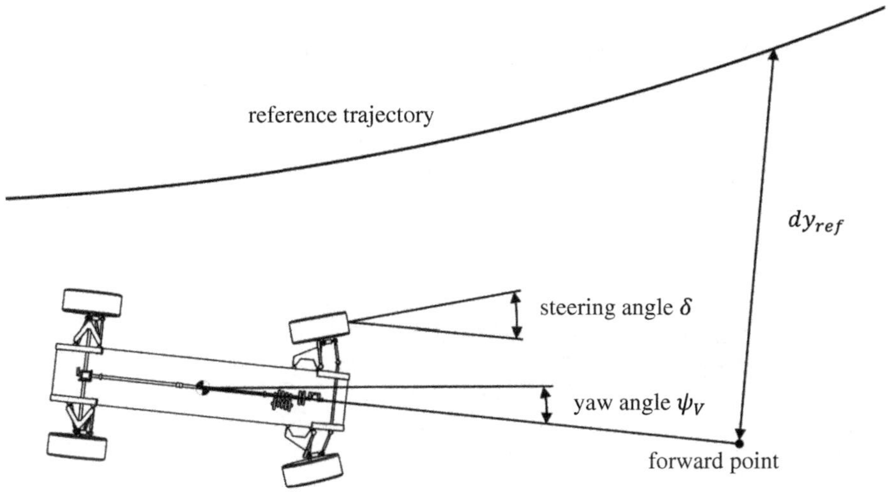

Fig. 11.13 Carriage shaft model

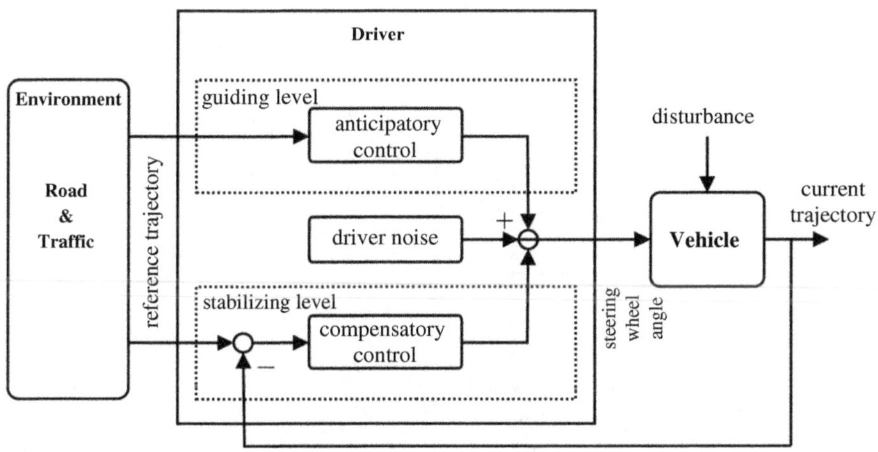

Fig. 11.14 Course control (Donges and Naab 1996)

For the steering controller, a linear predictive model is used as foundation. In this model it is assumed that the driver steers in such a way that an imaginary point at a predefined forward point coincides with the view point. Physically speaking, this means that the driver tries to minimize the lateral deviation. The model consists of an anticipatory control and a compensatory control, Fig. 11.14. The compensatory control corresponds with the experience of the driver, while the anticipatory control minimizes the deviation from the reference trajectory. The reaction time of the driver can be considered via a delay element.

As represented in Fig. 11.15, the input interface for course control of the driver model consists of the yaw angle and the position of the vehicle, as well as the reference course and the reference velocity. The output interface is defined by the steering angle and the throttle flap position.

The steering angle consists of two control components, Ref. Fig. 11.16. The forward control determines the difference between the vehicle orientation (yaw angle) and the tangent that results from the predefined reference course

$$\delta = \arctan dy_{ref} - \psi_V, \tag{11.194}$$

with the steering angle (forward control component) δ, the tangent of the reference trajectory of the vehicle position and the yaw angle ψ_V. The feedback control part is a second order controller, which minimizes the lateral deflection of the forward point from the corresponding reference trajectory point. The controller parameter identification is defined in Opgen-Rhein (1998).

Fig. 11.15 Interfaces of the driver controller

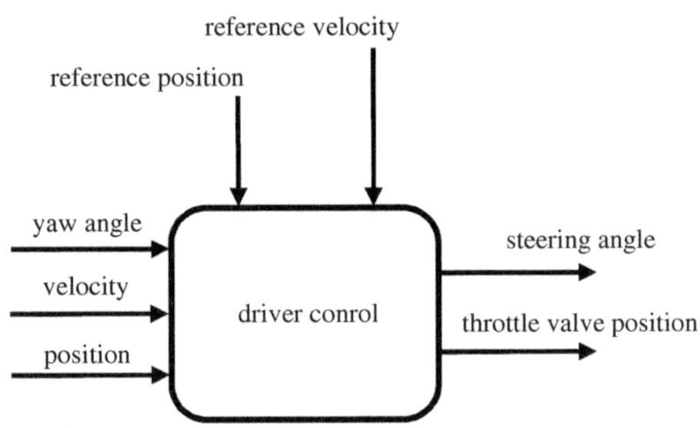

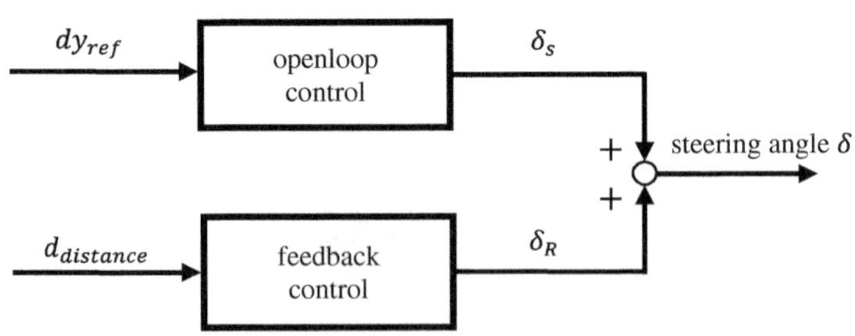

Fig. 11.16 Construction of a steering controller

11.4 Parameterization

Table 11.8 provides a few exemplary values for the described vehicle model. For more data sets an investigations on dynamic behavior of different types of vehicles see Unterreiner (2013).

Table 11.8 Example values

Data for the single track average model		
m—vehicle mass	1,506.5	kg
θ_{zz}—moment of inertia about the vertical axis w.r.t. S	2,585.6	kg m^2
l—wheel base	2.6897	m
l_v—distance between center of gravity—front axle	1.1853	m
l_h—distance between center of gravity—rear axle	1.5044	m
s_z—height of the center of gravity	0.54475	m
i_L—steering transmission ratio	16.3	
c_W—value	n.a.	
A—frontal area	2.14	m^2
r_{dyn}—(dynamic) tire radius	0.31075	m
J_r—moment of inertia of tire per axle	1.7	kg m^2
Date for twin track model of an average model		
m—vehicle mass	1,506.5	kg
θ_{xx}—moment of inertia about the longitud axis w.r.t. S	482.7	kg m^2
θ_{yy}—moment of inertia about the lateral axis w.r.t. S	2,394.3	kg m^2
θ_{zz}—moment of inertia about the vertical axis w.r.t. S	2,585.6	kg m^2
l—wheel base	2.6897	m
l_v—distance between center of gravity—front axle	1.1853	m
l_h—distance between center of gravity—rear axle	1.5044	m
$2S_{R_1}, v$—track width front	1,492.2	mm
$2S_{R_1}, h$—track width rear	1,487.4	mm
h_s—height of the center of gravity	0.54475	m
i_L—steering transmission ratio	16.3	–
c_w—value	0.33	–
A—frontal area	2.14	m^2
r—(dynamic) tire radius	0.31075	m
θ_r—moment of inertia of tire per axis	1.7	kg m^2
c_{f,v_i}—spring stiffness front	34,900	N/m
c_{f,h_i}—spring stiffness rear	32,600	N/m
d_{v_i}—damper front	4,360	Ns/m
d_{h_i}—damper rear	3,870	Ns/m
c_{s_i}—anti-roll bar stiffness front/rear	2,882	Nm/rad

References

Adamski D (2001) Komponentenbasierte Simulation mechatronischer Systeme [Dr.-Ing.]. Dissertation, Universität Duisburg-Essen, Düsseldorf

Donges E and Naab K (1996) Regelsysteme zur Fahrzeugführung und-stabilisierung in der Automobiltechnik. Automatisierungstechnik 44. S 226–236

Opgen-Rhein P (1998) Modellierung eines einfachen Fahrermodells für die Fahrdynamiksimulationsumgebung FASIM C++ [Studienarbeit]. Gerhard-Mercator-Universität Duisburg, Duisburg

Unterreiner M (2013) Modellbildung und Simulation von Fahrzeugmodellen unterschiedlicher Komplexität, Duisburg-Essen, Duisburg

Chapter 12
Three-Dimensional Complete Vehicle Models

To state the equations of motion for a complete road vehicle, or in particular a passenger car, in a first step the mechanical setup of the real vehicle has to be replaced by a mechanical substitute system, consisting of rigid bodies, ideal joints, and external forces (including its force transfer direction). Here, this mechanical system will be analyzed, using the method of multibody systems (MBS). The basic ideas of this approach will be explained, and in particular, this modeling technique will be using the ideas presented in Chaps. 3 and 4, as well as in Chaps. 5, 6, 7, 8 and 9. The method contains particular ideas for the analysis of vehicle components. In this context, a detailed description of the nonlinear kinematics of wheel suspensions plays a major role, since these effects are responsible for a strong influence on the dynamic behavior of the complete vehicle (see in particular Chap. 6).

12.1 Modeling of the Complete Vehicle

An important aspect, which will be analyzed in more detail in Sect. 12.2, is related to the modularization of the modeling process of the complete vehicle. In particular, this means that mechanical, as well as non-mechanical subsystems and components —like wheel suspensions, drivetrains, wheels, and tires, but also electronic elements, controllers, or ECUs (Electronic Control Units)—have to be modeled and prepared, so that they can be put or "clicked" together into the overall mechatronic complete vehicle model in a simple and straightforward manner.

In the subsequent sections, the mechanical modeling will be discussed in more detail for different vehicle types with different drivetrain concepts, like rear-wheel, front-wheel, and four-wheel drive, starting with a rear-wheel drivetrain of a BMW 5-series (E60), Fig. 12.1 (see also Chap. 8). In this modeling

D. Schramm et al., Vehicle Dynamics,
DOI 10.1007/978-3-662-54483-9_12

process kinematics play a key role, as shown in Chap. 3. The particular equations of the global kinematics, as required later for the derivation of the dynamic equations (see Chap. 4), are put together by combining the equations of relative kinematics with those of the equations of absolute kinematics. While the equations of the absolute kinematics (as shown earlier in Sect. 3.5) can always be stated in a fully recursive manner, the situation for the relative kinematics is much more complicated due to the arising closed kinematic loops within the system. Also, one has to give special attention to a suitable choice and description of the kinematic loops, arising in particular in the wheel suspensions. It is the goal to find a mathematical structure of the appertaining relative kinematics equations, which is as explicit as possible.

12.1.1 Kinematics of a Rear-Wheel Driven Complete Vehicle Model

The chassis or vehicle body respectively will be modeled as a rigid body with mass, and as having a full tensor of inertia (Chap. 5). The front-wheel suspensions, implemented as MacPherson front suspension struts with dissolved lower wishbone, as well as the rear-wheel suspensions, implemented as Integral–IV rear suspensions, will be represented by systems of rigid bodies, interconnected by joints (constraints!), and force elements, respectively. The wheels are pivoted to the wheel carriers. The modeling of the complete vehicle is closely connected to the real vehicle (see Fig. 12.1). The representation of the complete vehicle as a complex multibody system is shown in Fig. 12.2.

MacPherson front suspension strut with dissolved lower wishbone Starting with the representation of the real system as shown in Fig. 12.3, the corresponding multibody system will be derived in a further step. Each of the two wheel carriers—also modeled as rigid bodies with mass and a full inertia tensor—is rigidly connected to the lower half of the shock absorber tube. The piston rod of the damper is sliding in the wheel carrier fixed tube of the damper and it is simultaneously connected to the car body by a spherical joint. The supporting spring is located between car body and outer part of the damper guide. The connecting joint between the wheel carrier, the lateral suspension arm, and the push rod, respectively is also a spherical joint (Figs. 12.3 and 12.4). Additional details can be seen in Fig. 12.4. The steering motion of the wheel carrier is initiated via steering rod and drag link, respectively, and its instantaneous steering axis is given by the connection of the spherical joint (located on the vehicle side at the spring strut) and the spherical joints of lateral rod and push rod, respectively (located at the wheel carriers). The spherical joints in this design have to be in alignment (Fig. 12.4) (see also Frik

Fig. 12.1 BMW 5-series—real vehicle (by courtesy of BMW AG 2003)

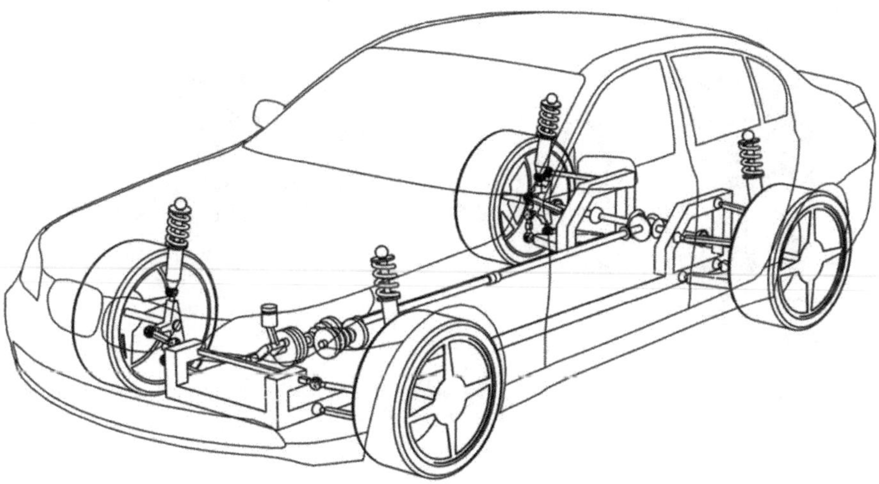

Fig. 12.2 BMW 5-series as a complex multibody system including car body

1994). The steering mechanism of the front axle of the BMW 5-series is modeled by a rack, which can move with respect to the car body, and is connected to the drag links by spherical joints (Fig. 12.3). Even though the rack is a single rigid body, it will be separated into two halves for modeling reasons in the left and right wheel suspension. The connection to the steering wheel, which is not modeled here, is realized via steering gear and steering shaft (see Figs. 12.3 and 12.4). The given

Fig. 12.3 BMW 5-series—MacPherson front suspension strut with dissolved lower wishbone—with, and without steering (by courtesy of BMW AG 2003)

time-history of the steering wheel deflection is defining a "set point" value of the left half of the rack. The extremities of the drag links are realized as spherical joints, which enable an isolated and free rotation about the connecting line of the two ball centers, and which are of no further interest in the modeling process.

In the modeling process, the following parts are elements or subsystems with mass: the two drag links, the steering mechanism, and the lateral suspension arm. Furthermore, the longitudinal compliance in the front axle will not be considered in the modeling (see also Sect. 12.1.2). In a further step, the closed kinematic loops (including their coupling), which typically occur in wheel suspensions, will be

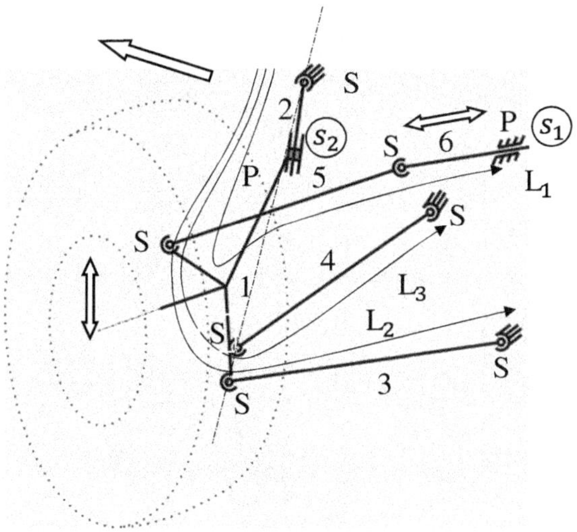

Total number of DoF

$$f = 6n_B - \sum_{i=1}^{n_G}(6 - f_{G_i})$$
$$= 6 \cdot 6 - \underbrace{7 \cdot (6-3)}_{\text{7 spherical joints}} - \underbrace{2 \cdot (6-1)}_{\text{2 prismatic joints}}$$
$$= 36 - 21 - 10$$
$$= \underbrace{1}_{\text{spring deflection}} + \underbrace{1}_{\text{steering}} + \underbrace{3}_{\text{isolated d.o.f.}}$$

Wheel carrier kinematic loops

$$n_L = n_G - n_B = 9 - 6 = 3$$

1. wheel carrier
2. piston rod
3. strut
4. transverse link
5. tie rod
6. steering rack

Fig. 12.4 BMW 5-series–MacPherson front suspension strut with dissolved lower wishbone (front *left*)–structure

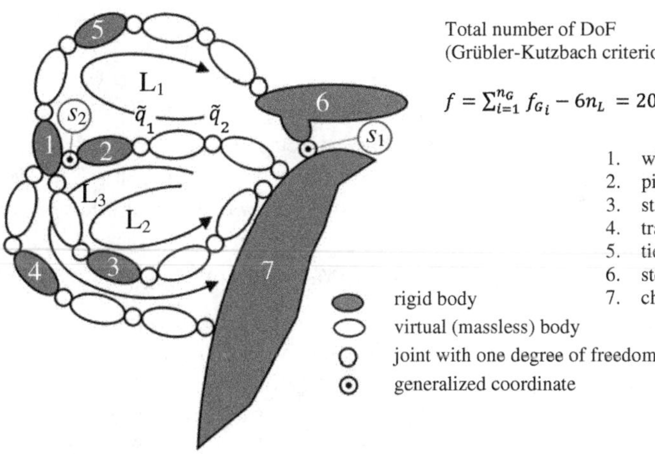

Total number of DoF
(Grübler-Kutzbach criterion)

$$f = \sum_{i=1}^{n_G} f_{G_i} - 6n_L = 20 - 6 \cdot 3 = 2$$

1. wheel carrier
2. piston rod
3. strut
4. transverse link
5. tie rod
6. steering rack
7. chassis

Fig. 12.5 BMW 5-series—MacPherson front suspension strut with dissolved lower wishbone (front *left*)—topology

regarded. This analysis is based on the mathematical elements introduced in Chap. 3. In particular, the symbolism contained in Sect. 3.5 is very helpful, and one achieves the topological representation of the MacPherson front suspension strut with dissolved lower wishbone as shown in Fig. 12.5.

In a subsequent modeling step–again based on the fundamentals from Chap. 3, combined with its adapted application to the analysis of wheel suspensions Chap. 6–

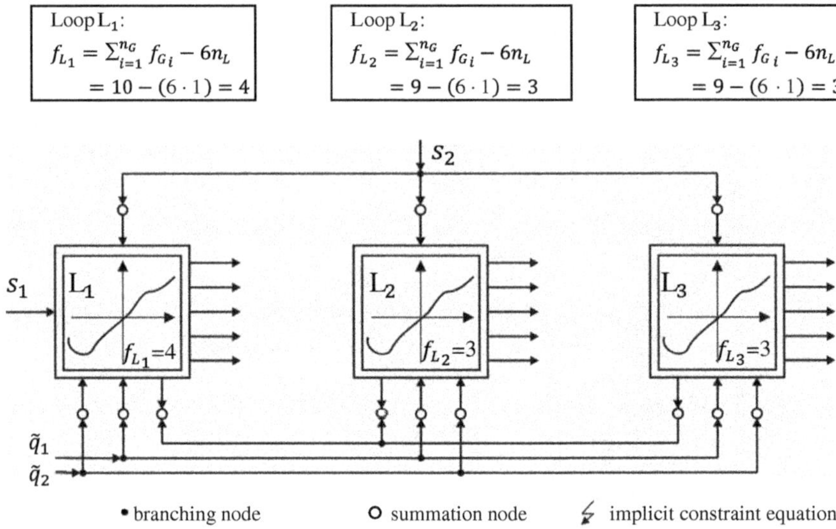

Fig. 12.6 BMW 5-series—MacPherson front suspension strut with dissolved lower wishbone (front *left*)—block diagram and solution flow

one obtains the block diagram for the solution flow of the relative kinematics of the wheel suspension, as shown in Fig. 12.6.

Now, the solution flow of the relative kinematics of this front suspension is no longer fully recursive and explicit. As an alternative, in the block diagram of Fig. 12.6, at the positions marked by the symbol "↯", two implicit algebraic equations have to locally be solved numerically. In order to create these equations, one has to introduce two so-called pseudo-inputs $\tilde{q}_1$ and $\tilde{q}_2$ (Figs. 12.5 and 12.6). For more details, see (Hiller et al. 1986–1988; Hiller 1995), as well as Chap. 6.

Integral IV–Rear suspension The two rear-wheels of the BMW 5-series are suspended and guided individually and separately by two integral IV-wheel suspensions, which are combined to an integral IV-rear axle (Fig. 12.7).

The integral IV-wheel suspension belongs to the group of the so-called multilink wheel suspensions. Its kinematic topology is as follows: The wheel carrier is connected to the vehicle body by means of a lever, a guiding rod, a transverse link, and the spring strut–including the piston rod (which does not affect the kinematics!). A–from a kinematic point of view–particular element is the additional supporting integral rod. Due to the five geometric constraints in this wheel suspension, one degree of freedom remains for the suspension mode (coordinate α) of this wheel suspension. The calculation can be carried out by means of the GRÜBLER criterion (Eq. 3.3, Fig. 12.8). The kinematic structure of the multibody system of this wheel suspension is given by Fig. 12.8, (Pichler 1999).

Note: The longitudinal compliance in horizontal direction of the system is not taken into account (see also Sect. 12.1.2)!

Fig. 12.7 BMW 5-series—integral IV-rear axle (by courtesy of BMWAG 2003)

Analogously to the procedure applied to the kinematics of the front axle, the topology and block diagram, including solution flow, can be created (Figs. 12.9 and 12.10). In case of the integral IV-rear suspension, a fully recursive solution of the relative kinematics of the system is no longer possible. Introducing the two pseudo-inputs $\tilde{q}_1$ and $\tilde{q}_2$ (Fig. 12.9), one obtains the results, as shown in Fig. 12.10 (see also Chap. 6).

The DoF of the complete vehicle may be assigned as follows (Table 12.1): the car body has six DoF with respect to inertia. Each front wheel has one suspending degree of freedom, which is completed by a common degree of freedom for the steering motion. The rear-wheels combined have two DoF for the suspension. The non-powered front wheels, as well as the powered rear-wheels, are combined with

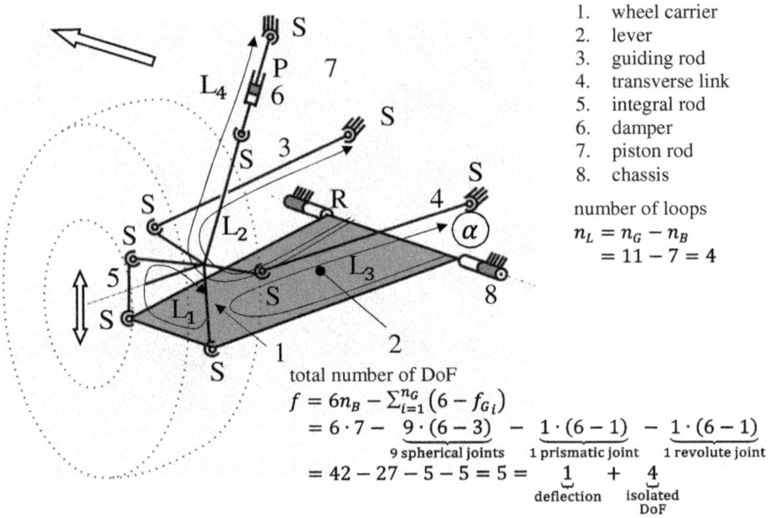

Fig. 12.8 BMW 5-series–integral IV-rear suspension–rear *left* (by courtesy of BMW AG 2003)–structure

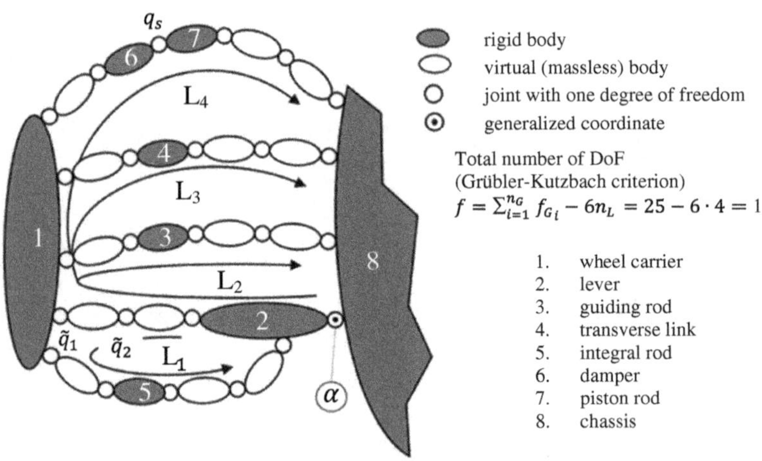

Fig. 12.9 BMW 5-series–integral IV-rear suspension (by courtesy of BMW AG 2003)–topology

two additional DoF in the drivetrain (relative rotation of the prop shaft $\Delta\varphi_H$, and the relative rotation of the clutch $\Delta\varphi_K$, respectively) thus creating the subsystem "drivetrain" (Schmitz 1994). It needs to be mentioned that there are slight differences with respect to the real vehicle: the engine bearing has two DoF—one for the lift and one for the roll motion (variables s_M and φ_M) of the engine with respect to the vehicle body (see also Chap. 8). The DoF for the complete vehicle, combined with their assignment to the vehicle components are collected in Table 12.1.

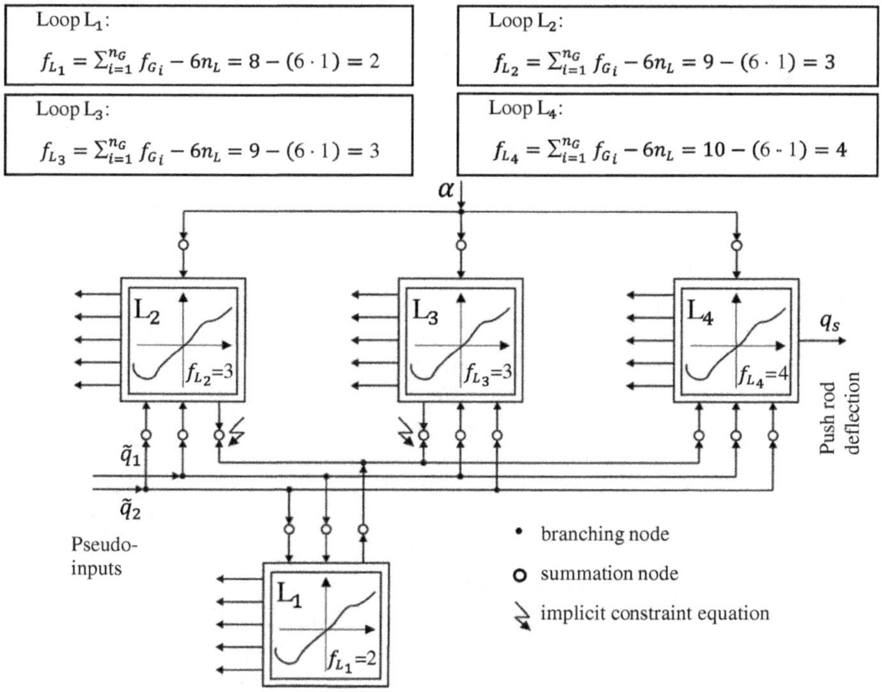

Fig. 12.10 BMW 5-series–integral IV-rear suspension (by courtesy of BMW AG 2003)–block diagram and solution flow

Table 12.1 Kinematic structure of the BMW 5-series (E60)

Vehicle component	DoF	Body	Loops
Chassis	6	1	0
Front-wheel suspension	$2 \times 1 + 1$	11	2×3
Rear-wheel suspension	2×1	14	2×4
Engine bracket	2	1	0
Drivetrain (including all wheel rotations)	6	12	3
Σ	19	41	17

For further analysis of this wheel suspension, one can allocate one generalized coordinate (representation of the complete system in minimum coordinates) to each degree of freedom presented in Table 12.1, to describe position, velocity, and acceleration of the complete system. In the next step, the topology of the complete vehicle will be derived, but the particular element of the sub-frame bearing will not be considered in this context. The six DoF of the vehicle body with respect to inertia can be interpreted as an open kinematic chain, located between car body and inertia, and consisting of three prismatic joints with one degree of freedom each for the translational DoF, and three revolute joints with one degree of freedom each for the rotational DoF of the vehicle. The front and rear-wheel suspensions are hinged with respect to the car body. The connection of each rear-wheel carrier to the chassis is realized in this vehicle by the kinematic structure of the Integral IV-suspension presented above. In this presentation however it is assumed that in the connecting rods, which are hinged by two spherical joints (three DoF) to the car body, one of these spherical joints can be replaced by a universal joint with two DoF to avoid a so-called isolated degree of freedom (Chap. 3).

The scheme of the topology for the complete vehicle is shown in Fig. 12.11, which can be read as follows: Every joint in the system with more than one degree of freedom will be modeled by a group of monovalent joints with one degree of freedom each, represented by small circles in Fig. 12.11. The corresponding joint coordinates are either angles or linear displacements.

Examples: A spherical joint with three DoF is represented by three revolute joints with intersecting axes, and two massless bodies in between; a universal joint with two DoF by two revolute joints with intersecting axes, and one massless body in between (see also Sect. 3.5).

Of the total number of joint coordinates in the topological model of Fig. 12.11, 19 are—corresponding to the number of DoF of the complete vehicle—independent generalized coordinates, all other joint coordinates in the system are dependent! The topological structure of Fig. 12.11 contains 17 independent kinematic loops, designated as L_1 to L_{17}, while it partially also contains tree-type or open substructures. Examples for this can be found at the front-wheels, or at the car body. The spatial substructures, containing the closed kinematic loops, can mainly be found in the wheel suspensions, and partly in the drivetrain (kinematic loops L_{15}, L_{16} and L_{17}). A particular advantage of the approach presented here results from the fact that a major part of the kinematic subsystems under consideration (wheel suspensions and further kinematic structures within the model) can be solved, as already mentioned above, in an explicit or almost explicit manner. A further advantage results from the fact that a large number of the dependent joint coordinates in the model are not needed for further calculations, and therefore do not have to be calculated (see also Sects. 3.5 and 3.6).

Finally it should be emphasized again that the analysis of the kinematics of wheel suspensions and drivetrains is presented in references (Schnelle 1990; Frik 1994; Schmitz 1994; Hiller 1995; Schuster 1999; Pichler 1999; Bardini 2008).

The modeling steps for the kinematic analysis of a complete rear-wheel drive vehicle

- real vehicle,
- multibody system (MBS),
- topological representation and
- individual analysis of vehicle components,

as summarized in Fig. 12.12, can therefore be regarded as an important prerequisite for an efficient analysis of the forthcoming dynamics of the multibody system of the vehicle, to be shown later in this Chapter.

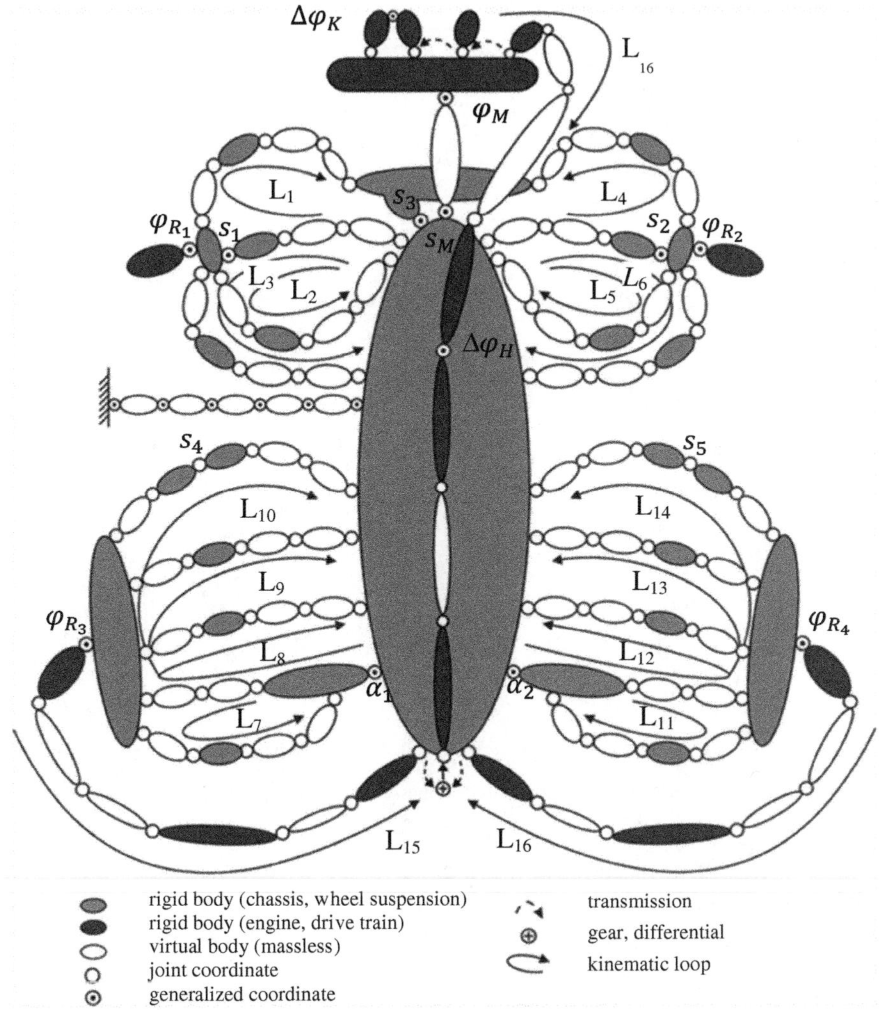

Fig. 12.11 BMW 5-series (E60)—topology complete vehicle

12.1.2 Kinematics of Front- and Four-Wheel Driven Complete Vehicle Models

Similar to the approach developed in the previous section for a rear-wheel drive vehicle, the modeling process for a front-wheel driven, as well as for a four-wheel drive vehicle will be derived. It will be shown that the basic ideas for the two drivetrain types—besides some specific characteristics—are the same as those in Sect. 12.1.1.

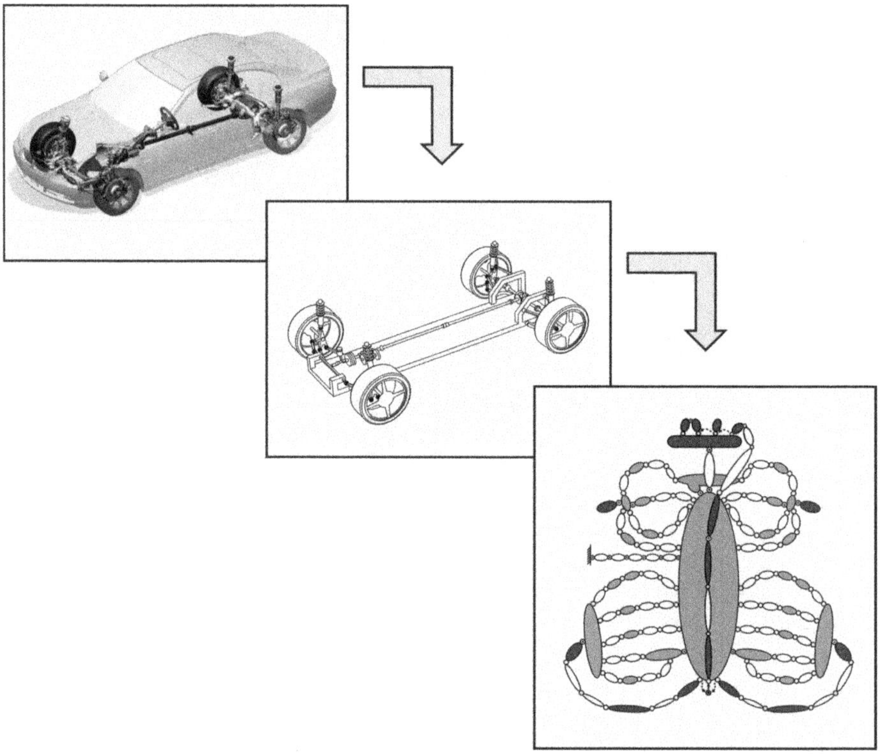

Fig. 12.12 BMW 5-series (E60)–steps of the kinematic modeling process

Kinematics of a front-wheel drive complete vehicle model The front-wheel drive vehicle under consideration is a Volkswagen VW Golf VI, as shown in Fig. 12.13. Figure 12.14 gives the corresponding rigid body model as a basis for the later multibody system.

The front-wheel suspension of this vehicle is realized using MacPherson front suspension struts (Fig. 12.15). This suspension type has been analyzed in detail in Sect. 6.5, though as a special factor, the misalignment due to comfort of the rear hinges of the lower wishbone is considered below.

To maintain the modeling idea, which is based on a multibody approach with rigid bodies, the compliance can be substituted by a set of prismatic joints with integrated spring, combined with a revolute joint for the kinematic compatibility and a compliance in only one direction (and therefore only one additional degree of freedom in the system). This numerically efficient approach is different from other methods in which the compliance is modeled by springs, leading to more DoF in the system and to stiff differential equations in the eventual dynamic equations. The drawback of this modeling idea may be the higher complexity of the structure of the rigid body kinematics (one additional kinematic loop). The kinematic modeling steps: structure, topology, block diagram, and solution flow for the MacPherson

Fig. 12.13 VW Golf VI–real vehicle (by courtesy of ATZ 2008)

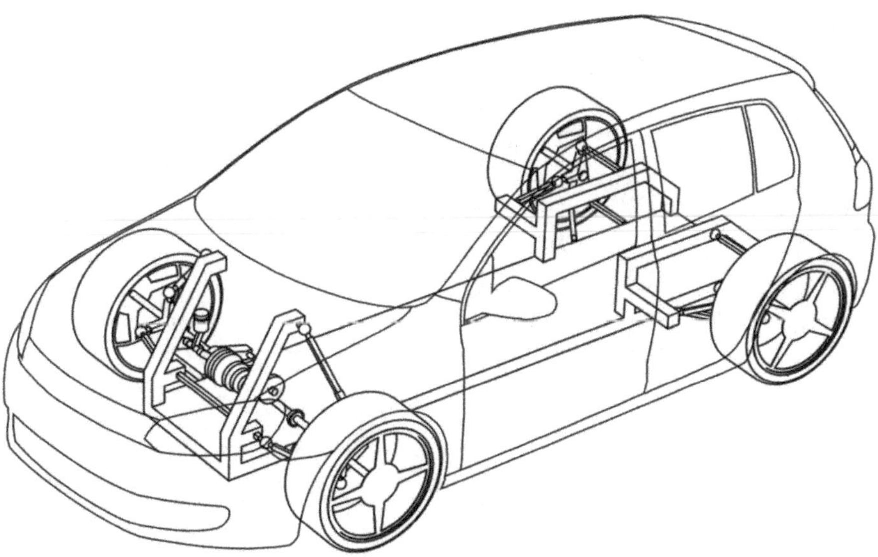

Fig. 12.14 VW Golf VI–multibody model with car body

Fig. 12.15 VW Golf VI—MacPherson front suspension axle (ATZ 2008)

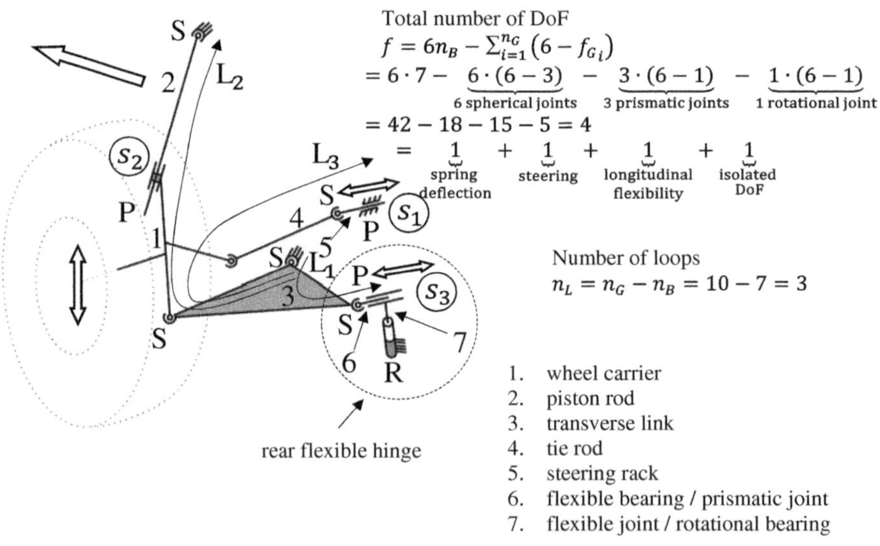

Fig. 12.16 VW Golf VI—MacPherson front suspension strut (with elastic hinge)—structure

front suspension axle with compliant hinges are presented in Figs. 12.15, 12.16, 12.17 and 12.18. It has to be mentioned that this kinematic structure leads to a recursive and explicit solution of the corresponding kinematic equations (see also Figs. 6.26, 6.27, and 6.28).

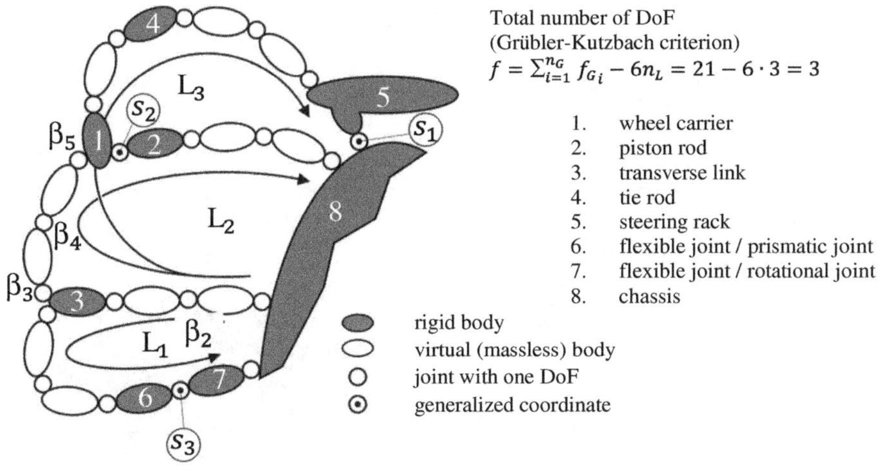

Fig. 12.17 VW Golf VI—MacPherson front suspension strut (with elastic hinge)

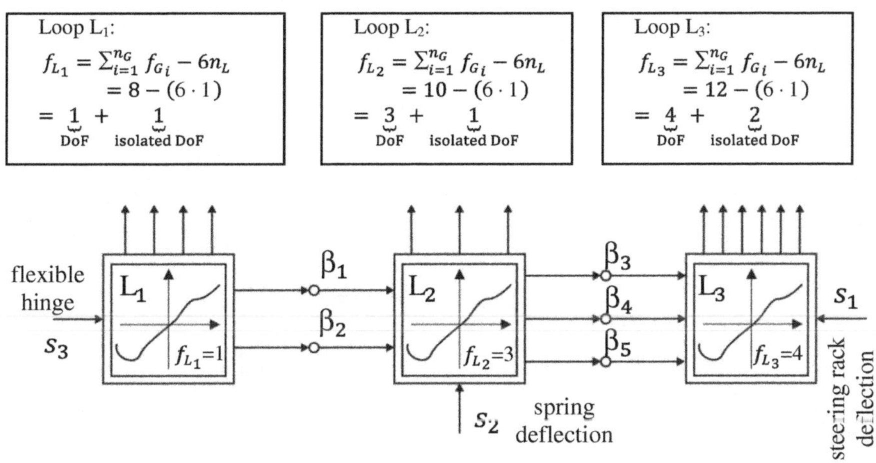

Fig. 12.18 VW Golf VI—MacPherson front suspension strut (with elastic hinge)—block diagram and solution flow

The rear axle of the VW Golf VI consists of two four-link wheel suspensions following the well-established principle of individual multi-link wheel suspensions (Fig. 12.19). It shall be noted that the longitudinal rod is elastically hinged, and also rigidly connected to the wheel carrier (Fig. 12.20), with the purpose of compensating the braking torques. The steps of the complete modeling process could be chosen similar to the integral-IV-suspension of the previous section. But a better alternative is offered by a method presented in Sect. 6.6. The efficiency of this approach is illustrated by the derivation of the equations for the relative kinematics

Fig. 12.19 VW Golf VI—four-link rear suspension (ATZ 2008)

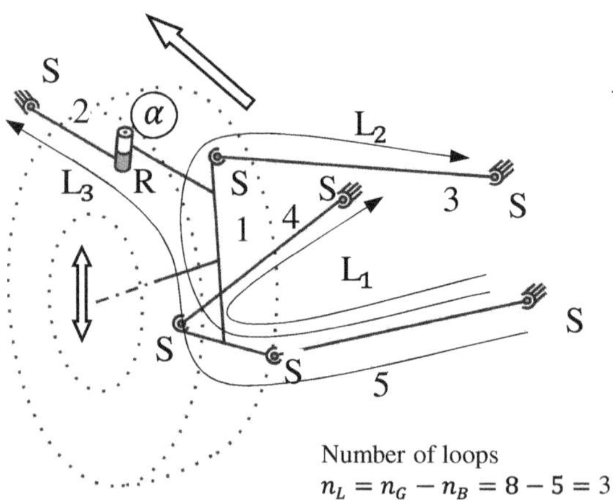

Total number of DoF

$$
\begin{aligned}
f &= 6n_B - \textstyle\sum_{i=1}^{n_G}\left(6 - f_{G_i}\right)\\
&= 6\cdot 5 - \underbrace{7\cdot(6-3)}_{\text{7 spherical joints}} - \underbrace{1\cdot(6-1)}_{\text{1 rotational joint}}\\
&= 30 - 21 - 5\\
&= \underbrace{1}_{\text{spring deflection}} + \underbrace{3}_{\text{isolated DoF}}.
\end{aligned}
$$

1. wheel carrier
2. trailing arm
3. transverse link
4. drag link
5. rear drag link

Number of loops
$n_L = n_G - n_B = 8 - 5 = 3$

Fig. 12.20 VW Golf VI—four-link rear (*left*) suspension (with compliant hinge)—structure

of a five-point wheel suspension—not using the kinematic loops, but rather an iterative solution of the corresponding geometric constraint equations. The kinematic structure and the topology of the four-link wheel suspension are presented in Figs. 12.20 and 12.21, respectively.

The scheme of the topology of the complete vehicle is presented in Fig. 12.22.

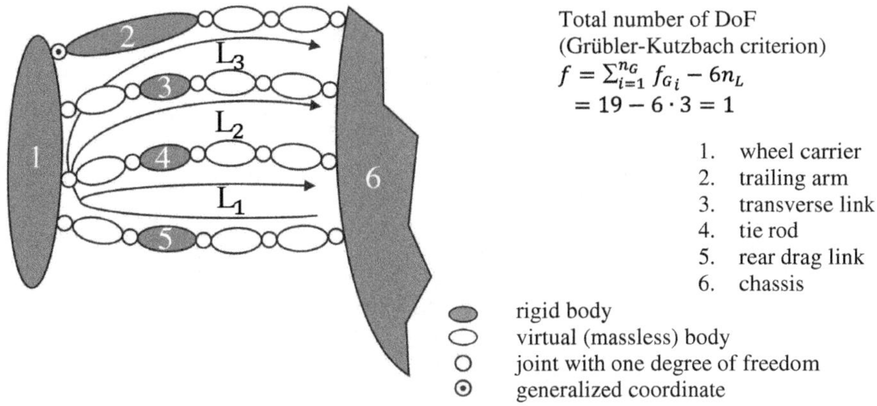

Fig. 12.21 VW Golf VI—four-link rear suspension—topology

The grouping of the vehicle components with their number of bodies, number of kinematic loops, and DoF is given by Table 12.2. The corresponding coordinates of the drivetrain are: the relative rotation of the clutch $\Delta\varphi_K$, and the four-wheel rotations $\varphi_{R_1}, \ldots, \varphi_{R_4}$.

The steps of the complete modeling process of the kinematic analysis of a front-wheel drive vehicle—real vehicle, multibody system, topology, and individual analysis of vehicle components—are summarized in Fig. 12.23. They represent —as before for the rear-wheel drive vehicle of Sect. 12.1.1—a major step in a compact and efficient dynamical analysis of the complete vehicle.

Kinematics of a four-wheel drive complete vehicle model As an example for a four-wheel drive car the Audi A5 (8T) has been selected, and Fig. 12.24 shows the real vehicle. The corresponding multibody system as a rigid body model is represented by Fig. 12.25.

The front-wheel suspensions are realized as five-link suspensions, following the previously mentioned multi-link principle (Fig. 12.26). The structure of the front suspension, its topology, as well as its block diagram, and kinematic solution flow are given by Figs. 12.27, 12.28 and 12.29.

Note: The kinematic principle of the multi-link wheel suspension under consideration has been discussed in detail in Sect. 6.6, proposing an alternative kinematic analysis, already applied for the front-wheel driven VW Golf earlier in this section (the closed kinematic loop approach is replaced by a method, which is making use of the equal structure of the constraint equations (Sect. 6.6.2)).

The main elements of the rear axle of the Audi A5 are two trapezoidal-link suspensions (Fig. 12.30), a topology which is very close—except the coupling to the damping element (without any kinematic influence)—to the integral-IV-wheel suspension of the BMW of Sect. 12.1.1. The modeling steps—structure, topology, and block diagram with solution flow—are put together in Figs. 12.31, 12.32 and 12.33.

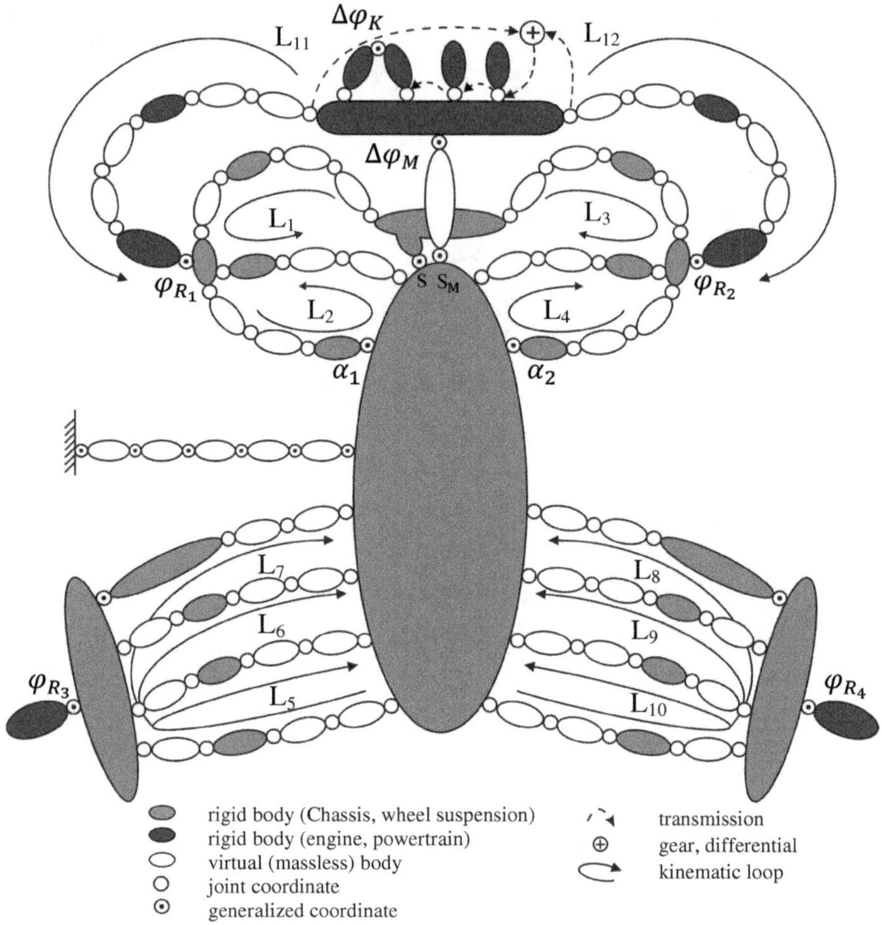

Fig. 12.22 VW Golf VI—topology of the complete vehicle

Table 12.2 Kinematic structure of VW Golf VI

Vehicle components	DoF	Bodies	Loops
Chassis	6	1	0
Front-wheel suspension	$2 \times 1 + 1$	13	2×2
(el. wishbone joints)	2×1	0	2×1
Rear-wheel suspension	2×1	10	2×3
Engine bracket	2	1	0
Drivetrain (including all wheel rotations)	5	15	2
Σ	18 (20)	40	12 (14)

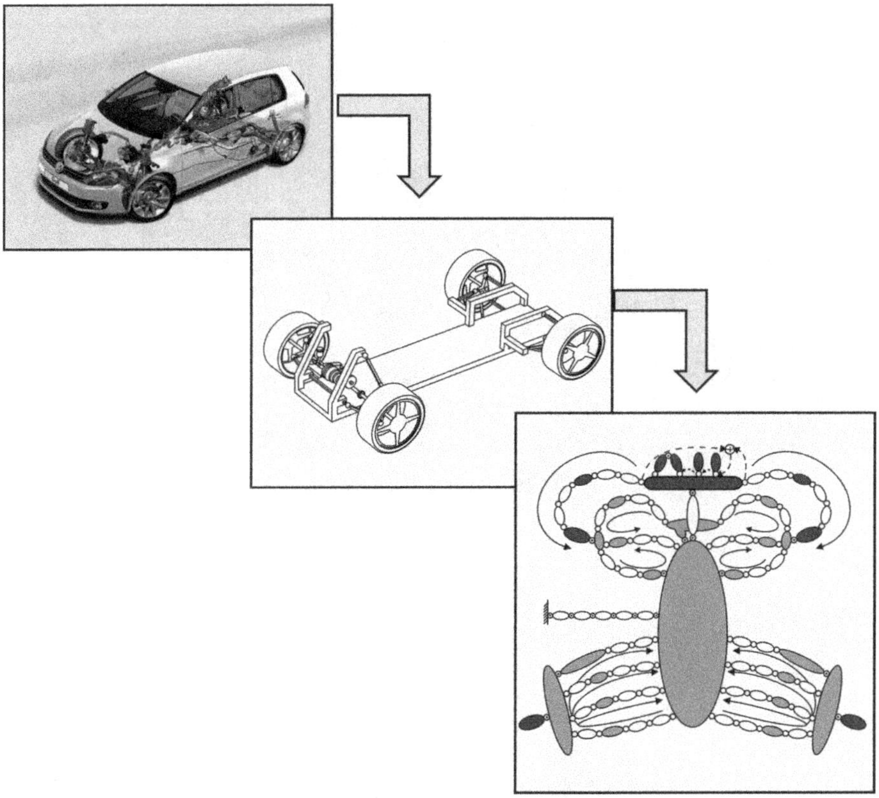

Fig. 12.23 VW Golf VI—steps of the kinematic modeling process

The scheme of the topology of the complete vehicle is presented in Fig. 12.34. The corresponding coordinates of the drivetrain are as follows: the rotary displacement of the clutch $\Delta\varphi_K$ and of the front and the intermediate shaft, $\Delta\varphi_V$ and $\Delta\varphi_H$, as well as the four wheel rotations $\varphi_{R_1}, \ldots, \varphi_{R_4}$.

Table 12.3 gives the grouping of the vehicle components, with their number of bodies, number of kinematic loops, and DoF.

The steps of the complete modeling process of the kinematic analysis of a four-wheel drive vehicle–real vehicle, multibody system, topology, and individual analysis of vehicle components, are summarized in Fig. 12.35. They represent–as before for the rear-wheel and front-wheel drive vehicle, respectively–the key concept in a compact and efficient dynamical analysis of the complete vehicle.

Fig. 12.24 Audi A5 (8T)—real vehicle (by courtesy of Audi AG, 2008)

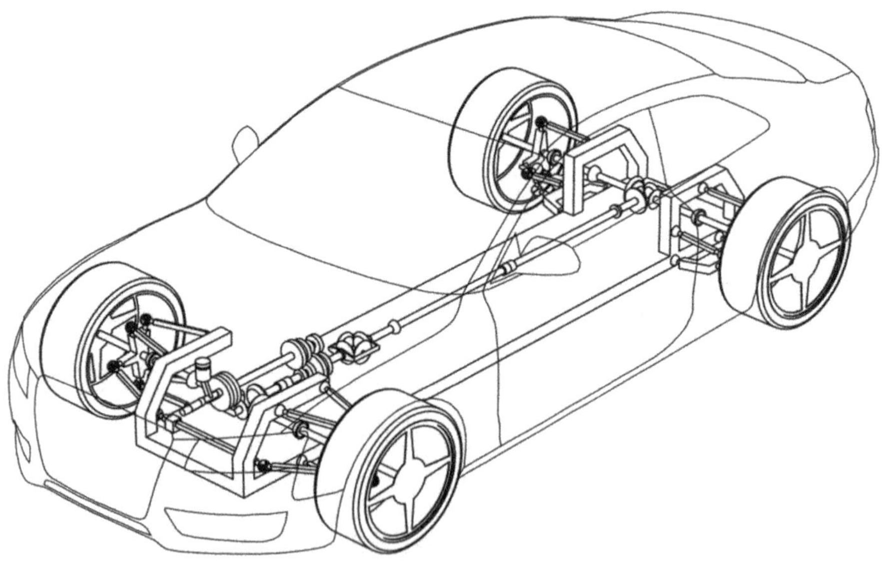

Fig. 12.25 Audi A5—multibody model with car body

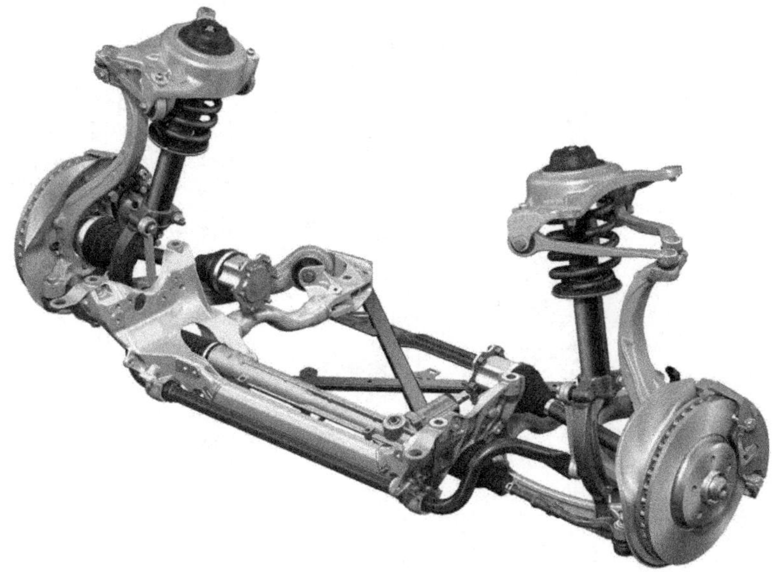

Fig. 12.26 Audi A5—five-link front axle (by courtesy of Audi AG, 2008)

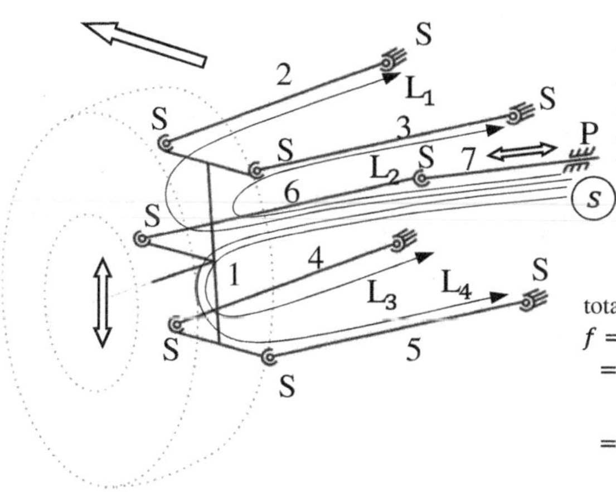

1. wheel carrier
2. suspension arm front top
3. suspension arm rear top
4. suspension arm front bottom
5. suspension arm rear bottom
6. tie rod
7. steering rack

number of loops

$$n_L = n_G - n_B = 11 - 7 = 4$$

total number of DoF

$$f = 6n_B - \sum_{i=1}^{n_G} \left(6 - f_{G_i}\right)$$

$$= 6 \cdot 7 - \underbrace{10 \cdot (6-3)}_{\text{10 spherical joints}} - \underbrace{1 \cdot (6-1)}_{\text{1 prismatic joint}}$$

$$= 42 - 30 - 5$$

$$= \underbrace{1}_{\text{spring deflection}} + \underbrace{1}_{\text{steering}} + \underbrace{5}_{\text{isolated DoF}}$$

Fig. 12.27 Audi A5—five-link front suspension (*left*)—structure

12.1.3 Dynamics of the Complete Vehicle Model

Based on the mathematical analysis of the kinematics of a complete vehicle as presented in the first sections of this chapter, the mathematical equations of the

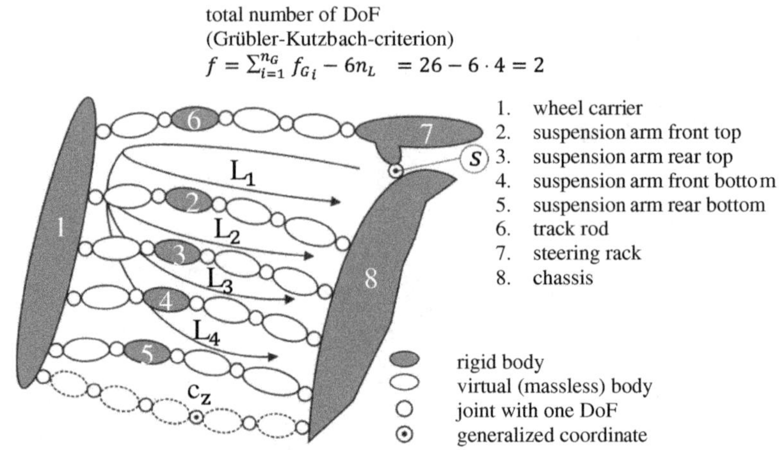

Fig. 12.28 Audi A5—five-link front suspension—topology

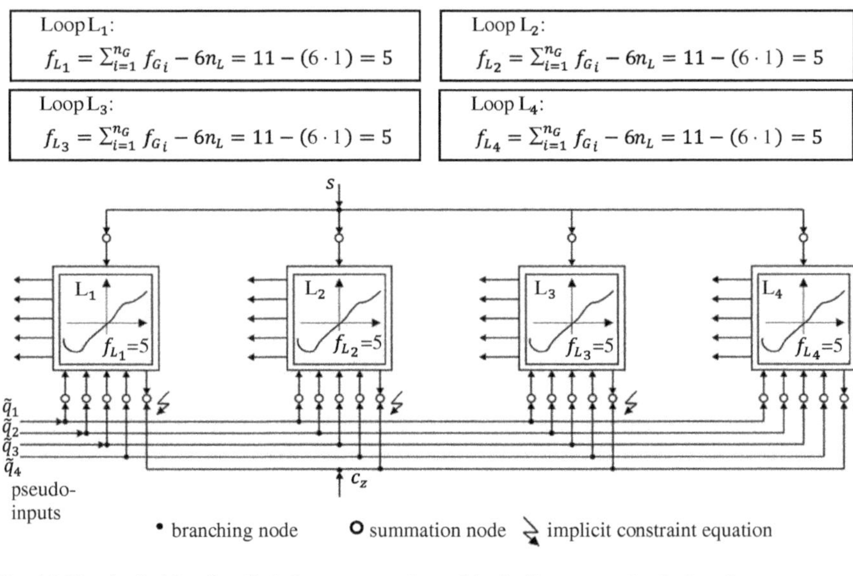

Fig. 12.29 Audi A5—five-link front suspension—block diagram and solution flow

dynamics of this system have to be derived next. As a very efficient tool, the method of kinematic differentials can be used, as developed by Kecskemethy, in collaboration with Hiller (see Hiller et al. 1986; Hiller and Kecskeméthy 1987; Kecskemethy 1993; Hiller 1995).

The method of the kinematic differentials was already presented in Chap. 4, and its numerical efficiency can be further improved by two optimization steps: By making use of the structure of the topology, as it arrives in the discussed vehicle

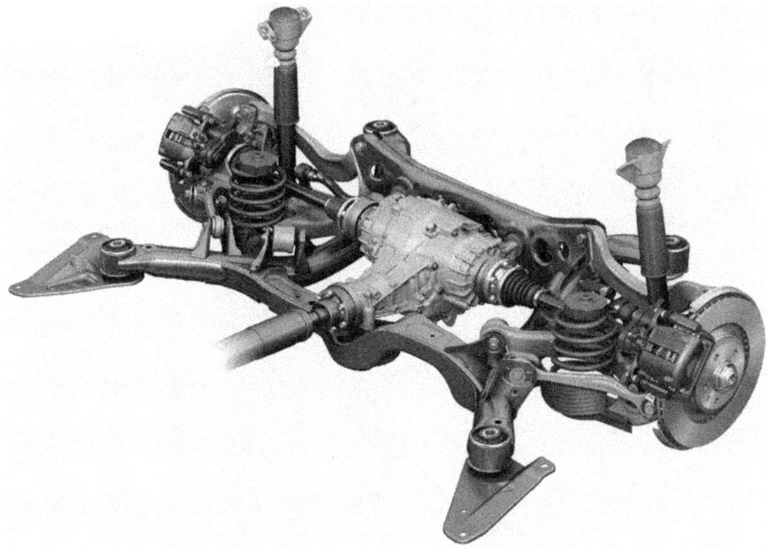

Fig. 12.30 Audi A5–trapezoidal-link rear suspension (by courtesy of Audi AG, 2008)

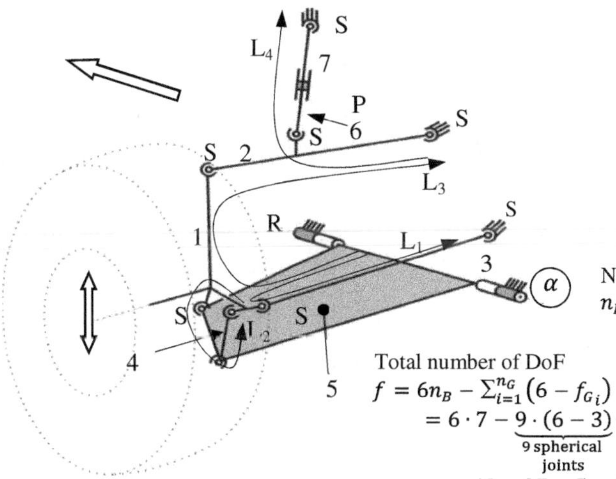

1. wheel carrier
2. upper suspension arm
3. tie rod
4. coupling link
5. trapezoidal link
6. damper
7. piston rod
8. chassis

Number of loops

$n_L = n_G - n_B = 11 - 7 = 4$

Total number of DoF

$$f = 6n_B - \sum_{i=1}^{n_G} (6 - f_{G_i}) = 6 \cdot 7 - \underbrace{9 \cdot (6-3)}_{\text{9 spherical joints}} - \underbrace{1 \cdot (6-1)}_{\text{1 revolute joint}} - \underbrace{1 \cdot (6-1)}_{\text{1 prismatic joint}} = 42 - 27 - 5 - 5 = 5 = \underbrace{1}_{\text{vertical deflection}} + \underbrace{4}_{\text{isolated DoF}}$$

Fig. 12.31 Audi A5–trapezoidal-link rear suspension–structure

models, a large number of zero-operations can not only be avoided, but mathematical terms can be efficiently comprised. In particular, for this last aspect, the mathematical coupling of mechanical subsystems needs to be analyzed. The vehicle model, as presented in Fig. 12.12, can illustrate this. It can be observed, that some

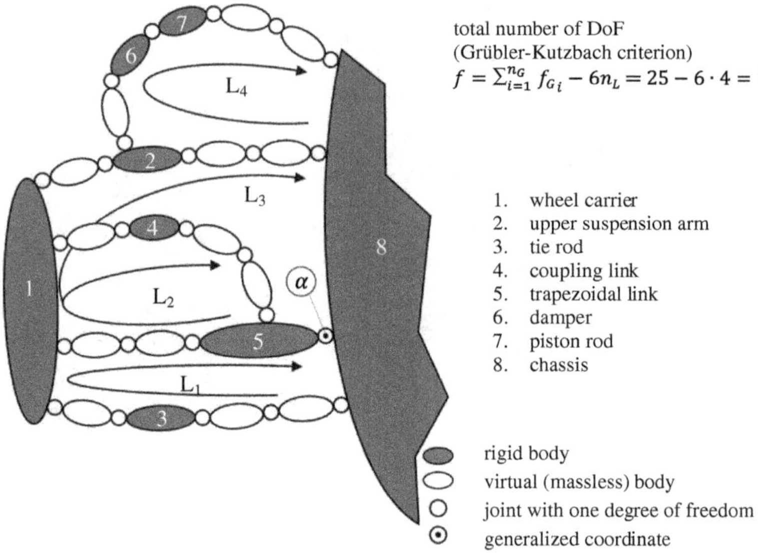

Fig. 12.32 Audi A5–trapezoidal-link rear suspension–topology

Loop L_1: $f_{L_1} = \sum_{i=1}^{n_G} f_{G\,i} - 6n_L = 9 - (6 \cdot 1) = 3$	Loop L_2: $f_{L_2} = \sum_{i=1}^{n_G} f_{G\,i} - 6n_L = 8 - (6 \cdot 1) = 2$
Loop L_3: $f_{L_3} = \sum_{i=1}^{n_G} f_{G\,i} - 6n_L = 9 - (6 \cdot 1) = 3$	Loop L_4: $f_{L_4} = \sum_{i=1}^{n_G} f_{G\,i} - 6n_L = 9 - (6 \cdot 1) = 3$

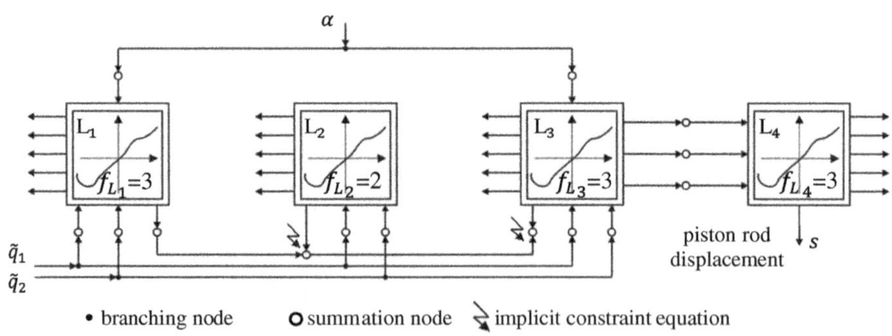

Fig. 12.33 Audi A5–trapezoidal-link rear suspension–block diagram and solution flow

of the subsystems can be kinematically completely decoupled. This means, that changes in individual generalized coordinates of particular subsystems do not have any influence on the motion of other subsystems.

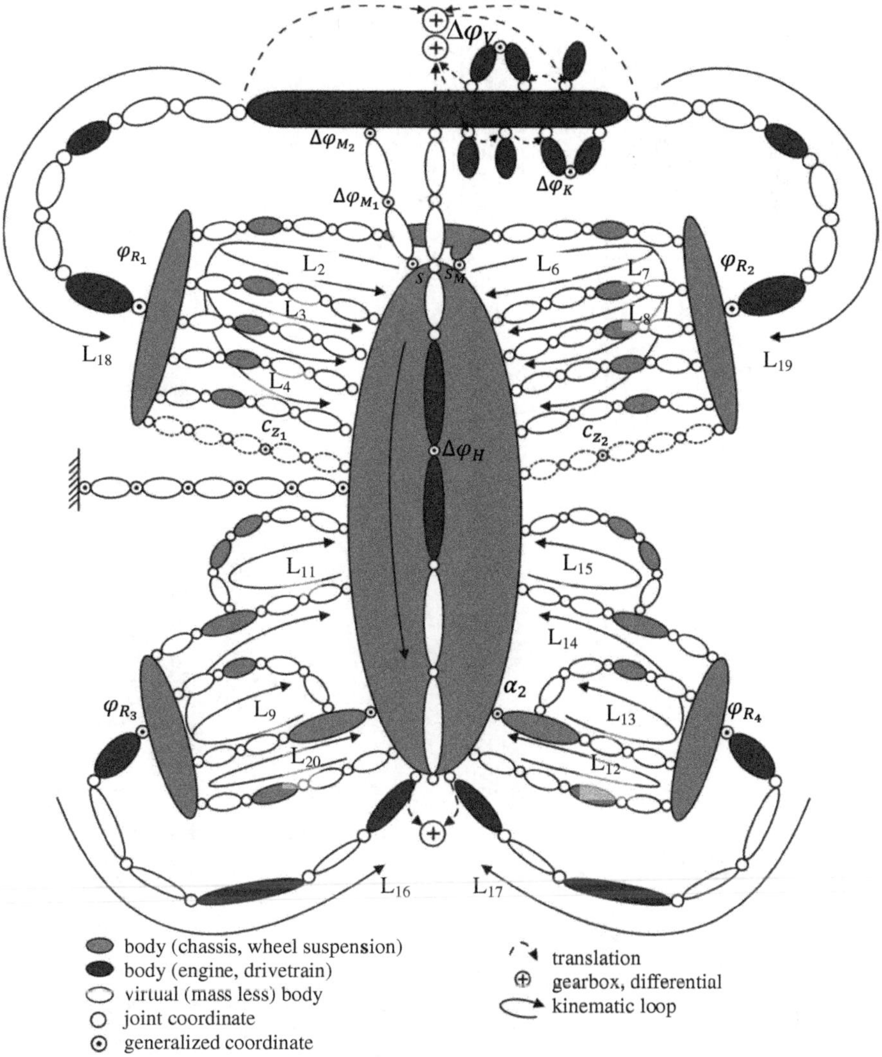

Fig. 12.34 Audi A5–topology of the complete vehicle

Table 12.3 Kinematic structure of the Audi A5

Vehicle component	DoF	Bodies	Loops
Chassis	6	1	0
Front-wheel suspension	$2 \times 1 + 1$	13	2×4
Rear-wheel suspension	2×1	14	2×4
Engine bracket	3	1	0
Drivetrain (including all wheel rotations)	7	20	4
Σ	21	49	20

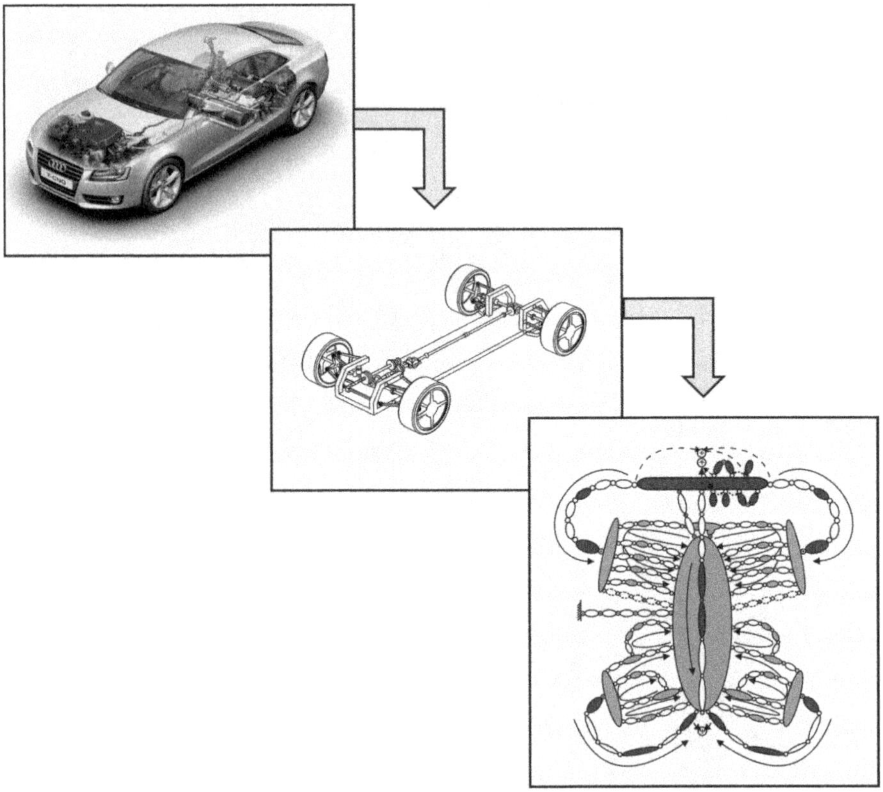

Fig. 12.35 Audi A5–steps of the kinematic modeling dynamics of the model of the complete vehicle

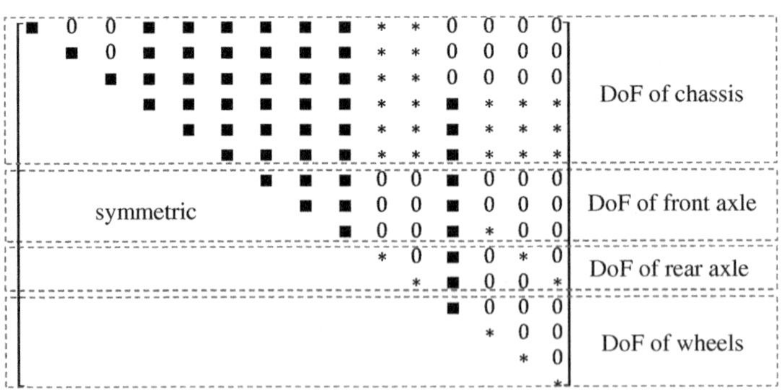

Fig. 12.36 Coupling of subsystems of the mass matrix—example: influence of the left front wheel

As a demonstrative example for the coupling of elements, components, and subsystems, the mass matrix of a vehicle model with 15 DoF is presented in Fig. 12.36. Due to the symmetry of the mass matrix, only the upper triangular part of the mass matrix is shown. (The presented vehicle corresponds to the vehicle of Fig. 12.12, but without considering the DoF of engine support, and the DoF of compliance in the drivetrain). One can now recognize the different influences of individual subsystems on the elements of the mass matrix, as illustrated in this example by the influence of the left front-wheel: The influenced element are characterized by the symbol ■, while the elements which are not influenced are displayed by the symbol *.

In addition to this potential for an optimization, a concentration of mass matrix elements is possible, which again reduces the required number of numerical operations. For further information see (Schnelle 1990; Schmitz 1994; Pichler 1999).

12.2 Simulation of Motor Vehicles

Depending on the task for which the complex vehicle models of Sect. 12.1 are designated it is necessary to modify the respective models according to the different simulations that are to be conducted. It therefore becomes the goal to create a suitable environment for the simulations. The needed simulations are in the present case mostly concerned with processes of driving dynamics such as the emerging systems in the context of different driving safety systems, comfort systems or the ever-growing range of advanced driver assistance systems (ADAS). Such simulations include for example the active antilock braking systems (ABS), traction control systems (TCS), electronic stability programs (ESP), adaptive cruise control (ACC), as well as the passive systems such as in example the automatic supplemental restraint systems (SRS) for the occupant protection in the event of rolling. Behind these applications there is a close cooperation of more than two decades in the form of concrete cooperation and development projects with the automotive industry suppliers as well as the vehicle manufacturers.

In order to prevent accidents, the control systems used in vehicles become more and more complex. This in example includes the electronic stability program ESP. Not only does it regulate the longitudinal dynamics (ABS and ASR), but it also incorporates the lateral dynamics by observing the side slip angle and the yaw rate (Erhardt and van Zanten 1995). Using the electronic stability program ESP, the stabilization of the vehicle shall be ensured in situations of driving- or physical thresholds. This way, an excessive demand of the driver is supposed to be avoided. ABS, ASR and ASP are already progressed extensively and today are part of the basic configuration of most vehicles. In order to shorten the development time and to reduce the development costs, the tool of driving dynamics simulation is usually used. In cooperation with the Robert Bosch GmbH, the development environment FASIM_C++ for three dimensional driving dynamic simulations was developed by

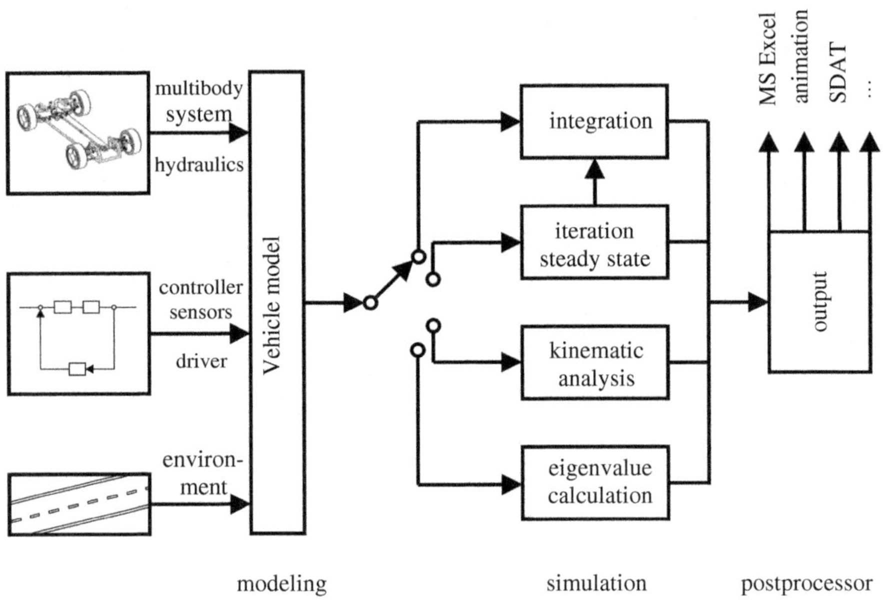

Fig. 12.37 Program concept of FASIM_C++

the chair of mechatronics at the University of Duisburg from 1990 to 1995 for the first production-line ESP development for the then-current S-Class (W140) of the Daimler Benz AG. This tool was used for the development of ESP in combination with commercial software packages in various application projects (Erhardt and van Zanten 1995).

12.2.1 Setup and Concept of FASIM_C++

On the one hand, the simulation program FASIM_C++ offers the possibility of a generic insight into the dynamic behavior of the system of the vehicle and on the other hand supports especially the elaborate development of controllers of the driving dynamics. The program fundamentally consists of two parts: the actual model of the vehicle and the frame of the simulation (Fig. 12.37).

The model of the vehicle can be understood as a combination of

- a multibody system for the depiction of the mapping of the mechanical features,
- the hydraulics for the brake system,
- the driver,
- time-discrete systems such as rules and sensors, and
- the external influence of the environment.

The model of the vehicle can be used for simulations, or for the iteration of conditions of equilibriums, or for the analysis of kinematics, or the linearization including the calculation of an eigenvalue. There are different possibilities available to present the results. The whole program can be operated via the graphic user interface (Adamski 2001) for Microsoft Windows.

For the structure of such a simulation program there are numerous requirements:

- Substantial driving dynamic effects of a vehicle need to be displayed.
- A separation between the frame of the simulation (including the numerical integration process) and the system description needs to be available.
- The frame of the simulation needs to support the connection of fast and slow subsystems (i.e. hydraulics and vehicle) as well as the incorporation of discrete sensor-controller systems.
- The program shall be transparent in its structure so that changes and additions to it are easy to implement.
- The exchange of models of components needs to be simple. Examples: transition from simple to complex models of components, exchange of types of wheel suspensions. Here, the quantity of the respective DoF needs to be modifiable.
- During the simulation, a simple access to any system parameters (i.e. sensor signals) needs to be possible.
- The input of control inputs from the controller into the system needs to be simple and flexible as well.
- Low computing requirements that can be achieved by establishing the equations of motion efficiently.

As a possible approach the concept of the modularly constructed simulation environment FASIM_C++ was developed. It was created in the object oriented programming language C++ (Pichler 1999). The equations of motion of the mechanical components are set up in fractional coordinates using the process of kinematic differentials (Chap. 4). Via a special formulation, which will be discussed in further detail below, this access allows the decomposition of the vehicle into modules (chassis, front axle, rear axle, wheels drivetrain, engine bracket) with standardized interfaces. The remaining non-mechanical components (braking hydraulics, environment, driver, combustion engine) are also described in a modular form.

The modular structure is an essential advantage of the simulation program FASIM_C++ because a comparably simple replacement of single system components is possible. In the section below, an overview over the modules created so far is given; the user thus has his choice of suitable components for his simulation.

Obviously, the following fundamental principles of the structure of FASIM_C++ can be derived:

- Every module is encapsulated externally and contains a separate data record as well as methods for the initialization, for the kinematic and dynamic calculation of the module (for mechanical components).

- The internal communication between the modules as well as with a superior frame constructing the equations of motion is created via interfaces, which are not dependent on the complexity of the module. An alteration of the number of DoF only becomes valid for the dimension of a few interface quantities, and can therefore be processed easily.
- The flow of calculations follows the build of the model. With the generalized coordinates as an entry point, at first the kinematics of the multibody system is calculated in the following order: chassis, axles, drivetrain, followed by the wheels. Subsequently, using the now known kinematic values, the non-mechanical modules (driver, engine, brake hydraulics) are determined. Finally, the dynamics of the mechanical components of the multibody system is calculated in reverse direction. Afterwards, the equations of motions are created using the frame of simulation. They are then written as a state vector in the form $\dot{x} = f(x, t)$. After the numerical integration of this first order differential equation system, it is then partitioned into the corresponding generalized coordinates inside the modules and the flow of calculation is restarted.

For the implementation of a simulation project, the following work sequence is predetermined for the user. At first, he chooses the modules from the module library (Ref. Sect. 12.2.2) which best match his vehicle model. Thus, she could for example after choosing a chassis, pick a model of a MacPherson wheel suspension and a double wishbone axle. By adding a front-wheel drive and a dynamic wheel model, the mechanical components of the vehicle model are completed. In order to complete the whole model, a model for the driver, the brake hydraulics as well as an engine model is introduced. Finally, an ABS or ESP controller can be installed. The user is now free to change parameters within the data set or to create individual simulations (cornering, driving over shifted bumps etc.).

12.2.2 Modular Structure of a Vehicle Model

The program FASIM_C++, which has primarily been developed for road vehicles, creates the equations of motion for complex, modular multibody systems with active elements in minimal coordinates. The systems are depicted as rigid body models with differing complexity that, in example, considers the non-linear kinematics of the wheel suspension which is expressed especially via numerous kinematic loops which are essential for the correct illustration of the dynamic behavior of the vehicle.

In developing the program, especially the mostly automatic generation of the equations of motion at high modularity of the complete program and low time requirements for the calculation have been considered. Furthermore the individual modules were to be created quickly and exchanged easily. The following section explains how this was made possible.

In order to realize a higher degree of modularity, a systematic structuring of the multibody system vehicle is necessary. The interfaces between the modules are appearing in a way that the multibody system is separated at selected rigid links. This means that any descriptions of the joints are integrated in the models in such a way that they are offering the advantage that those links can be treated uniformly from the outside. A multibody system can therefore be divided into descriptive, realistic assemblies (Pichler 1999).

The assembly of the single modules into a complex multibody system can be interpreted as putting together the single components to create the overall system "vehicle". For every type of a component, a module is implemented and stored in a library. Since the coupling elements are automatically determined while creating the equations of motion, the calculation of the kinematics and dynamics of a module can mostly be implemented independently from all other modules.

In order to put enough emphasis on the modularity and flexibility of this concept, the construction of different types of vehicles shall be presented in addition to a simple passenger car: a car/trailer combination as well as an articulated truck. Their implementation in FASIM_C++ will be shortly outlined below.

Passenger car In a first step, a vehicle is separated into the modules chassis, front and rear axle respectively as well as the four wheels (Fig. 12.38).

The differentiation of the types of vehicles (vehicles of different classes and manufacturers) can be achieved by using adapted modules. As an example for the module type front axle, an axle of the following type can be chosen:

- solid axle,
- double wishbone axle,
- MacPherson axle,
- double-link spring strut,
- four-link front suspension,
- five-link front suspension.

A special data set is connected with every type, which contains the manufacturer-specific dimensions and parameters.

A principal example for a vehicle is the modeling of a Mercedes Benz W140 (Fig. 12.39), which was the target vehicle of the previously mentioned cooperation with the automobile supplier industry. In its mechanical concept and its topological structure it is very similar to the BMW 5-series which was introduced earlier in this chapter. The front axle is a double wishbone axle with three DoF while the steering is executed as a four-bar linkage with steering damper. As the rear axle, the model of an independent five-link suspension (one DoF each) is used. For the drivetrain, a rear-wheel drive with an automatic transmission is available. The wheel is subdivided into a rim, belonging to the drivetrain, and a tire, being linked with each other rigidly. The rim assumes the degree of rotational freedom as well as the features of mass of the wheel. The wheel model can be chosen from different models with differing complexity and scopes. For standard applications, a wheel model is chosen after HSRI (Ref. Sect. 7.4.6). A more precise description of the model of the

Fig. 12.38 Structure of a passenger car

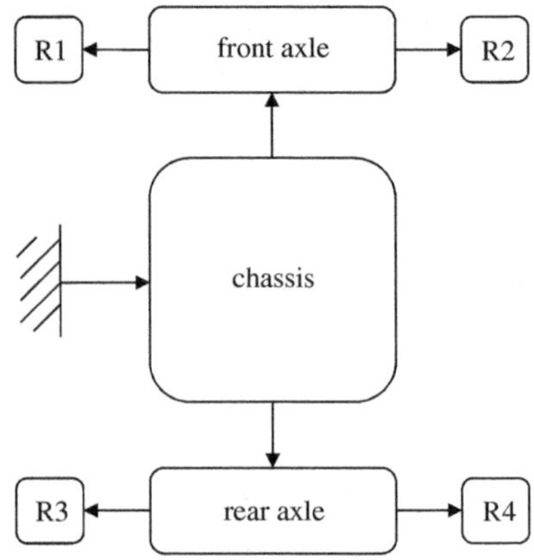

Fig. 12.39 Complex multibody system of a MB W140 (Mercedes Benz S-Class in the 1995 years)

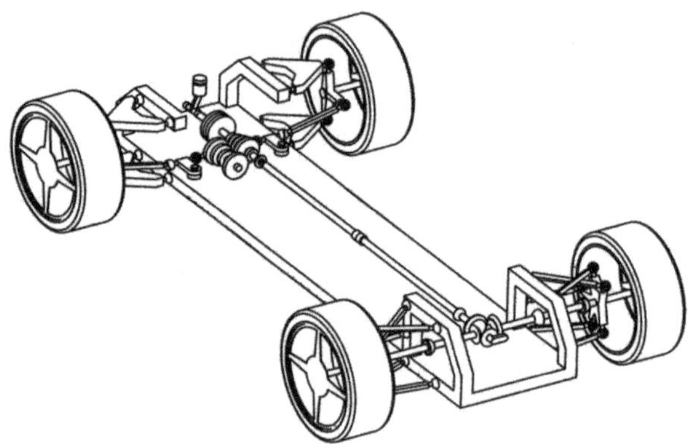

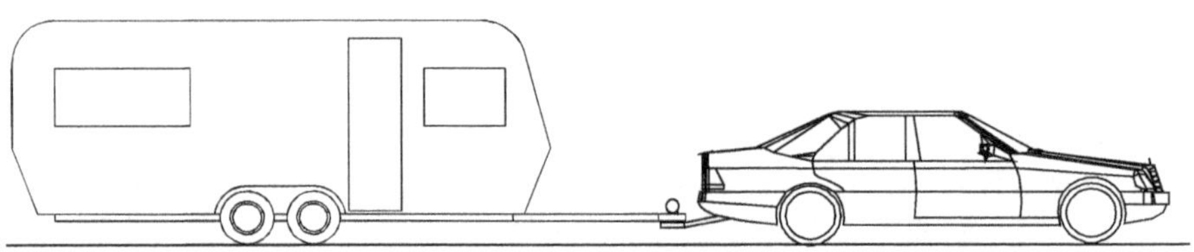

Fig. 12.40 Outline of a car/trailer combination

vehicle can be found in Schmitz (1994). The kinematic structure is similar to the rendered vehicle of Sect. 12.1.1.

Caravan By consequently using the full range of the programming language C++, in FASIM_C++, it is possible to simulate a wide variety of vehicle types without having to create always a new program. This way it is possible for example to simulate a passenger car with a trailer (Fig. 12.40). The corresponding modular structure is depicted in Fig. 12.41.

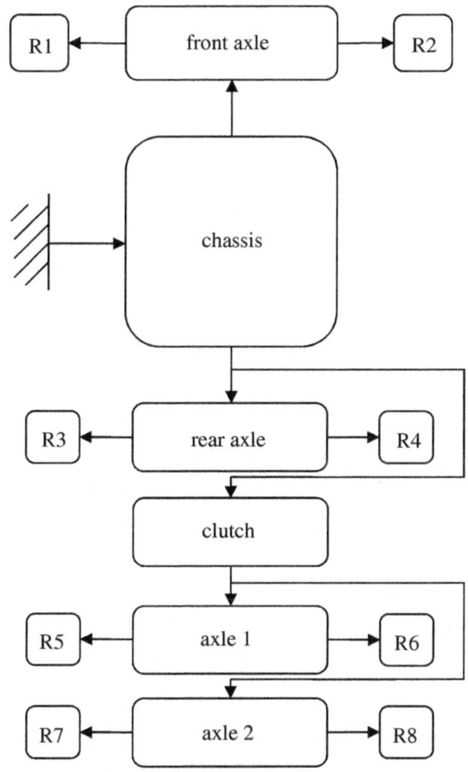

Fig. 12.41 Modular structure of a car/trailer combination

Since the major part of trailers uses an overrun brake (Uffelmann 1980), the caravan has been equipped with such a brake as well. Since the overrun brake is an example for a structure variant system, it shall be examined more closely below. A torsion bar spring axle is used as the axle of the caravan.

The overrun brake automatically creates a braking torque, which originates from the seizing process of the trailer when the drawing vehicle brakes. The drawbar of the trailer is longitudinally moveable in a certain area so that when braking, the trailer pushes the drawbar against a spring due to the thus created longitudinal force. Due to the relative movement between trailer and drawing vehicle, the brake is activated. The drawbar contains two stops. If one of the stops is reached, the translational DoF needs to be turned off. The conditions of turning on and off respectively, is determined in this simple model using a simple estimate. While the development environment FASIM_C++ is able to calculate in time all needed sizes in every position at almost every point. It has been observed that an estimate of the required forces leads to quicker results that are just as good. For the assessment whether the DoF needs to be engaged or not, on the one hand the longitudinal force of the trailer F_A is used

$$F_A = m_A a_{abs} \tag{12.1}$$

with

F_A longitudinal force of the drawbar in N,
m_A mass of the trailer in kg,
a_{abs} acceleration in coordinates of the inertial system in m/s^2, (only the component in the x, y-level is considered),

and on the other hand the spring force F_F of the spring built into the drawbar

$$F_F = c_F x_{B_{max}} \tag{12.2}$$

with

F_F spring force / N,
c_F spring constant / (N/m),
$X_{B_{max}}$ range of spring up to the stop /m

As long as $F_A \leq F_F$ is valid, meaning as long as the force resulting from the delay is smaller than the spring force, the DoF is turned on. The stops of the brake are however chosen as such, that they remain in the back stop for rides with low acceleration, or deceleration respectively due to the aerodynamic force of the trailer and the friction of the wheels. As a result, the DoF of the overrun brake is turned off and the computing time is reduced.

Semitrailer vehicle Different from the previous example of the car/trailer combination, the coupling device (fifth wheel coupling) of an articulated vehicle like e.g. a semitrailer (Fig. 12.42) is attached in front of the rear axle of the towing vehicle. The model of a semitrailer vehicle shows that the possibilities of FASIM_C ++ are not limited to passenger cars. In order to show to what extend this is possible, a simple vehicle model was designed. Even the complete vehicle model however is able to reproduce essential effects well. Since a vehicle with such magnitude (length approx. 15 m, weight approx. 30 t) cannot be considered to have rigid bodies, two concentrated elasticities were introduced in the chassis of the tractor unit. The implemented modules "elasticity" allow for a rotation about the longitudinal axis of the vehicle. This way, the torsional flexibility of the frame is reproduced. The model structure of the articulated vehicle is depicted in Fig. 12.43. Overall, this model contains 32 DoF. The kinematic topology of the articulated vehicle is depicted in Fig. 12.44. It is kept very simple and only contains one kinematic loop (steering).

12.2.3 Construction of the Equations of Motion

When calculating the kinematics of the system, established concepts are used (Hiller 1989, also Ref. Chap. 3). Here, the calculation of the global kinematic is

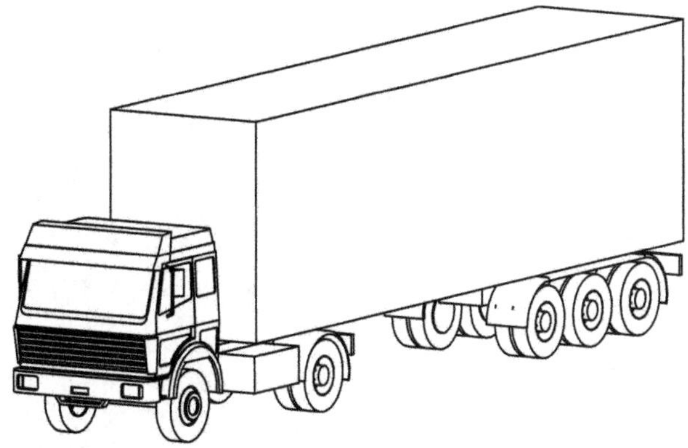

Fig. 12.42 Outline of an articulated vehicle (semitrailer)

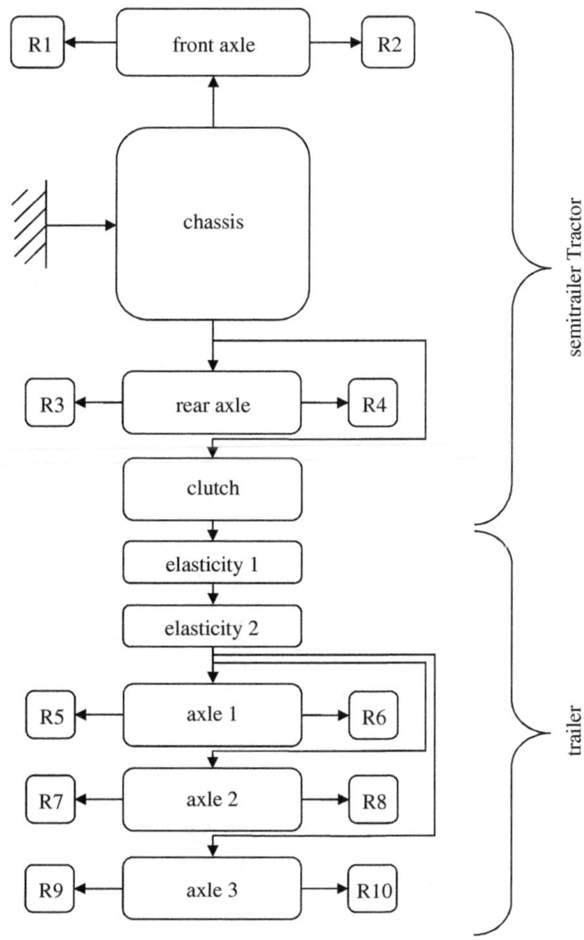

Fig. 12.43 Modular structure of the articulated vehicle

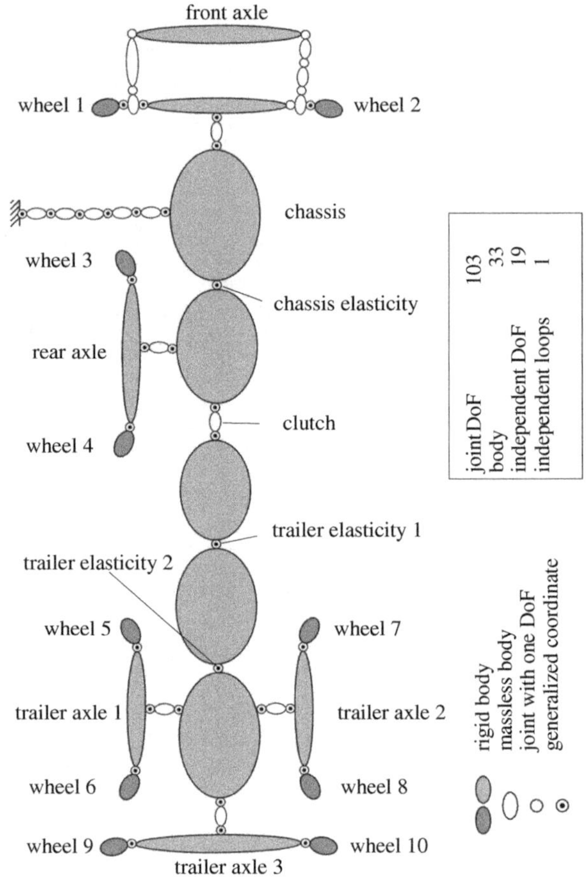

Fig. 12.44 Joint-body-structure of an articulated vehicle—topology

divided up into relative kinematic—meaning the solving of the constraint equation—and an absolute kinematic—meaning the determination of the absolute position, velocity, and acceleration of the bodies. In order to topologically structure the multibody system, the concept of kinematic transformers is used to display the response characteristics of a single kinematic loop and its connection to a kinematic network (Hiller et al. 1986), where only one kinematic loop exists. The kinematic of the individual loops can be solved using the process of the characteristic pair of joints in recursive, or in mostly recursive form (Woernle 1988). In order to calculate the kinematics, the program package "MOBILE", which was developed in the object oriented programming language C++, can additionally be applied (Kecskemethy 1993).

The equations of motion of the system are presented in minimal coordinates in which the needed partial derivations of the body coordinates are obtained using the concept of kinematic differentials (Hiller and Kecskeméthy 1989). This approach is

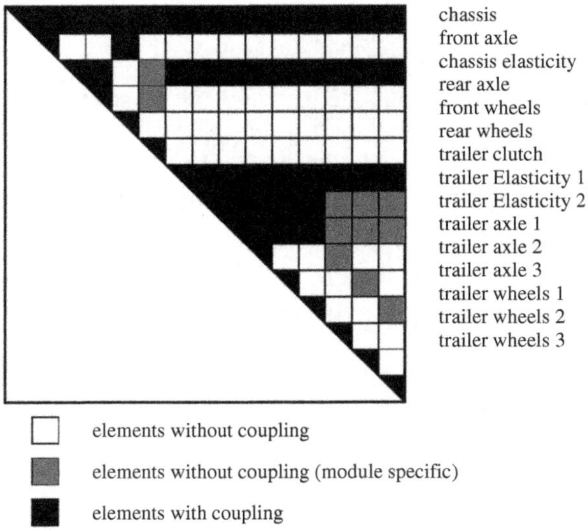

Fig. 12.45 Schematized display of the mass matrix for the articulated vehicle

based on the analysis of the kinematics mentioned above. The equations of motion can mostly be set up automatically if the single modules of the system, as already illustrated, are coupled with each other in a tree-type structure and are initialized, beginning at the roots and going towards the leafs. Here, the interfaces are built independently. By partitioning the mass matrix and the JACOBIANs consistently, the zero-operations, occurring due to the topology of the system, can be avoided (Pichler 1999). The influence of the coupling of the sub systems on the mass matrix has already been developed extensively for a passenger car in Sect. 12.1.3 (Ref. Fig. 12.36). The modular structure, the joint-body structure of the kinematic chains and the schematic presentation of the mass matrix of an articulated vehicle is presented in Figs. 12.43, 12.44 and 12.45.

The calculation is overall divided into an initialization, a kinematics part (position, velocity and kinematic differentials), in which the system is calculated starting at the root going towards the leaves, and a dynamics part (applied forces and torques respectively) and creating of the equations of motion, in which the system works in reverse order.

The used equations of motion are based on d'ALEMBERT's principle in the LAGRANGE version (see Chap. 4).

$$\sum_{i=1}^{n_B} \left[(m_i \ddot{\boldsymbol{r}}_{S_i} - \boldsymbol{F}_i)^T \delta \boldsymbol{r}_{S_i} + (\boldsymbol{\Theta}_{S_i} \dot{\boldsymbol{\omega}}_i + \boldsymbol{\omega}_i \times \boldsymbol{\Theta}_{S_i} \boldsymbol{\omega}_i - \boldsymbol{T}_{S_i})^T \delta \boldsymbol{\varphi}_i \right] = 0, \qquad (12.3)$$

using the terms:

n_B	number of bodies with mass,
$m_i, \boldsymbol{\Theta}_{S_i}$	mass and inertia tensor with respect to the center S,
$\ddot{\boldsymbol{r}}_{S_i}$	acceleration of the center,
$\boldsymbol{F}_i, \boldsymbol{T}_{S_i}$	applied force and torque with respect to the center of mass,
$\delta \boldsymbol{r}_{S_i}, \delta \boldsymbol{\varphi}_i$	virtual displacement of translation and rotation

The virtual displacements from Eq. (12.3) are not independent from each other due to their coupling. In order to be able to state the equations of motion of the multibody system in minimal coordinates, the introduction of f independent generalized coordinates $q_1, \ldots, q_f$ is useful. In this, f corresponds to the number of DoF and therefore determines the dimension of the mass matrix. The equations of motion of the mechanical system in minimal coordinates result in

$$\boldsymbol{M}\ddot{\boldsymbol{q}} + \boldsymbol{b} = \boldsymbol{Q}, \tag{12.4}$$

with the terms and dimensions:

$\boldsymbol{q}$	$: [f \times 1]$	Vector of the generalized coordinates,
$\boldsymbol{M}$	$: [f \times f]$	Generalized mass matrix,
$\boldsymbol{b}$	$: [f \times 1]$	Vector of the generalized coriolis-, centrifugal-, and gyroscopic forces,
$\boldsymbol{Q}$	$: [f \times 1]$	Vector of the generalized applied forces.

Using the approach of kinematic differentials, and using the pseudo velocities $\tilde{\dot{r}}_{S_i}^{(j)}$, $\tilde{\omega}_i^{(j)}$ and the pseudo accelerations $\tilde{\ddot{\boldsymbol{r}}}_{S_i}$, $\tilde{\dot{\boldsymbol{\omega}}}_i$, the elements of the equations of motion are calculated (Hiller and Kecskeméthy 1989; Kecskemethy 1993). The elements thus obtain the form (Sect. 4.6)

$$M_{j,k} = \sum_{i=1}^{n_B} \left[m_i\, \tilde{\dot{\boldsymbol{r}}}_{S_i}^{(j)} . \tilde{\dot{\boldsymbol{r}}}_{S_i}^{(k)} + \tilde{\boldsymbol{\omega}}_i^{(j)} . \left(\boldsymbol{\Theta}_{S_i} \tilde{\boldsymbol{\omega}}_i^{(k)} \right) \right], \tag{12.5}$$

$$b_j = \sum_{i=1}^{n_B} \left[m_i\, \tilde{\dot{\boldsymbol{r}}}_{S_i}^{(j)} . \tilde{\ddot{\boldsymbol{r}}}_{S_i} + \tilde{\boldsymbol{\omega}}_i^{(j)} . \left(\boldsymbol{\Theta}_{S_i} \tilde{\dot{\boldsymbol{\omega}}}_i + \boldsymbol{\omega}_i \times \boldsymbol{\Theta}_{S_i} \boldsymbol{\omega}_i \right) \right], \tag{12.6}$$

$$Q_j = \sum_{i=1}^{n_B} \left[\tilde{\dot{\boldsymbol{r}}}_{S_i}^{(j)} . \boldsymbol{F}_i + \tilde{\boldsymbol{\omega}}_i^{(j)} . \boldsymbol{T}_i \right]. \tag{12.7}$$

In order to transfer the kinematic differentials on to the newly introduced modular structure of the vehicles correctly, the equations of motion need to be transformed. The advantage of the then chosen approach can be found in the modularization of the needed matrices and vectors which is now possible in a way in which the single elements of the corresponding modules can be calculated. Therefore, the Eqs. (12.8)–(12.10) are divided up into an inner sum inside one module l including all its bodies n_B and an outer sum including all modules n_M (Pichler 1999).

$$M_{j,k} = \sum_{l=1}^{n_B} \sum_{i \in l_i} \left[m_i \, \tilde{\dot{\boldsymbol{r}}}_{S_i}^{(j)} . \tilde{\dot{\boldsymbol{r}}}_{S_i}^{(k)} + \tilde{\boldsymbol{\omega}}_i^{(j)} . \left(\boldsymbol{\Theta}_{S_i} \tilde{\boldsymbol{\omega}}_i^{(k)} \right) \right], \tag{12.8}$$

$$b_j = \sum_{l=1}^{n_B} \sum_{i \in l_i} \left[m_i \, \tilde{\dot{\boldsymbol{r}}}_{S_i}^{(j)} . \tilde{\ddot{\boldsymbol{r}}}_{S_i} + \tilde{\boldsymbol{\omega}}_i^{(j)} . \left(\boldsymbol{\Theta}_{S_i} \tilde{\dot{\boldsymbol{\omega}}}_i + \boldsymbol{\omega}_i \times \boldsymbol{\Theta}_{S_i} \boldsymbol{\omega}_i \right) \right], \tag{12.9}$$

$$Q_j = \sum_{l=1}^{n_B} \sum_{i \in l_i} \left[\tilde{\dot{\boldsymbol{r}}}_{S_i}^{(j)} . \boldsymbol{F}_i + \tilde{\boldsymbol{\omega}}_i^{(j)} . \boldsymbol{T}_i \right]. \tag{12.10}$$

The vectors with the symbol "~" are the pseudo-velocities, or pseudo-accelerations respectively, introduced in Sect. 4.6 which are defined as follows:

$$\tilde{\dot{\boldsymbol{r}}}_{S_i}^{(j)} = \frac{\partial \boldsymbol{r}_{S_i}}{\partial q_j}, \tilde{\ddot{\boldsymbol{r}}}_{S_i} = \sum_{j=1}^{f} \sum_{k=1}^{f} \frac{\partial^2 \boldsymbol{r}_{S_i}}{\partial q_j \partial q_k} \dot{q}_j \dot{q}_k, \tag{12.11}$$

$$\tilde{\boldsymbol{\omega}}_i^{(j)} = \frac{\partial \boldsymbol{\omega}_i}{\partial \dot{q}_j}, \tilde{\dot{\boldsymbol{\omega}}}_i = \sum_{j=1}^{f} \frac{\partial \boldsymbol{J}_{\omega_i}}{\partial q_j} \dot{\boldsymbol{q}} \dot{q}_j; \boldsymbol{J}_{\omega_i} = \frac{\partial \boldsymbol{\omega}_i}{\partial \dot{\boldsymbol{q}}} \tag{12.12}$$

with

$m_i, \boldsymbol{\Theta}_i$ mass and inertia sensor of the body *i*,
$\boldsymbol{F}_i, \boldsymbol{T}_i$ applied force, applied torque
$\boldsymbol{\omega}_i$ angular velocity

By using the kinematics of the vehicle model, the pseudo velocities and pseudo accelerations can be calculated. Due to the tree structure of the model, this analytical calculation using the kinematic differential is considerably faster than the conventional analytical calculation by means of differentiation of the particular relationships. The modular structure becomes apparent from the mass matrix of the already described model of an articulated vehicle, depicted in Fig. 12.45.[1]

It becomes evident from Fig. 12.45 that the mass matrix is symmetrical, and that therefore only the first half of the matrix needs to be interpreted. The calculation of the single elements of the mass matrix is now done in blocks. Since the path from the inertial system (root) to the module (leaf) is known due to the interfaces between the modules, only those couplings need to be considered that are on this path. All other couplings would systemically only display zero-operations. The calculation of the blocks is separated into two steps:

[1]The modules have been summarized since the presentation of all 32 DoF would be too unclear.

- calculation of the blocks of the main diagonal,
- calculation of the blocks on the minor diagonal.

Next to the elements that are structurally defined as zero, there are elements in which a coupling is existent in the standard case, but in which for special kinematic zero-operations may exist. In this specific case of the considered articulated vehicle, the wheels are perpendicularly placed on the axle, so that the coupling term between the axle and the wheel disappears (due to the scalar product). This cannot be detected automatically and therefore needs to be treated accordingly inside the module so that an optimization of the computing time can be achieved.[2]

For the elements of the vector of the generalized coriolis-, centrifugal-, and gyroscopic forces $\boldsymbol{b}$ and the vector of the generalized forces $\boldsymbol{Q}$, only those calculations are executed that provide a contribution to the end result different from zero. Considering for example the rotating motion of the wheels of the third axle of the trailer, it becomes evident that while the coordinates of the wheels are influenced by the motion of the corresponding axle, both elasticities of the trailer, the coupling device, the elasticity of the chassis as well as of course the chassis itself, it is not influenced by the motion of the other modules. It therefore becomes obvious that the rotation of a different wheel does not have an impact on the observed wheel.

The outer structure of the model does not contain kinematic loops. Within the single modules, there may very well exist loops (i.e. steering, wheel suspensions, etc.). These loops are solved within the module using suitable mechanisms. The calculation of the kinematics and dynamics is done for each module individually. At first, the kinematics with respect to the tree structure, beginning at the root, is evaluated. Then, the dynamics are calculated in reverse order.

With the used processes, the single modules can be linked to create a multibody system with any number of DoF and modules using a tree structure. Notably, the modules can be replaced and the structure of the whole system can be varied. The developed concept can be expanded to include the case in which the modules are connected as loops. This structure of loops of the models is seen in vehicles for example when introducing axle carriers and when modeling the drivetrain.

12.2.4 Numeric Integration

In this section different approaches to integration methods are used depending on the task and the system that is to be simulated will be discussed. In detail, the following approaches are available in FASIM_C++:

Single-step method

- HEUN- approach

[2]A calculation of these terms would not change the overall results but increases the calculation time.

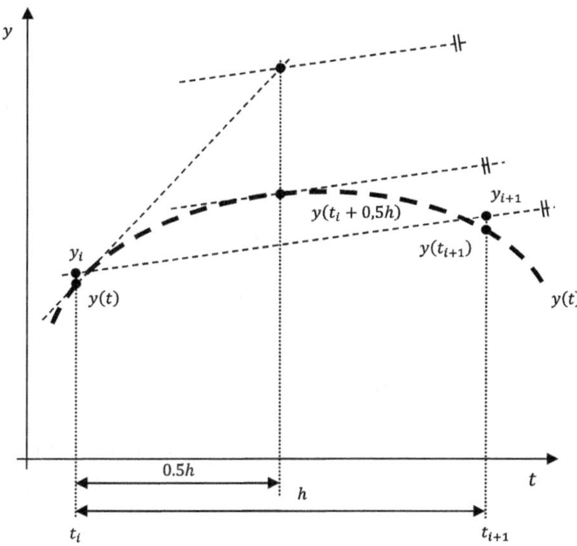

Fig. 12.46 Explicit one-step method according to HEUN

Multi-step method

- Predictor-correction approach by ADAMS
- LSODAR-integration approach (reversible between explicit ADAMS- and implicit BDF methods)

One-step method For a simple one-step method, the approach of EULER is used:

$$y_{i+1} = y_i + hf(t_i, y_i). \tag{12.13}$$

This method usually needs very small increments h in order to provide good approximate values (Schwarz and Köckler 2009; Gottschling and Schramm 2014). Therefore, an approach according to HEUN is used which evaluates the function $f(t, y)$ in two places. Geometrically this approach can be interpreted via the gradients (Fig. 12.46). The consistency order is consequently increased from 1 to 2.

$$y_{i+1} = y_i + hf(t_i + \frac{h}{2}, y_i + \frac{h}{2} f(t_i, y_i)) \tag{12.14}$$

The advantage of this method lies in its short computing time. For more complex driving maneuvers such as for example the lock-up or rollover simulations, the attainable accuracy often is not adequate, in which case more complex approaches need to be used.

A great advantage of this approach is its real time capability, which is a requirement for hardware-in-the-loop-simulations. For the calculation of an additional value at a certain point in time, only the sizes that are available at the current time instant are needed (explicit one-step method).

Multi-Step method Since components with very different eigen-frequencies may be implemented in the modeling of a vehicle as a mechatronic system, the probability is high to obtain a system of stiff differential equations. Integration approaches with automated step size control try to keep the local error as low as possible. If the error exceeds a fixed maximum error value, the step size is reduced. If the error drops significantly below the maximum error, the step size is increased respectively. For non-stiff systems, the step size can usually be chosen depending on the largest eigenvalue according to amount (the smallest time constant respectively) without leaving the stable region of the integration process. For stiff systems this causes problems as a step size control depending on the largest eigenvalue according to amount can lead to numeric problems with rounding errors for the components with substantially smaller eigenvalues and will additionally results in considerable increased computing times. Therefore it is necessary to equip FASIM_C++ with numeric integration methods that are able to integrate stiff, as well as non-stiff differential equations.

For the numerical integration of the differential equations, linear multi-step methods in the form of

$$y_i = \sum_{n=0}^{K_1} \alpha_{i,n} y_{i-n} + h \sum_{n=0}^{K_2} \beta_{i,n} \dot{y}_{i-n} \tag{12.15}$$

are used which are described extensively in Gear et al. (1985). Here, $\alpha_{i,n}$ and $\beta_{i,n}$ are constant coefficients. The step size h is internally controlled via the integrator. In this formula, the index i describes the last determined value which was achieved through numeric integration.

Depending on the size of the quantities K_1 and K_2, which are dependent on the used order of integration q, two special groups of integration methods can be derived. The explicit Adams methods can be derived from Eq. (12.15), for which $K_1 = 1$ and $K_2 = q - 1$. Using the framework of the here discussed methods, integration is done with an order of q between one and twelve (Shampine and Gordon 1975; Radhakrishnan and Hindmarsh 1993). Adams formulae are applied to solve non-stiff differential equations. Through rearranging, the following relationship can be derived

$$y_i = y_{i-1} + h \sum_{n=0}^{q-1} \beta_n \dot{y}_{i-n}. \tag{12.16}$$

The BDF-formulae (BDF = Backward Differentiation Formula), which are used for solving stiff differential equations, are multi-step methods that result immediately from Eq. (12.15). Among others, their advantage is the significantly higher numeric areas of stability. Implicit approaches are more complex per step, but allow for larger step sizes, an important factor for stiff differential equations. It is notable however that the influence of the dynamics of the eigenvalues with the largest amounts is numerically reduced.

For $K_1 = q$ and $K_2 = 0$ results after short conversion

$$y_i = \sum_{n=1}^{q} \alpha_n y_{i-n} + h\beta_0 \dot{y}_i. \tag{12.17}$$

These approaches need however a much higher computing time. Since the stiffness of a system is a local characteristic, which is determined by the JACOBIAN matrix, the system matrix of the local linearization, it doesn't change for the observed, nonlinear system during the simulation. If all eigenvalues of the JACOBIAN matrix are within the same magnitude, the system does not need to be approached with the methods for stiff systems and a shift towards methods for non-stiff systems would shorten the computing time immensely.

Therefore, FASIM_C++ uses the LSODAR integration method in order to solve stiff and non-stiff differential equations. By analyzing an algorithm described in Petzold (1980), it is able to recognize stiff, or non-stiff systems respectively and to treat them differently. Additionally, a root-finding option is integrated which will be discussed in more detail in Sect. 12.2.5.

12.2.5 Treatment of Events

The numerical integration of the equations of motion of multibody systems requires a suitable analysis of results in which the differentiation between time-, step-, and state incidences is made (Otter 1995). Figure 12.47 illustrates the control of the step size and incident control implemented in FASIM_C++.

In the vehicle model, changes of the state occur for example at wheel lock-up, while the control of the step event depends on time-discrete systems such as sensors, controllers, or output modules.

Structurally variant systems For some systems it may be required to change the number of DoF or the transmission behavior of some modules during the simulation. This necessitates a stop of the simulation and a subsequent reboot of the integrator. The number of DoF for example changes in a braking maneuver with wheel lock-up. As soon as one wheel is locked, the degree of rotational freedom ceases to exist and the system needs to be reconfigured. As soon as the wheel turns again, the degree of freedom is subsequently added and the system again needs a reconfiguration.

Therefore, an event-driven function $\boldsymbol{g}(\boldsymbol{q}, t)$ is defined which is dependent on the state variables and time. The elements of an event vector represent single incidents (wheel lock-up, shifting in an automated gearshift, etc.). This function is transferred to the integrator so that the sign change of elements of the state vector can be detected. Depending on the event, the number of DoF is adjusted and a

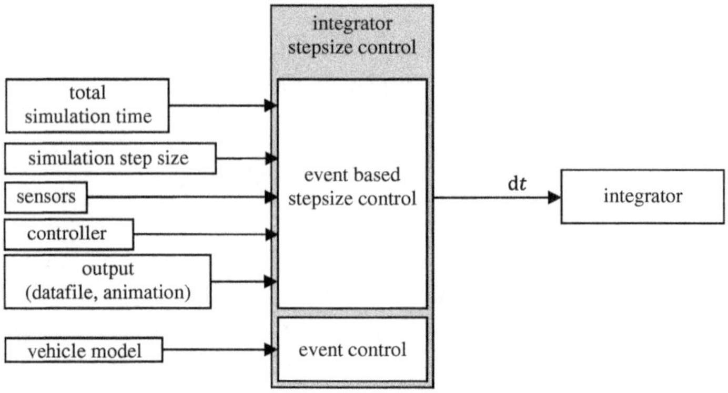

Fig. 12.47 Control of the integrator

re-initialization of the system with a subsequent reboot of the integrator is performed. A simple example for a structurally important system is an overrun brake that was designed for the model of a caravan as it was introduced in Sect. 12.2.2. A further example for a reconfiguration is the treatment of impacts.

Time-discrete time models The development environment FASIM_C++ was designed for the development of complex control systems. Since the controllers and sensors used are time-discrete systems, the integration needs to be interrupted depending on the sampling time of the controllers and sensors. For that reason it needs to be ensured that the integrator interrupts the corresponding time interval of the integration in time so that the corresponding time-discrete system can be called up. In order to implement this control, a list of time-discrete systems is created in FASIM_C++, which communicates the next exit point to the integrator. After interference with the time-discrete system, the control is given back to the integrator. This list contains not only the controllers and sensors, but also the output mechanisms for the simulation data and for the animation of results.

References

Adamski D (2001) Komponentenbasierte Simulation mechatronischer Systeme [Dr.-Ing.]. Dissertation, Universität Duisburg-Essen, Düsseldorf

ATZ (2008) ATZ Extra: Der VW Golf VI. *ATZ* D 58922

Bardini R (2008) Auslegung von Überschlagschutzsystemen für Personenkraftwagen mithilfe der Simulation [Dr.-Ing.]. Dissertation, Universität Duisburg-Essen Düsseldorf: VDI-Verlag

Erhardt R and van Zanten AT (1995) Die Regelung der Fahrdynamik im physikalischen Grenzbereich. *VDI-Berichte* 1224. S 423–38

Frik S (1994) Untersuchungen zur erforderlichen Modellkomplexität bei der Fahrdynamiksimulation [Dr.-Ing.]. Dissertation, Universität—Gesamthochschule Duisburg

Gear C, Leimkuhler B and Gupta G (1985) Automatic integration of Euler-Lagrange equations with constraints. *Journal of Computational and Applied Mathematics* 12, ISBN 0377-0427. S 77–90

Gottschling J and Schramm D (2014) Numerische Methoden für Ingenieure. Vol. 2, Klartext Medienwerkstatt GmbH, Essen, Universität Duisburg-Essen. ISBN 978-3-944339-08-5

Hiller M (1989) Modeling the dynamics of a complete vehicle with nonlinear wheel suspension kinematics and elastic hinges. In IUTAM-Symposium on Dynamics of Controlled Mechanical Systems, Springer-Verlag, ISBN 978-3-642-83583-4

Hiller M (1995) Multiloop Kinematic Chains, and Dynamics of Multiloop Systems in Kinematics and Dynamics of Multi-body Systems

Hiller M and Kecskeméthy A (1987) A computer-oriented approach for the automatic generation and solution of the equations of motion for complex mechanisms. In *Proceedings of the 7th. World Congress on Theory of Machines and Mechanisms*, Vol. 1, S 425-30, IFToMM, Pergamon Press., Sevilla, Spain

Hiller M and Kecskeméthy A (1989) Equations of motion of complex multibody systems using kinematical differentials. In *Transactions of the Canadian Society of Mechanical Engineers*, Vol. 13, S 113–21

Hiller M, Kecskeméthy A and Woernle C (1986−1988) Computergestützte Kinematik und Dynamik für Fahrzeuge, Roboter und Mechanismen. In *Carl-Cranz-Kurs*, Vol. 1.16, Carl-Cranz-Gesellschaft, Oberpfaffenhofen

Hiller M, Kecskeméthy A and Woernle C (1986) A loop-based kinematical analysis of complex mechanisms. In *Presented at the Design Engineering Technical Conference*, Columbus, Ohio

Kecskemethy A (1993) Objektorientierte Modellierung der Dynamik von Mehrkörpersystemen mit Hilfe von Übertragungselementen [Dr.-Ing.]. Dissertation, Universität Duisburg, Düsseldorf: VDI-Verlag

Otter M (1995) Objektorientierte Modellierung mechatronischer Systeme am Beispiel geregelter Roboter [Dr.-Ing.]. Dissertation, Düsseldorf: VDI-Verlag

Petzold L (1980) Automatic selection of methods for solving stiff and non-stiff systems of ODE. *Sandia National Laboratory Report SAND*. S 80–8230

Pichler V (1999) Modellbildung der Dynamik von Kraftfahrzeugen unter Anwendung objektorientierter Konzepte [Dr.-Ing.]. Dissertation, Universität Duisburg, Düsseldorf: VDI-Verlag

Radhakrishnan K and Hindmarsh AC (1993) Description and use of LSODE, the Livermore solver for ordinary differential equations. National Aeronautics and Space Administration, Office of Management, Scientific and Technical Information Program

Schmitz T (1994) Modellbildung und Simulation der Antriebsdynamik von Personenwagen [Dr.-Ing.]. Dissertation, Universität—Gesamthochschule Duisburg, Duisburg

Schnelle K-P (1990) Simulationsmodelle für die Fahrdynamik von Personenwagen unter Berücksichtigung der nichtlinearen Fahrwerkskinematik Dissertation, Universität Stuttgart, Stuttgart

Schuster C (1999) Strukturvariante Modelle zur Simulation der Fahrdynamik bei niedrigen Geschwindigkeiten [Dr.-Ing.]. Dissertation, Universität—Gesamthochschule Duisburg, Duisburg

Schwarz HR and Köckler N (2009) Numerische Mathematik. Springer DE, ISBN 978-3-8348-0683-3

Shampine LF and Gordon MK (1975) Computer solution of ordinary differential equations: the initial value problem. WH Freeman San Francisco - ISBN 978-0-7167-0461-4

Uffelmann F (1980) Berechnung des Lenk- und Bremsverhaltens von Kraftfahrzeugzügen auf rutschiger Fahrbahn [Dr.-Ing.]. Dissertation, TU Braunschweig, Braunschweig

Woernle C (1988) Ein systematisches Verfahren zur Aufstellung der geometrischen Schließbedingungen in kinematischen Schleifen mit Anwendung bei der Rückwärtstransformation für Industrieroboter [Dr.-Ing.]. Dissertation, Universität Stuttgart, Düsseldorf: VDI-Verlag

Chapter 13
Model of a Typical Complex Complete Vehicle

This chapter shows the modeling of a typical mid-range vehicle in FASIM_C++ (Adamski et al. 1997; Opgen-Rhein 1998). Additionally, the methods used for the verification and validation of the model are described. This results in the verified model of a typical mid-range passenger car. This vehicle model is examined concerning its significance using the real data from experiments with four different mid-range vehicles. Using this driving dynamics simulation model, the analyses, presented in Chap. 14 for example for the design of rollover protection systems can be executed.

13.1 Modeling of the Complete Vehicle

Just as it was described in Sect. 12.2.3, FASIM_C++ is based on the formulations of the equations of motion of complex modular multi body systems using kinematic differentials. This method enables a division of the vehicle into larger units (modules) with standardized points of intersection. The definition of the points of intersection as well as the modular notation of the equations of motion can be found in Pichler (1999). The dynamic behavior of the mechanic modules such as chassis, front axle, rear axle and drivetrain is described in the general system using differential equations of the second order. The behavior of the non-mechanical components such as braking hydraulics, environment, driver and engine is though described using differential equations of the first order.

Before all modules can be combined to a total system as described in Sect. 12.2.2, the sub-modules describing the kinematics need to be implanted into FASIM_C++. The kinematics can either be calculated explicitly as it was done in example in Sect. 6.5.2, or described using the program package MOBILE (Kecskeméthy 1996).

For the complete vehicle model introduced here, the kinematics of the wheel suspension was calculated using MOBILE. For the modeling of multi body

D. Schramm et al., Vehicle Dynamics,
DOI 10.1007/978-3-662-54483-9_13

systems, basic building blocks such as reference coordinate systems, joints (prismatic and revolute joints), stiff connections as well as objects for the definition and solving of kinematic loops in the programming language C++ are already implemented in MOBILE.

In order to include the kinematics sub-modules, FASIM_C++ uses gateways, so-called "shells", which complete the calculation of the global kinematic, which is necessary for the creation of the equations of motion. For the presentation of the integrating of the MOBILE-wheel suspensions into the axle module, Fig. 13.1 includes an exemplary sub-modular structure of the front axle. For the calculation of the axle kinematics, the turned-on MOBILE-modules are hereto activated from the steering and from the chassis first. Then, the wheel suspension kinematics is calculated inside the MOBILE-modules. At last, the calculation between the anti-roll bars in between both wheel suspensions is calculated. The calculation of the kinematics of the complete vehicle is done according to the flow depicted in Fig. 13.1, starting at the inertial system.

If all non-mechanical and mechanical modules (see Sect. 12.2.2) of the vehicle are available, they can be combined to a complete model. The building of the single

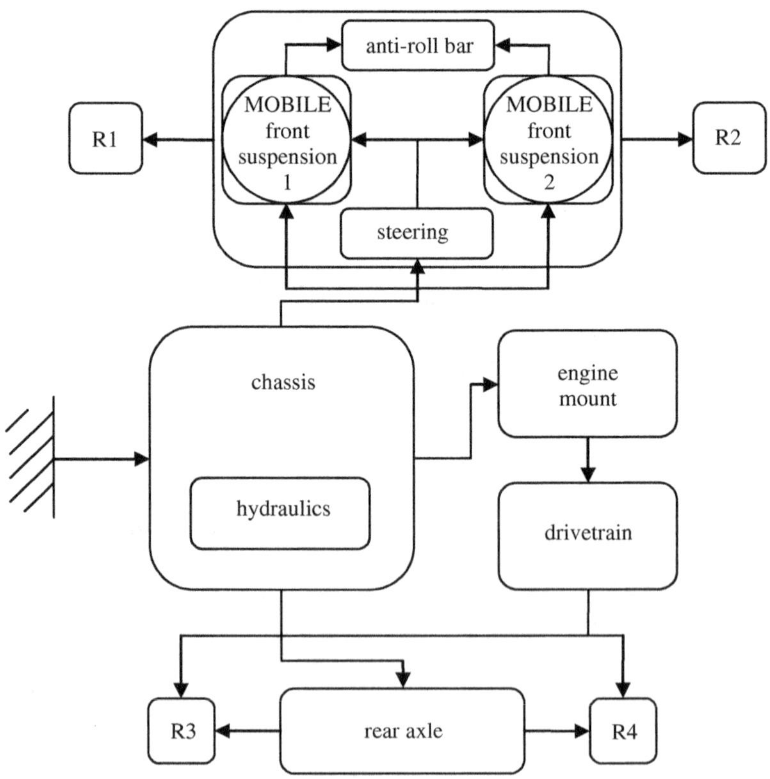

Fig. 13.1 Module of the complete vehicle in FASIM_C++

modules is done in a text file, the so-called vehicle pointer file. Additionally, the names of the needed data files for the initialization of the modules are entered in this control file. When starting the simulation, in the first step (initialization), the reading of the model parameters from these data files, the management of this data as well as a possibly necessary pre-calculation of constant parameters is done.

In the second step (kinematics), the calculation of the local kinematic of the single modules is done. The last step (dynamics) includes the determined applied forces and torques along with the variables needed for the creation of the equations of motion.

The combination of the modules in the vehicle pointer file enables the fast modeling of different mid-range vehicles with differing wheel suspension systems. For this, the needed kinematic modules and their corresponding vehicle specific data sets need to be available. Thus, in the model library of FASIM_C++ has implemented for example the wheel suspension kinematics depicted in Table 13.1. In combination with the corresponding data sets, with this, four different mid-range vehicles (A–D) can be simulated.

For the modeling of the mid-range vehicles A–D, already known modules are used for the modules driver, engine, chassis, drive train, hydraulics, and wheel. Here, only adaptions of the available data sets are implemented. For the modules engine, hydraulics, as well as drive train this was however only partially possible since not all data is available for these. Here, data sets of other comparable validated vehicles can be used. The influence of these components is in the roll-over maneuvers of Sect. 14.2 only of minor significance, as only the vehicle with open clutch is observed. In the sub modules steering and stabilization for the front and back axle, already available modules could be used as well. For this, depending on the type of vehicle, the stabilization is executed using a simple torsion beam with or without connected end link, and the steering is modeled as a rack and pinion version.

The wheel is of great importance in all maneuvers as the contact element between the road surface and the vehicle. It is allocated in FASIM_C++ as a sub-module of the drive train and great effort was put into finding detailed, vehicle specific values for the data set of the wheels. Thus, in all vehicles, the exact specifications of for example the wheel sizes are considered. This data is important for the calculation of the exact kinematics of the road-wheel-contact. The kinematic wheel model, which FASIM_C++ is based upon, originates from the mechanical replacement model for the calculation of the contact patch, which was introduced in Sects. 7.4.1, 7.4.2 and 7.4.3. From this, the movement of the contact patch in

Table 13.1 Types of wheel suspension of typical mid-range vehicles (see Sect. 6.2)

Vehicle	Front wheel suspension	Rear wheel suspension
A	MacPherson (damper strut)	Five-link
B	MacPherson (spring strut)	Twist-beam
C	MacPherson (spring strut)	Four-link
D	Four-link suspension	Double wishbone

correlation with the toe angle as well as the rotation- and translational velocity of the wheel carrier as well as the center of the wheel can be calculated. This then enables the calculation of the tire slip angle and longitudinal slip. The calculation of the dynamic wheel forces is done using the HSRI-wheel models (see Sect. 7.4.6) (Highway Safety Research Institute). This model allows the determination of the longitudinal and horizontal forces on the wheel from slip angle and slip. For the calculation of the wheel forces with this physical model, next to wheel specific coefficients for the calculation of the sliding friction forces, the vertical spring stiffness and the longitudinal and horizontal stiffness of the normal force of the wheel is needed. These parameters are not, or only partially, available for some wheel models, so that parameters are used from other, comparable wheel models from validated vehicle models. This approach leads to adequate exact vehicle models for the relevant driving maneuvers for the roll-over simulation.

13.2 Model Verification and Validation

For an evaluation of the models described in Sect. 13.1 concerning its simulation results, different approaches can be used [see hereto also (Pelz 2001; Kramer and Neculau 1998)]. In practice, the earliest possible execution of simulation runs, especially before the phase of model building has emerged as a very important corrective, in order to find modeling mistakes early and to improve the model quality in steps. This approach has led to qualitatively meaningful models in the complete vehicles of the mid-range vehicles A, B, C and D, even before the implementation of validation and verification. Additionally, the vehicle models were validated for roll-over models on ramps, or embankments respectively. The general approach for the evaluation of the model quality can fundamentally be divided up into the two parts verification and validation. In doing so, the model verification asks the question whether the model fulfills the requirements both as a complete model and in its components. The task of model validation on the other hand consist of showing, using the data obtained from real experiments, that the executable simulation model is suitable for a certain application [see also (VDI-Richtlinie-2206 2004)]. The steps for validation and verification of the simulation models that were executed for the vehicles A-D can be divided up into the following tasks.

13.2.1 Verification

1. Plausibility tests
   - Testing of the toe- and camber characteristic curves of front- and back axle of the respective wheel suspension sub modules.
   - Testing of stationary vehicle conditions of the complete vehicle model based on the following maneuvers:
     - slip-free straight ahead cruising at 30 km/h,
     - steady state circular test at constant 50 km/h.
2. Visual inspection and graphical animation of the wheel suspensions and the complete vehicle.

13.2.2 Validation

1. Exact adjustment of the real experiment scenarios inside the simulation environment.
2. First visual comparison of real experiment videos and animated simulation results of the experiment.
3. Removal of model mistakes and fine-tuning of the not exactly available model parameters.
4. Graphical comparison of the relevant measured and simulated data of the roll angular velocity ω_x and the lateral and vertical acceleration a_y and a_z.

Because a complete discussion of these verification and validation tasks for all examined vehicles would exceed the frame of this book, for the presentation of the method, only a few examples shall be presented for the relevant aspects substituting the whole process.

As a first plausibility test for the verification, for all four modeled vehicles, the toe angle and camber angle curves of the front- and rear axles were calculated. In Fig. 13.2, the comparison of the calculated and the simulated toe angle and camber angle curves at the example of vehicle A is depicted. The deviation between simulation and measurement are based on the elasto-kinematics of the bearings which were neglected in the model. Yet, the curves produced by the simulation have proven fairly exact results even despite of the simplifications of the model, so that the behavior of the wheel suspensions can be accepted as verified.

The second part of the plausibility test is mostly aimed at the testing of the complete vehicle model regarding the basic functions of wheels, wheel suspensions, anti-roll bars, steering and drive train.

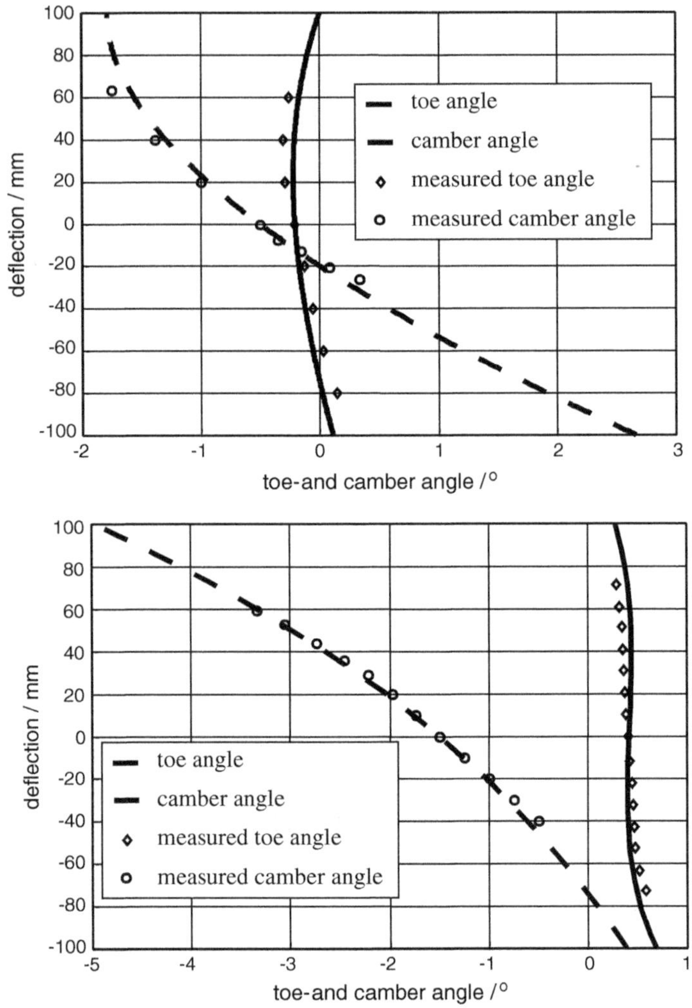

Fig. 13.2 Verification of the toe and camber angle curves for vehicle A

Constant driving at 30 km/h shows the correct simulation of a stationary driving state free of acceleration, steering, and surface stimulation. In the unclutched state, the vehicle needs to stably roll in a quasi-stationary state. Possible model imperfections could result in oscillating states, non-plausible drive paths as well as simulation cancellations due to instabilities.

The steady state circular test on the other hand verifies the complete vehicle for acceleration, and steering influences in this maneuver, after a certain time period, a stationary circle needs to be established as presented in Fig. 13.3 at the example of vehicle D. For the testing of a correct function of the anti-roll bars, for stationary circular drive with deactivated anti-roll bars, an increased roll angle should result.

Fig. 13.3 Verification of the model behavior at a circular test at constant 50 km/h

Generally, the visual inspection of the sub models, or the complete model, respectively, as depicted in Figs. 6.40 and 13.3 is of great value. Using animations at an early stage, mistakes can be avoided in the model building already. They also impressively and quickly support the plausibility testing and give hints to possible mistakes that wouldn't be possible to see from data material without further analysis.

First step within the framework of the validation is the exact reconstruction of the respective experiments. Hereto, the real experimental data of driving maneuvers on ramps and embankments are available. The available driving maneuvers are summarized in Table 13.1. In order to be able to recreate the experiments inside the simulated environment as exactly as possible, an extensive documentation during the experimentation phase is inevitable. This holds especially true for the parameters that are varied within an experimental series. In all experiments, these parameters are in equal measure the respective weight of the vehicle and possible larger added vehicle loads for example in the form of measurement engineering and dummies, and the road condition as well as the weather condition during the implementation of the experiment. Since the experiments are partially conducted outside, this information is necessary in order to estimate the friction coefficient between the wheel and the road. If the conditions of the environment are known, the kinetic coefficients of friction can be estimated using literature (Wallentowitz and Mitschke 2006; Bosch 2011).

Furthermore, the configurations depending on the kind of experiment need to be known. Figure 13.4 presents the set-up of the validation experiments with vehicle B. Contrary to the curved guide rail proposed in SAE-policy SAE J857, in this experiments a straight guide rail in combination with automated steering has been used. The vehicle is accelerated using the guide rail and rolls freely after the uncoupling from the guide rail. At a predefined point, the front wheels are turned in an angle of 35° within 0.2 s to the left (Fig. 13.4), so that the vehicle crosses the ramp with the wheels inside the curve. The ramp is 5 m long and 1 m broad. Its height was varied between 40 and 50 cm for the experiments so that depending on the speed of the vehicle (50–80 km/h), a roll-over maneuver, or almost a roll-over maneuver is achieved. In the experiments that lead to a roll-over maneuver, the tendency to roll over was heavily dependent on a "dig-in effect" of the wheels in the area of the topsoil.

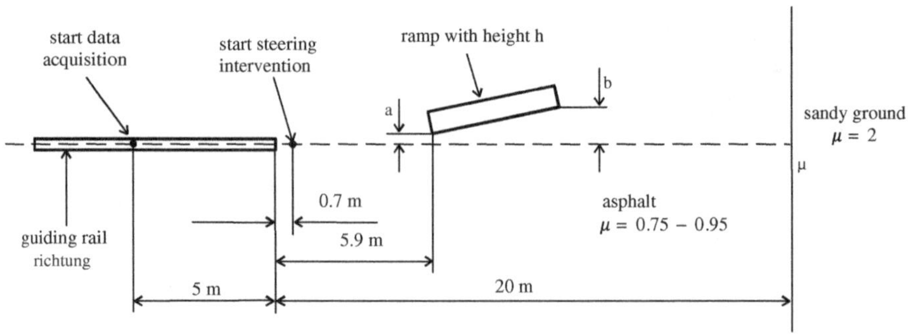

Fig. 13.4 Top view of ramp test area similar SAE J857 (dimensions *a* and *b* dependent on test vehicle)

Since for this effect there is no specific wheel-ground-model available in FASIM_C++, this can only be approximated using a sudden change of friction (Fig. 14.9). This leads to a strong increase of the roll angular velocity when entering the topsoil in the simulation. Since in reality, the vehicles have already reached a roll angle of 50°, the validation has only been observed up until this range (Fig. 13.10).

The executed ramp crossing experiments for the validation of vehicles C and D are comparably simple in their set-up. These experiments have been executed using a steel ramp analogous to Fig. 13.5. Here, the vehicle rolls over the ramp with declutched drive train on dry asphalt using the left wheels and without steering. The height of the ramp has been varied in steps of 10 cm between 50 and 100 cm. The examined velocity range lies between 18–72 km/h.

For the validation of vehicles A and D, embankment crossings were performed (Bardini et al. 1996). Figure 13.6 shows the embankment maneuver on dry, hardened clay, adjusted in FASIM_C++. In this experiment, the vehicle rolls freely at a velocity of 57 km/h at a crossing angle of 11° straight onto the edge of the embankment. In the embankment maneuver using the artificially adjustable embankment presented in Fig. 13.7, there is only one level of embankment, which is inclined at the angle β. Here, the vehicle rolls freely towards the edge of the embankment of 1.80 m at an approach angle α of 15° or 20°. The tested range of velocities is between 18 and 45 km/h. The kinetic coefficient of friction on the

Fig. 13.5 Adjustable ramp (length 5 m) for validation tests similar ADAC-screw test

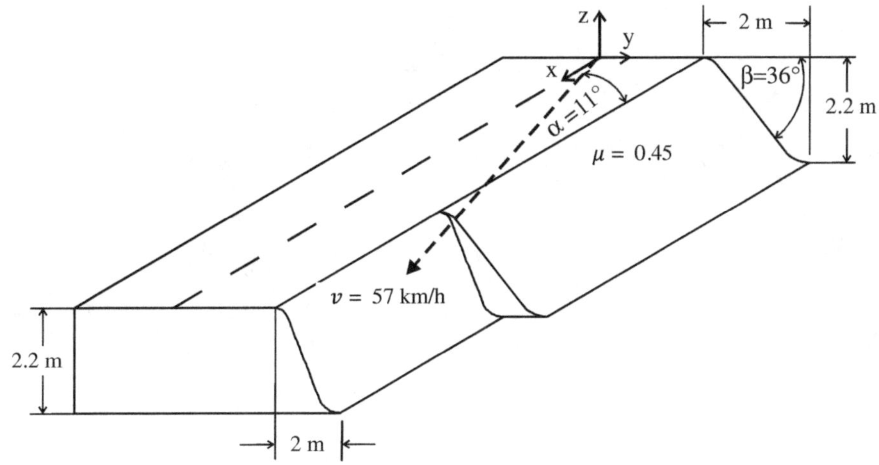

Fig. 13.6 Experimental setup and parameters for the embanking maneuvers for vehicle A

embankment which is especially coated with a layer of a material similar to asphalt and is approximated as $\mu = 1.0$. In order to obtain more reliable roll-overs on the relatively flat embankment, a steering angle is added on the upper edge of the embankment. This is automatically activated as soon as the middle of the vehicle reaches the upper edge of the ramp.

The second step of the validation is illustrated using the embankment maneuver with vehicle A. Using the video validation partially displayed in Fig. 13.8, it is possible to adjust the kinematics of the vehicle to the real behavior. Here, the following discrepancies in the model behavior can be recognized:

1. deflection and hopping movements of the wheel suspensions are too strong (cause: not defined bump stops and deflection limit of the suspension),
2. deformation of wheels too strong (cause: only linear vertical spring stiffness),
3. unwanted movement of steering (cause: steering system not stiff enough) and
4. roll movement of the chassis too strong (cause: anti-roll bar stiffness too low).

These abnormalities that have been observed from video analysis directly lead to step 3 of the validation. The removal of model mistakes and the execution of the

Fig. 13.7 Experimental setup for the artificial, adjustable embankment

Fig. 13.8 Comparison between simulation and experiment (Bardini et al. 1996)

fine-tuning can for all modeled vehicles be applied at these four points. The aspects 1 and 2 are caused on a definition of the operating range of the vehicle data that is too narrow. Hence for the simulation of normal maneuvers of driving dynamics on a plane, often spring characteristics for wheel suspensions and wheels are sufficient. For the simulation of extreme driving maneuvers at the threshold, such as roll-over maneuvers on the other hand, also require the upper and lower confines of the operating range. In Fig. 13.9, the non-linear progressive, respectively digressive spring characteristics of the wheels and the wheel suspensions are exemplary shown for one of the modeled vehicles. For the wheels, the area of wheel stiffness was expanded to include the range of rim stiffness. According to the rim sizes used at the corresponding vehicles, for the vertical wheel stiffness, the material constant for steel or aluminum respectively is used. Similarly, the compression and depression travel of the wheel suspensions was restricted. In order to achieve this, the linear characteristics of the normal operating range are complemented with the measured characteristics of the so-called bump-and rebound stops.

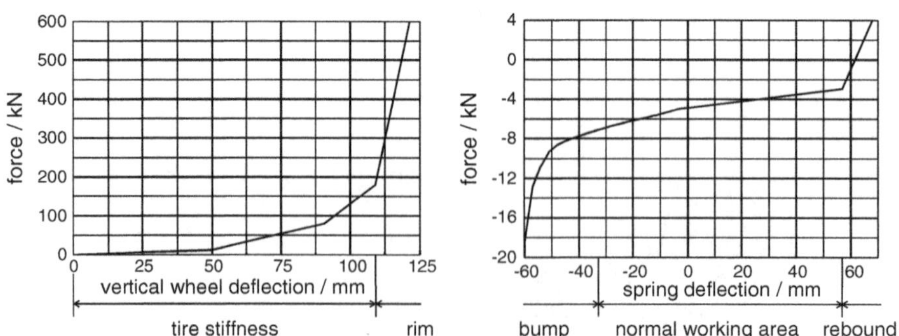

Fig. 13.9 Spring characteristics of the wheels and wheel suspensions

The points 3 and 4 can be ascribed to the manner in which the model parameters of the steering mechanism and the anti-roll bars are not directly available from measurements of the real vehicle. The anti-roll bar model (see Fig. 9.9), used in FASIM_C++ concentrates the characteristics of the whole anti-roll bar including the soft vehicle-mounted bearing in rubber sleeves in a U-shaped mechanism with a turn suspension rate, so that the anti-roll bar model only acts as a simplified force element on the wheel suspensions. Hence, the real value of the anti-roll bar stiffness, available from measurements, needs to be adjusted corresponding to the real vehicle behavior. A similar manner applies for the idealized steering mechanism in which the suspension and bumper constants depict the behavior of the complete steering as concentrated coefficients.

During the process of fine-tuning the complete model, the previously named, but not exactly available model parameters are adjusted so that a vehicle behavior can be displayed which is as close to the real behavior as possible. In order to test the success of the adjustments, in step 4 of the validation, as presented for example in Fig. 13.10 for vehicle B, the relevant measured and simulated data is compared graphically. The data of roll angles and roll velocity of lateral and vertical acceleration stem from a ramp crossing at a velocity of 68 km/h and a height of the ramp of 40 cm, which is executed without roll-over. In Fig. 13.11 on the other hand, the corresponding graphs of a roll-over on a ramp of one meter at 45 km/h using vehicle D is depicted.

Figures 13.10 and 13.11 shall be representative for all listed driving maneuvers that are listed in Table 13.2 and shall convey an impression of the quality of the

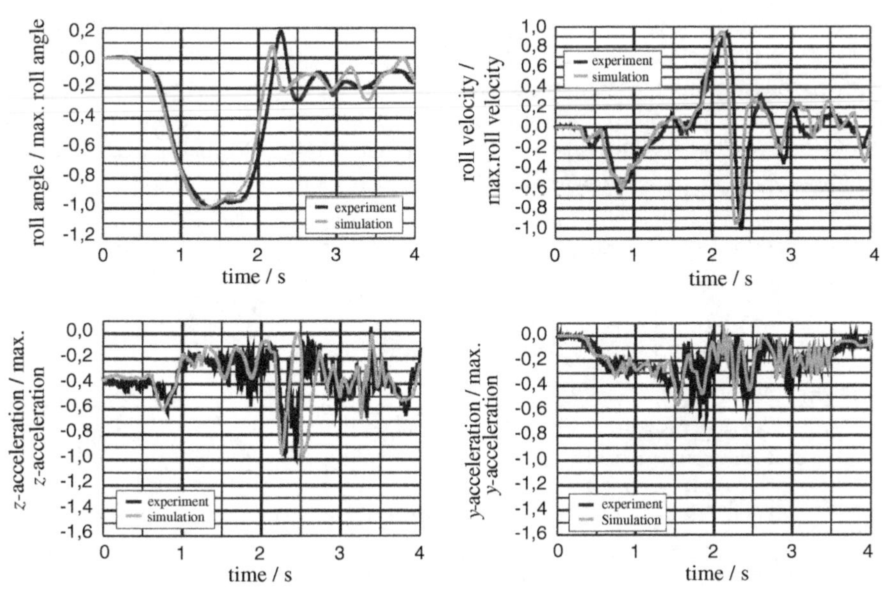

Fig. 13.10 Validation of a ramp crossing similar to SAE J857 without roll-over

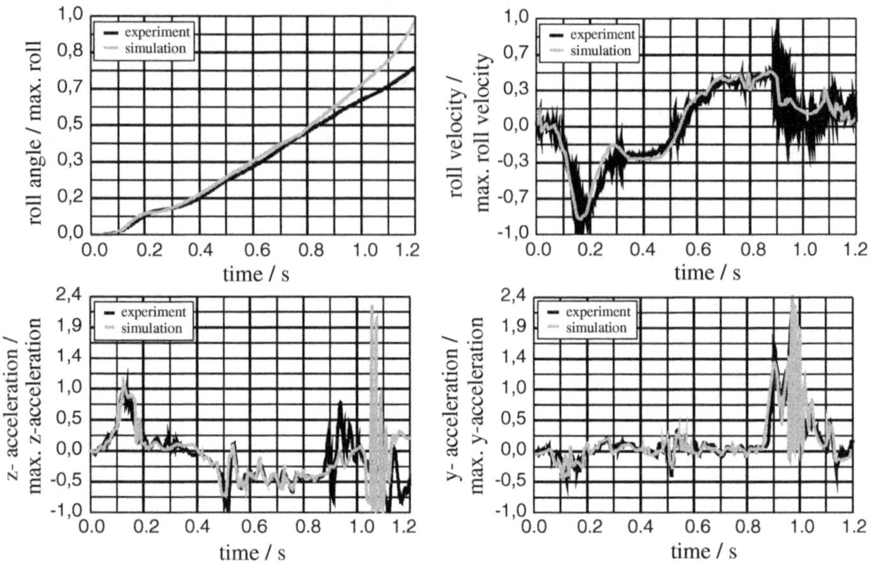

Fig. 13.11 Validation of a ramp crossing similar ADAC-screw test with roll-over

created models for the simulation of roll-over maneuvers. Over all, all models present a high quality for the simulation of almost-roll-over maneuvers. This statement however needs to be restricted for complete roll-over maneuvers. For roll angels larger than 60°, the model behavior cannot be seen as validated, as for such large roll angels on the one hand the wheel model cannot calculate useful vertical and lateral forces (see Fig. 13.12 from 1.1 s) and also the contacts of the chassis with its environment become relevant. These contacts are not considered in the models because for roll-over simulation, a validated calculation up to a roll angle in the size of the static tilt angle of mid-range cars (50–55°) is sufficient. Hence, a reliable statement can be made about the occurrence of roll-overs can be expected for the validated roll-over scenarios. Up to this angular range, the validation confirms a qualitatively and quantitatively good accuracy for the signals relevant for the

Table 13.2 Drive maneuvers available for validation of the vehicle models

Vehicle	Maneuver
A	Embankment crossings with dry hardened clay with roll over (see Fig. 13.8)
B	Ramp crossings similar to SAE J857 with various velocities and ramp heights (Fig. 13.4)
C	Ramp crossings similar to ADAC-screw tests with various velocities and ramp heights (Fig. 13.5)
D	Ramp crossings similar to ADAC-screw tests with various velocities and ramp heights (Fig. 13.5) and drive over embankments with artificial, adjustable embankment angles (Fig. 13.7)

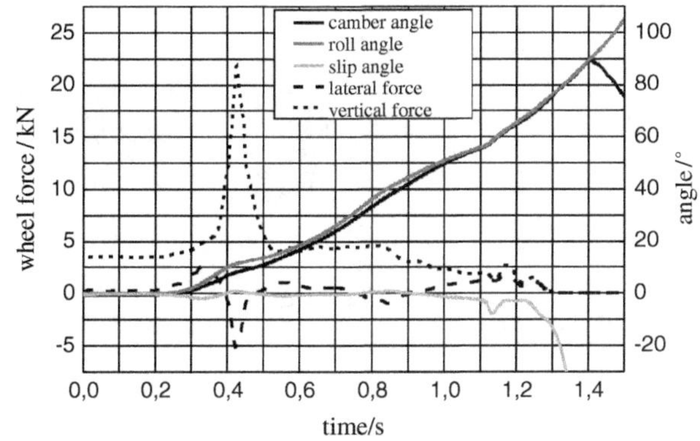

Fig. 13.12 Wheel forces at the rear wheel averted from the ramp during roll-over

interpretation of roll-over protection systems, so the angular range relevant for the triggering of roll-over protection system is completely covered.

13.3 Parameterized Vehicle Model

As the previous sections in this chapter show, the remodeling of a vehicle takes a considerable effort. Furthermore, a large quantity of vehicle specific data is necessary. In order to not depend on the availability of detailed vehicle data for the roll-over simulation of vehicles other than the already validated vehicles it was examined which of the vehicle parameters are actually relevant for the roll-over behavior of mid-range cars. The examinations are divided up into three steps:

1. Creation of an average mid-range vehicle model as reference model for the parameter examination.
2. Gradual examination of all model parameters of the average model concerning their impact on the roll behavior of the vehicle.
3. Verification of the behavior of the parameterized models using the validated original models.

13.3.1 Definition of a Reference Model

In a first step, the average vehicle parameters presented in Table 13.3 were determined based on the registrations statistics (Kraftfahrt-Bundesamt 2006) for mid-range cars. The thereby considered vehicles represent a market share of approximately 95 % (Fig. 13.13).

As a model for the axle kinematics, for the average model, the kinematics of vehicle C (see Table 13.1) is used. In Fig. 13.14, the corresponding topological

Table 13.3 Parameters of the average mid-range vehicle

No.	Parameter		Average model	Max. deviation	No.	Parameter		Average model	Max. deviation
Chassis:					*Rear axle (multi-link)*				
1.	Mass chassis	kg	1.296.5	255.5	11.	Mass of wheel carrier	kg	8.21	1.79
2.	Center of gravity	m	−1.1855	0.2353	12.	Inertia tensor of the wheel carrier	kg m^2	0.0426	0.0574
			0	0				0.103	0.147
			0.2337	−0.0337				0.1033	0.1667
3.	Inertia tensor	kg m^2	482.7	−122.7	13.	Spring characterization		Figure 13.13	n. A.
			2.394.3	−194.3	14.	Damper characterization		Figure 13.13	n. A.
			2.585.6	−225.6	15.	Half-track width	m	0.7437	0.0353
4.	Seat position of occupants front	m	−1.3902	0.0902	16.	Wheel base	m	2.6897	−0.1377
			±0.3699	±0.0114	*Anti-Roll Bar rear*				
			0.2735	-0.0385	17.	Twist spring constant	Nm/°	7.3686	−3.4586
Front axle (MacPherson w. rack and pinion steering)					*Engine bracket*				
5.	Mass of the wheel carrier	kg	10.33	2.67	18.	Position vector to the bearing point	m	0.0031	−0.1431
6.	Inertia tensor of the wheel carrier	kg m^2	0.033	0.067				0	0
			0.083	0.167				0.0857	0.0713
			0.09	0.18	*Wheel*				
7.	Spring characterization		Figure 13.3	n. A.	19.	Wheel radius	m	0.3108	−0.0123
8.	Damper characterization		Figure 13.13	n. A.	20.	Longitudinal moment of inertia	kg m^2	1.375	0.05
9.	Half-track width	m	0.7461	0.0436	21.	Vertical moment of inertia	kg m^2	0.85	0.1
Anti-Roll Bar front					22.	Linear vertical tire spring stiffness	N/m	239.333	−10.667
10.	Twist spring constant	Nm/°	6.4315	2.173	23.	Wheel Characterization		Figure 13.9	n. A.

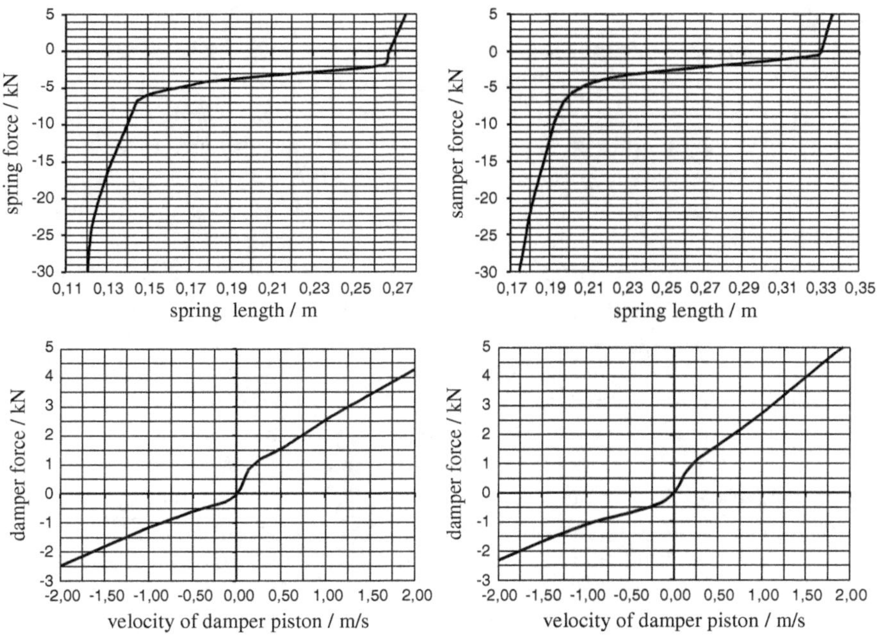

Fig. 13.13 Characteristic curves for springs and dampers of the front wheel (*left*) and rear wheel suspension (*right*) of the average mid-range vehicle model

structure of the complete vehicle is depicted. The MacPherson front wheel suspension in spring strut version and the multi-link wheel suspension with four levers are at the time of the data assessment (Kraftfahrt-Bundesamt 2006) most prevalent in the axle kinematics of mid-range cars. The adjustment of the respective parameters from Table 13.3 is in most cases achieved by a simple adaption of the corresponding numerical values in the data files. That way, the extension or constriction of the track can for example be achieved by changing the y-coordinate of the center of the wheel. The changes of the spring and bumper characteristics of the wheel suspensions as well as the characteristics of the wheels on the other hand, are more complex. In order to determine the average characteristics, the data of the validated vehicles were used as a baseline, since no further detailed data was available. For the determination of the average spring characteristics, the characteristics of the validated vehicles are shifted about the length of the length of the suspension spring. The shifted characteristics therefore operate at the same point. Subsequently, the spring characteristics are divided up into their linear and non-linear aspects and the average slopes and lengths of the linear range are calculated.

There upon, the addition of the averaged non-linear part, contingent on the bump-stop and the rebound-stop (see Fig. 13.9 right). Finally, the characteristics need to be shifted corresponding to the length of the suspension spring of vehicle C, and while considering the spring forces at balance (preload in installation position), towards the operating point of the average vehicle. A much simpler approach can be chosen for the determination of the average damper characteristics. For this purpose, only the averaging at the respective points of support of the characteristics is necessary. The determination of the average mean vertical spring stiffness of the wheels is done the same way. Here, too the averages are calculated at the respective points of support. The determined characteristics are shown in Fig. 13.13.

In a second step, the 23 parameters listed in Table 13.3 need to gradually be examined concerning their effects on the roll behavior of the average mid-range vehicle model. Hereto, for each of the 23 parameters, simulations were conducted with one altered value each, and were then compared to the reference behavior of the average model. The valuation is done based on the basis of the simulation of a drive over a 5 m long and 0.5 m high ramp at a velocity of 54 km/h. As an evaluation criterion, a deviation corridor of 10 % of the reference behavior is determined. If the results of a parameter change in the complete signal curve remains inside the corridor, the parameter is graded as not dominant and can is disregarded for the needed parameter model. As a variation of the 23 parameters, the maximum deviations, presented in Table 13.3 are plugged in. The maximum deviation of each parameter marks the largest recorded change of every single value found in the data set from the corresponding mean of the average model. Exemplary, Fig. 13.15 presents the result of the parameter variation for the location of the center of mass (parameter No. 2 in Table 13.3). From the executed complete analysis of the influencing parameters, the following eight vehicle parameters have been observed as being dominant:

- mass of chassis (parameter No. 1),
- mass center location of the empty complete vehicle (parameter No. 2),
- inertia tensor of the chassis (parameter No. 3),
- spring characteristics of the front axle (parameter No. 7),
- track width of the front axle (parameter No. 9),
- spring characteristics of the rear axle (parameter No. 13),
- track width of the rear axle (parameter No. 15) and
- wheel base (parameter No. 16).

In a third step of the formulation of the parameterized vehicle model it is tested, whether a vehicle model with 15 not dominant, constant parameters and 8 dominant, adjusted parameters can recreate the behavior of a corresponding, more detailed vehicle model.

13.3.2 Comparison of Parameterized Versus Validated Models

For the evaluation of the quality of the parameterized vehicle models, they are used to simulate two different ramp maneuvers each. By comparing the corresponding

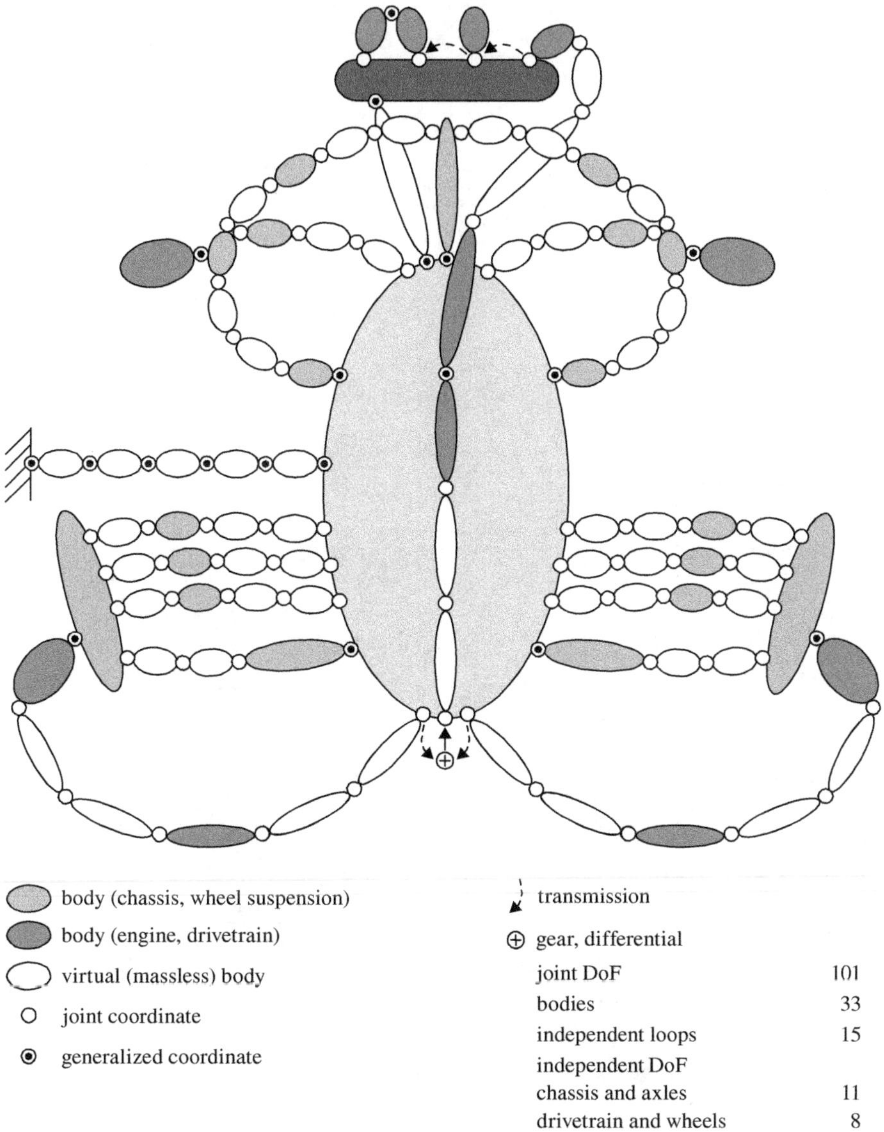

Fig. 13.14 Topological structure of the average mid-range vehicle model

simulations with the validated models from Sect. 13.2, the quality of the parameter models can be presented easily. The comparison is based on the following two ramp maneuvers:

- Non-critical maneuver with low danger of roll-over: Height of ramp 0.5 m, velocity 54 km/h.

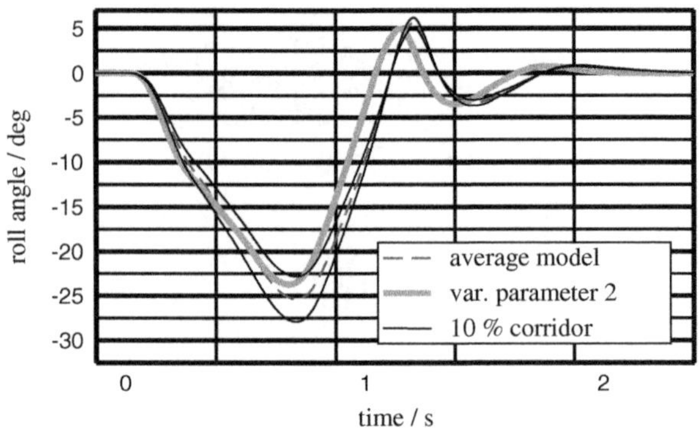

Fig. 13.15 Roll angle curve for variation of parameter No. 2

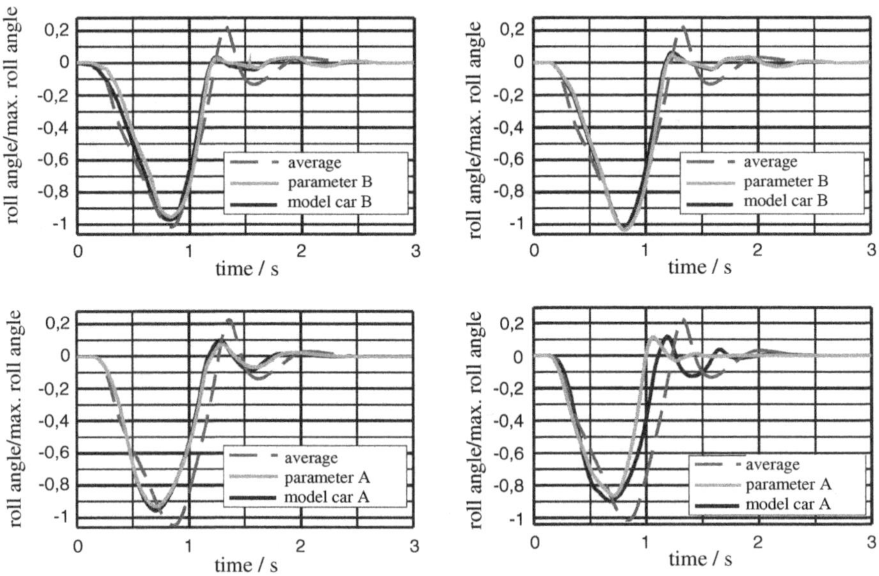

Fig. 13.16 Comparison of the parameterized models with the validated models in a driving maneuver with low danger of roll-over

- Critical maneuver with high danger of roll-over: Height of ramp 0.7 m, velocity 65 km/h.

For the critical maneuver, a configuration was especially chosen that was at the threshold of roll-over. Using this it can be ensured that the parameter models also provide reliable prognoses for the roll-over behavior. In Figs. 13.16 and 13.17, the simulated roll angle of the parameter models of the vehicles A, B, C and D are presented which are each adjusted in their eight parameters. Additionally, each diagram contains the roll angle curve (dotted line) of the correct model. For the non-critical maneuvers, for all parameter models a distinct alignment for all

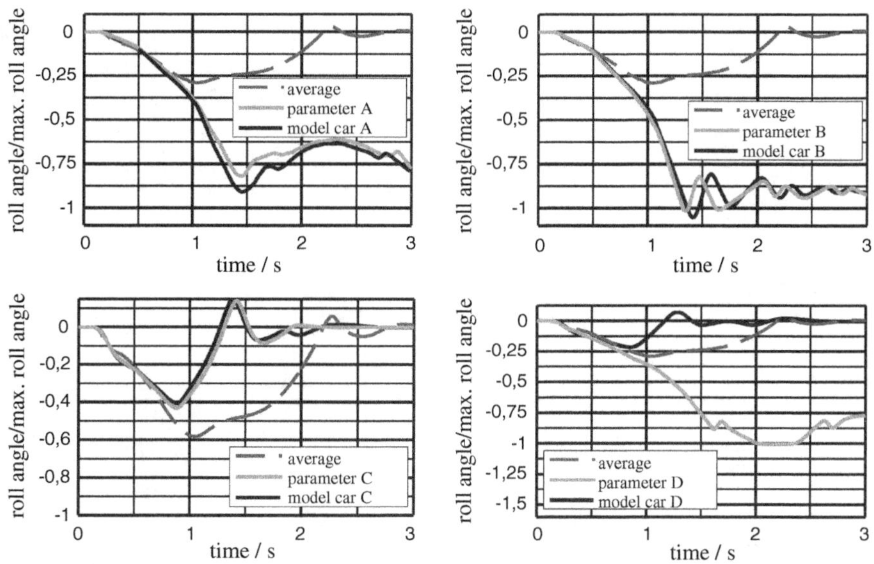

Fig. 13.17 Comparison of the parameterized models with the validated models in a driving maneuver with high danger of roll-over

parameter models can be reached. In the critical maneuver close to the roll-over threshold, this statement can on the other hand only be confirmed for the parameter models of the vehicles A, B, C and D. For vehicle D the parameter model diverges distinctly from the reference behavior, even causing a roll-over, which does not occur for the reference model. The axle kinematics of the front wheel suspension (see Table 13.1), which differ from the other models, could be responsible for these deviations in Vehicle D.

Further simulations using gradual changes of the parameters towards the original parameters in vehicle D have shown that by adopting the correct front axle kinematics and front axle spring characteristics, a substantial improvement of the model can be achieved. Therefore, the front axle kinematics as well as the front axle spring characteristics need to be added to the dominant parameters listed in Sect. 13.3. The front axle also needs to be adjusted for mid-range vehicles that do not have a MacPherson front wheel suspension. The influence of the rear wheel suspension on the behavior of the vehicle on the other hand is much smaller. When using different rear wheel suspensions, the results of all models only show slight deviations from the corresponding reference behavior.

All in all, the parameter model for the simulation of the behavior of mid-range vehicles with McPherson front wheel suspension can be used with high quality. For other vehicles, an individual examination is needed. The necessity of the implementation of the correct front axle kinematics in the parameter model does not completely lower the modeling effort. The parameters summarized in Table 13.4 can however still be kept as average values. Of course it is generally advisable to

Table 13.4 Generally valid parameters of the mid-range vehicle model (See Table 13.3)

No.	Parameter	Dimension	Value
Chassis			
4.	Seat position passengers front	m	−1.3902
			±0.3699
			0.2735
Front axle (with rack and pinion steering)			
8.	Damping		see Fig. 13.13
Anti-roll bar front			
10.	Rotation spring value	Nm/°	6.4315
Rear axle (multi-link)			
11.	Mass of wheel carrier	kg	8.21
12.	Moment of inertia of wheel carrier	kg m^2	0.0426
			0.103
			0.1033
14.	Damping		see Fig. 13.13
Anti-roll bar rear			
17.	Rotation spring value	Nm/°	7.3686
Engine bracket			
18.	Vector to bearing	m	0.0031
			0
			0.0857
Wheel			
19.	Wheel radius	m	0.3108
20.	Longitudinal moment of inertia	kg m^2	1.375
21.	Transversal moment of inertia	kg m^2	0.85
22.	Linear vertical tire stiffness	N/m	239.333
23.	Wheel characteristics		Figure 13.9

use all data available for a vehicle that is to be simulated in a model. But if these are not available, for example in an early stage of the vehicle development, for a mid-range vehicle the parameters presented here can be used. With the aid of such a model, both parameters for the execution of first real roll-over application tests can be estimated, and an ex ante calibration of the trigger parameters for first test vehicles can be performed (compare Sect. 14.2).

References

Adamski D, Schuster C and Hiller M (1997) Fahrdynamiksimulation mit FASIM C++ als Beispiel für die Modellierung mechatronischer Systeme. *VDI-Berichte* 1315. S 117-41

Bardini R, Hiller M, Schuster C, Breunig V, Lang H-P and Mattes B (1996) Developing a rollover sensing concept using multibody simulation. In *Poster Presentation at 3rd International Symposium on Sophisticated Car Occupant Safety Systems - Airbag 2000*, Karlsruhe, Germany

Bosch R (2011) Automotive Handbook. In *VDI-Verlag GmbH*, 8 (ed.), S 1265. Wiley-ISBN 978-1-119-97556-4

Kecskeméthy A (1996) MOBILE - Version 1.2 - User's Guide and Reference Manual. Fachgebiet Mechatronik, Gerhard-Mercator-Universität - GH Duisburg

Kraftfahrt-Bundesamt (2006) Pressemitteilungen - Neuzulassungen von Pkw nach Segmenten und Modellreihen

Kramer U and Neculau M (1998) Simulationstechnik. Hanser, München u.a. - ISBN 978-3-446-19235-5

Opgen-Rhein P (1998) Modellierung eines einfachen Fahrermodells für die Fahrdynamiksimulationsumgebung FASIM C++ [Studienarbeit]. Gerhard-Mercator-Universität Duisburg, Duisburg

Pelz G (2001) Modellierung und Simulation mechatronischer Systeme: Vom Chip- zum Systementwurf mit Hardwarebeschreibungssprachen. Hüthig-Verlag, Heidelberg. - ISBN 978-3-7785-2848-8

Pichler V (1999) Modellbildung der Dynamik von Kraftfahrzeugen unter Anwendung objektorientierter Konzepte [Dr.-Ing.]. Dissertation, Universität Duisburg, Düsseldorf: VDI-Verlag

VDI-Richtlinie-2206 (2004) Entwicklungsmethodik für mechatronische Systeme. (ed.), Beuth-Verlag GmbH, Berlin

Wallentowitz H and Mitschke M (2006) Dynamik der Kraftfahrzeuge (VDI-Buch). Springer-Verlag, Heidelberg. - ISBN 978-3-540-42011-8

Chapter 14
Selected Applications

The constantly growing complexity of motor vehicles also in the field of driving dynamics makes the use of the vehicle dynamics simulation in an efficient development process indispensable. Next to the evaluation of basic driving characteristics such as comfort and stability, the simulation supports the design and application process of basic vehicle dynamics controllers such as the anti-lock braking system (ABS), drive slip control (ASR) or the electronic stability program (ESP) and advanced driving dynamics functions like e.g. Torque Vectoring (Rahimi Fetrati et al. 2016). It is the task of the simulation tool to map a vehicle in the computer in order to use it as a test environment in the controller application. This is exemplified in Sect. 14.1, on the three basic maneuvers of steering angle, stationary circular travel and double lane change. Three of the vehicle models described in Chaps. 10 and 11 are used.

In addition, the vehicle dynamics simulation is also used in the development and application of passive safety systems such as the rollover detection (Sect. 14.2). A further application example is given by the regulation of the roll dynamics by active stabilizers in Sect. 14.3.

14.1 Simulation of Test Maneuvers

14.1.1 Simulation of a Step Steering Input (ISO 7401)

A so-called open-loop maneuver (open-loop control) is analyzed as an example of a test procedure, which is used to evaluate the vehicle's overall system without the driver. A step steering input per (ISO_7401 2003) is simulated using the overall vehicle model described in Chap. 13. In this maneuver, during a stationary straight-ahead traveling at constant speed, the steering wheel is abruptly turned to a stationary value $\delta_{H,stat}$. Since, for physical reasons, it is not possible to produce an

D. Schramm et al., Vehicle Dynamics,
DOI 10.1007/978-3-662-54483-9_14

ideal step input, the standard only requires that the time duration during which the steering angle assumes values between 10 % and 90 % of its steady-state value is less than 0.15 s. To take account of errors or inaccuracies in the evaluation, the point in time t_0, from which the response times of the vehicle are measured, is only defined when 50 % of the stationary steering wheel angle is reached.

The size of the steering wheel angle is to be determined in preliminary experiments. It must be selected in such a way that a stationary transverse acceleration of 4 m/s^2 is established after the settling process. The standard rate is in this case 100 km/h. The drive pedal position must be kept constant during the entire test period.

With a steering angle step input of 12° (see Fig. 14.1) the required transverse acceleration of 4 m/s^2 (Fig. 14.2) results under the conditions mentioned above for the simulation model used here. Thus, the response of the vehicle to the steering angle can now be evaluated. A fast and well-damped vehicle reaction is desirable.

The speed of the vehicle reaction is quantified by so-called response times. The most frequently used response times are the response time $T_{\dot{\psi}}$ and the peak response time T_{max} of the yaw angle speed. The response time is defined as the duration of time from t_0 to reaching 90 % of the stationary value. The response time $T_{\dot{\psi}}$ results to 0.25 s in Fig. 14.3. The peak response time can be determined for the transverse acceleration as well as for the yaw angle speed. However, priority is normally given to the yaw angle speed for evaluating the speed. The peak response time $T_{\dot{\psi}}$ is 0.32 s, see Fig. 14.3.

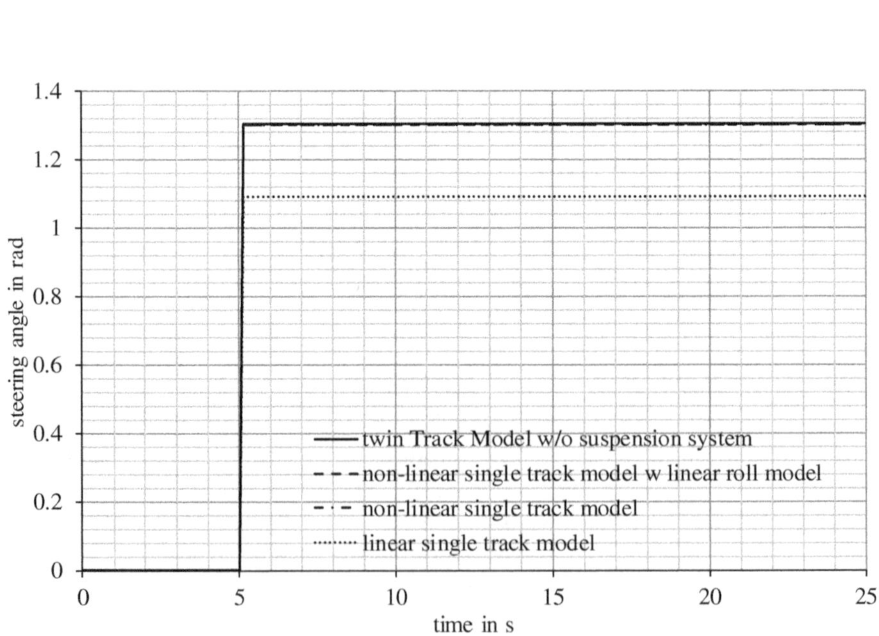

Fig. 14.1 Steering angle for step steering input

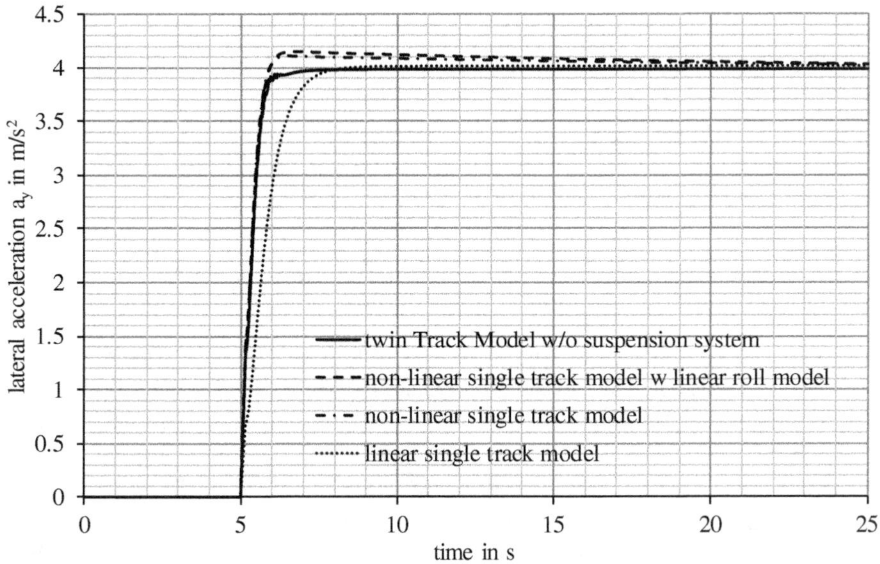

Fig. 14.2 Lateral acceleration for step steering input

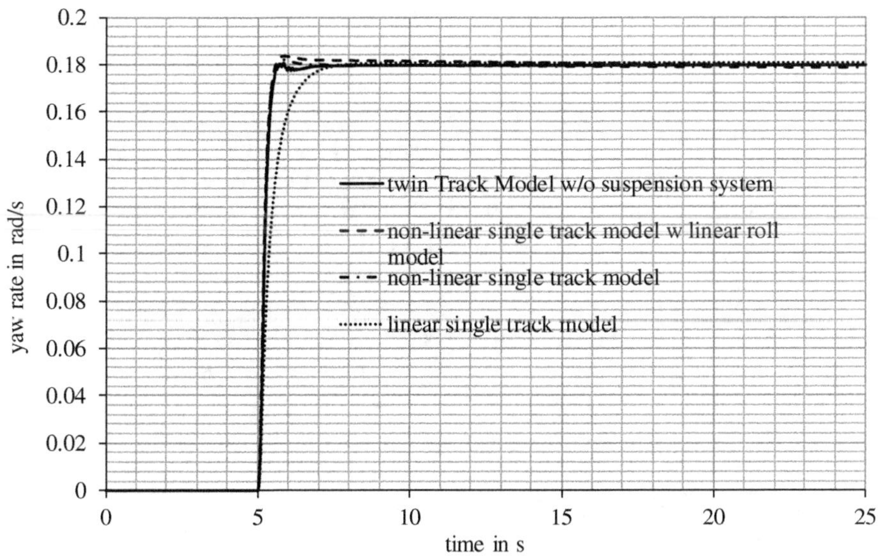

Fig. 14.3 Yaw rate for step steering input

The evaluation of the damping of the vehicle reaction can be judged by the ratio of the maximum value to the stationary value of the yaw angle speed or alternatively by the normalized overshooting range used in the control technology

$$U_{\dot{\psi}} = \frac{\dot{\psi}_{max} - \dot{\psi}_{stat}}{\dot{\psi}_{stat}}. \tag{14.1}$$

The overshoot range $U_{\dot{\psi}}$ of the simulated vehicle model is 2.3 %. This value indicates an under-steering, stable driving behavior.

14.1.2 Simulation of Stationary Circular Travel

The "stationary circular travel" maneuver according to ISO_4138 (2004) is one of the most important so-called open loop standard test methods. It is used both in driving tests and in simulation. With this maneuver, important vehicle dynamics variables, such as self-steering gradient, yaw gain and steering tendency, and thus driving behavior and stability of motor vehicles, can be determined right down to the dynamic limit range. For this purpose, variables such as steering wheel angle, vehicle speed, float angle, roll angle, transverse acceleration and yaw angle speed are recorded.

There are, in principle, four methods for obtaining the stationary circulating values (ISO_4138 2004), one of which is explained here. This method uses a circle with a constant radius. It can therefore be performed well even with limited driving dynamics surfaces. During the execution, a left and a right circle are driven on a path radius of at least 30 meters while increasing the vehicle speed slowly and uniformly. This results in a transverse acceleration, which thus increases slowly and monotonously up to a value from which the steady state can no longer be maintained by the driver. While increasing the speed, the driver must adjust the steering angle so that the vehicle remains on the orbit. If the maneuver is executed in a simulation environment instead of a real test track the human driver is replaced by an artificial driver, see Sect. 11.3. Due to the quasi-stationary test conditions, this maneuver is well reproducible for an experienced driver and the driver has little influence on the system behavior (Rau 2007).

However, a different method will be used here. In this case, the driver model can be avoided. Starting with a large radius, the steering wheel is for this method always pushed further until the vehicle can no longer be held on the track (Figs. 14.4–14.6).

14.1.3 Simulation of a Double Lane Change

One of the standard test methods ISO (1975) and ISO (2000) in the non-stationary range is the double lane change (ISO flare test). The aim of this experiment is to evaluate the driving behavior of a motor vehicle in the closed control loop in a practical driving situation. For this purpose, the test simulates an evasive maneuver together with the subsequent deflection to the right lane. In a slightly modified form,

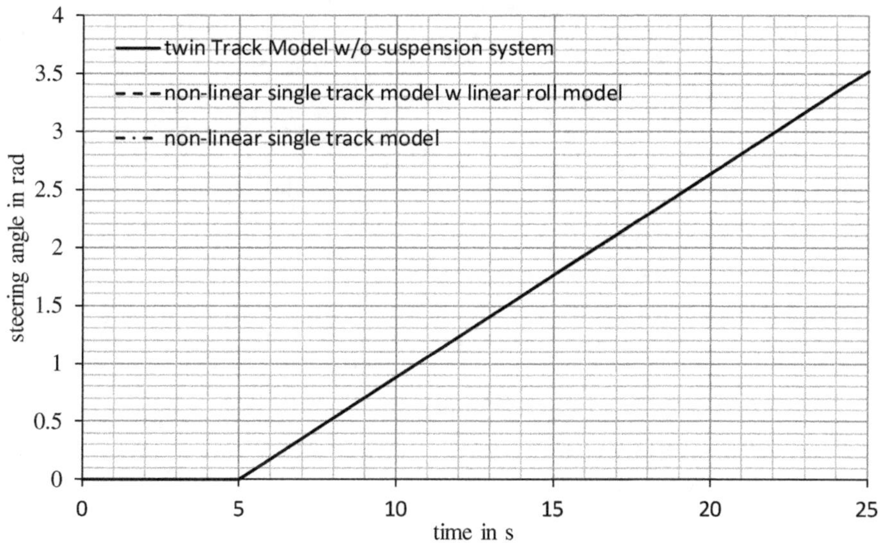

Fig. 14.4 Steering angle for stationary circular travel

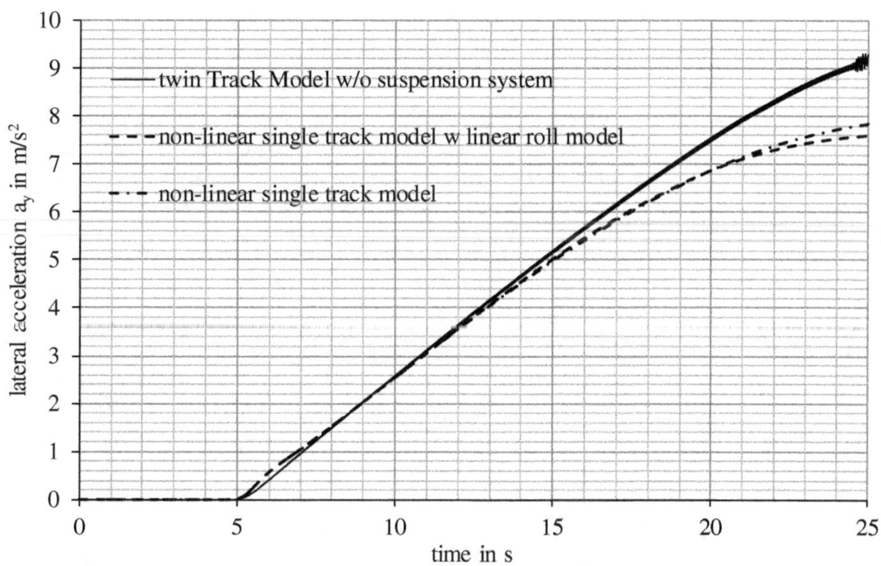

Fig. 14.5 Lateral acceleration for stationary circular travel

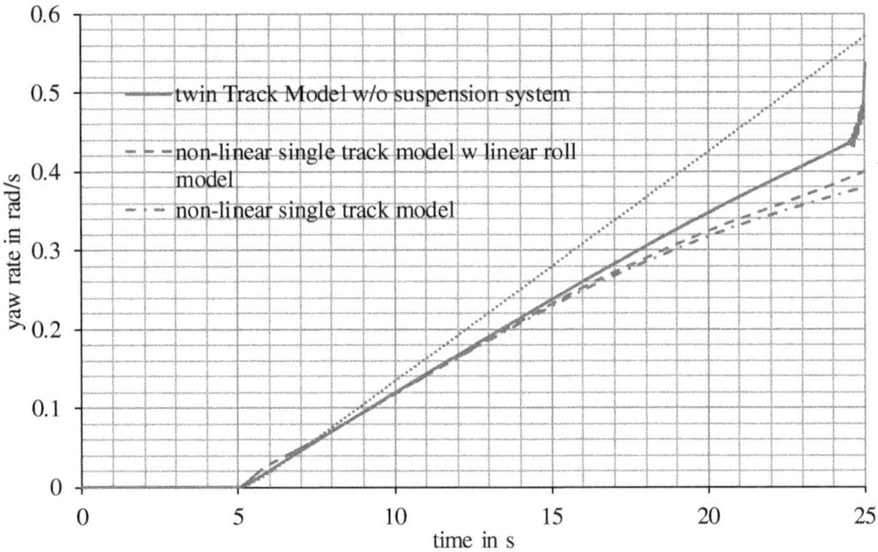

Fig. 14.6 Yaw rate for stationary circular travel

this test was known in a modified form under the name "Elk Test" in 1997 to a broader public in Germany. Today, it is an integral part of the test program of automotive companies, test magazines and organizations.

The dimensions of the test section are determined by a length of the measuring section of 110 m and a lateral offset of 3.5 m (Fig. 14.7). The track width is adjusted to the respective test vehicle. The respective transit time is measured with light

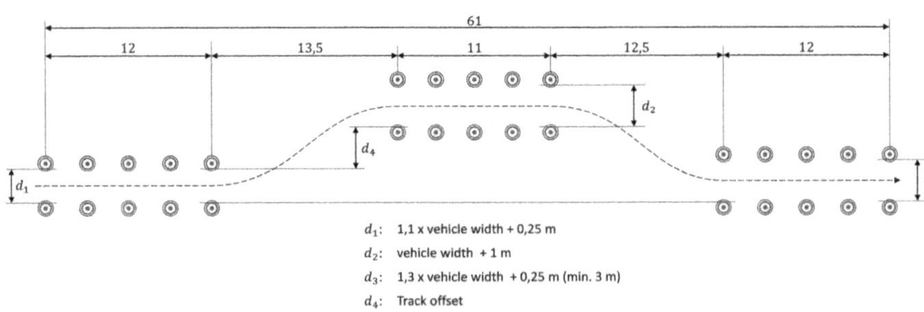

Fig. 14.7 Double lane change (all dimensions in m) according to ISO (1975), see also Reński (2001)

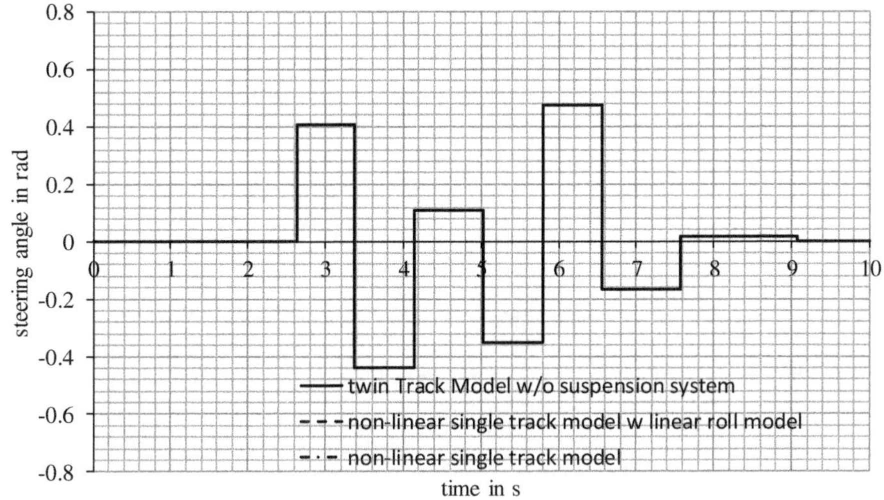

Fig. 14.8 Steering angle for double lane change

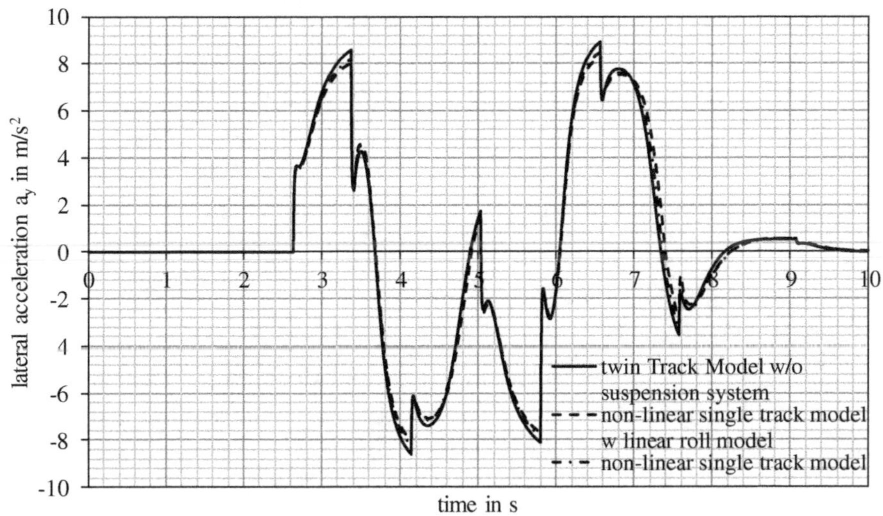

Fig. 14.9 Lateral acceleration for double lane change

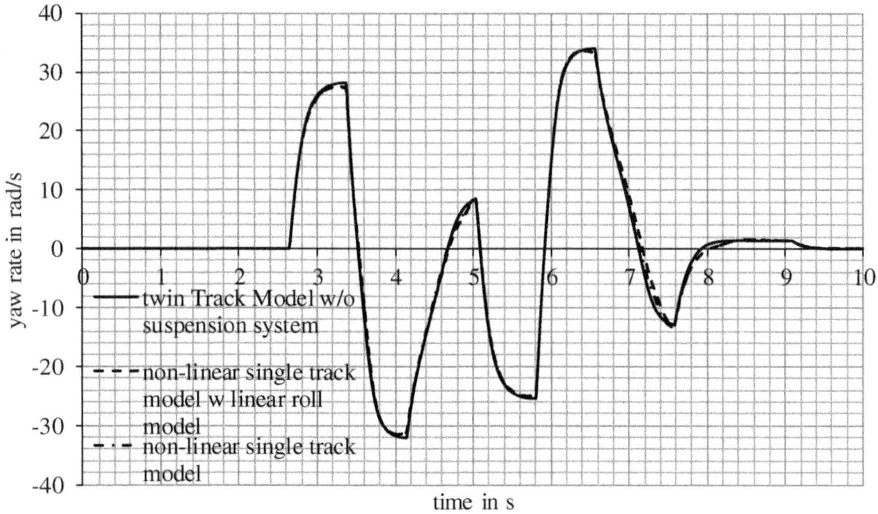

Fig. 14.10 Yaw rate for double lane change

barriers at the beginning and end of the measuring distance. The assessment criterion is the mean value of transit times from at least three error-free journeys in which pylons are not to be touched. To investigate the non-stationary driving characteristics, other driving dynamics are often recorded in addition to the time measurement, e.g.

- steering wheel angle,
- steering wheel torque,
- lateral acceleration,
- yaw rate,
- driving speed and
- sideslip and roll angles.

To evaluate the quality of the driving behavior of the vehicle being tested, the required steering wheel motions and the transverse acceleration which have occurred are recorded (Figs. 14.7–14.10).

14.2 Simulation of Vehicle Rollover

This section describes the intricate example of the support of the application process for the rollover detection function in controlling restraint systems in passenger cars supported by multi body simulation. Many of the following implementations are taken from Bardini (2008). In the context of the entire vehicle development process, the application process for a rollover protection system is secondary. It primarily depends on the availability of prototype vehicles and parts. The earlier this is accessed in the course of product development, the more costly the process becomes. On the other hand, beginning the process as early as possible minimizes the development risk, increasing the maturity and the quality of the application at launch. Against the background of this conflict it is important to find the best possible application process. To this end, the simulation can make a decisive contribution.

In order to ensure the function of rollover detection in Fig. 14.11, accident scenarios are presented to derive those representative experimental configurations that reflect real accident data as comprehensively as possible. These must then be

Rollover Cause	Accident Scenario	Field Relevance	Test Setup
lateral sliding in conjunction with varying ground (trip-over)	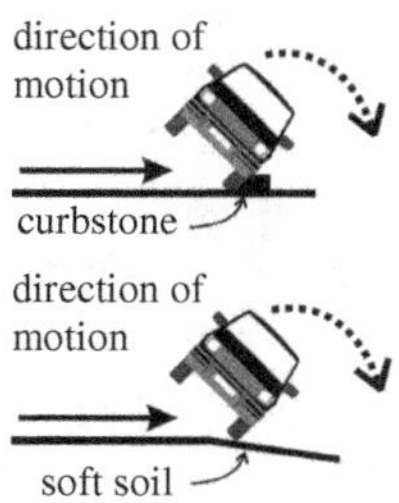	56%	curbstone  soiltrip 
losing roadway onto a sloping plane (fall-over)	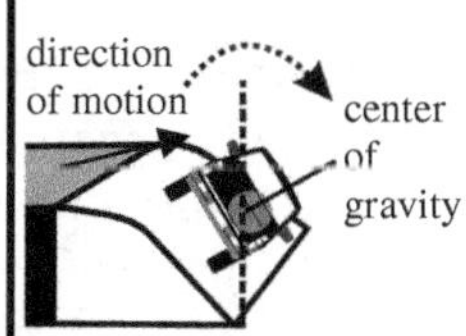	13%	embankment 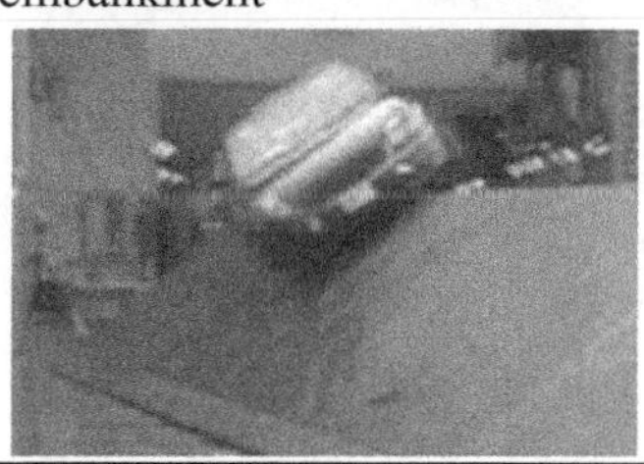
driving over a ramp-like object (flip-over)	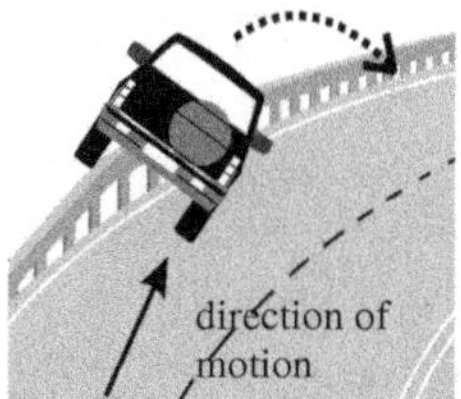	12%	ramp 
	total	81%	

Fig. 14.11 Dissipation of field relevant test scenarios (Bardini 2008)

simulated using a simulation of vehicle dynamics. According to Fig. 14.11 up to 81 % of the relevant field rollover accidents are represented by four test scenarios. To ensure a reliable activation in the majority of the actual rollover accidents, the curb, the sand bed, the embankment and the ramp test at various speeds during the development process will be executed using prototypes.

In order to minimize the effort required for this purpose, tests that are usually already contained in the development programs of the vehicle manufacturer are considered first, if they are appropriate for the application of the rollover protection system of passenger cars. From these established experiments, for example, in the lateral curb impact and in the VDA-slalom test and in the VDA lane change test, synergies can be used for the interpretation of the rollover detection. In standard side crash misuse test maneuvers such as curb side impact, the no-fire threshold is usually applied to impact velocities between 10 and 15 km/h. At higher speeds, the acceleration sensor-based systems will deploy the restraint systems using the side crash algorithm.

Due to physical constraints the rollover of a mid-range car cannot take place at speeds below 17 km/h. Thus it can be assumed that at any case the side crash detection takes place before the rollover detection. This usually already happens at a roll angle of the vehicle below 5°. Because of this synergy there is no need to apply the rollover detection at the curb stone with an additional test in the development process. This also applies in the context of chassis development where standard vehicle dynamics tests like slalom and lane change (moose test) are conducted. The measured data obtained in these experiments during application of the rollover detection have to be fulfilled as robust no-fire tests.

As shown in Fig. 14.5, the product development, can be divided up into three phases both temporally and in terms of product maturity. In the first phase, which begins about 3 years before the series starts, there the early prototypes are available. In terms of significance, they are of reduced value compared to the later series product. Materials, package and design details may change in later development stages. Therefore the data included in this phase for later application could in the worst case be unsuitable. In addition, in this stage of construction, vehicles are very expensive and thus only a small number of destructive tests are available. Against this background, a waiver of the implementation of a complete test program for the first data record is obvious. However, it also increases the development risk, because problems can then only be discovered later in the next prototype phase, which begins about 2 years before the series starts. But if it is the aim to already equip the vehicles in this next stage of development with triggering enabled systems, experimental data must be provided for the first data collection.

As this first application must not meet the stringent requirements for a suitable performance for the end user, it is possible to favor a conservative interpretation of high misuse robustness. This also increases the tolerance to the first application for the experimental data to be provided, so that the predecessor model can also be used for the acquisition of experimental data. This assumes that from the eight identified dominant vehicle parameters from Sect. 13.3 there are no major deviations between predecessor and successor. As a practical matter for a maximum allowable deviation in this case a value about 5 % is acceptable. In addition, it is also possible to perform an adaptation of vehicle weight, center of gravity and inertia within certain limits by the introduction or movement of weights inside the vehicle. If such measures do not sufficiently solve the problem, or if, for example no previous model exists in the development of a new vehicle, then the data can only be obtained from the simulation. The model construction can begin in the concept stage, so at first availability of early prototypes, verification of the model using non-destructive maneuvers, such as circular or slalom runs, can be made. This model can simulate the maneuvers summarized in Sect. 14.2.1, so that a first data set to apply the rollover detection can be provided.

If either simulation or experimental data in the early prototype phase are available during the second prototype phase, then it is necessary to generate a complete data set to reduce the development risk as shown in Fig. 14.12. If not, both the data acquisition as well as the functional acknowledgment must be done during the preproduction phase.

A guaranteed achievement of the series start can only be achieved under the assumption of a zero error process flow. If there is prototype simulation data in the second phase, a reduction in crash-testing is possible. On the one hand test parameters of the individual previous experiments can be set more accurately using

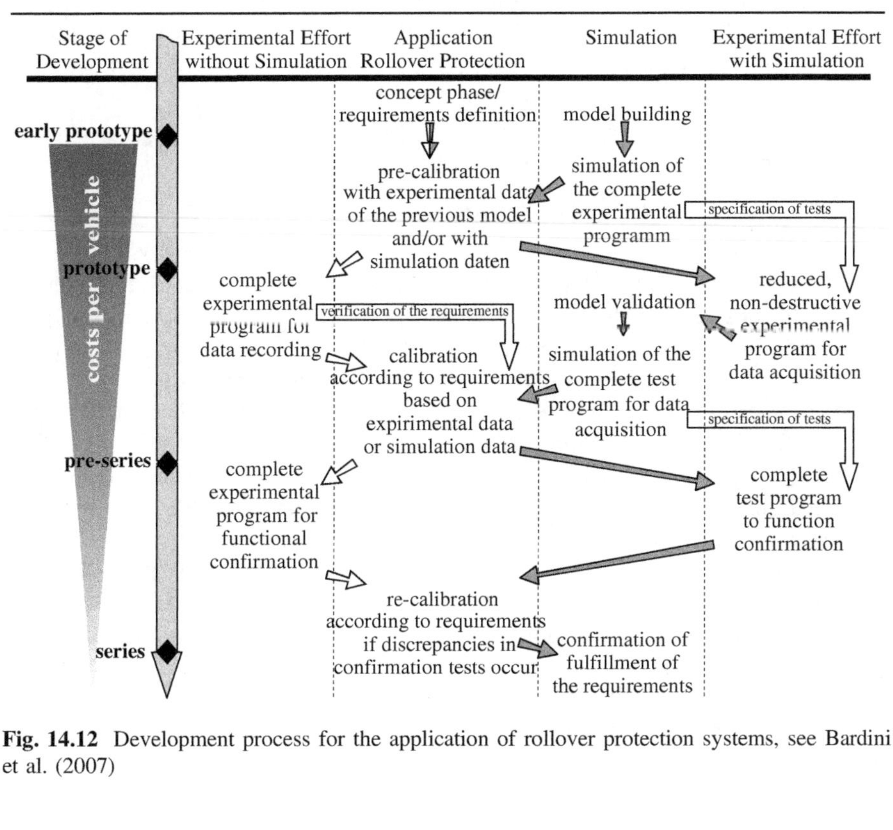

Fig. 14.12 Development process for the application of rollover protection systems, see Bardini et al. (2007)

simulation, and on the other hand the destructive tests can be omitted. Since for a robust algorithm application for each of the fire maneuvers from Fig. 14.11 three trials are required (see Sect. 14.2.1), a limitation of real experiments to only the impact free experiments is possible. Apart from sand bed experiments, all experiments that lead to the rollover of the vehicle must too be regarded as destructive. In experiments just below the stability limit it is possible to repair of the vehicle for further testing. Thus, in the absence of destructive tests, the need of test vehicles is reduced. By using impact-free tests in order to validate the simulation models, tests with rollover event can be replaced by simulations.

The tests for functionality confirmation during the pre-series (from about 16 months before the start of series) have less potential for savings when using simulations. Because the functional acknowledgment by definition serves to demonstrate the functioning of the entire system (hardware, software and restraint systems) the absence of real experiments is generally not possible for the real vehicle. Here, the simulation can make a contribution to conduct the experiments by specifying them with the help of the experimental parameters verified by simulation. If however deviations from the desired behavior occur in the confirmation tests, simulations using a more detailed analysis of the wrong behavior will be performed. If a recalibration of the application is required, a simulation provides the additional data of the critical maneuver, e.g. in adjacent speed ranges. Using occupant simulation, the effects of the adopted trigger behavior must be checked again to meet the requirements of the occupant protection requirements. It needs to be decided in each individual case whether the extent of the changes even requires a repeat of the entire test program for confirmation.

14.2.1 Virtual Proving Grounds

In this section, the virtual tests used for the application of a rollover protection system are described. All necessary applications for such experiments are summarized in Table 14.1. The experiments are divided up into two groups. One group consists of the so-called misuse attempts, which include extreme driving situations in which a triggering of the non-reversible components of the occupant protection system is unacceptable. The other group includes the so-called use-testing in which the triggering of the restraint systems is expected, but only if it actually comes to a rollover in the course of such a driving maneuver. Therefore, the trigger (fire) and the non-tripping experiments (no fire) must be distinguished in this case. The latter experiments are used to distinguish between the thresholds for each algorithm provided in the decision path for the fire and no fire-cases. This means that, for the use-tests in each case, a test should be carried out below the triggering threshold (roll angle $\leq 40°$), just right below the stability limit (roll angle 50°–60°) and one above the stability limit (>60°).

It is discussed here to what extent a simulation of this maneuver is possible and practical. Except for the curb impact, all misuse test maneuvers can be simulated

Table 14.1 Experiments necessary for the recognition of rollovers

Experiment	Application
Misuse testing	
VDA-Slalom test (10 pylons at a distance of 18 m each on a successive line)	Safeguarding of the resonance behavior (escalation)
VDA-lane change test (ISO 3888 Part 2)	Robustness of the application in standard maneuvers
lateral curb impact at 10–15 km/h	Testing of the influence of lateral acceleration impulses and definition of the upper border of the no fire range
Drive over two successive ramps with the right and left wheels each (alternating embankment)	Robustness at high roll angular velocities
Jump over a ramp with all wheels (simulation of a jump)	Safeguarding of the free-flight phase
Use testing	
Lateral curb impact at ≥ 15 km/h	Safeguarding of the trigger behavior at the curb brunt (lateral sliding with large acceleration impulse)
Lateral sliding into a sand bed	Trigger behavior for lateral sliding at low acceleration impulse
Drive over a ramp with two wheels	Trigger behavior for the vertical spin (ascending vehicle)
Drive over an embankment	Trigger behavior for the vertical spin (descending vehicle)

with high reliability using a multibody model as presented in Sect. 13.3. For the curb impact, as mentioned earlier in this section, the side-crash application data can be used for rollover detection. Since test type data is available during the development process of this, so the simulation of this maneuver in the application process for rollover is not necessarily required.

The situation is similar with the two VDA tests (VDA 2006). Since these are so-called closed-loop tests, the results are strongly influenced by the driver, or in the simulation, by the quality of the driver model. Depending on this the results for example of lane-change test as shown in Fig. 14.13 can be of very different quality. Here, the maximum feasible transit speed of an experienced driver has higher values than that for the simulation and thus in the application higher values have to be considered. Therefore it is advisable to take account of real measurement data.

The basic structure of the two remaining misuse tests of Table 14.1 is shown in Fig. 14.14. These maneuvers can be simulated, with the help of the necessary ramp scenarios. In order to simulate the appropriate change of the slope, two 8.5 m long ramps are placed in a row at between 30 and 60 cm adjustable height, so they can be run over by the vehicle one after the other. The parameters of ramp height, ramp spacing and vehicle speed are selected so that the highest possible roll-speeds are generated, without the vehicle jumping over the ramps uncontrollably. This is for

Fig. 14.13 Virtual VDA-lane change test [(matches ISO 3888 Part 2 (ISO))]

Fig. 14.14 Virtual testing grounds for misuse testing alternating embankment and ramp test

mid-size vehicles only at speeds below 70 km/h. For the simulation of the jump hill there are theoretically possible many different configurations with using ramp set up. The aim of this maneuver is the generation of a free flight phase with duration of at least 300 ms. To avoid damage to the vehicle, the use of a lower ramp with a maximum height of 35 cm which is traveled at speeds ranging from 60 to 80 km/h has also proven to be suitable.

For the use testing from Table 14.1, for the simulation of the curb impact the same conditions are valid as the ones already described for the case of misuse. The lateral sliding into a sand bed however is different (see Fig. 14.15).

In this maneuver, no trigger via the side crash recognition can occur because the strong lateral acceleration impulse at the beginning of the rollover accident does not exist. Therefore, at the sand bed, as well as the maneuvers ramp and embankment, the generation of the signals necessary for the application is necessary (1st roll angle $\leq 40°$; 2nd roll angle 50–60°; 3rd roll angle >60°).

Fig. 14.15 Virtual lateral sliding into a sand bed

At this, the sand bed maneuver occupies an exceptional position. As opposed to the others, these experiments can be done with only minor damages to the vehicle. Here, due to the gradual sink-in of the wheels in the sand and the subsequent rollover of the vehicle in the sand, larger plastic deformations can be prevented, so a further use of the test vehicle is thinkable. The simulation of these maneuvers is however not possible without creating a previous verification test using the multi body simulation. For a meaningful simulation, the parameters presented in Fig. 14.16 for the sink-in depth as well as the length of the sliding distance in the sand bed need to be determined. Using these measurements that need to be determined from a real rollover experiment in a sand bed, an empirical model approach for the adjustment of the process of the function depicted in Fig. 14.16 of the kinetic friction of the sand bed maneuver can be created. Here, the digging into ground, arising from the sinking-in of the wheels in the sand, can be recreated by continuously increasing the friction on a lowering surface.

The basic setup of the remaining use-tests from Table 14.1 is depicted in Fig. 14.17. Which parameters are suitable for the height of the ramp, the approach angle, the geometry of the embankment and the velocity of the mid-range vehicle will be discussed in Sect. 14.2.2 on the basis of simulation results.

In order to be able to determine the corresponding behavior for of the passengers, the use of a passenger simulation tool seems practical which contains already validated models of crash test dummies (i.e. MADYMO, Coo et al. 1991). Hereto, the position and orientation data of the chassis, which were calculated in the

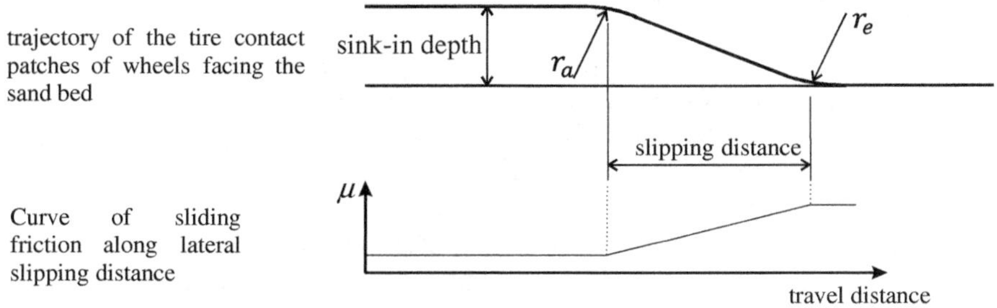

Fig. 14.16 Ground modeling for lateral sliding into a sand bed

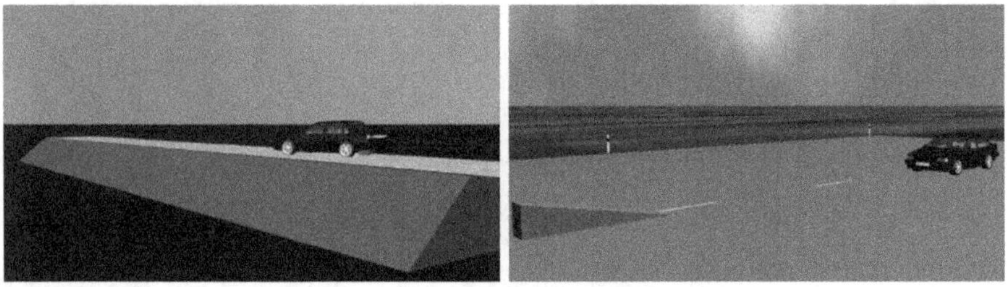

Fig. 14.17 Virtual testing grounds for the use-experiments embankment and ramp

simulation of driving dynamics, can be used as movement guideline for the interior of the vehicle. These need to be available relating to the inertial system. In experiments described in this section, using the simulation of driving dynamics, the data can be displayed readily in the needed framework. If the data however stems from real crash experiments, a significant effort for the corresponding processing of the measuring data is needed.

14.2.2 *Results of the Simulation*

Using the misuse and use maneuvers summarized in Table 14.1, an algorithm can be verified for the rollover recognition in an early stadium of the vehicle development and can be supported in the application as shown in Fig. 14.12. The characteristics of the vehicle- and passenger movements that occur during the different maneuvers determine the requirements of the algorithm and the passenger protection. To this end, appropriate parameters are determined so that the critical parameters for the rollover protection can be set up. Especially in the maneuvers with rollover, the simulation of the passenger behavior is necessary for the estimation of the trigger requirements. Below, simulation results chosen to this end from different maneuvers are presented.

14.2.2.1 Misuse Testing

As discussed in Sect. 14.2.1, the curb impact takes a special role and will therefore not be considered further in this discussion of misuse testing. Instead, the two maneuvers VDA lane-change test (Fig. 14.13) as well as Alternating embankment (Fig. 14.14) will be examined representatively as a less critical and a more critical maneuver.

The VDA lane-change test (VDA 2006), also known as moose test, does not present a rollover risk for modern passenger cars. Extensive and elaborate testing with passenger cars sold on the US market show that Pick-Ups and SUVs can more

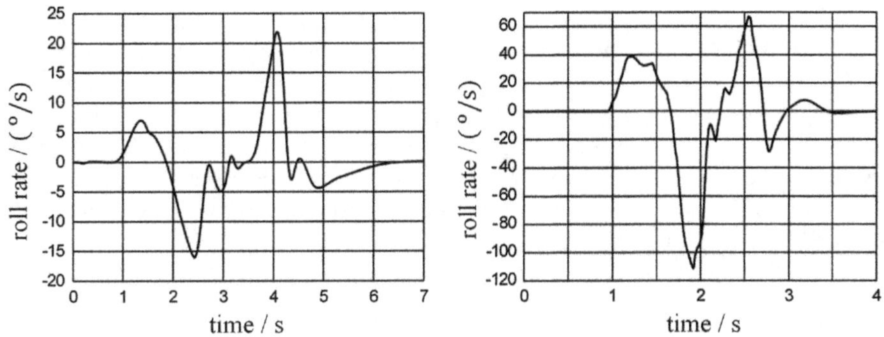

Fig. 14.18 Roll rate of the average mid-range vehicle model for the VDA lane-change test (*left*) and for a drive over an alternating embankment (*right*) each at 60 km/h

easily roll over in this situation. Most passenger cars however are very tolerant against lane-change concerning the rollover inclination. This is also confirmed here in the simulation conducted with the average mid-range vehicle. In Fig. 14.18, the course of the roll rate of the lane-change maneuver at 60 km/h is depicted. As a maximum angular rate about the longitudinal vehicle axis, 22°/s can be observed. This angular rate contains enough rotation energy in order to lift the center of gravity of the vehicle about an additional maximum of 3 mm. Since in this maneuver, the maximum roll angle is below 5°, this is uncritical. Reaching the threshold of plausibility of the algorithm is also unlikely as the maximum lateral delay is at 0.7 g. Therefore, the VDA lane-change test concerning the misuse-safety can overall be classified as not dangerous.

The alternating embankment maneuver is however much more critical. An angular roll rate of 112°/s combined with a lateral acceleration of 1 g and a vertical acceleration of 1.2 g can in case of an insufficient application of the algorithm quickly lead to a false triggering. This maneuver creates high roll rates and accelerations only at the moment of rebound of the vehicle from the roll movement of the first embankment while going up the second embankment. In Fig. 14.18, this moment corresponds with the point in time of 1.85 s. The corresponding moment of the animation of the maneuver is depicted in Fig. 14.19.

Fig. 14.19 Drive over an alternating embankment at 60 km/h (t = 1.85 s)

Here it can be registered that at this point in time (the trigger algorithm was embedded into the simulation model and its state is presented in the animation as a colored ball above the hood) the algorithm presents the plausibility status. The trigger is however not released as the Roll rate of 122°/s at that point in time does not suffice for a tip-over generating elevation of the center of mass. In the framework of a complete application for all use- and misuse cases tests need to be conducted questioning in what way the plausibility thresholds could be increased without negatively affecting the reach of the use-requirements.

The following three sections analyze these use- requirements more precisely.

14.2.2.2 Ride Over a Ramp

As already described in Sect. 14.2.1, different barriers need to be set up for the algorithm application of the use-tests. That is, tests need to be conducted each below the action level (roll angle 30°–40°), just under the statistical action level (50°–60°) and above the statistical action level (>60°). In order to actually reach these thresholds in relevant and reproducible conditions during the tests, the parameters ramp height and vehicle velocity need to be defined individually for each vehicle model. Here, the simulation can support the planning of the experiments very efficiently.

The simulations can prevent extensive pre-trials, or hazardous attempts by calculating in advance the matrix depicted in Fig. 14.20.

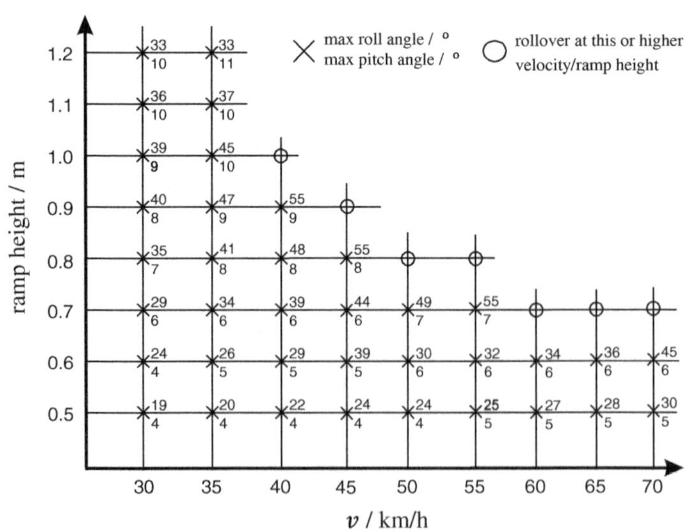

Fig. 14.20 Maximum roll and pitch angle, dependent on ramp height and velocity of the average mid-range vehicle model (ramp length constant at 5 m)

The matrix is based on findings from accident research stating that rollover accidents usually do not occur at speeds above 70 km/h and that obstacles similar to ramps are usually not higher than 1.2 m (Ajluni 1989 and Harkey 1999).

From the here executed examination of the ramp tests follows an increasing rollover tendency of the vehicles with lower weight, lower inertia as well as lowering of the track width and the wheel base. For an average mid-range vehicle model, the data needed for the application in dependence on Fig. 14.20 can be created for example using a ramp length of 5 m and a ramp height of 0.8 m at velocities of 30, 40 and 60 km/h.

The corresponding roll angle processes of these maneuvers are depicted in Fig. 14.21. Depending on the requirements of passenger safety, the minimum triggering level for the rollover maneuver on the ramp can be applied. For the requirements of passenger safety during rollover it is principally true that an early triggering of the restraint systems is desirable. This must however not be at the expense of the misuse robustness. Therefore, the latest point of time for the triggering of the system needs to be specified from the perspective of passenger safety. This can be simulated for the window bag using passenger simulations. For the trigger of the belt tensioner however there is no point in time that is too late for the trigger for the purpose of an OOP-problem (Out of Position). Here, only a reduction of the protective effect can be observed in a later triggering. Furthermore, the passenger simulation needs to decide whether a certain accident situation even justifies the activation of irreversible restraint systems. This is only useful when a corresponding protection potential can be expected, or when a rollover actually occurs.

The screen shots of a simulation of driving dynamics (*left*) and their corresponding MADYMO-passenger simulation (*right*) depicted in Fig. 14.22 illustrate the use of a simulation using the example of a ramp maneuver. In this maneuver in

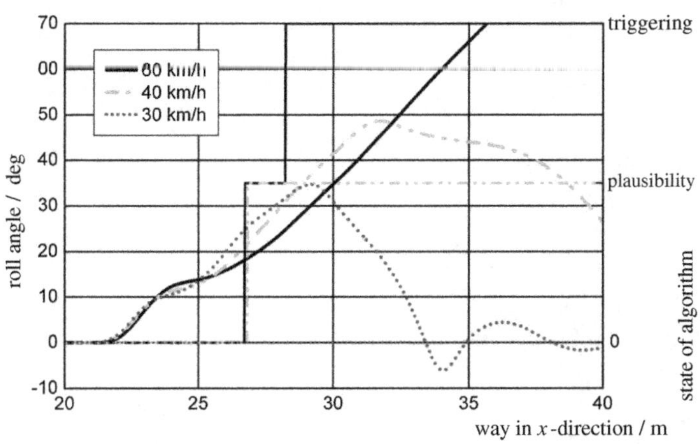

Fig. 14.21 Roll angle and trigger status for ramp crossings with different velocities at 80 cm ramp height and a ramp length of 5 m

which the vehicle drove at a velocity of 80 km/h over a ramp of 35 cm, the recognition algorithm does not trigger the system. The correctness of this trigger behavior is confirmed by the unbelted 50 % male Hybrid III Dummy that was used in this passenger simulation. All characteristic load values in this simulation are significantly below the legal thresholds, so that an execution of protective measures

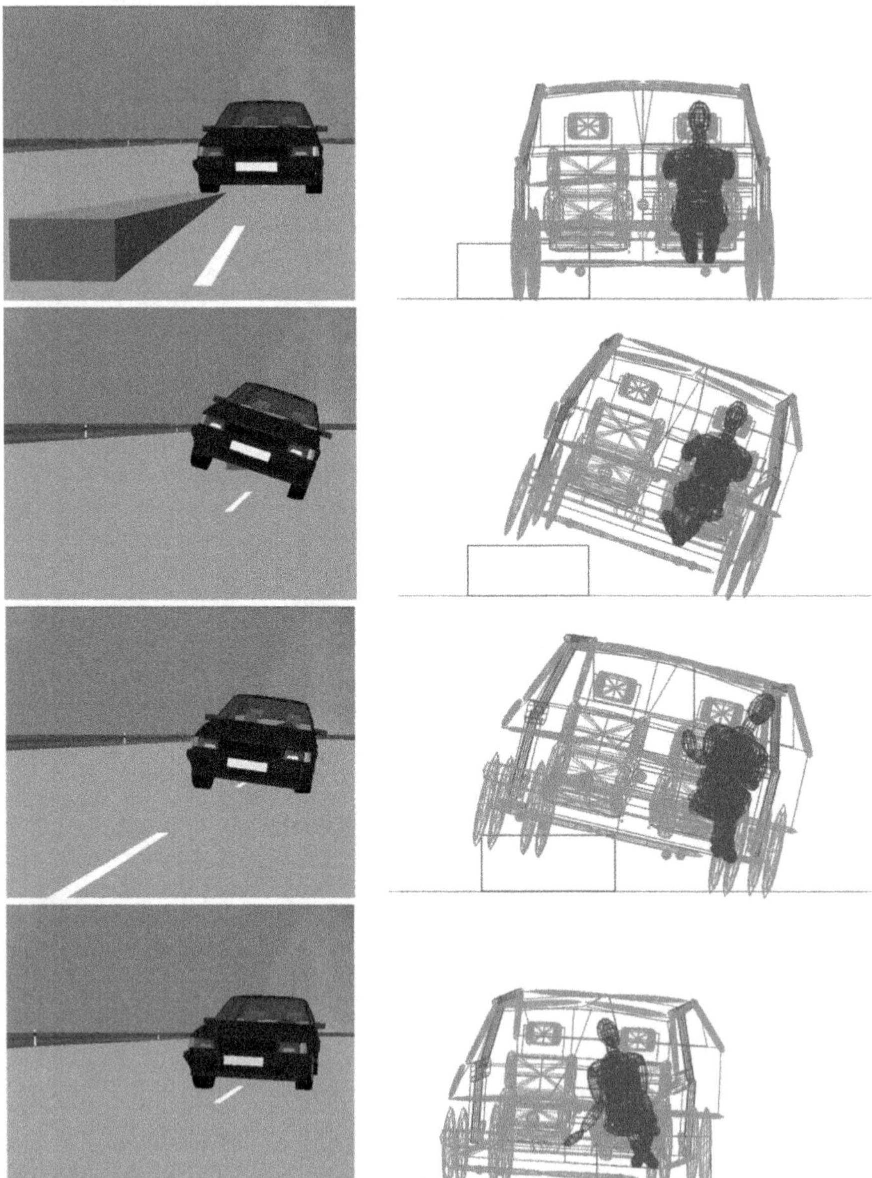

Fig. 14.22 Ramp maneuver with 35 cm ramp height at 80 km/h

is not necessary. Furthermore, during the journey over the ramp, almost no relative motion between the head of the dummy and the interior of the vehicle can be observed so that the OOP-risk in ramp maneuvers can be considered rather low. Only the strong dislocation of the passenger at the end of the simulation is noticeable. A real person however would not be expected to be dislocated in such a fashion due to the actively executed retention forces. When executing this simulation with a seat belt, the situation is even less precarious. Here too, similar to the misuse maneuvers on the Alternating embankment (see Sect. 14.2.2.1), the need for safety measures could not be substantiated.

Further passenger simulation shows that a critical dislocation of the dummy head of about 60° occurs. This means that for the calculation of a robust triggering decision, compared to the curb impact or sand bed (see Sect. 14.2.2) there is a relatively high amount of time available. A trigger for a roll angle of 25° as is provided by the application depicted in Fig. 14.14 therefore largely eliminates the OOP-risk and furthermore facilitates an early link of the passengers to the vehicle movement using the belt tightener.

14.2.2.3 Passing Over Embankment

Analogously to the approach described in the previous section for the definition of suitable testing parameters, a matrix is created for the embankment which shows the roll and pitch angles in relation to the velocity. In the embankment maneuvers however, the approach angle α to the edge of the embankment needs to be added as an additional variation parameter. In order to find out which approach angle leads to the largest possible roll angle buildup, simulations were executed for an embankment with a decline angle β of 35° at a velocity of 40 km/h with approach angle α between 5° and 32.5°. The results of this simulation are summarized in Table 14.2. The presented maximum roll angles have all been measured within a distance of eight meters (length of the inclined embankment in lateral direction). The largest roll angles then occur between 20° and 30°.

As the evaluation of different accident statistics shows (Mohamedshah and Council 2007 and Otte 2005), the vehicles usually deviate from the road at angles between 5° and 25°. Considering this, an approach angle of 20° is chosen for further examinations. Another influencing factor which is relevant for the real accident occurrence is a potential steering movement of the driver (Ajluni 1989 and Mohamedshah and Council 2007). Usually the driver tries to steer the vehicle back

Table 14.2 Roll and pitch angle dependent on approach angle α when running over an embankment (v = 40 km/h, β = 35°, μ = 0.9)

Approach angle/°	5.0	7.5	10.0	12.5	15.0	17.5	20.0	22.5	25.0	27.5	30.0	32.5
Pitch angle/°	4	5	7	10	12	13	14	15	16	18	20	22
Roll angle/°	38	39	39	40	45	50	55	63	64	62	60	57

onto the road. Such a steering movement leads to an increase of the likeliness of a rollover if the vehicle is already driving on the embankment. Therefore, the simulations for the creation of the matrix presented in Fig. 14.16, is based on a full steering angle against the embankment. This is conducted when the front wheel on the side facing away from the embankment moves onto the obstacle.

Following Fig. 14.23, for the average mid-range vehicle model the parameters angle of embankment and velocity of the vehicle can be chosen appropriately. The data needed for the application of embankment rollovers can for example be created at an angle of embankment of 35° at velocities of 25.38 and 43 km/h. Here, a minimum height of 3.3 m and minimum length of 20 m is necessary because otherwise the vehicle gets back to its horizontal level too early and the desired roll angle cannot be reached. If the height cannot be achieved due to constraints of the experiment, configurations with larger embankment angles should be chosen.

The use of smaller approach angle also allows for lower heights of the embankment, though the needed length of the embankment then increases. In Fig. 14.24, the roll angle developments due to the two previously named parameters. Depending on the requirements of passenger safety, the minimum triggering level for the roll over maneuver on the embankment can be administered.

Similarly to the application of the ramp maneuver, the ideal compromise between a sufficient triggering and misuse robustness needs to be found for the embankment maneuver. Therefore, the latest possible moment of triggering needs to be calculated that allows for the unobstructed deployment of the window bag, using the passenger simulation.

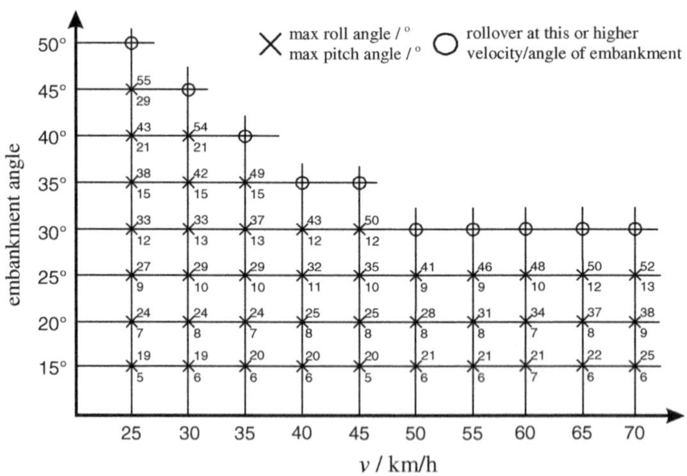

Fig. 14.23 Maximum roll and pitch angle, dependent on the angle of the embankment and the velocity of the average mid-range vehicle model (with steering motion and constant approach angle of 20°)

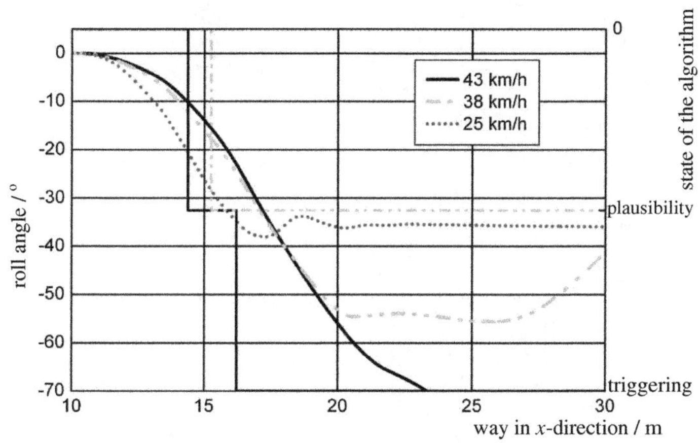

Fig. 14.24 Roll angle and trigger status for embankment crossings with different velocities for an approach angle of 20° and an angle of the embankment of 35°

The screen shots depicted in Fig. 14.25 of a simulation of driving dynamics (*left*) and the corresponding MADYMO-passenger simulation (*right*) illustrate the typical passenger kinematics of an embankment maneuver. In this maneuver in which an Approach angle of 15° and a velocity of 50 km/h is used for an embankment with an incline of 55°, the triggering occurs at a roll angle of 30°.

In Fig. 14.25, the moment of the roll over recognition via the trigger algorithm is visualized using the upper of the two balls above the vehicle. The lower ball signalizes the moment in which the plausibility barrier is recognized. In the presented maneuver, the recognition happens at a roll angle of 25°. The timeliness of this trigger behavior is confirmed by the belted 50 % male Hybrid III Dummy that was used for the passenger simulation.

Numerous passenger simulations have shown that during the ride over the embankment, almost no relative movement between the head of the dummy and the interior of the vehicle can be detected. Therefore, the OOP-risk for the embankment can also be indicated as low. Just like for the ramp, stronger relative movements of the passenger only occur when the wheels hit the ground after the roll, or when the chassis has direct contact with the ground respectively ($\varphi > 90°$).

It could be determined for all observed embankment maneuvers that a relevant OOP-situation can only occur when the vehicle roll angle of about 35° is reached. This means that for the calculation of a robust trigger decision, similar to the ramp roll over, there is relatively much time available. A trigger behavior provided by the application with a good misuse robustness presented in Fig. 14.24 therefore fulfills the requirements for the avoidance of OOP-risks and additionally enables an early link of the passengers to the vehicle movement using the belt tightener.

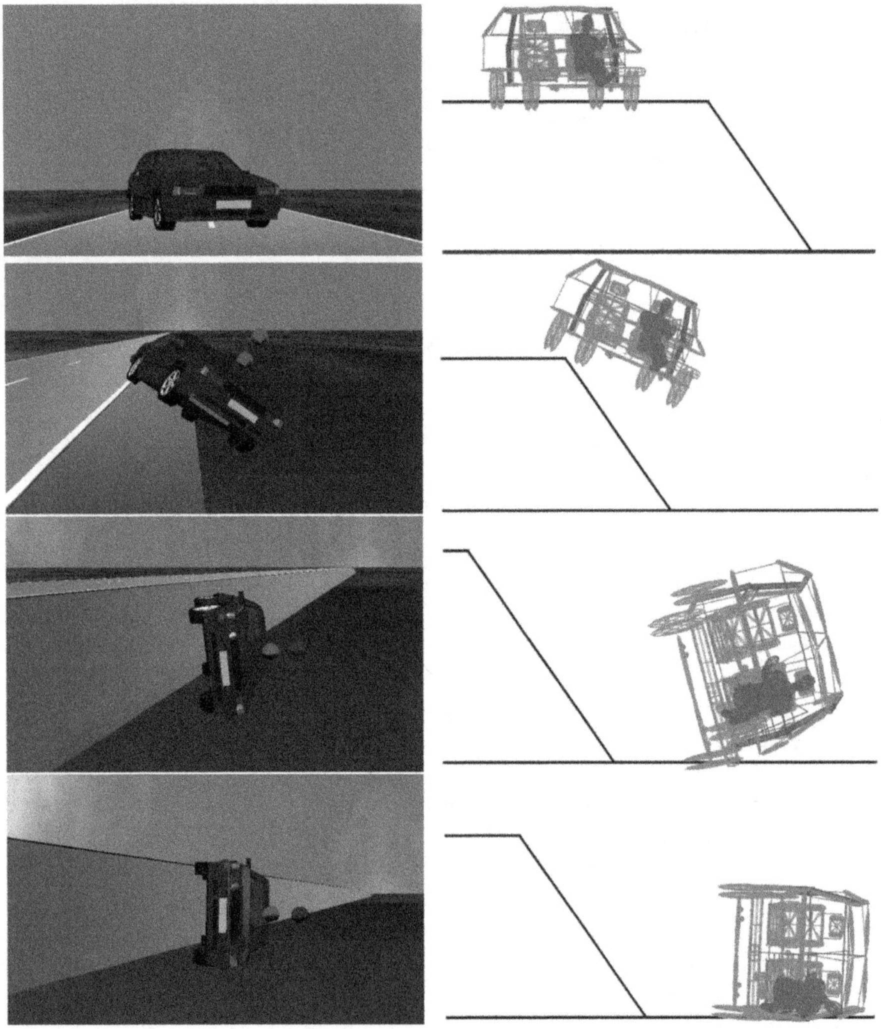

Fig. 14.25 Embankment maneuver with 2.2 m height of the embankment, 55° angle of the embankment, 15° approach angle and 50 km/h velocity

14.2.2.4 Sand Bed

The roll over scenarios of the trip-over category (see Fig. 14.11) that almost exclusively occur in the lateral direction of the vehicle are represented using the curb brunt and the sand bed. The definition of suitable testing parameters for the sand bed maneuver via simulation is however afflicted with certain limitations. The empirical model parameters of the wheel-ground contact of the sand bed maneuver are not transferable to other vehicles without another validation based on

Table 14.3 Maximum reached roll angle depending on the sliding velocity upon entering the sand bed

Lateral sliding velocity/(km/h)	17	20	21	22	23	24	25	27	30	33
Roll angle/(km/h)	12	19	25	33	46	52	100	180	180	180

experimental data (see Sect. 14.2.1). The model only holds validity for the dig-in behavior achieved in the framework of the validation testing. If the wheels or the composition of the ground change, an adaption of the empirical testing parameters (Fig. 14.16) is necessary. Therefore, the results presented here are not directly transferable to other mid-range vehicles. The model does however depict tendencies and the magnitudes of testing parameters for mid-range vehicles as indications.

In the sand bed roll over, just like for the ramp and the embankment, tests need to be conducted each below the action level (roll angle 35°–40°), slightly below the statistical action level (roll angle 50°–60°) and above the statistical action level (>60°). To this end, simulations with different lateral sliding velocity were executed using the average mid-range vehicle. The results of these simulations are listed in Table 14.3. On the basis of these results, the relevant action levels, for example at velocities of 22.4 and 33 km/h can be determined. In Fig. 14.16, the roll angle development resulting from the two previously named parameters are presented. Depending on the requirements for passenger safety, the action level for the roll over maneuver in the sand bed can be administered.

Similarly to the application of the ramp- and embankment maneuvers, a compromise between a sufficient triggering and misuse robustness needs to be found for the sand bed maneuver. Therefore, again the latest possible moment of triggering needs to be calculated that allows for the unobstructed deployment of the window bag, using the passenger simulation. The screenshots depicted in Fig. 14.27 of a simulation of driving dynamics (left) and the corresponding MADYMO-passenger

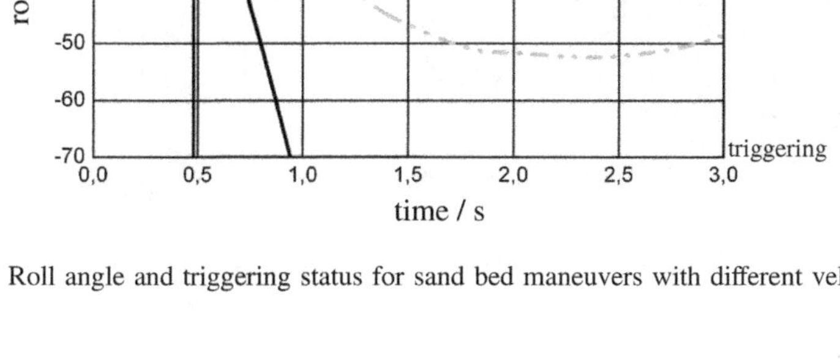

Fig. 14.26 Roll angle and triggering status for sand bed maneuvers with different velocities

simulation (right) illustrate the typical passenger kinematics of a sand bed maneuver. In this maneuver with a lateral sliding velocity of 33 km/h, a roll angle of 11° is sufficient for a triggering. In Fig. 14.27, the moment of the roll over recognition is visualized using the upper sphere above the vehicle. The lower sphere signals the moment in which the plausibility barrier is recognized. For the presented maneuver, recognition already occurs at a roll angle of 6°. The timeliness of this trigger

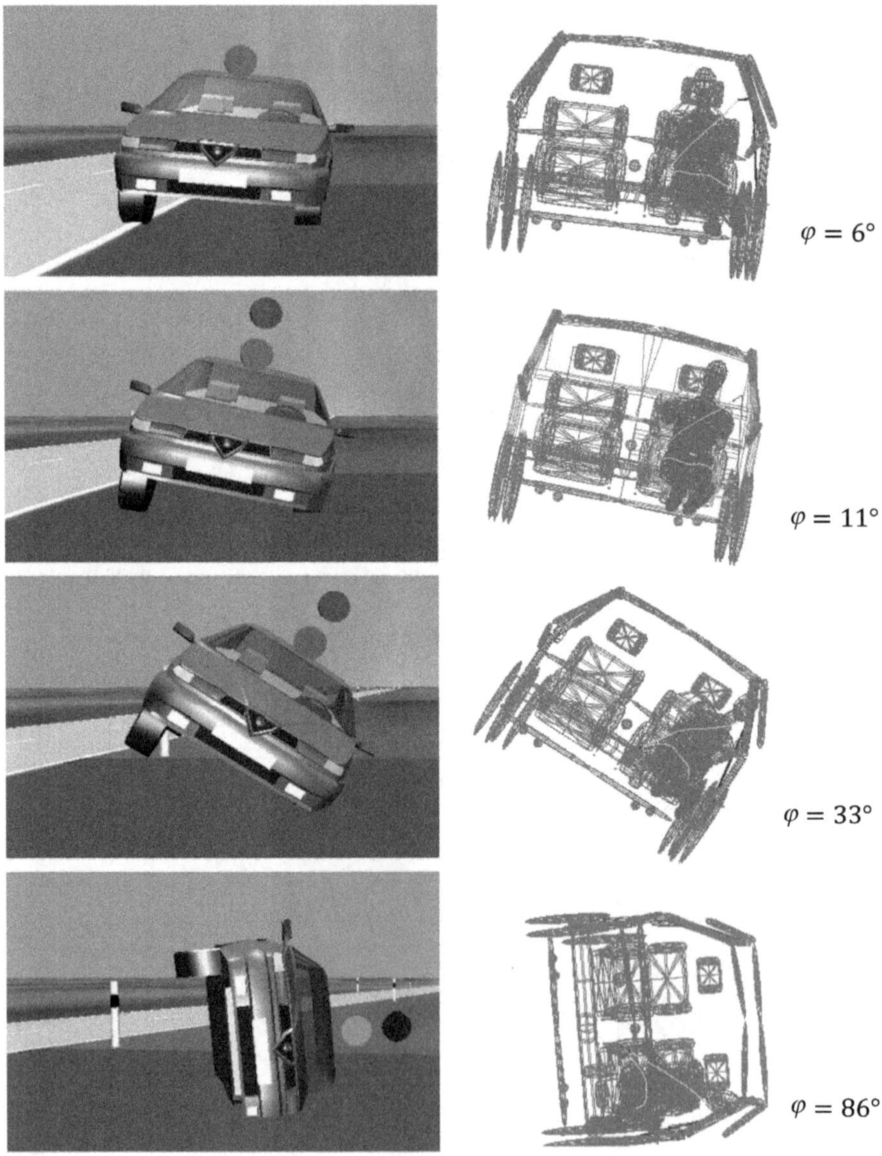

Fig. 14.27 Sand bed maneuver with 33 km/h lateral sliding velocity

behavior is confirmed by the belted 50 % male Hybrid III Dummy which was used for the passenger simulation. Multiple passenger simulations have shown that during the sliding of the vehicle through the sand bed, relative movements between the head of the dummy and the interior of the vehicle occur even for low roll angles. They are caused by a continuously acting lateral delay during the process of the wheels digging in. This lateral delay typically lays between 1 and 2.5 g (Fig. 14.26).

Therefore, the out of position (OOP)-risk needs to be classified as rather high for sand bed rollover maneuvers. As opposed to the maneuvers on ramps and embankments, the dummy head is in contact with the glass very early in the sand bed roll over. In all of the sand bed maneuvers examined here it could be concluded that an OOP-relevant situation can already occur at a vehicle roll angle of about 12°. This means that for the calculation of a robust triggering decision there is relatively little time available. The trigger behavior shown in Fig. 14.27 with a triggering decision at a roll angle of 11° therefore fulfills the requirements for the avoidance of OOP-risks while at the same time guaranteeing good misuse robustness.

14.3 Control of the Roll Dynamics Using Active Anti-Roll Bars

Next to the chassis components in the narrow sense such as wheels, tires, brakes, wheel trunk, wheel carrier, steering system, suspension and insulation as well as (passive) anti-roll bars, many vehicles have nowadays implemented active components into the chassis. These active components are used in order to realize for example chassis control systems. In this section, the modeling procedure of a single track model with an added linear roll dynamic as described in Chap. 10 is used to simulate a chassis control system with active anti-roll bars.

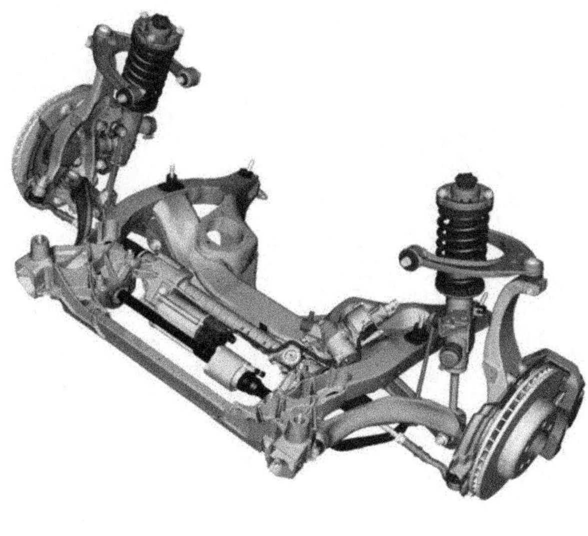

Fig. 14.28 Realization of an active anti-roll bar (by courtesy of the BMW Group 2009)

Figure 14.28 exemplary shows a double wishbone axle with integrated anti-roll bar. A detailed presentation, including a discussion of possible realizations is contained in (Öttgen 2005).

14.3.1 Passive Anti-Roll Bar

Anti-roll bars are used for axles with independent wheel suspension in order to reduce the slope of the chassis during cornering on the one hand, and to influence the tendency to under- or over steer (resonant intrinsic steering behavior) on the other. As already outlined in Chap. 9, the anti-roll bar holds the function of a torsion bar that is pivotally mounted to the chassis. By different spring deflections of the wheel suspensions of an axle, torsion of the rod arises which stamps forces to the wheel suspension via a leverage mechanism. This collectively leads to an induration of the chassis suspension and therefore lowers the roll angle. The coupling of both wheel suspensions of an axle in one-sided road excitations does however have a negative effect on driving comfort.

The sum of the anti-roll bar stiffness of both axles decides the resulting roll angle reduction. The wheel vertical load differences, and subsequently the self-steering behavior of the vehicle can be influenced at each axle using the allocation of the stiffness on the front- and rear axles.

14.3.2 Stiffness Distribution Between Front- and Rear Axle

Below, the simulation results of (Öttgen 2005) are used that were won using a linear roll model. The roll model is described by the equation derived in Chap. 10:

$$\begin{aligned} &\Theta_{xx,A}\ddot{\varphi} + 2\left(s_{D,v}^2 d_v + s_{D,h}^2 d_h\right)\dot{\varphi} \\ &+ 2\left(s_{F,v}^2 c_{F,v} + \frac{c_{St,v} l_{St,v} s_{St,v}}{2b_v^2} + s_{F,h}^2 c_{F,h} + \frac{c_{St,h} l_{St,h} s_{St,h}}{2b_h^2}\right)\varphi \\ &= m_A a_y (h_S - h_W). \end{aligned} \tag{14.2}$$

In order to make visible the influence of the stiffness distribution of the chassis, and to simplify the identification of the physical parameters, the additional parameter λ_{FD} is introduced which describes the distribution of the spring rigidity and the damping rate between front and rear axle by approximation. With this, simple correlations between the front and rear anti-roll bar can be achieved. One has:

$$c_{F,v} = \lambda_{FD} c_F, \quad c_{F,h} = (1 - \lambda_{FD}) c_F \tag{14.3}$$

for the roll elasticities and

$$d_v = \lambda_{FD} d, \quad d_h = (1 - \lambda_{FD}) d, \tag{14.4}$$

with $\lambda \in [0 \quad 1]$ for the roll damping values. By plugging Eqs. (14.3) and (14.4) in Eq. (14.2), one obtains, first without consideration of the anti-roll bars, the equation of motion:

$$\Theta_{xx,A}\ddot{\varphi} + 2\left(s_{D,v}^2 \lambda_{FD} + s_{D,h}^2 (1 - \lambda_{FD})\right) d\dot{\varphi} + 2\left(s_{F,v}^2 \lambda_{FD} + s_{F,h}^2 (1 - \lambda_{FD})\right) c_F \varphi$$
$$= m_A a_y (h_S - h_W). \tag{14.5}$$

For a typical mid-range vehicle, (Öttgen 2005) determines the following exemplary values from reference point measurement, using an identification process:

$$\lambda_{FD} = 0.69, \quad c_F = 36,390 \frac{\text{N}}{\text{m}}, \quad d = 5,401 \frac{\text{Ns}}{\text{m}}. \tag{14.6}$$

Similarly to the parameters of the wheel suspension, the distribution of the anti-roll bar stiffness between front and rear axle is described using a scalar parameter $\lambda \in [-1 \quad 1]$:

$$c_{St,v} = \frac{1}{2}(1 + \lambda) c_{St}, \quad c_{St,h} = \frac{1}{2}(1 - \lambda) c_{St}. \tag{14.7}$$

For the given exemplary vehicle, here an exemplary value for the combined anti-roll bar stiffness of $c_{St} = 2,882 \frac{\text{Nm}}{\text{rad}}$ could be found.

In order to illustrate the different distributions of stiffness between front and rear axle, Fig. 14.29 shows the respective trajectories of a vehicle when driving into a stationary circle of the x_E, y_E—level with identical steering movement, a velocity of (v = 15 m/s) and a factor of adhesion of ($\mu = 1$). The used vehicle configurations differ however from in the distribution of the combined stiffness of the anti-roll bars at the front- and rear axle by the distribution parameter $\lambda \in [-1 \quad 1]$. When $\lambda = -1$, the complete stiffness lies at the rear axle, while $\lambda = 1$ represents the case in which the complete stiffness lies at the front axle. The trajectories of the center of gravity of the vehicle in Fig. 14.29 show that a vehicle exhibits a stronger under steering behavior when the distribution of stiffness leans towards the front axle ($\lambda = 1$) compared to the case in which it is further towards the rear axle ($\lambda = -1$). This consequently leads to a larger curve radius.

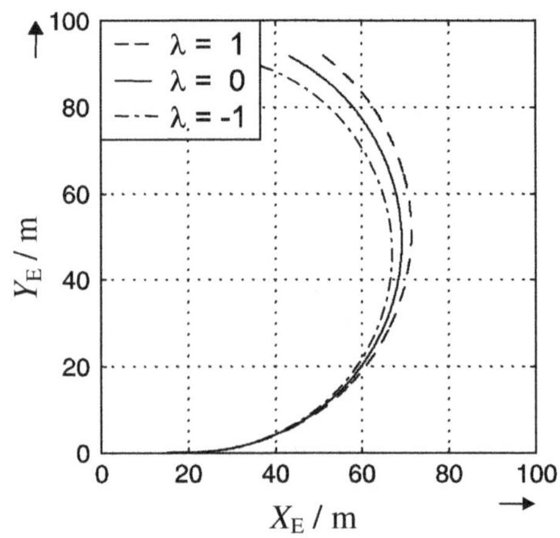

Fig. 14.29 Trajectories with different distributions of the anti-roll bar stiffness

In Fig. 14.30 the roll angles φ and the wheel steering angle δ_R are plotted from circular driving under static conditions that unfolds, or that is set up, to drive in a circle with the radius $R = 50$ m for an identical vehicle longitudinal velocity. One can observe in Fig. 14.31 that the distribution λ of the anti-roll bar stiffness has little impact on the roll angle. The increase of the combined anti-roll bar stiffness c_{St} however achieves a significant reduction of the roll angle.

In order to drive through the same circle, Fig. 14.30 exhibits an increasing wheel steering requirement for a stronger distribution of the stiffness towards the front axle. Here, the total stiffness only causes an increase of the influence of the parameter λ for larger values.

In a passive chassis, an under steering setting is usually intended in order to give the driver the chance to control the vehicle better by acting in an intuitive way (turning the wheel more). On the other hand, a strongly under steering vehicle loses agility and is perceived as less maneuverable. From these conflicting goals, the wish

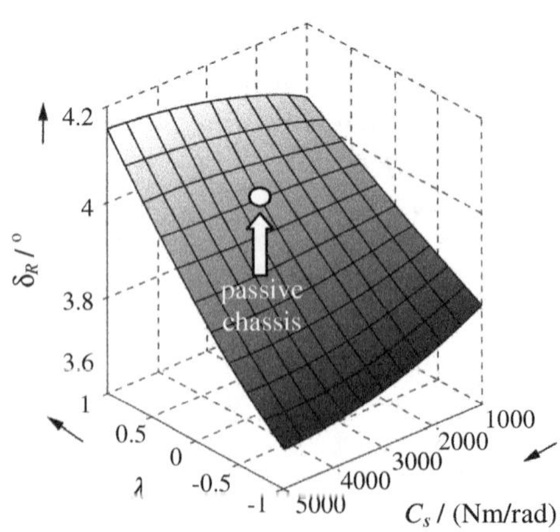

Fig. 14.30 Wheel steering angle for circular driving under static conditions

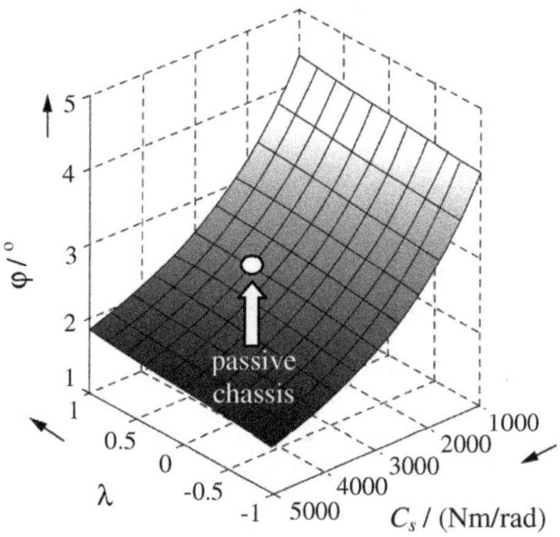

Fig. 14.31 Roll angle for circular driving under static conditions

for a generally over steering vehicle in regular vehicle operations and an under steering vehicle in critical vehicle operations arises. The approach for the solution of this conflict of objectives discussed in this section is the use of active anti-roll bars in order to realize situation-dependent torques at the front and rear axle. This way, other than in vehicles with passive chassis, almost all of the operating range presented in Fig. 14.31 is available. Additionally, a reduction of the roll motion and the resonant intrinsic steering behavior wished for in each case can be set up almost independently from each other and depending on the particular driving situation.

14.3.3 Adjustment of the Roll Dynamics by Means of Active Anti-Roll Bars

As regulator control elements for the roll stabilization active anti-roll bars are used (see Fig. 14.28). To this end, torsion rods of the passive anti-roll bars are replaced by electrical or hydraulic actuators, which actively apply torques $M_{St,v}$ on the front and $M_{St,h}$ rear axle, see Chap. 9. The applied torques are supported using the column hinged onto the wheel trunk and therefore initiate torque about the longitudinal axis of the vehicle into the platform of the vehicle using stress. This way a roll stabilization is possible and by partitioning the applied torques to the front and rear axle, the resonant intrinsic steering behavior is influenceable.

14.3.4 Control Unit Design

The control unit design is achieved by (Öttgen 2005) using the reduced vehicle model in a continuous time domain, because this way standard interpretive

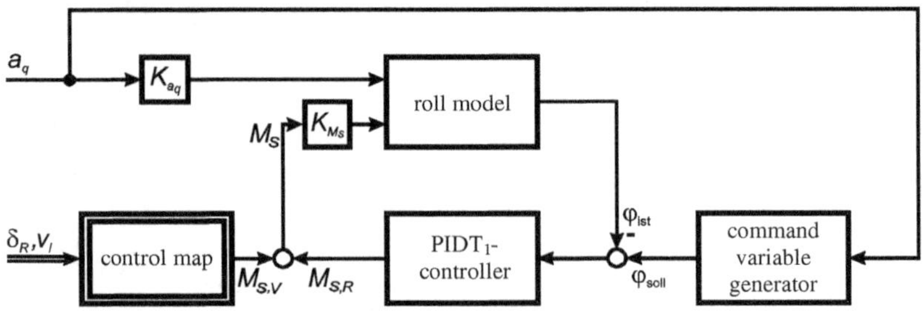

Fig. 14.32 Concept for the control system of the roll angle

processes for continuous systems can be used. In doing so a control system is proposed which on the one hand uses the combined anti-roll bar torque M_{St} to influence the roll dynamic and on the other hand actively influences the self-steering dynamic due to the distribution of the anti-roll bar torques at the front and rear axles. The control circuit for the control system of the roll dynamic used by (Öttgen 2005) is displayed in Fig. 14.32. Here, the roll angle needs to be determined using appropriate estimation procedures such as for example a angular velocity sensor as shown by Lich and Breitmaier (2003) for the over-roll sensing.

The roll angle φ_{ist} is the control variable, and the cumulative anti-roll bar torque M_{St} is the correcting variable. The metrologically measurable lateral acceleration a_y represents a disturbance value from the point of view of control engineering. Furthermore, the command variable φ_{soll} and the feed forward torque $M_{S,V}$ dependent on the wheel steering angle δ_R and the longitudinal velocity of the vehicle v_l are considered whose determination is covered later. The description of the roll dynamics of the reduced model changes with the combined anti-roll bar torque M_S and the distribution $\lambda \in [-1 \quad 1]$ as follows:

$$\ddot{\varphi} + \left(\frac{d_v s_{d,v}^2 + d_h s_{d,h}^2}{2\Theta_{xx,A}}\right)\dot{\varphi} + \left(\frac{c_v s_{f,v}^2 + c_h s_{f,h}^2}{2\Theta_{xx,A}}\right)\varphi = K_{a_y} a_y + K_{MS} M_{St}, \qquad (14.8)$$

with the related excitation by centrifugal forces

$$K_{a_y} a_y = \frac{m(h_S - h_W)}{\Theta_{xx,A}} a_y \qquad (14.9)$$

and the actively relative roll torque

$$K_{MS} M_{St} = \left(\frac{\frac{1}{2}(1+\lambda)s_{f,v}}{2a_v b_v \Theta_{xx,A}} + \frac{\frac{1}{2}(1-\lambda)s_{f,h}}{2a_h b_h \Theta_{xx,A}}\right) M_{St}. \qquad (14.10)$$

The transfer function of the linear roll model from Eq. (14.8) as control path in the image section is:

$$G_s(s) = \frac{1}{\Theta_{xx,A}s^2 + \frac{d_v s_{d,v}^2 + d_h s_{d,h}^2}{2} s + \frac{c_v s_{f,v}^2 + c_h s_{f,h}^2}{2}}. \quad (14.11)$$

After the transfer function from Eq. (14.11) the distance corresponds with a PT_2-element. The Eigen angular frequency and damping of the control path amount to:

$$\omega_0 = 6.318 \frac{rad}{s}, \quad D = 0.47. \quad (14.12)$$

The frequency response of the linear roll model from Eq. (14.18) is depicted in Fig. 14.33. For the choice of the controller structure, the characteristics stability and stationary accuracy of the control circuit are vital. Furthermore it is attempted to create a control circuit that is as fast as possible in order to create a high damping. These requirements do however contradict each other which is why a good compromise is needed. To this end, a PID-controller can be used.

Concerning the realizability, a $PIDT_1$-controller—also known as real PID-controller—is used for the control of the roll dynamics which has the following transfer function (Unbehauen 2008):

$$G_R(s) = K_R\left(1 + \frac{1}{T_I s} + T_D \frac{s}{1 + T_1 s}\right). \quad (14.13)$$

For the controller design, the frequency characteristic method is used (Öttgen and Bertram 2003), in which an inference can be made from the behavior of the open system to the closed system (Lunze 2013). Next to the stability behavior, also

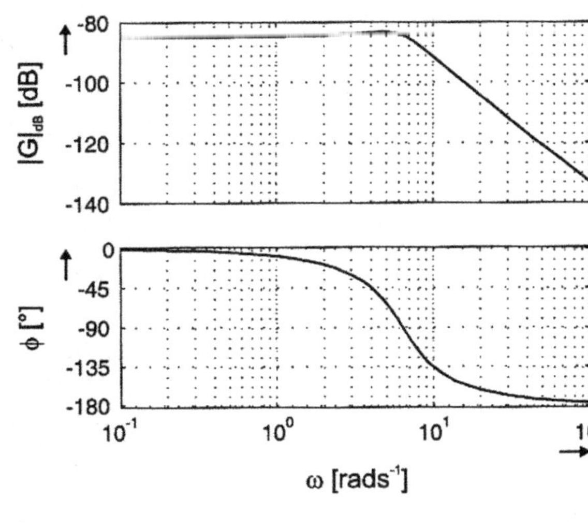

Fig. 14.33 Frequency response (Bode-diagram) of the control path

a qualitative statement can be made about the transient response, meaning damping and quickness. The frequency response characteristic of the open system has the following form:

$$G_0(s) = K_{MS} G_s(s) G_R(s), \tag{14.14}$$

with

$$K_{MS} = \frac{\frac{1}{2}(\lambda + 1) s_{f,v}}{2 a_v b_v} + \frac{\frac{1}{2}(1 - \lambda) s_{f,h}}{2 a_h b_h}. \tag{14.15}$$

For the stability of the closed control circuit with degenerative feedback it is necessary that the phase response does not pass below the 180° line for the gain crossover frequency ω_T of the amplitude response through the 0 dB line. The gain crossover frequency is furthermore a measurement for the quickness of the system. In Eq. (14.16), the controller gain and time constants are compiled while an overall gain of the open control circuit of $K_O = 10$ is chosen:

$$K_R = 28155, \quad T_I = 0.306\,\text{s}, \quad T_D = 0.072\,\text{s}, \quad T_1 = 0.01\,\text{s}. \tag{14.16}$$

The frequency response characteristics of the open control circuit from Eq. (14.14) is presented in Fig. 14.34. From this it can be seen that the system is stable in the whole area of operation as the phase response does not go below the 180°-line. The system contains a phase margin of around $\phi_R \approx 68°$, which hints towards a high damping. The average frequency of about $\omega_T \approx 13.5$ rad^{-1} suggests a short response time.

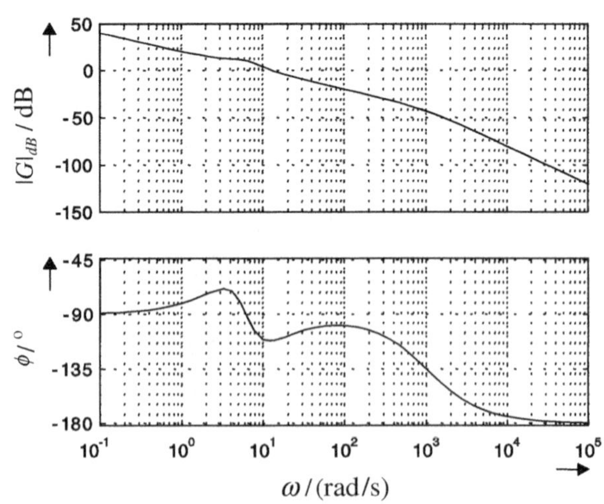

Fig. 14.34 Bode-Diagram of the open control circuit

14.3.5 Response and Disturbance Reaction

The control unit for the roll stabilization of a vehicle is evaluated in Öttgen (2005) based on reference value and disturbance values steps. To this end, simulations were executed using a time-discrete linear roll model (Öttgen 2005) for a more detailed description and discussion.

14.3.6 Roll Torque Distribution with Fuzzy Logic

Up until this point a control unit for the roll stabilization of a vehicle was used with the help of active anti-roll bars. To this end, the combined anti-roll bar torque of the vehicle was appropriately controlled first. Additional active anti-roll bars of course also influence the self-steering behavior of a vehicle. This is then discussed using the single track model with linear roll dynamics (Fig. 14.35). Now the driver input in the form of the steering angle δ_H, which leads to a wheel steering angle $\delta_R = \frac{1}{i_L}\delta_H$ via the steering transmission i_L, is used as an input variable. The longitudinal velocity of the vehicle is treated combined with the factor of adhesion μ as an environment value is treated as an additional parameter. The roll angle φ_{ist} is controlled using the PIDT$_1$-controller explained earlier. Additionally, the correcting variable combined anti-roll bar torque M_{St} is now distributed between front and rear axle (Öttgen and Bertram 2004), depending on the lateral acceleration a_y, the factor of adhesion μ and the longitudinal velocity of the vehicle v_L using a linguistic fuzzy algorithm (Zadeh 1973 and Mamdani and Assilian 1975).

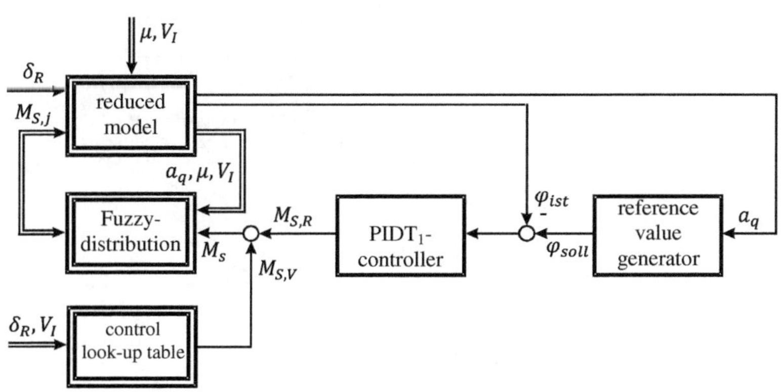

Fig. 14.35 Concept for the active distribution of the roll torques

14.3.7 Active Principle

In this section, the active principle of active anti-roll bars in order to influence the self-steering behavior using the shift of the wheel vertical forces at one axle is explained in more detail. The side slip stiffness used for the determination of the lateral wheel force, is non-linearly dependent on the respective vertical force of the particular wheel. As an approach for the determination of the side slip stiffness in dependence on the vertical force, a polynomial of the third order is chosen in Öttgen (2005). The identified correlation for the wheels at the front axle at a factor of adhesion of $\mu = 1$ is presented in Fig. 14.36. The wheel vertical force and the side slip stiffness in straight driving with symmetrical mass distribution between the left and right side of the vehicle is depicted using the drawn through lines. The driving in a circle then results in a wheel load difference of $2\Delta F_v$. Due to the non-linear correlation, the wheel with decreasing vertical force loses more side slip stiffness than the other wheel with increasing vertical force gains

$$\left|\Delta c_{\alpha,z}\right| < \left|\Delta c_{\alpha,a}\right|. \tag{14.17}$$

A larger stabilizing torque at this axle during cornering leads to a further increase of the wheel load difference Eqs. (14.18) and (14.19):

$$F_{v,VL} = \left(\frac{l_h}{l}\frac{mg}{2} - \frac{ma_y h_W}{s_R} + \frac{\Theta_{xx,D}}{s_R}\ddot{\varphi}\right) - \frac{d_v s_{d,v}^2}{2s_R}\dot{\varphi} - \frac{c_{f,v} s_{f,v}^2}{2s_R}\varphi - \frac{M_{S,v} s_{f,v}}{2b_v s_R}, \tag{14.18}$$

$$F_{v,VR} = \left(\frac{l_h}{l}\frac{mg}{2} - \frac{ma_y h_W}{s_R} + \frac{\Theta_{xx,D}}{s_R}\ddot{\varphi}\right) - \frac{d_v s_{d,v}^2}{2s_R}\dot{\varphi} - \frac{c_{f,v} s_{f,v}^2}{2s_R}\varphi - \frac{M_{S,v} s_{f,v}}{2b_v s_R}. \tag{14.19}$$

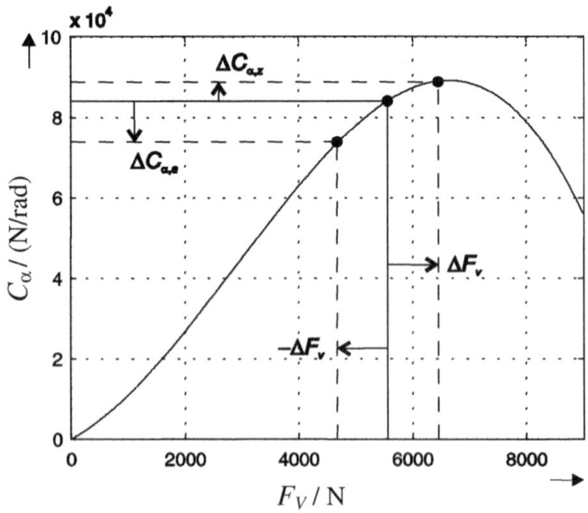

Fig. 14.36 Identified tire characteristics of the front axle at a factor of adhesion of $\mu = 1$

Fig. 14.37 Distribution of the anti-roll bar torques between front and rear axle

Thus, the sum of the side slip stiffnesses at this axle decreases. For the composition of the corresponding wheel lateral force, a larger side slip angle is needed. This means that an increase of the anti-roll bar torque at one axle leads to a decrease of the side slip stiffness and an increase of the side slip angle. With a larger torque at the front axle, an understeering behavior can be reached, and with an increase of the torque at the rear axle, a more over steering vehicle behavior can be achieved. By using an active distribution of the anti-roll bar torques between front and rear axle, a diverse under or over steering vehicle behavior can therefore be achieved. The distribution of the torques is done using the parameter $\lambda \in [-1 \quad 1]$. That way, $\lambda = -1$ describes the complete torque at the rear axle, while $\lambda = 1$ describes the applied torque completely at the front axle Fig. 14.37.

14.3.8 Potential of a Roll Torque Distribution

After having already displayed the influence of the stiffness distribution of passive anti-roll bars between front and rear axle, now the potential of the active roll torque distribution concerning the self-steer behavior of a vehicle using active anti-roll bars is presented. In Fig. 14.38, three trajectories from rides through a circle with

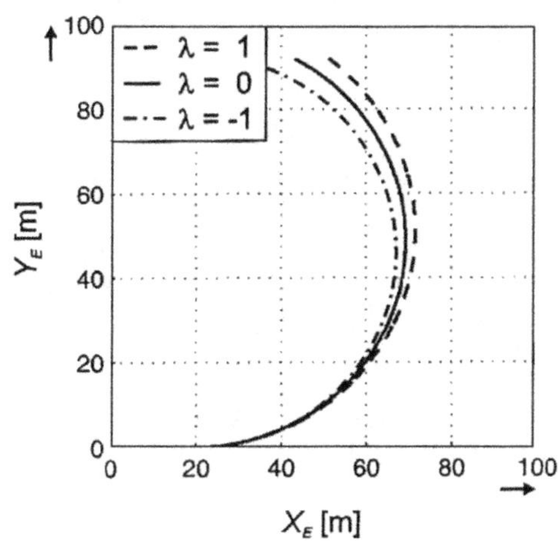

Fig. 14.38 Trajectories with different distributions of the combined anti-roll bar torque

the reduced vehicle model are shown at a longitudinal velocity of the vehicle of v_l = 15 m/s and a factor of adhesion of $\mu = 1$.

The three vehicle configurations only differ due to different distributions of the combined anti-roll bar torque of $\lambda = -1$, $\lambda = 0$ and $\lambda = 1$. Here, the combined anti-roll bar torque is determined depending on the described control unit design. In comparison with the distribution of stiffness of the passive anti-roll bars, the influence with active anti-roll bars is even larger due to the higher anti-roll bar torques due to amount. Thus, the vehicle show a much more neutral driving behavior at a distribution of $\lambda = -1$ with a radius of gyration of R = 44.9 m, than at the distribution of $\lambda = 1$ with R = 52.1 m. Thus, a distribution towards the rear axle equals a gain in the area of agility due to the decrease of the needed effort of steering for the driver in order to make the vehicle drive in the desired path. In Fig. 14.39, the lateral acceleration and roll angle processes are presented. Using the stationary lateral accelerations in Fig. 14.39a it can be recognized that the vehicle with the distribution towards the front axle ($\lambda = 1$) has a significantly higher required steering angle, in order to drive through the same circuit. The roll angle curves in Fig. 14.39b illustrate again the dependence of the reference input on the lateral acceleration via the passive roll model which was also simulated in the algorithm.

An under steering vehicle has clear advantages in instable and critical driving situations due to the push via the front wheels. The vehicle is manageable more easily because the driver needs to summon a larger steering wheel angle following the bend. In a stable and secure vehicle operation however, a neutrally synchronized vehicle is much easier to steer and possesses a much larger lateral guidance potential for e.g. suddenly occurring critical driving situations. The advantages of both vehicle configurations are united in Öttgen (2005) by a distribution of the anti-roll bar torques independent from the driving condition, as well as Fuzzy Logic.

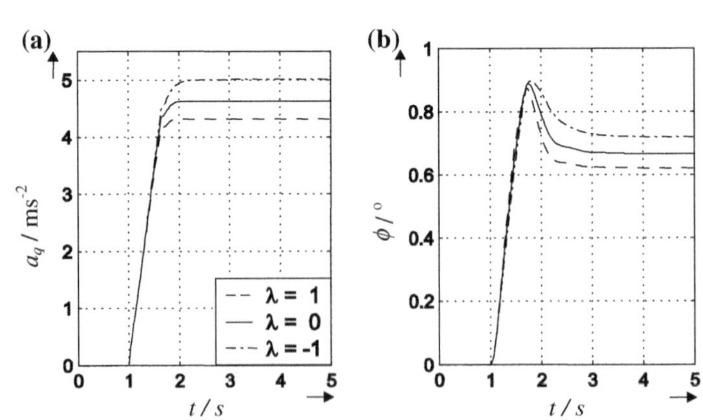

Fig. 14.39 Lateral acceleration **a** and roll angle **b** with different distributions of the combined anti-roll bar torque

14.4 Driving Simulators

Driving simulators of various types and complexity have become an indispensable tool for research and development, especially in the field of automotive engineering. They are suitable for describing the complex interaction between driver, environment and vehicle in a realistic and efficient manner. Since the middle of the 20th century, simulators have been used in research and development as well as a tool for risk-free training in many other industrial sectors. For example, simulators have also been used in the aerospace industry for a long time. The first driving simulators used in automotive engineering emerged in the 1970s. However, research into the field of vehicle dynamics could not be executed satisfactorily in the beginning. With the hardware available at that time, the vehicle dynamics could not yet be simulated with the required degree of realism. And also, the development of real-time simulation algorithms, including numerical solving techniques, had not yet reached a sufficient level of development. Only the massively increasing CPU computing power and new, powerful numerical solution algorithms have made it possible to compute sufficiently complex models of the vehicles and traffic situations in real time. At the same time, modern methods of visualizing virtual realities enable completely new possibilities. In addition, there was the development of algorithms that allowed the calculated simulation results to be converted into movements of the simulator in such a way that a realistic driving feeling was generated. Thus it was possible to steadily increase the complexity of the simulated systems together with the degree of immersion (reality feeling). This provided the basic prerequisites for driving simulators in the field of driving dynamics.

14.4.1 Areas of Application and Implementation of Driving Simulators

Important applications of driving simulators are the simulation of driving dynamics and investigations in the area of the human-machine interface. The latter is possible since the test persons can immerse themselves into environments that are at least close to reality through the driving simulator. They provide an excellent tool for the development of systems that represent a complex interface between driver and vehicle (e.g. in the field of driver assistance). They are always used when analyses based on classic off-line simulations are not sufficient i.e. because the driver is a significant and acting part of the overall system. The simulator is often the only means for investigating questions in the field of automotive technology, especially when tests with real vehicles are too risky in public road traffic, or even on test tracks, or when prototypes are not yet available.

The advantages of at least partly avoiding real physical tests with prototype vehicles through the use of driving simulators are similar to those of other simulation techniques described in this book:

- Reproducibility and comparability of test series.
- Full control and wide potential range of variation in ambient conditions and the avoidance of environmental irritation that may affect the test persons in real driving tests.
- Cost savings in comparison to real tests during the trial, and the absence of cost-intensive prototypes.
- Efficiency improvement in the development process, since the driving simulator can be used in early development phases, if no prototype is available.
- Safe conduction of experiments even during testing of critical situations.

On the other hand, a ride in the simulator can never completely replace vehicle tests with a real vehicle. Depending on the experimental scenario and the simulator used, some limitations may arise:

- Perceptual physiological restriction of the driving simulation by the imaging and motion system used.
- Scenario design and the representation of vehicles and systems.
- Drop out of test persons due to kinetosis ("simulator disease" or "motion sickness").
- A changed and possibly unrealistic driving behavior of the test persons, since they often drive more risk-free and less conform in a simulator.

Driving simulators must meet a wide range of requirements, depending on the specific application. Thus the performance of the motion system of the simulator is decisive for the study of the driving dynamics discussed in this book. Especially for this purpose, it is indispensable to convey a realistic driving impression. However, with increasing complexity, size of the movement space to be realized, and the universality of the simulator the construction and operating expenditure of the motion system increases disproportionately.

In all cases, driving simulators should ensure the most realistic behavior of driver and vehicle as well as their combination in different test scenarios. For this, various and partly contradictory features of the simulator are required:

- Precise simulation models for describing the physical behavior of the involved systems as well as their interdependency's,
- The possibility to simulate all components and systems involved in real-time,
- The immersion of the driver in the driving situation,
- The driving task, possibly together with secondary activities, must be executable as far as possible like in the real environment.

The methods for the specific modeling and simulation of the vehicle or individual components were described in detail in chaps. 10 and 11, after establishing the foundations in the first 9 chapters. These ensure a simulation of the vehicle movement in real time upon implementation on a suitably powerful computer hardware (Unterreiner 2013, Unterreiner et al. 2013). The methods described in Chaps. 12 and 13 are only applicable to a limited extent or may even be not applicable at all as of now. This holds in a similar form for commercial simulation programs such as, e.g. ADAMS.

The driving task and the resulting relevant requirements for driving simulators are described below, see also (Maas 2017).

The structuring of the action strategies in solving tasks in general, but also in the tasks of driving a motor vehicle in particular, can be represented according to (Donges and Naab 1996) in three behavioral levels, Fig. 14.40.

The task of vehicle guidance without further secondary activities can be divided into three temporal planes, see Fig. 14.41. The variation of duration and complexity of tasks within the levels of vehicle management are very great. However, the processing of tasks in the control circuit driver-vehicle-environment is always based on sensory stimuli. On the other hand, the nature of the recording (sense-channel) as well as the processing can differ considerably. The operator (driver) of the driving simulator generally works with acquired knowledge. Depending on the task, the human being reverts to different stimulus-reaction mechanisms, Fig. 14.40:

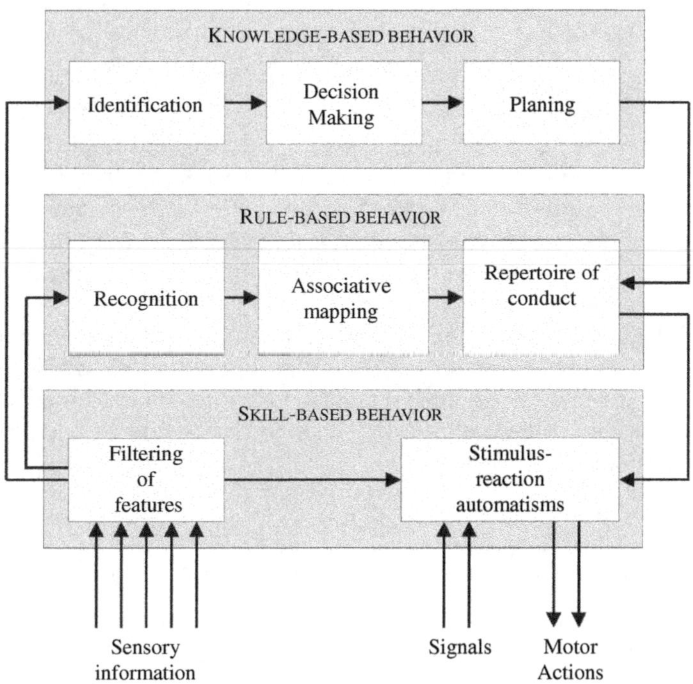

Fig. 14.40 Different levels of the driving task according to (Rasmussen 1983)

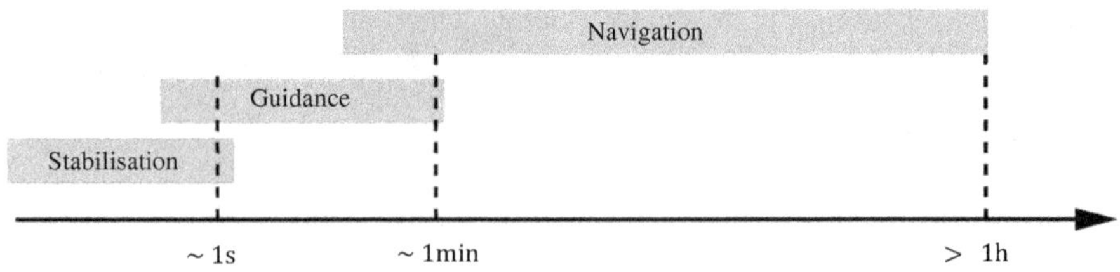

Fig. 14.41 Time model presentation of the vehicle management task according to (Donges and Naab 1996)

- In the case of stabilizing actions (for example, the compensation of suddenly occurring lateral winds), this is done on the basis of pure stimulus reactions. This usually trained behavior is very fast and is directly linked to a stimulus. There is no further conscious interpretation of the situation required by the driver in this case.
- The rule-based behavior relies on solution strategies based on trained rules such as the proper recognition of and compliance with road signs.
- For orientation in unknown areas, a driver will collect information from the environment (such as the location of prominent buildings or landmarks). By linking this information, he will then consciously initiate an action.

Particularly important for the development of driving simulators with consideration of the three-level model shown in Fig. 14.40 is the interaction of the sense channels (stimuli) with reactions at a certain level of vehicle guidance. In order for a driving simulator to provide sufficiently realistic action patterns, the stimuli expected and required by the driver must be made available. Thus, in the presentation of highly dynamic scenarios, it is indispensable to provide the necessary stimuli for haptic sensory impressions if the interaction of the driver and the vehicle is to be analyzed on the stabilization level. Driving simulators, which are used for the investigation of highly dynamic processes, are therefore technically much more complex than those used for long-term driving studies with secondary tasks. Investigations which have little or no dynamic driving behavior can often be successfully carried out with very simple simulators.

14.4.2 The Control Circuit Driver-Vehicle-Environment

The consideration of the perception and the resulting action when driving a car can be described well with a simplified three-phase model. This model includes the processing stages:

- perception (gathering of information),
- cognition (processing of information) and
- motoring (information transfer).

For the selection of the relevant information sources and the further processing of the information obtained, a human being will distribute his attention to the three phases of the information processing process. A suitable attention distribution is usually learned. The quality of the learning performance determines how well a task can be performed and whether errors can occur in extreme cases.

Different forms of attention can be distinguished in terms of selectivity and intensity. The concept of selective attention is described by the fact that humans have to choose between different and, as a rule, competing sources of information.

Furthermore, the human being must usually divide his attention to various stimuli simultaneously perceived by him. In a change of attention, he will turn away from a stimulus and then turn to another. The level of activation is a measure of the intensity of attention (alertness) and thus the ability to respond to stimuli. Here the reduced vigilance (constant attention at monotonous stimulus frequency, for example an experienced driver on long, monotonous highway sections) and the constant attention (high proportion of relevant stimuli) is important. For the perception (reception of sensory impressions) the visual sense (visual stimuli), the acoustics (auditory perception), as well as the movement and position perception (haptics and vestibular perception) play an important role.

From these considerations, the necessary information sources can be derived for a driving simulator:

A fusion of optical, auditive, and haptic stimuli is generally used for driving a motor vehicle. Exactly the same channels can also be used to transmit information to the vehicle via input devices and sensors.

While the driver receives most of the information via the visual sense, most of the "inputs" are still executed via haptic controls. The most important controls are

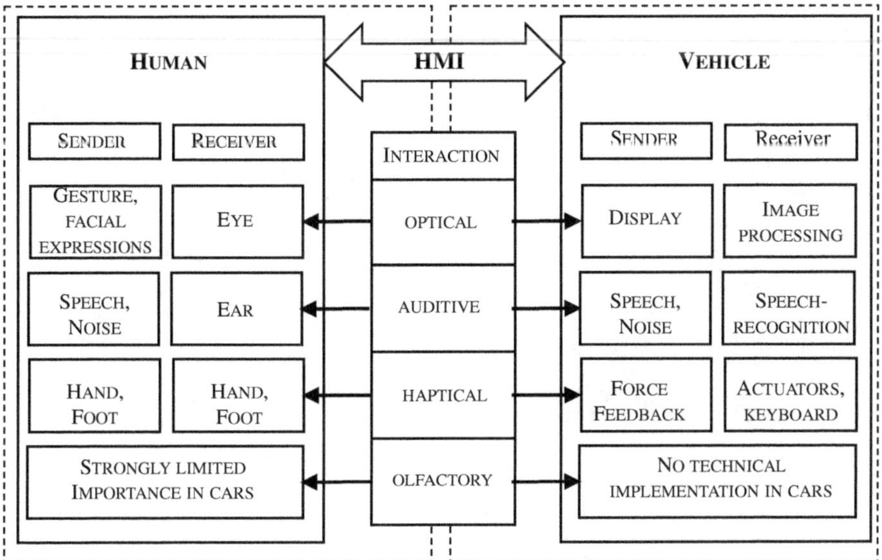

Fig. 14.42 Channels of human-machine interaction in the driving simulator (Hiesgen 2011)

the steering wheel and the pedals, which determine the movement of the vehicle. A wide range of additional control elements is used in various vehicle-specific versions, from voice control to image-processing sensors, which make it possible to make inputs via gestures.

14.4.3 Implementation of Driving Simulators

The driving simulator can, irrespective of the degree of complexity of its construction, in principle be divided into the subareas:

- simulation,
- driver's workplace,
- motion system, and
- environmental representation.

Their specific characteristics often have to be adapted to the requirements of the planned simulator experiment, Table 14.4.

The most important subsystems of driving simulators are described below. For further details on the construction of driving simulators the reader is referred e.g. to the contributions of (Negele 2007, Capustiac et al. 2011b, Capustiac 2011, Hiesgen 2011, Maas 2017).

14.4.4 Simulation Models and Interfaces

The simulation of the vehicle and its surroundings are the central functional elements of a driving simulator. The simulation core must simulate all the physical effects relevant to the simulator experiment and make the interaction with the driver's inputs as realistic as possible. In particular, the vehicle dynamics must be

Table 14.4 Structuring the elements of R & D driving simulation according to (Hiesgen 2011)

Simulation	Driver's workplace	Motion System	Environmental representation
Driving dynamics	cockpit	Number of degrees of freedom	Projection / visualization
Driver assistance systems	Man-machine interfaces	Acceleration	Field of view resolution
Surrounding traffic	HIL elements	Max. speed	2D / 3D
Roadway / underground	Ergonomics …	Frequency bands	…
Surroundings weather		Motion Cueing…	
…			Sound sources
			3D sound…

modeled with the minimum required accuracy. Sufficiently realistic models should additionally be used for other road users and the environment. The required modeling depth is decisive here. In typical urban driving situations, 20 or more vehicles may be involved. There may be additional traffic such as pedestrians or cyclists. Too high a modeling depth would make the real-time simulation required for use in a simulator more difficult or even impossible. It is therefore sensible to simulate the own vehicle (EGO vehicle) with a very high physical detail, while at the same time calculating the other traffic participants in a very simplified manner. The simulator core of a driving simulator must always communicate with the driver workstation as well as the environment display in real time. Actions of the driver must produce reactions of the system, the timing of which corresponds to that in real-time and which can also be immediately perceived by the driver. Only if this condition is fulfilled, a simulator experiment will map the control circuit driver, vehicle, and environment so accurately that the relevant findings may be obtained.

Since the simulation of the vehicle dynamics must take place in hard real-time, it is imperative to use a real-time vehicle dynamics model. For this purpose, the linear and non-linear vehicle models discussed in chap. 10 can be considered, with corresponding roll-model extensions. However, it is possible to use optimized variants of the models discussed in chap. 11. The methods for modeling and simulating a vehicle as well as individual systems of the vehicle are described in Chaps. 10 to 13 of this book.

The driver's workplace in the driving simulator must be build-up in such a way that the test person can interact with the vehicle while driving in the simulator as one would in a real vehicle. This includes, in the first instance, the control elements, which are absolutely necessary for vehicle guidance, such as the steering wheel and pedals. In addition, further input and output devices at the driver's workstation may be required for the simulator test in order to process a predetermined experimental plan. For example, novel control elements can also be tested in the simulator. There may also be additional installations such as sensors for monitoring the driver. A simple example of this are sensors, with which the movement of the pupils or the viewing direction can be detected.

For a successful simulator study, it is also important that the driver takes the "ride" as a serious task and not as a game, and acts accordingly. This includes, for example, the prevention of major risks, as often taken in a computer game, since nothing serious can happen in the simulator. This is accomplished by a number of externalities, e.g. the getting into a realistic cockpit by a vehicle door, but also by a clear instruction of the subjects and clearly formulated tasks.

The spectrum of the different variants of simulator designs used today range from simple steering wheels via reconstructed cockpits to full vehicles, which are mounted onto a motion platform. Simple designs have the decisive advantage over full vehicles, that both handling and integration into a simulation is comparatively simple.

The central control element and decisive for the lateral guidance and stabilization of the vehicle is the steering wheel. It serves as a bidirectional interface between driver and vehicle. The steering wheel serves, on the one hand, to determine the direction of movement, but on the other hand also has the important function of giving the feedback of the road and vehicle condition. This is also called a steering feeling and has a decisive effect on the driver's behavior. The representation of a realistic steering feeling is very complex in a Simulator. One reason for this is that the driver sometimes perceives and interprets the same stimuli differently in the simulator than in the real vehicle. The steering is often perceived to be more realistic when it has a significantly greater hand force than in a real vehicle because additional forces that are normally occurring at the same time from the vehicle's lateral acceleration are missing in many simulators. This shows that at least some of the lateral acceleration can be compensated for by corresponding interventions in the steering torque.

The visualization and auditory stimuli are also part of the surroundings of the vehicle. Haptic signals, such as vibrations due to an uneven road, must be realized by the motion system. Both the nature of the representation of the environment as well as the motion systems are very diverse in the various simulators.

The most important part of the environment is the visual presentation of the simulated scenario. In addition to the degree of reality of the displayed objects, the

- temporal offset to simulation,
- viewing angle (optical flow), and
- recognizable image margins

are of vital importance regarding the scope of visualization in general. The majority of the information required to drive a motor vehicle is recorded visually. Therefore, even a small temporal offset (dead time) between simulation and visual representation prevents a safe guidance of the vehicle in the simulator. In that case, the simulated vehicle is only poorly or not at all drivable, meaning that in extreme cases, the control circuit driver, vehicle, and environment becomes unstable.

The optical flow gives the driver the impression of speed in the field of view by means of the temporal angular change to objects ("motion blur"). The lack of optical flow (for example, when using a flat screen in front of the simulator) prevents a reliable assessment of the speed of the vehicle, thus destroying the realistic driving impression. The speed impression can be increased by means of speed-dependent road excitations (vibrations). The periphery field of vision is also necessary when road surfaces with a strong curvature are to be safely driven on.

In order to allow for the best possible immersion of the subjects into the simulation, an appropriate infrastructure for the driving simulator must be installed. This applies particularly to the visualization of the environment. Therefore, simulators in automotive development and research use a so-called projection cave. These caves surround the driver's workstation of the simulator in such a way that the driver can see the projected environment in all directions. The external shape of the cave can take the form of a cuboid, a cylinder, or even a dome. Cylindrical but especially spherical projection surfaces have the great advantage that the driver always looks perpendicularly at the projection surface. A disadvantage of this solution is the greater technical effort that has to be invested.

Fig. 14.43 Toyota driving simulator

14.4.5 Motion Systems

Motion systems for simulating a real vehicle movement are necessary to give the test persons a realistic feeling for the motion of the vehicle. Suitable motion systems are generally very cost-intensive because of the complex hardware. However, they must be used at least when tests are to be carried out in which the driving dynamics and their feedback to the driver are the subject of the investigation. The Toyota simulator shown in Fig. 14.43 is one of the highest design stages of this type of simulator design.

In this case, the entire visualization system, which is installed in the white dome, along with the fully contained vehicle, is set in motion by a hexapod. The hexapod can move the dome in all six spatial directions. However, it is mainly used for rolling and pitching the vehicle around the x- and y-axis, respectively. The yawing of the vehicle is represented by a free rotation of the dome about its vertical axis. In this case, longer lateral or longitudinal accelerations can also be simulated. For this purpose, the entire structure is moved translatory over a large x-y table. For the realistic propagation of the driving speed to the driver, the high-frequency vertical excitation, caused by the rolling off of the tires on the road, is decisive. For this simulator, active plungers are installed on the four wheels. These excite the wheels along the respective wheel-fixed z-axis.

Superstructures with similar functionalities are used in almost all large vehicle manufacturers and in some research facilities.

When initiating the movement into the simulator, a distinction must be made between two basically different approaches:

- The “real” movement (acceleration), as obtained from the simulation of vehicle dynamics, is initiated directly and unchanged or
- The movements (accelerations) are modified by a clever control of the motion system in such a manner, that even small intensities of excitation are sufficient to convey a realistic movement impression.

The second method is called motion cueing. A motion-cueing algorithm converts the motion of a vehicle model into simulator input commands so that a realistic movement impression is produced for the subject, although the movements which are present in reality are in fact not carried out. The underlying principle is understandable by using the example of a sustained curve ride and the lateral accelerations that the driver perceives as centripetal forces. If the real movements of the vehicle were mapped directly and unchanged by the motion system of the simulator, the rolling of the vehicle could e.g. easily be constituted via a hexapod. The lateral acceleration could be achieved by moving of the entire structure translatory. Due to the limited working space of the sled system, however, the simulator would quickly reach the limits of its movement capabilities. This can be avoided by using motion cueing algorithms. In this case, the acceleration due to gravity is used to give the driver the feeling of the lateral acceleration. For this purpose, the entire dome is strongly tilted by the motion system. Parallel rotation of the image gives the driver the impression that he moves almost horizontally. Due to the strong inclination of the simulator, the acceleration due to gravity gets a strong component in the vehicle transverse direction, which is interpreted by the driver as a lateral acceleration. The application of this technique requires that the driver has no optical reference point to unmoving parts. The determination of the set values for the control of the actuators of the motion system is a very complex mathematical task for this type of motion simulation. More information on the topic of motion cueing (the wash-out filters to be used) can be obtained from specialized research on this area, e.g. (Capustiac et al. 2011a, Capustiac et al. 2011b, Capustiac 2011).

14.4.6 Conducting Experiments with Driving Simulators

Experiments with driving simulators are extensions to the widely used development methodology based on (real-time) simulations and hardware-in-the-loop (HiL) simulations. The actual simulation algorithm must be capable of mathematically mapping all relevant physical effects and executing numerically in real time. Further information on the simulation of the vehicle as well as the various systems in the vehicle can be found in Chaps. 10 and 11 as well as further relevant literature.

A prerequisite for the successful execution of a simulator test is that the driver can accept the given driving task as real. This depends primarily on how realistic the stimuli presented to the driver are. However, simple simulator setups are already sufficient for many experiments.

The execution of a simulator experiment is essentially divided into the steps described below.

Preparation of the simulator set-ups and definition of the study results

A simulator study almost always consists of several stages. Prior to the actual simulator study, several preliminary studies are usually performed to validate the set-up of the simulator and, if necessary, to optimize it.

Before carrying out the series of tests, the examination objective must be precisely defined and a simulation scenario should be developed. At the same time, it is important to define exactly which "measured variables" should be included in the experiment. For this purpose, it must be ensured that all desired measurements can be obtained with the available models and the available simulator configuration.

Preparation of the test person

The preparation phase is used to explain the upcoming task upon which the test person is familiarized with the driving task. Often the test persons are additionally occupied with secondary driving tasks. Before starting the experiment, it is necessary to clearly define what is expected of the test person.

The use of driving simulators often deviates from the operation of real vehicles, or especially prototypical operating devices are tested. Therefore, the simulator and its operation must be briefly explained before the first ride starts. This phase can additionally be used to carry out a first data collection with the subject.

Familiarization of the test person with the simulator

The operation of a driving simulator generally deviates from the operation of a real vehicle. For this reason, the operation of the vehicle must be learned. The familiarization phase is used for driving in the simulator. In addition, many actions take place while driving a vehicle subconsciously. Adequate time must be provided for getting used to it. The ability to intuitively operate the simulator is particularly important when driving highly dynamic maneuvers that are based on stimulus response patterns (behavior-based behavioral patterns), Fig. 14.40. The habituation to the simulator can take some time depending on the structure and the planned investigations.

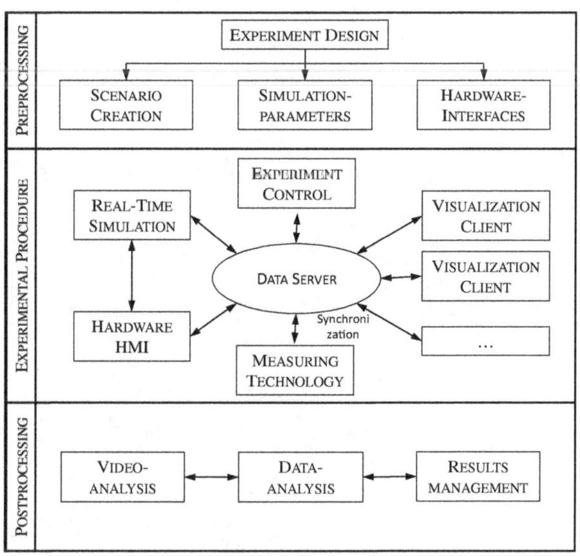

Fig. 14.44 Complete sequence of a driving simulator study according to (Hiesgen 2011)

Experimental Procedure
During the simulator study, the desired measured variables are recorded and stored. Many signal sequences can be obtained very easily from the simulation models. This also applies to signals that can only be measured with great effort in reality, such as, for example, the side slip angle. In a real vehicle this can only be recorded with complex approaches, but can directly be taken from the simulation model in a driving simulator. Apart from these, additional objective values are usually recorded (see below), which are to be evaluated or further processed according to the study. In addition, there may be other objective personal data that may provide information about the condition or reactions of the trial subject, e.g. ECGs or skin conduction values, see e.g. (Hesse 2011), but also the movement of the pupils, respectively the direction of view. It is Crucial for the successful conduction of the experiment that the test person is able to experience the ride unaffected.

Further data collection
After the test, further data can be collected as required. This relates to e.g. an anew survey of the test persons concerning subjective impressions after the test.

14.4.7 Recording of Measured Values in Simulator Tests

Measurement data can be captured much more easily in a simulator experiment than in reality. All state variables, for example, are already known from the simulation and must only be recorded. Other variables, which would not be measurable during a real journey or only with high effort, are also very easy to capture, Table 14.5. In addition, in the ideal case there is no measuring noise.

Further advantages of the laboratory conditions prevailing in the simulator test are the possibilities to observe the behavior of the driver comparatively simply.

14.4.8 Implementation of Simple Driving Simulators

The following explanations describe exemplary concepts for simulators that are realized at the Department of Mechatronics of the University of Duisburg-Essen, partly in cooperation with the Westphalian University of applied science. The

Table 14.5 Possible measured values during driving simulator test

Measurement of vehicle and traffic	Driver-related measured values	Survey
Objective vehicle parameters	Observation of the behavior (e.g., by video)	Questionnaire at different times during the experiment
Longitudinal, transverse and vertical dynamics	Physiological values	Free survey
Individual simulation variables for the vehicle (system)	E.g. Heart rate, skin conductance etc.	Subjective evaluation

presented stages of extension are split based on the mechanical design of the cockpit and the movement platform.

The key component of each simulator is the installed simulation environment. This can be largely identical for simulators of different configuration stages irrespective of the configuration of the hardware. Experiments on different simulators can therefore be conducted without having to modify the simulation software every time.

Desktop Simulator

For less demanding applications, such simulator setups are often sufficient, which can be operated directly at the developer's desk, Fig. 14.45. Such simulators provide a very simple human-machine interface. Steering wheels and pedals from the domain of entertainment electronics are often used for these simple simulators. The visualization can be carried out on a commercially available PC monitor. Both the human-machine interface and the visualization are easily scalable and adaptable to the requirements of the simulation.

Without a vehicle environment and also because of the usually small monitors, the degree of immersion in such a simulator is rather low. Direct conclusions about the behavior in a real vehicle are therefore not always possible. However, such a simple construction can prove itself useful in many applications. In particular, subsequent large-scale simulator studies can be prepared well and efficiently. In the development of systems or the design of simulator studies, the developer can make changes and test directly at his workplace. With these "preliminary studies", the desired study can then be carried out more precisely with a more complex simulator.

Simulator with cockpit

A first increase in the feeling of reality can be achieved by the use of a seat box (a cockpit). Here the typical man-machine interfaces (pedals, steering wheel and indicators) of a vehicle are integrated into a specially designed cockpit, Fig. 14.46. As a rule, this extension results in a much better immersion in the driving experience.

Such constructions are very popular for use on trade fairs or for demonstration purposes because of their good transportability. But even simple studies can be performed efficiently with such a simulator. The visualization ranges from simple PC monitors to elaborate 360° projections.

Fig. 14.45 „Desktop" Driving Simulator © Lehrstuhl für Mechatronik, Universität Duisburg-Essen

Fig. 14.46 Trade Fair construction (seat box) © Lehrstuhl für Mechatronik, Universität Duisburg-Essen

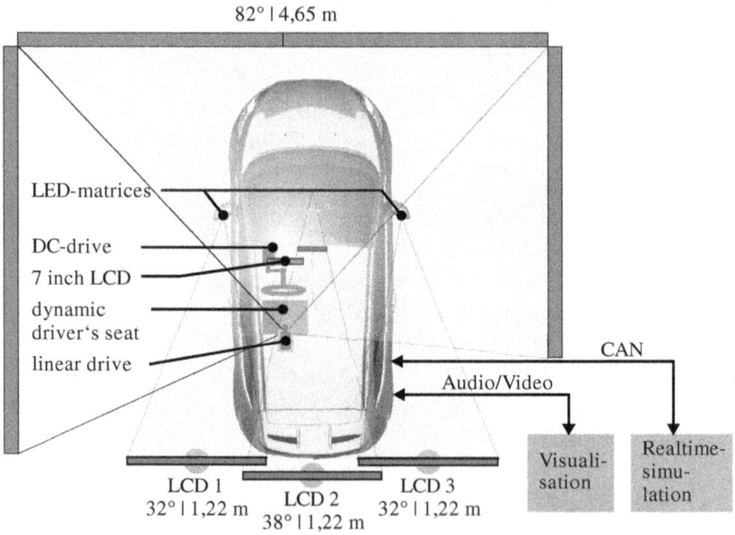

Fig. 14.47 Example of a static simulator using the example of the Ford Fiesta driving simulator at the Chair of Mechatronics at the University of Duisburg Essen © Lehrstuhl für Mechatronik, Universität Duisburg-Essen

Static full-vehicle driving simulator

The use of a full vehicle is another big step towards the reality of a driving simulator Fig. 14.47. The driver is located here in the vehicle environment familiar to him with the familiar operating elements, Fig. 14.48 and Fig. 14.49. Also, the getting into a complete vehicle supports the immersion in the driving experience.

Fig. 14.48 Exemplary driver's view of the Ford Fiesta driving simulator of the Department of Mechatronics at the University of Duisburg-Essen © Lehrstuhl für Mechatronik, Universität Duisburg-Essen

Fig. 14.49 Interior of the dynamic driving simulator at the Chair for Mechatronics at the University of Duisburg Essen © Lehrstuhl für Mechatronik, Universität Duisburg-Essen

However, the use of a complete vehicle as a driving simulator also considerably increases the effort. In particular, the requirements for the spatial conditions are significantly increased. Usually a separate room is needed for the simulator into which the entire vehicle can be placed, Fig. 14.47. This room often also needs to be fully air-conditioned. In addition, there must be a possibility to visualize the vehicle environment, Fig. 14.48. Usually, projectors and the so-called front projection are used for this purpose. This requires a minimum distance between the vehicle and the projection surface in order to avoid shadow and reflection effects caused by the vehicle itself. The human eye can reach conclusions on the position of an object in the close range (up to about 2m away from the eye) by the necessary focus on a distance. This results, in particular, in a minimum distance between the driver's eyes and the projection surface. If the projection surface is too close to the driver's workplace, the driver's eyes would focus on the near area. Thus, all the objects shown on the screen appear to be very close at first. If an object is displayed in the virtual reality that is not in the immediate vicinity of the eye point, the image does not fit the focus point of the eye. As a result, the simulation can be regarded as unrealistic (Negele 2007). Furthermore, it is significantly less tiring if the projection surface is at a distance from the eye which can be viewed with a relaxed eye.

It follows from these considerations that a space with a base area of 5 x 5 m or more is required for a simulator with a full vehicle.

Dynamic driving simulators

Driving simulators equipped with motion systems can have different stages of extension of the driver's workplace. Depending on the requirements and circumstances, seat boxes, simple cockpit constructions, or full vehicles can be used. A motion system must be adapted to the payload in the process of implementation. Motion systems can be categorized according to possible excitations. With some exceptions, the design stage for the driver's workplace, the visualization and a motion system are coordinated so that the degree of reality of the subsystems is not fundamentally different.

Dynamic driving simulator with complete cockpit

For the project DesignStudio NRW, a compact and cost-effective form of a dynamic driving simulator was developed at the Chair for Mechatronics at the University of Duisburg-Essen (Proff et al. 2015), Fig. 14.50. In this case, electromechanical actuators were installed in order to simulate the vehicle dynamics realistically but at the same time also cost-effectively, Fig. 14.51. The overall goal is to increase the degree of immersion (the feeling of reality) and thus open up further fields of investigation.

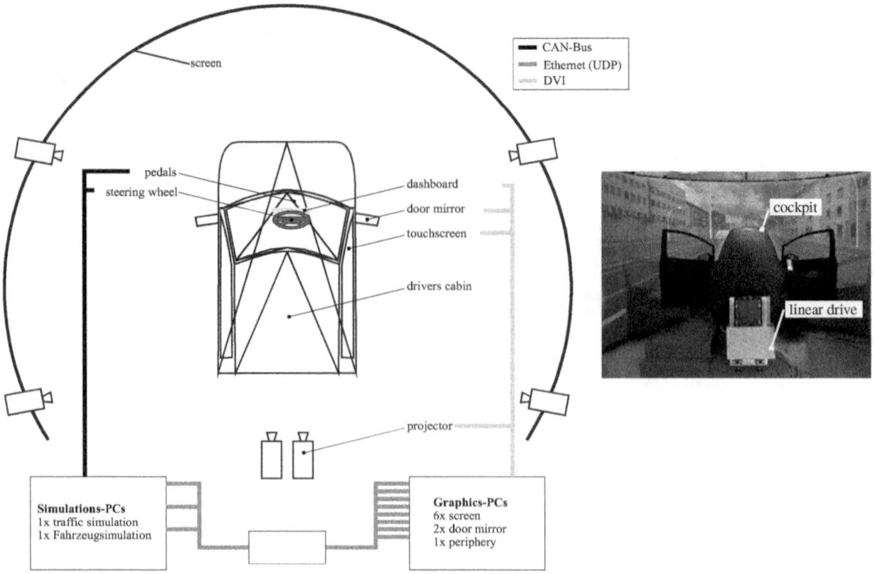

Fig. 14.50 Exemplary structure of the dynamic simulator at the Chair of Mechatronics at the University of Duisburg-Essen © Lehrstuhl für Mechatronik, Universität Duisburg-Essen und (Capustiac 2011, Maas 2017)

Table 14.6 Technical data of the movement platform

Size	Values
Base area	2,50 m × 2,00 m × 1,50 m (L × W × H)
Roll angle	10 Deg
Actuator stroke	380 mm vertical
Maximum frequency	40 Hz
body weight	350 kg
overall weight	1300 kg

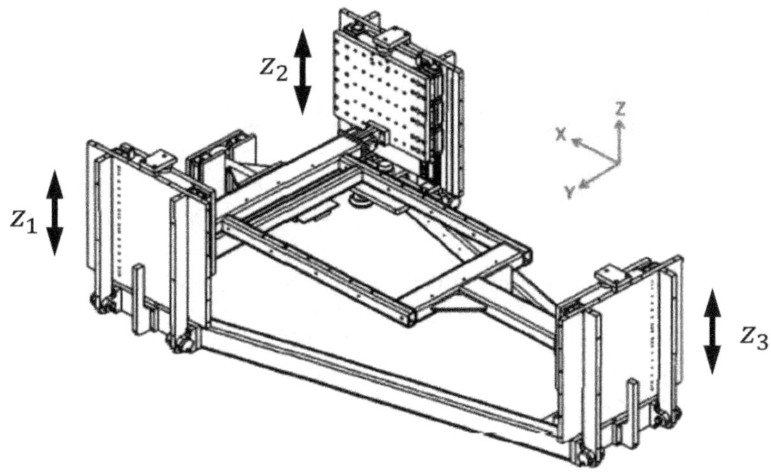

Fig. 14.51 Model of the motion platform © Lehrstuhl für Mechatronik, Universität Duisburg-Essen

This simulator is a very good compromise between the degree of realism and the complexity of the simulator. In particular, the very compact, modular design is advantageous in comparison to the systems shown in the following sections.

The cockpit is located at the center of a cylindrical screen, Fig. 14.50. Several high-performance projectors represent the environment. The images of the individual projectors are transformed in real time in such a way that a distortion-free representation is produced on the curved screen, and the edges of the images overlap without this being perceived by the observer. This results in a continuous image that covers approximately 252° of the driver's field of vision (Proff et al. 2015).

This area is sufficient for carrying out a normal trip. Also sharp turns as well as the support of the shoulder sight when turning or changing the lane are realistic. The area behind the vehicle is represented by several monitors, which are then seen in the rearview mirror. Alternatively, small displays can be attached directly to the cockpit as a rear-view mirror. Additional display elements for the combined instrument as well as a further freely configurable monitor in the interior are used for man-machine interaction.

The driver's workplace is located in a monocoque specially designed for this simulator. Because of the equipment with a normal vehicle seat and a corresponding instrument cluster, the cockpit gives the impression of a very realistic driving feeling, which is very close to that of a full vehicle.

The electromechanical motion platform used with its three powerful linear motors offers the possibility to move the structure with the three degrees of freedom, rolling, pitching, and lifting. In contrast to other simulators, the actuators are used to reproduce only the "real" rigid body movements of the vehicle body. No motion cueing is used here. The electromechanical actuators are designed for the generation of highly dynamic movements and enable excitation frequencies of up to 40 Hz and accelerations of up to 20 m / s² in the vertical direction. As a result, it is also possible to recreate road excitations, such as those caused by normal road surfaces, but also by ground swings or cobblestone pavements. This is achieved by the combination of powerful actuators and the low mass of the cockpit. Due to the realistic roadway feedback, a very realistic driving feeling is created (Proff et al. 2015).

Dynamic full-vehicle simulators

The motion platform described in Fig. 14.51 was also used with some extensions of the actuator system to build a driving simulator for an entire vehicle, see Fig. 14.52.

All of the described simulators have in common that a modified driver workstation from a real vehicle or a full vehicle is used. The driver workplace must be partially modified for the simulator, depending on the application. However, a force feedback on the steering as well as actuators or other aids are essential to create the right pedal feel. In addition, a correspondingly developed computer hardware must ensure that the vehicle inputs are transferred to the simulation environment and processed there in real time. In the above-described driving simulator with linear motors, two computers for simulating the vehicle and the environment are used as well as a further six computers for generating the graphical representation and the transformation on the curved projection surface. In addition, there are a further two computers for the rear-view mirrors as well as a further computer for the peripherals.

In the case of full vehicles with an electric servo steering, it is also possible under certain circumstances to load the necessary force feedback for the simulation directly via the already existing actuator of the servo steering system. In this case,

Fig. 14.52 Motion platform with complete vehicle © Foto: Annette van de Locht

no additional actuator is needed. Otherwise, it must be ensured that the steering (as one of the central elements of the man-machine interaction) offers the familiar haptic interface to the vehicle.

Dynamic simulators using a full-vehicle as a driver's workplace generally move the vehicle along with the environmental presentation in the direction of all six rigid-body degrees of freedom of the vehicle. The available work space is, of course, limited by the mechanical design of the motion system. Apart from the necessary six actuators, other active axes (actuators) are often used to create redundancy in the generation of the movement. This can be utilized for the optimized design of the movement. For example, very complex workspaces of a translatory x-y carriage system are combined with the high dynamics of one-dimensional lifting actuators in complex simulators. However, since the workspaces of all the simulators are still limited in nature, it is not possible to directly map the predetermined acceleration profile with the superstructure during sustained accelerations along an axis. For this purpose, further measures are required in the controls of the actuators.

In dynamic full-vehicle simulators, the vehicle is usually positioned in the interior of a projection dome. The driver is offered a 360° view of the virtual surroundings, combined with a very high degree of immersion. Furthermore, by skillful control of the actuators, the acceleration due to gravity can be used to give the driver the feeling of a real acceleration in the lateral or longitudinal direction. In this case, the impressed gravitational force is used, for example, to produce the lateral accelerations occurring during a curve travel. For this purpose, the entire vehicle is inclined together with the view system in such a way that the acceleration of the earth has a correspondingly large component along the corresponding vehicle axis. Since the relative position of the driver to the projection screen changes just as little as the projected image, this acceleration component is interpreted as a longitudinal or lateral acceleration. This procedure is known as motion cueing, see also Sect. 14.4.5. In the case of a consistent implementation according to powerful algorithms, the driver experiences a realistic acceleration impression, which, if correctly executed, he cannot distinguish from the impression of the real movement. However, the actual movements of the system shall not be perceivable by the test person. Therefore, some of the movements must occur below the perception threshold of the driver.

Examples of commercial dynamic full-vehicle simulators

The dynamic driving simulator of the DLR in Braunschweig differs from other simulators by the special and very interesting kinematic and kinetic motion system. A hydraulically operated hexapod is again used for the motion system (Pattberg 2005) und (Stöbe 2006). However, in this case, the cabin holding the vehicle is suspended below the upper joints, Fig. 14.53. This unusual kinematic arrangement allows particularly great translational and rotational movements.

The simulator of Daimler AG is one of the most modern and complex vehicle simulators, Fig. 14.54. The vehicle with the driver's workplace and all elements (visualization and sound) necessary for the environmental presentation are installed

Fig. 14.53 DLR simulator in Brauschweig (Stöbe 2006)

Table 14.7 Workspace and dynamics of the DLR simulator (Stöbe 2006)

	Travel	Speed	Acceleration
longitudinal	±1,5 m	±2 m/s	±10 m/s^2
transverse	±1,4 m	±2 m/s	±10 m/s^2
vertical	±1,4 m	±2 m/s	±10 m/s^2
roll	−20 deg / +21 deg	±50 deg/s	±250 deg/ s^2
pitch	±21 deg	±50 deg/s	±250 deg/ s^2
yaw	±21 deg	±50 deg/s	±250 deg/ s^2

Fig. 14.54 Daimler-Driving Simulator | Basic mechanical design (left) and 360-Degree-Visualisation (right) (Source: Daimler AG)

in a large dome. The dome is again supported by a hydraulically operated hexapod so that movements around all six spatial degrees of freedom can be realized.

In addition to the motion system of the hexapod, this simulator uses a high-dynamic linear axis to move the complete structure over several meters. It enables this particular driving simulator to realistically impress sustained accelerations as occur, for instance, in long-drawn curves. The dome can be rotated by 90°, so that the linear axis can represent both vehicle movements in the lateral as well as in the longitudinal direction.

References

Ajluni KK (1989) Rollover Potential of Vehicles on Embankments, Sideslopes, and Other Roadside Features. *PUBLIC ROADS* 52. S 107–13

Bardini R (2008) Auslegung von Überschlagschutzsystemen für Personenkraftwagen mithilfe der Simulation [Dr.-Ing.]. Dissertation, Universität Duisburg-Essen Düsseldorf: VDI-Verlag

Bardini R, Nagelstraßer M and Wronn O (2007) Applikation, Test und Absicherung einer Überschlagsensorik am Beispiel des neuen BMW X5. *VDI Bericht* 2013. S 149–67

Capustiac NA (2011) Development and application of smart actuation methods for vehicle simulators. Dissertation, Universität Duisburg-Essen

Capustiac A, Banabic D, Schramm D, Ossendoth U (2011a) Motion cueing: from design until implementation. Proceedings of the Romanian Academy Series A—Mathematics Physics Technical Sciences Information Science 12(3):249–255

Capustiac A, Hesse B, Schramm D, Banabic D (2011b) A human centered control strategy for a driving simulator. The International Journal of Mechanical & Mechatronics Engineering IJMME-IJENS 11(01):45–52

Coo PJAd, Wismans J and Niboer JJ (1991) Advances in MADYMO Crash Simulations. *SAE Technical Paper* 910879. S 135–46

Donges E, Naab K (1996) Regelsysteme zur Fahrzeugführung und-stabilisierung in der Automobiltechnik. Automatisierungstechnik 44(5):226–236

Harkey DL (1999) The Effect of Roadside Design on Rollover. In *AE Passenger Car Rollover Toptec*, San Diego, USA

Hesse B (2011) Wechselwirkung von Fahrzeugdynamik und Kfz-Bordnetz unter Berücksichtigung der Fahrzeugbeherrschbarkeit

Hiesgen G (2011) Effiziente Entwicklung eines menschzentrierten, integralen Querführungsassistenzsystems mit einem Fahrsimulator Dissertation

ISO N (1975) TR 3888: Road vehicles-test procedure for a severe lanechange manœuvre. Genf: Intern. Organization for Standardization

ISO D (1989) 7401: Lateral transient response test methods. In Deutsches Institut für Normung eV, Berlin (ed.)

ISO (2000) ISO 3888-2: Passenger cars—Test track for a severe lane-change manoeuvre—Part 2: Obstacle avoidance.

ISO_7401 (2003) Road vehicles—lateral transient response test methods—open loop test method. Genf, International Organisation for Standardisation

ISO_4138 (2004) Passenger Cars—steady-state circular driving behaviour—open loop test methods. Genf, International Organisation for Standardisation

Lich T, Breitmaier B (2003) Optimierte Überrollsensierung zur frühzeitigen Überschlagerkennung. *Automotiv Electronics. Sonderausgabe ATZ/MTZ Automotiv Engineering Partners März* 14

Lunze J (2013) Regelungstechnik. Springer—ISBN 978-3-642-29532-4

Maas N (2017) Konzeptionierung, Auslegung und Umsetzung von Assistenzfunktionen für die Übergabe der Fahraufgabe aus hochautomatisiertem Fahrbetrieb. Dr.-Ing. Dissertation, Universität Duisburg-Essen

Mamdani EH, Assilian S (1975) An experiment in linguistic synthesis with a fuzzy logic controller. *International journal of man-machine studies* 7, —ISBN 0020-7373. S 1–13

Mohamedshah Y, Council F (2007) Synthesis of Rollover Research

Negele H-J (2007) Anwendungsgerechte Konzipierung von Fahrsimulatoren für die Fahrzeugentwicklung, Technische Universität München

Otte DK, C (2005) Rollover accidents of cars in the German road traffic—an in-depth-analysis of injury and deformation pattern by GIDAS. In *19th International Technical Conference on the Enhanced Safety of Vehicles*, Vol. Paper Number 05-0093, Washington, DC

Öttgen O (2005) Zur modellgestützten Entwicklung eines mechatronischen Fahrwerkregelungssystems für Personenkraftwagen Dissertation, Universität Duisburg-Essen, Düsseldorf: VDI-Verlag

Öttgen O, Bertram T (2003) Aktive Beeinflussung des Eigenlenk- und Wankverhaltens eines Pkws. *Automatisierungstechnik 51*

Öttgen O, Bertram T (2004) Entwicklung eines Sollwertgenerators für fahrdynamische Regelungssysteme. VDI/VDE GMA Fachtagung Steuerung und Regelung von Fahrzeugen und Motoren, AutoReg 2004. *VDI-Berichte*. S 485–95

Pattberg B (2005) Der Wirklichkeit Nahe: Autofahren in der Simulation—Assistenzfunktionen sicher und effizient erproben. Ignition: S. 60–63

Proff H, Brand M, Mehnert K, Schmidt JA, Schramm D (2015) Elektrofahrzeuge für die Städte von morgen: Interdisziplinärer Entwurf und Test im DesignStudio NRW, Springer

Rahimi Fetrati S, Teufel S, Schramm D (2016) Electrification of Torque-Vectoring system to improve the vehicle driving dynamics. 7th International Munich Chassis Symposium 2016, München

Rasmussen J (1983) Skills, rules, and knowledge; signals, signs, and symbols, and other distinctions in human performance models. IEEE transactions on systems, man, and cybernetics (3):257–266

Rau M (2007) Koordination aktiver Fahrwerk-Regelsysteme zur Beeinflussung der Querdynamik mittels Verspannungslenkung. Promotion, Stuttgart.

Reński A (2001) Identification of driver model parameters. International Journal of occupational safety and ergonomics 7(1):79–92

Stöbe M (2006) Die Fahrsimulatoren des DLR—Funktionen und Anwendungsmöglichkeiten Braunschweig, Deutsches Zentrum für Luft- und Raumfahrt e.V.

Unbehauen H (2008) Regelungstechnik I: Klassische Verfahren zur Analyse und Synthese linearer kontinuierlicher Regelsysteme, Fuzzy-Regelsysteme. Springer DE—ISBN 978-3-8348-9491-5

Unterreiner M (2013) Modellbildung und Simulation von Fahrzeugmodellen unterschiedlicher Komplexität. Doktor Dissertation, Duisburg-Essen

Unterreiner M, Schramm D, Ossendoth U (2013) Untersuchung des Immersionsgrades eines bewegten Fahrsimulators in Abhängigkeit der Fahrzeugmodellkomplexität. VDI VDE Mechatronik, RWTH Aachen University, Deutschland

VDA (2006) VDA Spurwechseltest. (ed), Verband der Automobilindustrie e. V. (VDA)

Zadeh LA (1973) Outline of a new approach to the analysis of complex systems and decision processes. *Systems, Man and Cybernetics, IEEE Transactions on*—ISBN 0018-9472. S 28–44

Index

D. Schramm et al., Vehicle Dynamics,
DOI 10.1007/978-3-662-54483-9